The Grevillea Book

Volume Two

Species A–L

The Grevillea Book

Volume Two

Species A–L

Peter Olde and Neil Marriott

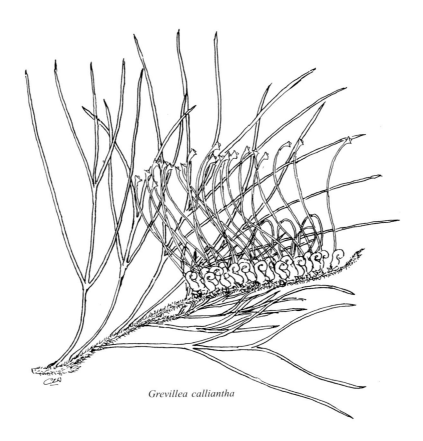

Grevillea calliantha

Timber Press

Portland, Oregon

Grevillea floribunda

Jacket front: *Grevillea annulifera* (N.Marriott)
Jacket back: *Grevillea fililoba*
Frontispiece: *Grevillea insignis* (P.Olde)

© Peter Olde and Neil Marriott 1995

First published in North America in 1995 by
Timber Press, Inc.
The Haseltine Building
133 S.W. Second Avenue, Suite 450
Portland, Oregon 97204, U.S.A.

ISBN 0-88192-305-2 (v. 1)
ISBN 0-88192-306-0 (v. 2)
ISBN 0-88192-307-9 (v. 3)
ISBN 0-88192-308-7 (set)

Printed in Hong Kong

Layout of the Species Texts

Botanical, ecological and horticultural information is provided for each taxon where possible. Botanical names indicate the rank at which we consider taxa should be recognised at the present time. Species are arranged in alphabetical rather than systematic order in order to facilitate finding a taxon.

For each species and infraspecific taxon the following information is provided: botanical name, author(s), year of first publication; journal or place of publication with relevant volume and page number(s); country of publication; the common name if appropriate; aboriginal name, relevant tribe and usage where known; the meaning or historical association of the specific or infraspecific epithet; a guide to pronunciation of the specific or intraspecific epithet; locality, date, collector, collector number, and current herbarium location of the type collection; synonyms; a full botanical description; the major distinguishing botanical features; closely related or confusing species as well as the key group to which closely related species belong; informal and formal variation; key to infraspecific taxa if relevant; natural distribution within State(s); ecological information including soil, associated vegetation type, flowering period, pollinator, regeneration mode and fire response; climate and average rainfall over the natural distribution; conservation status (following Briggs & Leigh 1989) with our own updated information for both species and subspecies; cultivation notes from our own experience and that of others, especially that provided by members of the Grevillea Study Group; propagation notes on cutting, seed, grafting and occasionally other means such as tissue culture; horticultural features providing a quick overview of the best features of a well-grown plant; historical information on cultivation if relevant; hybrid information including natural hybrids and those from cultivation where parentage is associated with the relevant species; usage by both European and Aboriginal cultures; general comments are sometimes given on the state of current botanical research or thinking about the status of a particular taxon.

Unless otherwise stated in the description, the development of the conflorescence is acropetal. For a number of taxa, measurements of floral parts and some observations on morphology rely on McGillivray (1993). Line drawings of the pistil, perianth and sometimes other features are provided to assist in identification. A small distribution map is given for each taxon. Colour plates usually show the flowers, the foliage and sometimes growth habit.

A key to species is given in Volume 1, which also contains a glossary of botanical terms.

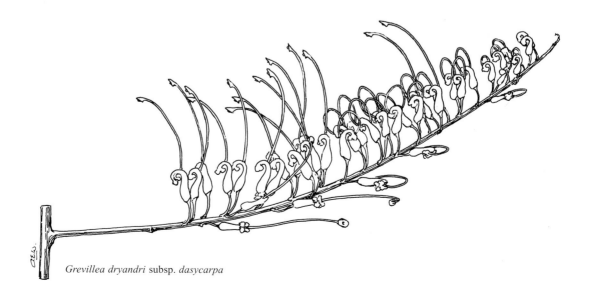

Grevillea dryandri subsp. *dasycarpa*

Abbreviations, Contractions and Symbols

C	Celsius
c.	Latin *circa*, about
cf.	Latin *confer*, compare
cm	centimetre
cult.	cultivated
e.g.	Latin *exempli gratia*, for example
et al.	Latin *et alii*, and others
etc.	Latin *et cetera*, and so forth
Hwy	Highway
i.e.	Latin *id est*, that is
km	kilometre
m	metre
mm	millimetre
NP	National Park
p./pp.	page/pages
pers. comm.	by personal communication
pH	a quantitative expression of the acidity or alkalinity of an aqueous solution, on a scale of 1–14; 7 is neutral.
q.v.	Latin *quod vide*, which see
R.	River
Ra.	Range
sp./spp.	species (singular/plural)
subsp.	subspecies
var.	variety
A.C.T.	Australian Capital Territory
N.S.W.	New South Wales
N.T.	Northern Territory
Qld	Queensland
S.A.	South Australia
Tas.	Tasmania
Vic.	Victoria
W.A.	Western Australia
±	*in species descriptions*, more or less
<	less than
>	more than
≥	more than or equal to

Herbaria where types are lodged are abbreviated in accordance with P.K.Holmgren *et al.* (1990), *Index Herbariorum* Part I, 8th edn, New York Botanical Garden.

The titles of books and journals and the authors of botanical names are not abbreviated.

Codes for conservation status follow J.D.Briggs & J.H.Leigh (1989), *Rare or Threatened Australian Plants*, 1988 revised edition, Special Publication [14], Australian National Parks & Wildlife Service, Canberra.

Grevillea alpina

Key to Line Drawings

1. Perianth just prior to anthesis
2. Flower just after anthesis
3. Pistil
4. Style end
4a. Pollen presenter
5. Enlargement of lower section of style
6. Fruit
6a. Seed
7. Cross-section of leaf
8. Nectary
9. Leaf
10. Flowers and foliage
11. Inner perianth surface
12. Tepal limb

Grevillea acacioides C.A.Gardner ex D.J.McGillivray (1986)   Plate 1

New Names in Grevillea *(Proteaceae)*: 1(Australia)

The specific epithet refers to the overall resemblance of the plant (but especially the leaves) to an *Acacia* (Greek *-oides*: resembling, like). AK-AY-SEE-OY-DEES

Type: E of Sandstone, W.A., 16 Aug. 1931, C.A.Gardner 2486 (holotype: PERTH).

A dense, rounded, sometimes suckering **shrub** 1–2 m tall. **Branches** usually erect; branchlets terete, silky. **Leaves** 3–8 cm long, ascending, sessile, simple, usually terete and 0.8–1 mm diam., rarely narrowly linear and c. 1.5 mm wide, longitudinally ridged, pungent, leathery, either silky but soon becoming glabrous or pubescent; midvein of flat leaves not evident. **Conflorescence** c. 1 cm long and wide, erect, axillary in upper axils or terminal on short branchlets, unbranched, umbel-like, basipetal; peduncle absent to 5 mm long, silky; rachis villous; bracts 4.5–6 mm long, ± ovate, papery, imbricate, falling before anthesis. **Flower colour:** perianth green in bud becoming greenish white or creamy white; style white, sometimes ageing pink. **Flowers** adaxially oriented; pedicels 4–7.5 mm long, glabrous; torus c. 1 mm across, oblique; nectary U-shaped to ± semicircular, entire or toothed; **perianth** c. 5 mm long, 1 mm wide, cylindrical to narrowly ovoid, glabrous outside, bearded inside, separating first along dorsal suture; tepals reflexed slightly before anthesis, afterwards free to base; limb revolute, ovoid to spheroidal, enclosing style end before anthesis; **pistil** 10–12.5 mm long, glabrous; stipe 1–1.3 mm long; ovary globular, somewhat retrorse; style exserted from near base on dorsal side and looped out and upward before anthesis, afterwards strongly incurved from about middle, dilating smoothly into style end; pollen presenter oblique, oblong, convex with a narrow, encircling rim. **Fruit** 9–13 mm long, 6–9 mm wide, 5–7 mm deep, erect, lateral to stipe, oblong-ellipsoidal with prominent transverse ridge just below middle, glabrous, glaucous, smooth to rugulose; pericarp 1.5–3 mm thick; valves separating completely. **Seed** 5.5–8 mm long, 1.5–2 mm wide, ellipsoidal, membranously and broadly winged all round.

Distribution W.A. Widespread in inland areas from near Wiluna south to Perenjori and Ravensthorpe and east to Queen Victoria Spring. *Climate* Hot, dry summer; cool to warm, occasionally wet winter. Rainfall 225–400 mm.

Ecology Grows in red or yellow sand and sandy granitic loam in dry mallee woodland, wattle shrubland or sandplain, sometimes in spinifex grassland; often associated with wattles, many of which have phyllodes similar to the leaves of this species. Flowers winter to spring. Pollinator unknown, but probably an insect. Perfume not noticeable.

Variation There is some foliage variation worthy of note in the following forms. There is also an herbarium record of a prostrate plant with pale red flowers from Newcarlbeon Soak, growing in granitic loam. Further survey is needed to establish whether this was an isolated plant or from a larger, distinct population.

Type form At the northern end of its range, from Sandstone to north of Wiluna, *G. acacioides* has leaves with a persistent variable indumentum especially prominent on young plants. These plants rarely exceed 1.2 m in height and are scattered in wattle shrubland or among spinifex in red sand. Young growth from cuttings of these plants propagated in 1988 bore, first, narrowly elliptic leaves without an apparent midvein, then leaves with the margins rolled inward, and finally terete leaves.

Glabrous form This form is commonly seen in the southern part of the species' range and is distinguished by glabrous, terete leaves. A flattish-leaved population of this form occurs near Carrabin. It has leaves that are elliptic in cross-section and up to 1.5 mm wide.

Major distinguishing features Leaves terete or almost so, without midvein; conflorescence umbel-like, basipetal, not exceeding foliage; bracts relatively large, papery; pistil glabrous; fruit wall 1.5–3 mm thick, transversely ridged in the middle, the valves separating completely after dehiscence.

Related or confusing species Group 17. *G. hakeoides* (Group 16) also appears close but may be readily distinguished by its leaves which have two distinct longitudinal grooves, either one down each side or both on the undersurface, and by its fruits that are rugose and thin-walled (c. 0.5 mm thick).

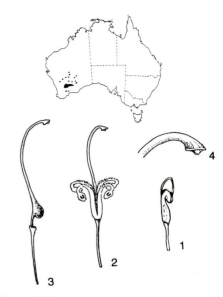

G. acacioides

1A. *G. acacioides* Close-up of flowers and foliage (N.Marriott)

1B. *G. acacioides* Shrub in spinifex habitat N of Wiluna, W.A. (P.Olde)

Conservation status Not presently at risk.

Cultivation G. acacioides is an extremely tough species, able to endure extremes of drought and cold, including heavy frosts. It has not adapted well to the conditions available in cultivation, however, and survives only in situations similar to its natural habitat. A hot, dry season appears to be essential. In the wild, the surface soil usually has an acidic pH with underlying alkaline sand or limestone rubble. The species has been cultivated in extremely well-drained, gravelly loam at Stawell in western Vic. but survived for only a few years. Nowadays, the only surviving plants in cultivation are grafted onto suitable rootstocks. An open, warm to hot, situation is preferred. Plants grown in cold or shaded sites have died from either root-rot or fungal attack on the foliage. Once established, grafted plants require no maintenance except occasional pruning. *G. acacioides* is usually unobtainable in nurseries although one specialist nursery occasionally offers grafted plants for sale.

Propagation Seed Sets plentiful seed that germinate readily if pre-treated by peeling or nicking. *Cutting* Rather difficult to strike. Some success has been achieved with firm new growth taken in late winter–spring. *Grafting* Difficult but has been grafted successfully onto *G. robusta*, G. 'Poorinda Anticipation' and G. 'Poorinda Royal Mantle' using the top wedge technique.

Horticultural features Under-rated horticulturally, *G. acacioides* has some potential in regions with a warm to hot dry season. Adult leaves vary from grey-green to bright green, the former in particular making it a striking foliage and screen plant. The white flowers are bright but tend to be hidden within the foliage, although sometimes a free-flowering plant may have flowers in almost every axil, forming a white mantle. The fruit can be a further interesting feature.

General comments Subject to more intensive collection and study, infraspecific ranking of the major variants is likely.

Grevillea acanthifolia A.Cunningham (1825)

Barron Field, *Geographical Memoirs of New South Wales*: 328 (England)

Named for the resemblance of its foliage to that of the genus *Acanthus* (Greek *akantha*: a thorn, prickle), and Latin *folium* (a leaf), in reference to the shape and pungent lobes of the leaves. AK-AN-THI-FOLE-EE-A

Type: peaty bogs on the Blue Mountains and banks of Cox's River, N.S.W., 1817, collected by A.Cunningham on Oxley's first expedition (lectotype: K) (McGillivray 1993).

An irregular, spreading to erect **shrub** with lignotuber, 1–3 m tall, 1–5 m wide, sometimes prostrate. **Branchlets** reddish, angular, tomentose but soon glabrous. **Leaves** 3.5–9 cm long, 3–7 cm wide, ascending to spreading, petiolate, stiff, ± ovate in outline, deeply bipinnatifid to bipinnatisect; ultimate lobes 0.5–2 cm long, usually divaricate, broadly triangular or linear to trifid, acute, pungent; margins recurved or refracted; upper and lower surfaces glabrous or sometimes with appressed hairs along veins; midvein prominent below. **Conflorescence** 2–9 cm long, terminal, erect, rarely decurved, unbranched, dense, conico- to oblong-secund; peduncle 0.1–1.5 cm long, tomentose to glabrous; rachis villous; bracts 2.5–5 mm long, ± ovate, imbricate, persistent. **Flowers** acroscopic; pedicels 1.5–2.5 mm long, villous; torus 1.5–2 mm across, oblique; nectary U-shaped, toothed; **perianth** 7–8 mm long, 2–3 mm wide, narrowly ovoid to S-shaped, ribbed, appressed-villous outside, glabrous inside, cohering except along dorsal suture; limb revolute, ovoid, woolly to villous; **pistil** 20–28 mm long; stipe 0.5–2.6 mm long; ovary villous; style glabrous, at first exserted below curve on dorsal side and looped upwards, straight to undulate after anthesis, ultimately reflexed strongly; style end hoof-like; pollen presenter oblique to erect, oblong-elliptic to round, convex to conical; stigma off-centre. **Fruit** 8.5–11 mm long, 6–8 mm wide, 5 mm deep, on incurved stipe, obovoid, villous, slightly ridged, with reddish brown striping; pericarp c. 0.5 mm thick. **Seed** 6–8 mm long, 2.5–3.5 mm wide, ellipsoidal; outer face convex, rough; inner face elliptic in centre, peripherally ridged, with a waxy margin drawn to excurrent elaiosome c. 1 mm long.

Major distinguishing features Leaves deeply twice-divided; lobes stiff, pungent, the undersurface exposed, glabrous or rarely silky; conflorescence secund; perianth zygomorphic, appressed-hairy outside with villous limb, glabrous inside; ovary shortly stipitate, densely hairy; fruit with reddish stripes or blotches.

Related or confusing species Group 35, especially *G. rivularis* which has the outer surface of the perianth glabrous.

There are three subspecies based on variation in the leaves, conflorescence and pollen presenter.

Key to subspecies

1 Pollen presenter broadly conical to convex, clearly shorter than wide, its base oblique to straight; leaf undersurface glabrous or with scattered hairs on midvein; leaf lobes triangular or linear; shrubby plant to 1.5 m tall

2 Conflorescence usually more than 5cm long, conico-secund; rachis much thickened after anthesis; leaves mostly bipinnatifid with triangular lobes — subsp. **acanthifolia**

2* Conflorescence usually less than 3cm long, oblong-secund; rachis not noticeably thickened after anthesis; leaves mostly bipinnatisect with linear lobes — subsp. **stenomera**

1* Pollen presenter conical, about as long as wide, its base straight; leaf undersurface silky beside mid-vein; leaf lobes linear; robust, erect shrub to 3m — subsp. **paludosa**

Grevillea acanthifolia A.Cunningham subsp. acanthifolia (1825) Plate 2

Shrub 1–1.5 m high, 2 m wide. **Branchlets** usually tomentose. **Leaves** 4–9 cm long, the undersurface mostly exposed. **Conflorescence** 4–9 cm long, erect; peduncle tomentose. **Flower colour**: perianth silvery to reddish grey outside; style mauve-pink to red. **Perianth** 2–3 mm wide, densely silky-villous outside; **pistil** 22–28 mm long; stipe 1.5–2.6 mm long; pollen presenter straight to oblique, convex.

Synonyms: *G. acanthifolia* Sieber ex K.Sprengel (1827); *G. acanthifolia* Sieber ex Schultes & Schultes f. 1827.

Distribution N.S.W., confined to the upper Blue Mountains from near Bell to Woodford. *Climate* Hot to cool, wet summer; cool, wet or dry winter, with infrequent snow falls. Rainfall 1000–1200 mm.

Ecology Grows in sandy or peaty soil in high, wet heath and hanging swamps, often with water running over the surface. Flowers spasmodically all year but with a spring to summer peak. Bird-pollinated.

Conservation status Not presently endangered. Conserved in the Blue Mountains NP, N.S.W.

Variation There is minor variation in habit, leaf division and flower colour. Prostrate plants with a spreading or mat-forming habit have been collected near Woodford in the lower Blue Mountains, west of Sydney. In other respects, these plants are similar to other populations of subsp. *acanthifolia*.

Cultivation One of the few grevilleas tolerant of poor drainage or 'wet feet' and is consequently regarded as a hardy species. It flourishes in cool, wet climates, but also does well in hot, dry climates provided there is adequate subsoil moisture during extended dry periods. It has been successfully established in most eastern States of Australia including Tas. It is cultivated in N.Z. and the U.S.A., and should do well in Europe. Summer humidity is not greatly to its liking even in coastal Sydney gardens, and flowering is poor in these conditions. Prolific flowering occurs only after winter-cold conditions, including frosts or even light snow,

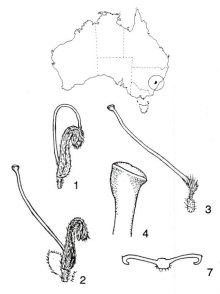

G. acanthifolia subsp. *acanthifolia*

2A. *G. acanthifolia* subsp. *acanthifolia* Pink-flowered plant (P.Olde)

which appear necessary for good bud initiation. Both full sun or partial shade are accommodated and it prospers in moist, well-drained, gravelly or sandy, acidic soils as well as heavy clay loams. In poor soils it becomes open and woody but responds with vigorous new growth to light fertilising as well as summer watering. Occasional pruning may be necessary to maintain shape. It is usually available in native plant nurseries on the east coast of Australia.

Propagation **Seed** Fresh seed, subjected to peeling pretreatment, gives good results. Even seed collected from the wild may be impure, however, as the species often hybridises with *G. laurifolia* which sometimes occurs nearby. **Cutting** Strikes well at most times of the year using firm, new growth. **Grafting** Generally unnecessary but may itself be suitable as a rootstock as it has proven resistant to *Phytophthora cinnamomi*, a root-rot fungus. It has been approach-grafted to *G. robusta* successfully.

Horticultural features Subsp. *acanthifolia* grows into an attractive, spreading shrub in cultivation with interesting, deeply divided, prickly, green leaves that are bronze or reddish when young. A profusion of conspicuous, bright pink 'toothbrush' conflorescences cover the shrub in late spring and summer, providing abundant nectar for birds and ornamental colour in the garden. The prostrate form develops into a vigorous, dense ground cover, while the typical form can be planted as a feature or low, screen plant. In cool climates it does well with summer watering.

Hybrids The natural hybrid, *G.* × *gaudichaudii*, which grows in disturbed areas around natural populations of *G. acanthifolia* and *G. laurifolia* near Clarence, N.S.W., is well known in cultivation. Cultivated hybrids are usually prostrate but erect shrubs can be found in the wild. *G. acanthifolia* is one parent of the garden hybrids G. 'Poorinda Peter' and G. 'Austraflora Copper Crest'.

History Subsp. *acanthifolia* was cultivated in stove-heated glasshouses in Britain as early as 1824 from seed sent back by A.Cunningham in 1823. From there it was distributed widely through Britain and the Continent. Hügel's conservatory in Vienna listed it in 1831 and it was recorded in cultivation at Dusseldorf (1834), Calcutta (1853–4) and Amsterdam (1857). The earliest recorded cultivation in Australia was in 1853 at Sydney.

Grevillea acanthifolia subsp. paludosa
R.O.Makinson & D.E.Albrecht (1989)

Plate 3

Muelleria 7: 89 (Australia)

The subspecific name, derived from the Latin *paludosus* (swampy, boggy), refers to the habitat. PAL-YOU-DOSE-A

Type: c. 2.5 km WNW of Mt Wog Wog trig, Nalbaugh NP, N.S.W., 22 Feb. 1987, D.E.Albrecht 3078 & P.Gilmour (holotype: MEL).

A spreading to erect **shrub** to 3 m tall and 5 m wide. **Branchlets** silky. **Leaves** 4–6 cm long, the undersurface mostly enclosed by margins, silky beside midvein. **Conflorescence** 3–5 cm long, erect, conico-secund; peduncle tomentose. **Flower colour:** perianth greyish fawn outside with a grey limb; style bright pink to red with green tip. **Perianth** 2 mm wide, silky to tomentose outside with villous limb; **pistil** 21–24 mm long; stipe 0.5–1.2 mm long; pollen presenter straight, conical.

Distribution N.S.W., confined to the Nalbaugh Plateau in the southern tablelands. *Climate* Summer hot, wet or dry; winter cool to cold, wet or dry. Rainfall 800–1100 mm.

Ecology Grows mainly on low hummocks in wet, peaty swamps surrounded by *Leptospermum lanigerum*, in dense thickets with *Boronia deanei*. Flowers spring to early summer. Regenerates from seed. Pollinated by birds.

G. acanthifolia subsp. *paludosa*

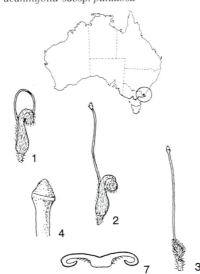

Variation A relatively uniform subspecies.

Conservation status Proposed 2EC-t. Although conserved in Nalbaugh NP, the single known population totals only c. 40 plants.

Cultivation Although recently introduced to cultivation, subsp. *paludosa* gives every indication of requiring similar conditions to subsp. *acanthifolia*. A similar, high altitude climate and poorly drained soils characterise the habitat in the wild where mists and light rain are frequent. Probably tolerates frost and snow. Summer watering appears necessary, and acidic, deep sand or clay soil with subsoil moisture as a minimum requirement. A position in either partial shade or full sun is indicated.

Propagation As for subsp. *acanthifolia*.

Horticultural features Subsp. *paludosa* has an extremely robust, exuberant habit and quite prickly, divided leaves. Short, bright pink toothbrush flowers abound in late spring and summer, making the subspecies extremely decorative. Its flowers produce a good nectar supply and freely attract birds. Use as a dense screen or barrier plant is indicated.

3A. *G. acanthifolia* subsp. *paludosa* Close-up of conflorescence (P. & J. Williams)

2B. *G. acanthifolia* subsp. *acanthifolia* Close-up of conflorescence, in cultivation, Shepparton, Vic. (P.Olde)

3B. *G. acanthifolia* subsp. *paludosa* Plant in natural habitat, Nalbaugh Plateau, N.S.W. (W.Molyneux)

Grevillea acanthifolia subsp. stenomera (F.Mueller ex G.Bentham) D.J.McGillivray (1975)

Plate 4

Telopea 1: 23 (Australia)

Based on *G. acanthifolia* var. *stenomera* F.Mueller ex G.Bentham (1870), *Flora Australiensis* 5: 439 (England).

The subspecific epithet is derived from the Greek *stenos* (narrow) and *meris* (a part), in reference to the narrow lobes of the leaves. STEN-O-MAIR-A

Type: head of the Macleay River, New England district, N.S.W., 1867, from C.Moore (holotype: K).

Shrub to 1.5 m tall and 3 m wide. **Branchlets** glabrous. **Leaves** 3–6 cm long, undersurface mostly enclosed by margins. **Conflorescence** 2–4 cm long, erect to decurved; peduncle glabrous. **Flower colour**: perianth greyish fawn outside with a grey limb; style bright pink to red with green tip. **Perianth** 2 mm wide, sparsely silky outside with villous limb; **pistil** 20–24 mm long; stipe 1.5–2 mm long; ovary villous; pollen presenter oblique, convex.

Distribution N.S.W., scattered along the northern tablelands from the Gibraltar Ra. S to about Armidale. *Climate* Summer hot, wet or dry; winter cool to cold, wet or dry. Rainfall 800–1100 mm.

Ecology Grows in peaty or sandy moist situations often on raised hummocks. Also occurs in dense thickets, in tall heath or woodland, beside streams and around the edges of rush swamps. Flowers spring to summer. Regeneration is probably from seed. Pollinated by nectarivorous birds.

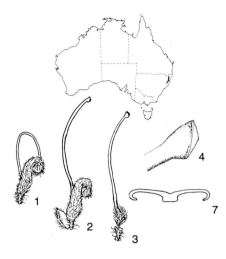

G. acanthifolia subsp. *stenomera*

4A. *G. acanthifolia* subsp. *stemomera* Shrub growing in sedge swamp, Gibralter Range NP, N.S.W. (P.Olde)

4B. *G. acanthifolia* subsp. *stenomera* Close-up of conflorescence (M.Keech)

Variation A relatively uniform subspecies.

Conservation status Not presently endangered.

Cultivation A reasonably hardy plant that tolerates a wide range of conditions, subsp. *stenomera* has been grown successfully in Sydney and Melbourne, as well as near its natural occurrence. Summer watering is necessary and the plant does not withstand extended dry periods, but although soil moisture should be high, it should not be stagnant. An acidic deep sandy loam or clay in an open, sunny position is ideal. Tolerates frost to at least -6°C and probably much lower, and the cold winters seem to induce a free-flowering spring display. Appearance is improved by pruning and by sparing application of low-phosphorus fertiliser. Has been sold commercially in Vic. but now rarely available.

Propagation As for subsp. *acanthifolia*.

Horticultural features In cultivation, subsp. *stenomera* forms a low, dense, spreading shrub with very prickly foliage, a feature that has contributed to its lack of popularity. Although its pink conflorescences are bright, they are quite short and relatively inconspicuous, lacking the impact of the more prolific and larger racemes of subsp. *acanthifolia*. Its spiky, rigid leaves would make it an effective 'people stopper' and it is suitable for planting as a low, screen in commercial and residential sites, especially in cool moist conditions. The flowers are strongly attractive to birds and the foliage offers suitable nesting sites for small birds.

General comments Botanists are currently considering if this taxon is sufficiently distinct to warrant specific rank. If so, a new name will be needed.

Grevillea acerata D.J.McGillivray (1986)

Plate 5

New Names in Grevillea *(Proteaceae)*: 1 (Australia)

The specific epithet alludes to the style end and is derived from the Greek *a-* (without) and *-ceras* (horned). The closest relative of this species (*G. buxifolia*) typically has a prominent horn on the style end. AY-SER-ART-A

Type: Gibraltar Range NP, 58 miles [c. 93 km] NW of Grafton on the Gwydir Highway, N.S.W., 8 Feb. 1973, M.D.Tindale 2069 (holotype: NSW).

A bushy, sometimes leggy **shrub** 0.6–1.3 m tall. **Branchlets** angular to terete, silky to tomentose. **Leaves** 1–3 cm long, 1–1.3 mm wide, ascending, sessile, simple, linear to narrowly elliptic, leathery; margins revolute, often enclosing lower surface; upper surface convex, granulate, sparsely hairy, soon glabrous, the midvein evident; lower surface often obscure, silky to villous, the midvein obscure. **Conflorescence** 0.5–1 cm long, 1–2 cm wide, erect, terminal, ± sessile, unbranched, barely exceeding foliage, dense, umbel-like; apical flowers opening first; rachis villous; bracts 2.5–6 mm long, linear-acuminate, persistent. **Flower colour:** perianth grey or pinkish grey with conspicuous brown hairs especially on the limb; style grey. **Flowers** adaxially oriented; pedicels 5–9 mm long, brown-villous; torus c. 1 mm across, ± straight; nectary semi-circular, toothed; **perianth** 3 mm long, 1–1.5 mm wide, oblong, strongly curved, grey-woolly inside and out, first opening along dorsal suture and strongly incurved, the ventral tepals separated but extending in a platform below limb c. level with top of ovary; limb strongly revolute, pyramidal, brown-villous, ± cohering after anthesis; **pistil** 9–12 mm long; stipe 0.5–1.3 mm long, silky to villous, almost glabrous on ventral side; ovary villous to woolly; style villous, hooked near apex, exposed to base except at limb before anthesis, afterwards strongly incurved, merging evenly into a disc-like style end with a narrow, glabrous rim; pollen presenter ± lateral, obovate, flat. **Fruit** c. 12 mm long, 5 mm wide, 4 mm deep, erect on straight pedicel, ovoid, loosely pilose; pericarp c. 0.3 mm thick. **Seed** 7 mm long, 2 mm wide, oblong–ellipsoidal, pubescent, with a subapical, cushion-like swelling; wing terminal, 1–2 mm long; outer face strongly convex; inner face flat, minutely tessellated, often concealed by margins; margins strongly revolute with a narrow, waxy wing along one side.

G. acerata

Distribution N.S.W., restricted to a small area near Glen Innes. *Climate* Summer warm to hot, wet; winter cool to mild, wet or dry. Rainfall 800–1000 mm.

Ecology Generally found in the shrubby understorey of dry sclerophyll forest growing in acidic, gravelly granite sand. Flowers normally spring to summer but flowers can often be found in winter. Insect-pollinated, possibly by native bees. After fire, regenerates from seed.

Variation A stable species with little variation.

Major distinguishing features Leaves simple, entire, linear, crowded, granular, the lower surface obscured by the margins; conflorescence umbel-like, sessile; flowers adaxially oriented; torus straight; nectary sometimes toothed; perianth zygomorphic, woolly inside and out; ovary densely hairy, shortly stipitate, the stipe sparsely hairy on the ventral side; style end disc-like; pollen presenter obovate, lateral.

Related or confusing species Group 22, especially *G. occidentalis* and *G. sphacelata*. *G. acerata* was once widely regarded as the northern form of *G. sphacelata* but this latter species is readily distinguished by its exposed leaf undersurface and the abaxial orientation of its flowers. *G. occidentalis* differs in its round pollen presenter, in its generally shorter floral bracts (< 2 mm long), in its more widely spaced leaves, and in its abaxially oriented flowers.

Conservation status 2RC-t. Restricted, but conserved over most of its distribution in the Gibraltar Ra. NP.

Cultivation *G. acerata* can be grown satisfactorily over a wide climatic range, thriving in coastal, tropical climates, inland, dry climates and cool, temperate climates. Although not cultivated by many enthusiasts to date, reports from successful growers in Vic., N.S.W. and Qld indicate that it is

5A. *G. acerata* Close-up of conflorescences (P.Olde)

both adaptable and hardy and tolerates limited dry conditions as well as frost down to at least -6°C. It prefers full sun to semi-shaded conditions in neutral to acidic well-drained sand, loam or gravelly loam, although plants grown in shady conditions may need pruning to maintain shape. Vigorous new growth is produced in response to summer watering and good garden conditions in general, including low phosphorus fertiliser. Not widely available in the nursery trade but can sometimes be obtained from nurseries specialising in grevilleas. Allow 1 m between plants.

Propagation Seed Sets numerous seeds that should germinate well if given the standard peeling pretreatment. *Cutting* Grows easily from cuttings taken from half-hardened new growth in spring or autumn. *Grafting* Untested.

Horticultural features A well-rounded, reasonably dense habit, crowded, dark green foliage with coppery new growth and terminal heads of bright

5B. *G. acerata* Different flower colour (M.Keech)

5C. *G. acerata* Shrub growing in natural habitat, Gibraltar Range NP, N.S.W. (P.Olde)

grey and brown, woolly flowers tinged with pink are the main characteristics of *G. acerata*. The umbel-like flower heads are conspicuous and mass on the shrub in great numbers in spring. In cultivation, especially in good conditions, blooms often continue for many months beyond the normal flowering period and provide an interesting contrast in the garden because of their unusual colour. It usually forms a compact, small shrub and, being relatively long-lived, is suitable for some applications in landscaping. In dry conditions, however, it tends to drop the lower leaves and become leggy, devaluing it for general landscape purposes. Pests and diseases do not seem attracted to it.

Grevillea acrobotrya C.F.Meisner (1855)
Plate 6

Hooker's Journal of Botany & Kew Garden Miscellany 7: 74 (England)

The specific name is derived from Greek *acros* (at the end, tip) and *botrys* (in botany, an inflorescence), a reference to the terminal conflorescences. AK-ROW-BOT-REE-A

Type: north of the Swan River, W.A., 1850–51, J.Drummond coll. 6 no. 185 (lectotype: NY) (McGillivray 1993).

Synonym: *G. vestita* var. *stenogyne* G.Bentham (1870).

An erect, open, prickly **shrub** 1–2 m tall, 1.5–2 m wide. Vegetative **branchlets** terete, silky to tomentose; floral branchlets ridged, ± glabrous, rarely tomentose. **Leaves** 1–4 cm long, 1–3 cm wide, ascending, shortly petiolate; lower leaves crowded, obovate to ovate, pungently toothed or slightly lobed around apical margin; leaves of floral branches sparse, usually deeply divided into 3, rigid, linear to subulate, pungent lobes; lobes occasionally again divided; upper surface glabrous or sparsely silky; margin recurved or refracted; lower surface silky, prominently veined. **Conflorescence** erect, pedunculate, axillary or terminal, simple or few-branched; unit con-florescence c. 2 cm across, globose, open; peduncle and rachis glabrous, sometimes silky; floral bracts 0.3–1 mm long, 0.3–0.6 mm wide, ovate-acuminate, falling early. **Flower colour:** perianth white or creamy white with chocolate buds; style white. **Flowers** regular, glabrous; pedicels 5–9 mm long, filamentous; torus 0.5–1 mm across, straight to slightly oblique; nectary inconspicuous, U-shaped, entire; **perianth** 2.5 mm long, 0.5 mm wide, cylindrical below limb; tepals splitting to base and rolling back at anthesis; limb 1.2 mm wide, erect, globular; **pistil** 3–4 mm long, glabrous; ovary globose; stipe 1–1.7 mm long; style 0.5 mm long, erect, constricted above ovary, slightly dilated above; pollen presenter erect, cylindrical with slight apical taper, its base scarcely broader than style. **Fruit** 8–11

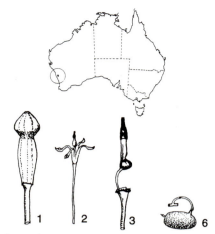

G. acrobotrya

mm long, 6–8 mm wide, 5–7 mm deep, on curved pedicel, attached horizontally to stipe, oblong, smooth; pericarp c. 0.5 mm thick. **Seed** 6.5–8.3 mm long, 3.3–4.2 mm wide, ellipsoidal, rimmed around border; wing lacking.

Distribution W.A., confined to a fairly small area between Eneabba and Badgingarra. *Climate* Summer hot, dry; winter mild to cool, wet. Rainfall 500–550 mm.

Ecology Grows in open low to tall heathland in well-drained shallow white or grey sand or sometimes in sandy gravel but always over laterite. Flowers all year, peaking in winter and spring. Insect-pollinated.

Variation Most populations have glabrous or mostly glabrous flower-bearing stems, but in several occurrences the stems are hairy (McGillivray, 1993).

Major distinguishing features Vegetative branchlets tomentose; leaves on basal branches and floral branches different in shape; leaf base wedge-shaped, the undersurface silky; flowers regular, glabrous; style slightly constricted above ovary, slightly swollen above; pollen presenter sub-cylindrical to conical, merging evenly into style, taller than broad at base; fruit smooth, oblong-ellipsoidal.

Related or confusing species Group 1, especially *G. amplexans*, *G. adpressa*, *G. erinacea*, *G. phanerophlebia*, *G. roycei* and *G. uniformis*. All differ in having leaves on the foliar branchlets similar to the basal, vegetative leaves.

Conservation status Despite its limited range, this species is not currently at risk.

Cultivation *G. acrobotrya* has proved both drought and frost resistant, and flourishes in both cool and warm temperate climates with low summer rainfall. Attempts at cultivating it have often foundered by not providing a well-drained situation in full sun, and it may be difficult to grow in humid coastal situations. It has, however, been successfully grown in Qld, western Vic. and at Burrendong Arboretum, N.S.W., in well-drained, sunny situations in acidic or neutral sands to heavy loams. Pruning may sometimes be required to reduce its somewhat untidy habit but it should be borne in mind that this will reduce the flowering potential of the plant for a season or two. Summer watering can induce root-rot disease and should be avoided. In summer-wet climates, pot culture has proved successful, but an open, free-draining mix with low phosphorus, slow-release fertiliser is essential. It is becoming available through specialist nurseries but may be difficult to find.

Propagation **Seed** Sets plentiful quantities of seed both in the wild and in cultivation which germinate well if fresh or pretreated. Seed germinates between 26 and 47 days after sowing (Kullman). *Cutting* Most readily propagated by cutting, taken from the lower, vegetative branchlets and not from the floral branchlets. Firm to semi-hard wood gives a good strike at most times of the year with best results in the warmer months. *Grafting* Has been grafted successfully onto *G. robusta* using the cotyledon or approach graft techniques.

Horticultural features *G. acrobotrya* flowers in a profusion of exquisite, chocolate-coloured buds followed by pleasantly perfumed cream flowers over a very long period. The almost leafless floral branches crowded with flowers above the basal foliage are rather novel, and while it can seem untidy in the eyes of some, the species is becoming increasingly popular in the gardens of native plant lovers. In landscaping, it would make a useful addition to the range of plants used for medium–dense screening in open, dry areas. It appears to be long-lived in cultivation, although its recent

6A. *G. acrobotrya* Fruits (P.Olde)

6B. *G. acrobotrya* Flowering among other plants at Burrendong Arboretum, N.S.W. (P.Olde)

6C. *G. acrobotrya* Floral branches (N.Marriott)

6D. *G. acrobotrya* Close-up of conflorescences showing chocolate buds (N.Marriott)

introduction to horticulture limits comment in this area. Unfortunately it is rarely available, even from native plant nurseries.

Grevillea acuaria F.Mueller ex G.Bentham (1870) **Plate 7**

Flora Australiensis 5: 452 (England)

Needle-leaf Grevillea

The specific name, derived from the Latin *acus* (a needle), alludes to the needle-like foliage in some forms. AK-YOU-ARE-EE-A

Type: south-western W.A., date unknown, J.Drummond (holotype: K).

Synonyms: *G. aculeolata* S.Moore (1899); *G. aculeolata* var. *longifolia* S.Moore (1899); *G. arida* C.A.Gardner (1923).

A rounded, bushy, usually lignotuberous **shrub** 0.2–1 m tall, rarely an erect, single-stemmed shrub to 1.5 m. **Branchlets** terete, glabrous, silky or tomentose. **Leaves** 0.5–3 cm long, 0.5–3.5 mm wide, erect to patent, subsessile, simple, subterete-trigonous to linear-subulate or linear, rarely narrowly elliptic, in some forms rigid, pungent, in others soft or leathery, non-pungent with oblique mucro; upper surface usually convex, sometimes flat, glabrous or hairy, silky to pubescent, dull or shiny, rarely punctate or with prominent ribs; margin entire, smoothly revolute, usually firmly abutting midvein below and obscuring lower surface, sometimes loosely rolled to recurved, the lamina below visible, silky, midvein prominent, glabrous, occasionally when dry the margin rolling over and obscuring even the midvein. **Conflorescence** erect, sessile or almost so, terminal or axillary, unbranched, umbel-like, flowers usually 4–6, sometimes solitary; rachis silky, sometimes pubescent; bracts 0.25–0.5 mm long, terete to triangular, villous, falling early. **Flower colour:** perianth scarlet, deep burgundy or cherry-red; style red or green with green tip. **Flowers** abaxially oriented, glabrous; pedicels 4–7 mm long; torus 1.5–2 mm across, very oblique, cup-shaped; nectary U-shaped to arcuate, the margin thick or thin, entire or with 2 teeth; **perianth** 7–10 mm long, 1.8–2 mm wide, ovoid-attenuate, ribbed, usually glaucous, bearded inside, separated at first along dorsal side to style end, after anthesis the tepals rolled down to curve; limb nodding to revolute, globular to spheroidal; **pistil** 14–21 mm long, glabrous; stipe 0.5–2 mm long; ovary usually glaucous, triangular with its base truncate and with a short basal protuberance dorsally; style at first looped out and upwards, afterwards gently incurved; style end convex to flat; pollen presenter lateral to very oblique, round to obovate, flat to convex. **Fruit** 8–11 mm long, 3–5 mm wide, usually ovoid to oblong–ovoid with attenuated apex, colliculose; pericarp c. 0.5 mm thick. **Seed** not seen.

Distribution W.A., widespread in the south-west between Lake Darlot, Brookton, Balladonia and Mt Ragged. **Climate** Hot, dry summer; mild, wet winter. Rainfall 150–600 mm.

Ecology Found in a variety of habitats; sandy depressions, winter-wet creek beds, open, red clay-loam woodland, rocky lateritic hills, near lakes and salt pans, in open to dense mallee and in heathland communities, in soils ranging from clay to sand. Flowers autumn–spring. Pollinated by nectar-seeking birds. Regenerates from seed or from lignotuber.

Major distinguishing features Leaves simple, entire, the upper surface usually convex, subshiny to dull, the venation obscure; conflorescence 1–6-flowered; torus very oblique, cup-shaped; perianth zygomorphic, ovoid, glabrous outside, bearded inside, the limb nodding in late bud; ovary glabrous, triangular, stipitate; fruit acuminate, colliculose.

Related or confusing species Group 15. *G. punctata* and *G. sulcata* differ in their broader perianth limb, the leaves with punctate upper surface and a prominent midvein, angularly refracted leaf margins, very prominent nectary and fruit 6–7 mm wide. *G. decipiens*, *G. oligantha* and *G. sparsiflora* differ in having the outer perianth surface hairy. *G. kennedyana* differs in its larger flowers (torus > 4 mm across).

Variation *G. acuaria* is here treated as a polymorphic species. Until current studies are completed, we recognise nine forms.

Needle-leaf form This form is found in the western sector of its range, in the wheatbelt from Brookton through Merredin, Cowcowing etc. to Coolgardie. It has leaves 1–3 cm long, patent to ascending, straight and rigid, subterete-trigonous, terminated by an aristate point c. 1 mm long. In this form the nectary is not prominent and has an entire, thick-walled margin; the style end is convex on the dorsal side; pistil 16–20.5 mm long. The type of *G. acuaria* is of this form.

Dagger-leaf form (represented by the synonym *G. aculeolata*) This form is a much-branched shrub very similar to the Needle-leaf form but differing in its broader, linear leaves, the midvein on the undersurface level with or slightly recessed below the margin. In many, but not all, collections of this form the flowers are much larger (perianth 10 mm long, pistil to 23 mm long) and have a sparser indumentum on the inner perianth surface. Some collections also have leaves ridged on the upper surface and angular margins. It is has been collected at Coolgardie, Norseman and Balladonia, extending north in a zone between Bullfinch, Laverton and Lake Darlot.

Widgiemooltha form (represented by the synonym *G. arida*). This differs from the first two in its more obtuse leaf and its leaf margins not firmly abutting the midvein on the undersurface. When dry, the leaf margins are so strongly rolled as to obscure even the midvein on the undersurface. Re-collection of this form at the type locality is needed. Specimens from the Toomey Hills closely resemble this form.

Parker Range form This very distinct form has a trim, low-growing habit with leaves like the above but pungent and with a spreading leaf, and also differs in its oblong–ellipsoidal fruits. It grows with the Soft-leaf form on the southern slopes of the Parker Range, indicating that at least one of these two should be recognised as a distinct species.

G. acuaria

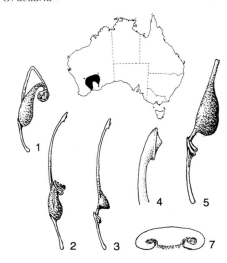

Kulin form A form with its leaf margins more closely abutting the undersurface and the leaf apex more blunt than the above two but differing from them in its pistil 14–16 mm long. It occurs between Kondinin and Corrigin.

Soft-leaf form This form usually has bright green, soft leaves 0.5–0.8 mm wide, c. 1 cm long, strongly ascending to erect and crowded, narrowly linear to trigonous, curved, non-pungent, the margins firmly abutting the midvein on the undersurface. It closely resembles the Needle-leaf form in some respects as well as the Glaucous-leaf form listed below, since some collections of this form near Varley have glaucous leaves. It also has a very conspicuous, toothed nectary and flat style end. It occurs in the Lake King to Bremer Ranges area, with collections from Mt Madden and Norseman. In this form, the styles are usually green.

Salmon Gums form In its leaf and floral morphology, this form resembles the Soft-leaf form but the leaves are stiffer, longer and less crowded. It occurs between Salmon Gums and Widgiemooltha. Its relationship to the Widgiemooltha form needs assessment as the two seem to merge in this area.

Shiny-leaf form This form has broad, obovate to elliptic green leaves (1.5–2 but up to 4.5 mm wide) with a glossy upper surface and obtuse-mucronate apex. At Maggie Hayes Hill on the breakaway, it is a single-stemmed shrub to 1.5 m with a few, erect branches. Elsewhere, it is a bushy shrub c. 0.5 m high. In its floral morphology, this form is the same as the Soft-leaf form, but the leaves differ in their leathery texture, width and degree of enclosure of the leaf undersurface. In this form, the undersurface is clearly exposed and silky. It occurs between Lake Cronin and the Bremer Range with disjunct collections from Mt Ragged and Dundas Nature Reserve south-east of Norseman. Its relationship to the Widgiemooltha form and Soft-leaf form requires further study. *G. punctata* has a similar leaf shape.

Glaucous-leaf form This form resembles the Shiny-leaf form in every respect and overlaps its distribution, but the distinctive blue-green glaucous leaves are not shiny. Some collections of the Soft-leaf form are also glaucous.

Conservation status Not presently endangered due to its occurrence in arid areas.

7A. *G. acuaria* Plant in natural habitat, N of Norseman, W.A. (C.Woolcock)

Cultivation Some forms have been extensively tested in cultivation, others have only been introduced recently. All forms appear to be adaptable, drought-tolerant and frost-hardy. Successful growers from Vic., Canberra, and N.S.W. have attested that, given a well-drained, sunny site, these forms grow readily into fairly dense, rounded shrubs in a wide range of soil types. However, a clay-loam with intermixed gravel is best. Light annual pruning will keep the plants compact and dense. Summer watering is not required if sub-soil moisture is available. Semi-shade is tolerated in warm, dry climates. Die-back can be expected in wet, cold conditions and in areas with high rainfall although pot culture is suitable in these areas. Some forms are available from native plant nurseries, mainly in Vic.

Propagation *Seed* Produces reasonable quantities of seed that germinate well if pretreated. *Cutting* Although often slow to put down roots, most forms give a high strike rate using firm to semi-hard material. Strikes readily from late spring to early autumn. *Grafting* The Shiny-leaf form grafts readily onto *G. robusta* and appears to be long-term compatible.

Horticultural features Most forms grow into neat, rounded shrubs with foliage dense to the ground. While the individual flowers are attractive, they are rarely prolific in cultivation, although the Salmon Gums form is particularly showy in the wild in full flower. In cultivation, flowering is prolonged with a winter peak and hence it is highly valuable to nectar-feeding birds. Plants are generally not affected by pests or diseases and are quite long-lived. Selected forms have some value as rockery or shrubbery plants, the most impressive being the Glaucous-leaf form.

General comments It is likely that several forms will be recognised as separate species or subspecies, but their clear separation from all other forms requires more field work.

7B. *G. acuaria* Dagger-leaf form near Kalgoorlie, W.A. (N.Marriott)

7C. *G. acuaria* Close-up (N.Marriott)

7D. *G. acuaria* Flowers and foliage, Maggie Hayes Hill, W.A. (P.Olde)

7E. *G. acuaria* A glaucous-leaf form, Round Top Hill, W.A. (P.Olde)

7F. *G. acuaria* Soft-leaf form E of Crossroads, Forrestiana, W.A. (P.Olde)

7G. *G. acuaria* Glossy-leaf form, Mt Day, W.A. (P.Olde)

7H. *G. acuaria* Flowers and foliage, Kulin, W.A. (C.Woolcock)

7I. *G. acuaria* Flowers and foliage, in cultivation, Royal Botanic Gardens, Sydney, ex Parker Range (N.Marriott)

Grevillea adenotricha D.J.McGillivray (1986) Plate 8

New Names in Grevillea *(Proteaceae)*: 1 (Australia)

The specific name is derived from the Greek *aden* (a gland) and *trichion* (a small hair), a reference to the glandular-pubescent indumentum covering the foliage and branchlets. AD-ENN-O-TRY-KA

Type: Manning Gorge, W.A., 1 Aug. 1973, B.Gill G20 (holotype: QRS).

A low, hairy **shrub** 0.8–2 m high, 0.8–2 m wide. **Branchlets** terete, glandular-hispidulous. **Leaves** 3–7 cm long, 1.5–2.5 cm wide, spreading to erect, petiolate, simple, oblong, sometimes narrowly ovate, sometimes undulate, toothed around the otherwise flat margins with 6–12 pungent lobes on each side; upper and lower surfaces concolorous, densely glandular-hispidulous with midvein and lateral veins conspicuous. **Conflorescence** 1.5–2.5 cm long, c. 1 cm wide, erect, terminal or axillary, simple, secund-globose; peduncle 1–3 cm long; peduncle and rachis glandular-villous; bracts 2 mm long, ovate-acuminate, villous outside, sometimes persistent at anthesis. **Flower colour:** perianth and style red. **Flowers** acroscopic; pedicels 2–2.5 mm long, glandular-villous; torus c. 1 mm across, ± straight to slightly oblique, extending far beyond dorsal side of pedicel; nectary prominent, U-shaped, entire and smooth; **perianth** c. 2.5 mm long, 2 mm wide, broadly ovoid, strongly curved, glabrous outside, bearded near base inside, cohering except along dorsal suture; limb revolute, spheroidal, the segments impressed at margin; **pistil** 4–6 mm long, glabrous, attached laterally on outer rim of torus (± at right angles to pedicel); ovary globose; stipe 1–1.3 mm long, lateral to torus; style not exserted before anthesis, refracted from dorsal line of perianth, afterwards straight to slightly incurved; style end flattened; pollen presenter lateral, obovate, umbonate. **Fruit** 9.5–11 mm long, 5 mm wide, erect to oblique on curved pedicel, persistent, oblong–ellipsoidal, glabrous; pericarp 0.5–1 mm thick. **Seed** 6–6.5 mm long, 3 mm wide, ellipsoidal with recurved margin and waxy, wing-like rim, smooth.

Distribution W.A., in the Kimberley at Manning Gorge, the Prince Regent River and Lushington Valley. ***Climate*** Hot, wet summer; mild, dry winter. Rainfall 900–1250 mm.

Ecology Grows on sandstone plateaus or bluffs, in skeletal, sandy soil. Flowers late autumn to late winter. Pollinator unknown. The role of the vegetative indumentum in the species' survival would be worth investigating.

Variation This species is known from only 4 collections which appear to be relatively uniform.

Major distinguishing features Vegetative parts glandular-hispidulous; leaves concolorous, oblong with flat, dentate margins; conflorescence sub-globose, pedunculate; perianth zygomorphic, glabrous outside, bearded within; pistil glabrous; ovary stipe lateral to torus; pollen presenter lateral; pericarp thin-walled.

Related or confusing species Group 10. Closely related to *G. longicuspis* which has rounded leaves with fewer (2–6) lobes per side and bearing a less obvious, non-glandular vegetative indumentum.

Conservation status 2RC. Possibly more widespread than presently known as the area is isolated and rarely visited. Conserved in Prince Regent River NP.

Cultivation *G. adenotricha* is unknown in cultivation but its natural habitat suggests that it would be hardy in arid-monsoonal, tropical climates and should succeed possibly as far south as Brisbane or Geraldton within Australia. Experience has shown that plants from these tropical regions do not adapt well to continuous cold conditions, hence cold areas are unlikely to be suitable. Its natural occurrence in sandy, skeletal soils in sandstone country usually in well-drained, sunny habitats suggests the primary requirements for introduction to cultivation, with adaptability to other situations to be tested later.

Propagation To date, propagation of this species has not been attempted. It is likely to germinate readily from seed.

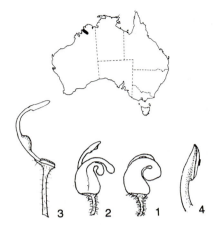

G. adenotricha

8A. Kevin Coate examining typical plant of *G. adenotricha*

8B. *G. adenotricha* Conflorescence. Note glandular hairs on stems (R.Gooddale)

Horticultural features *G. adenotricha* is unique in the decorative indumentum of curious glandular hairs that invest all but its flowers. Although it is a somewhat spindly shrub in the wild, it may become more shapely in cultivation, especially with regular pruning. Its red flowers, while individually extremely small, are borne on unusual long peduncles and are quite ornamental. The oblong leaves are also attractively shaped, being evenly and regularly toothed.

General comments The occurrence of this species in isolated country has resulted in very few collections. Several searches of the Manning Gorge have failed to relocate the species at that place. In May 1988, however, an expedition led by Kevin Coate from Perth to trace the footsteps of the explorer George Grey found the species near Hanover Bay. Two populations were discovered on ridgetops over a distance of several kilometres. One area, which had been burnt the previous year after a lightning strike, contained hundreds of young plants. The accompanying photographs from this trip are the first taken of the species. The exudate from the glandular hairs covering the plant emits a pervading, unpleasant odour similar to that of vomit. This attracts myriads of insects including native bees which feed on the exudate and visit the flowers (W.Molyneux, pers. comm.).

Grevillea adpressa P.M.Olde & N.R.Marriott (1993) Plate 9

Nuytsia 9: 250 (Australia)

The specific epithet is derived from the Latin *adpressus* (appressed), in reference to the indumentum on the leaf undersurface. AD-PRESS-A

Type: 5.6 km W of Arrino on Arrino West Rd, W.A., 16 Sept. 1991, P. Olde 91/112 (holotype: NSW).

Irregular **shrub** to 1–2 m high, 1–2 m wide. **Branchlets** white-silky or tomentose. **Leaves** 0.7–1.2 cm long, 1–1.5 cm wide, patent to spreading, sessile, stiff, stem-clasping, rhombic to angularly ovate, broadly dentate with basal lobes consistently retrorse; venation conspicuous, palmate, the primary and sometimes secondary veins tipped by excurrent, pungent points 1–5 mm long; reticulate venation evident, edge veins conspicuous; upper surface glabrous; lower surface white-silky to tomentose; margin shortly recurved to flat. **Conflorescence** axillary or terminal, pedunculate, simple or few-branched; unit conflorescence 1–3 cm long, globose, open; development acropetal; peduncle and rachis glabrous or almost so; floral bracts c. 2 mm long, ovate to lanceolate, ciliate, rarely persistent to anthesis. **Flower colour:** perianth white, the limb pale brown before anthesis; style white with pinkish tinges. **Flowers** regular, glabrous; pedicels 7–8 mm long, rarely to 12 mm; torus 0.5–1 mm across, straight to slightly oblique; nectary scarcely evident; **perianth** 3.5–5.5 mm long, 0.7–0.8 mm wide, white, oblong-obovoid to ellipsoidal below limb, a few trichomes sometimes at base inside; limb 1.5 mm wide, brown, globose to subovoid,

G. adpressa

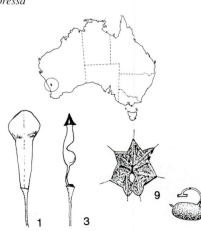

9A. *G. adpressa*, near Arrino, W.A. (N.Marriott)

9B. *G. adpressa* Flowering habit (P.Olde)

the segments ribbed; **pistil** 2.5–5.8 mm long; stipe 1–2.5 mm long, flexuose; ovary globose; style white, constricted immediately above the ovary, the zone of constriction 0.1–0.3 mm long, then dilated, the dilation globose to cylindrical, 0.5–1 mm wide; pollen presenter conical, its base 0.6–1 mm wide, straight, broader than style. **Fruit** 9–12 mm long, 5 mm wide, 5 mm deep, ± perpendicular to stipe on curved pedicels, oblong–ellipsoidal, smooth; pericarp 0.5mm thick. **Seed** 5.5–6 mm long, 2.5 mm wide, ellipsoidal, biconvex, with a membranous border on the inner face, otherwise smooth.

Distribution W.A., between Mingenew and Watheroo. *Climate* Hot, dry summer; mild, wet winter. Rainfall 330–460 mm.

Ecology Grows in brown, gravelly loam or yellow sand in low heath. Flowers spring. Pollinated by insects.

Variation A stable, uniform species.

Major distinguishing features Branchlets densely hairy; leaves stem-clasping, silky on undersurface; conflorescence globose; pedicels filiform; nectary obscure; flowers glabrous; perianth regular, a few trichomes sometimes on base inside; style with a globose dilation above a short constriction; pollen presenter conical; fruit smooth, oblong–ellipsoidal.

Related or confusing species Group 1, especially *G. amplexans* and *G. uniformis*. *G. amplexans* has glabrous leaves. *G. uniformis* differs in the leaf base patent to spreading, and a usually smaller, slightly less turgid style.

Conservation status Recommended 3VC.

Cultivation *G. adpressa* has been grown for some years in western Vic. and Qld where it has proved relatively hardy and reliable. It will tolerate both drought and frost reasonably well but appears to dislike summer humidity. Good drainage is essential but it is not fussy as to soil type, growing equally well in sandy soil and heavy, gravelly loam. Difficulties can be experienced in establishing this species, and protection from strong winds and extreme weather conditions is recommended in the early stages. The species is not generally available in nurseries.

Propagation *Seed* Should germinate readily from fresh seed with some pretreatment. *Cuttings* Cuttings of fresh, semi-firm material strike readily. *Grafting* Untested.

Horticultural features *G. adpressa* has some horticultural merit but is essentially a plant for the collector. Flowering can, in some instances, be quite showy but plants are often untidy and scrappy in the wild with little to recommend them over closely related species. In cultivation, however, plants are more compact, and the rounded grey leaves are distinctive. With prickly leaves it should not be planted near walkways. Landscapers could find it an effective contrast plant in the garden, for both its leaves and its white flowers.

General comments *G. adpressa* and *G. amplexans* are closely related, but grow in homogeneous populations over the same area and are almost certainly reproductively isolated. There is also a close relationship with *G. uniformis* (prior to 1993 a subspecies of *G. acrobotrya*), in many characters closer than that with *G. amplexans*. We consider it reasonable to maintain all four taxa at specific rank.

Grevillea agrifolia A.Cunningham ex R.Brown (1830) Plate 10

Supplementum Primum Prodromi Florae Novae Hollandiae exhibens Proteaceas Novas: 24 (England)

Blue Grevillea

Aboriginal: wurrgan (Kwini)

The specific name is derived from the Latin *aeger* (weak, ill) and *folium* (a leaf), in reference to the type collection having chewed, tatty leaves. AG-RI-FOLE-EE-A

Type: Lacrosse Is., [W.A.], 18 Sept. 1819, A.Cunningham (lectotype: BM) (McGillivray 1975).

Synonym: *G. agrifolia* var. *major* A.J.Ewart & B.Rees (1911).

A robust, single-stemmed **shrub** or small **tree** 2–6 m high, often with stout branches from near the base. **Branchlets** angular to almost terete, silky. **Leaves** 5–16 cm long, 3.5–9.5 cm wide, ascending to spreading, simple, petiolate, concolorous, obovate to almost round, conspicuously veined; upper and lower surfaces similar, densely and minutely silky; margins undulate, flat, entire or with 1–8 shallow pungent teeth in upper half. **Conflorescence** decurved, pedunculate, axillary or terminal on short branchlets, simple or branched; unit conflorescence 3–7.5 cm long, conico-cylindrical, dense; peduncle silky; rachis silky or glabrous; bracts 1–1.8 mm long, ovate, glabrous with ciliate margins to sometimes silky outside, falling before anthesis. **Flower colour:** perianth green in bud turning white at anthesis; style pale green to cream with green tip. **Flowers** adaxially acroscopic; pedicels 2.5–4 mm long, glabrous or silky; torus 1.5–2.2 mm across, oblique; nectary conspicuous, V-shaped, toothed or entire; **perianth** 5–6 mm long, 2–3 mm wide, strongly curved, ovoid, glabrous or silky outside, densely bearded inside, at first separated along dorsal suture, all tepals separating below limb before anthesis, the dorsal tepals reflexing and exposing inner surface, afterwards all free to base and loose; limb revolute, spheroidal, the style end partially exposed before anthesis; **pistil** 13.5–19 mm long; stipe 2.5–4 mm long; ovary glabrous or with glandular hairs in upper half; style at first exserted from below curve on dorsal side and looped upwards, strongly incurved in upper half after anthesis, sparsely glandular-pubescent usually over its entire length, rarely glabrous or almost so; pollen presenter very oblique, obovate, flat. **Fruit** 18–24 mm long, 14–17 mm wide, erect or slightly oblique, ellipsoidal to subglobose with obscure apiculum on dorsal side c. 5 mm from the obtuse apex, glabrous, rugose; pericarp 2–4.5 mm thick. **Seed** 7.5–10 mm long, 4–5 mm wide, oblong–ellipsoidal, biconvex, membranously winged all round.

Distribution W.A., widely distributed in the E Kimberley as far S as Halls Creek, and in the NW

G. agrifolia

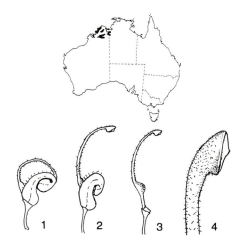

10A. *G. agrifolia* Cultivated specimen, Brisbane (M.Hodge)

10B. *G. agrifolia* Flowers and fruit, in cultivation (P.Olde)

part of the Victoria R. district, N.T., as well as on offshore islands. *Climate* Hot, wet summer; dry, warm winter. Rainfall 400–1300 mm.

Ecology Grows in yellow, sandy podsol or yellow-brown loam with lateritic gravel, in open, mixed vegetation of lowland woodland, often in rocky situations on sandstone or quartzite. Flowering dependent on the length of the wet season, but usually from autumn to spring. After low-intensity fire, regenerates from epicormic buds on the stems and branchlets. Small brown honeyeaters have been observed probing the flowers, but insect pollination is not ruled out.

Variation A relatively stable, uniform species. The outer perianth surface varies from glabrous to silky in a random fashion. Some plants, notably from the Drysdale River NP, have glabrous styles (usually hairy).

Major distinguishing features Leaves simple, concolorous, lightly toothed, minutely silky both sides, usually > 4 cm wide; conflorescence conico-cylindrical; perianth zygomorphic, bearded inside; ovary glabrous or sparsely hairy; style usually hairy, far exceeding perianth; pollen presenter oblique; fruit erect, 18–22 mm long; pericarp 2–4.5 mm thick.

Related or confusing species Group 11, especially *G. microcarpa* which has smaller fruits, narrower leaves and a smooth style end. Species in Group 18 are closely related, especially *G. velutinella*, which has similar leaves but differs in its inconspicuous leaf venation, its much longer conflorescences, its flowers borne on longer pedicels (7–9.5 mm) and a pistil that barely exceeds the perianth.

Conservation status Due to low development in the area of its natural habitat as well as its widespread distribution, this species does not appear to be at risk.

Cultivation *G. agrifolia* is reported in cultivation at the Darwin Botanic Gardens but grows readily at least as far south as Brisbane and has even been grown successfully outdoors in frost-free areas in Sydney. At this latitude, the leaves tend to spot with black fungus and look unhealthy during the winter, although they recover in the summer. The species tolerates hot, dry climates but performs best with copious summer rainfall. An acidic to neutral, sandy loam is favoured and it should be grown in a full sun, well-drained situation where it will respond by growing into a beautiful specimen. In more southerly climates, growing it among large rocks or against a north-facing wall is recommended. Native plant nurseries in Qld occasionally stock this species and it sometimes appears on seed lists.

Propagation *Seed* Sets plentiful seed that germinates well if pre-treated by soaking for 24 hours in hot water or by peeling. *Cutting* Grows readily from cuttings of firm, young growth taken in late spring–summer. *Grafting* Successfully grafted onto *G. macleayana*, *G. robusta* and *G.* 'Poorinda Royal Mantle' using the top wedge, whip and mummy techniques.

Horticultural features *G. agrifolia* grows into a robust, dense and handsome shrub with large, softly hairy, blue-grey leaves. New growth is bronze, gradually becoming paler before maturing to its adult colour. The axillary racemes of whitish green flowers tend to be obscured by the large leaves but are quite pleasing on close investigation and have a spicy fragrance. The large, green fruits are also attractive and persist on the plant for a long period. In the garden landscape, it makes a spectacular specimen or feature plant or can be used effectively as a screen or contrast plant. This is in contrast to the appearance in the wild where the leaves are often chewed and the plants are more open.

Early history Seed was sent to England by Allan Cunningham in 1820 to be grown in hothouses, but it does not appear to have persisted for long.

General comments *G. agrifolia* var. *major*, collected by G.F.Hill at Napier Broome Bay in Dec. 1909 and described in 1911, cannot be accepted because of the random nature of its occurrence. Plants with a silky perianth occur among populations of plants with a glabrous one but are clearly of the same species.

Grevillea albiflora C.T.White (1944)
Plate 11

Proceedings of the Royal Society of Queensland 55: 79 (Australia)

White Spider Flower

The specific name is derived from Latin *albus* (white) and *flos* (a flower), in reference to the colour of the flowers. AL-BI-FLORE-A

Type: Gilruth Plains, E. of Cunnamulla, Warrego District, Qld, 20 May 1939, S.T.Blake 14065 (holotype: BRI).

A spreading, irregular, somewhat hoary **shrub** or small **tree** 3–6 m high. **Branchlets** angular, tomentose. **Leaves** 8–30 cm long, ascending, ± sessile, pliable, leathery, usually pinnatipartite with 5–9 strap-like, linear lobes 1–2 mm wide, the lower lobes sometimes with secondary division, sometimes simple leaves also present; upper surface silky to pubescent becoming glabrous, midvein evident; margins smoothly revolute to midvein; lower surface silky, woolly in grooves, flattened midvein prominent. **Conflorescence** erect, paniculate, terminal, occasionally axillary in upper axils; unit conflorescences 5–10 cm long, 4 cm wide, cylindrical, medium to dense, the topmost flowers opening first or all synchronously; peduncle 2–5cm long, densely tomentose; rachis 2.8–3 mm wide at base, woolly

G. albiflora

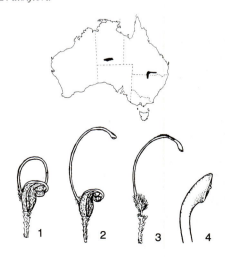

to villous; floral bracts 1.5–2 mm long, 1–1.5 mm wide, conspicuous in bud, ovate, glandular-villous, falling before anthesis. **Flower colour:** perianth green in bud, becoming white; style white to creamy yellow. **Flowers** horizontal with ventral suture extrorse; pedicels 3–6 mm long, woolly to villous, patent; torus c. 2 mm across, ± straight; nectary U-shaped; **perianth** 10–12 mm long, 1.5 mm wide, cylindrical, villous outside, glabrous inside except a few hairs at base, at first separated along dorsal suture, all tepals then free below limb, the dorsal tepals reflexing and exposing inner surface; limb revolute, velvety, spheroidal; **pistil** 15–21.5 mm long; stipe 1.6–2.4 mm long, appressed-villous; ovary woolly to villous; style at first exserted below curve on dorsal side, looped outwards before anthesis, strongly incurved to C-shaped after anthesis, villous near base then glabrous in upper 2/3; pollen presenter oblique, convex, obovate–elliptic; stigma central. **Fruit** 20–25 mm long, 14–17.5 mm wide, persistent, perpendicular to pedicel, lens-shaped, velvety; pericarp 2–3 mm thick. **Seed** 16–19 mm long, 9–15 mm long, ovate, membranously winged all round, smooth.

Distribution N.T., Qld and N.S.W. Widely distributed in two disjunct inland areas about 1400 km apart: one in the N.T. from near Ayers Rock east to Rainbow Valley, the other from east of Cunnamulla, Qld, to west of St George and southwest to near Bourke, N.S.W. *Climate* Summer hot and dry with occasional thunderstorms; winter warm to cool with many frosts. Rainfall 150–225 in N.T.; 300–400 mm in Qld and N.S.W.

Ecology Grows on exposed, deep, red sand ridges or sandplain, often in company with *G. juncifolia* which has similar foliage. Flowers late spring to summer. May be insect-pollinated, as the flowers have a fresh spicy scent that attracts many insects.

Variation A fairly uniform species, but some populations from the N.T. are more robust with broader leaf lobes and larger flowers than populations from Qld and N.S.W.

Major distinguishing features Leaves pinnatipartite; conflorescence paniculate with cylindrical units; floral rachis woolly–villous, broader than peduncle; perianth zygomorphic; ovary densely hairy, stipitate; style glabrous except at base; fruit persistent, lens-shaped.

Related or confusing species Group 30, *G. eriobotrya*, in which the whole style is hairy and the inside of the perianth is glabrous, and *G. pterosperma* which differs in its longer floral bracts and oblique torus.

Conservation status Not at risk in N.T. Rare in N.S.W. and Qld.

Cultivation *G. albiflora* has been successfully grown in inland areas of Vic. and Qld, including Toowomba and Glenmorgan, and superb specimens are cultivated on several inland properties in S.A.. Grafted plants have flowered in western Vic. and Brisbane. Once established, it is drought-hardy and requires little maintenance except the removal of old flowering canes to improve shape. Although not widely grown, success has been achieved in acidic to alkaline, gravelly loam, deep sand and sandy loam. It will not tolerate heavy clay, and demands well-drained soil in full sun. Hot, humid summers are not tolerated and its natural occurrence in dry, low rainfall, inland areas suggests a similar climate for cultivation. It will endure frosts to at least -6°C. Most specialist native plant nurseries do not stock the species, but it has been sold from time to time in Qld nurseries.

Propagation Seed Sets copious quantities of seed that remain on the shrub for a long period and are therefore easy to collect. Seed germinates readily with standard peeling treatment. Soaking seed in warm water for 1 or 2 days prior to sowing has also been effective. *Cutting* Experience has shown that species with a dense indumentum on the branchlets are subject to rotting and fungal attack and can rarely be induced to produce roots. This is true of *G. albiflora*. *Grafting* By choosing appropriate rootstocks, this species could undoubtedly be induced to grow in warmer, coastal climates provided summer humidity did not too seriously affect the foliage. Successful grafts with *G. robusta* have been made, but the combination may not be compatible long-term. It has been grafted successfully onto G. 'Poorinda Anticipation' using cotyledon grafts.

Horticultural features In flower, *G. albiflora* is an exceptionally showy plant with bright white, cylindrical racemes carried in dense clusters beyond the leaves. These are produced in summer and the species suggests itself as another useful out-of-season flowering plant for the garden. By contrast, the beautiful, hoary-white indumentum on its branchlets and the grey-green foliage seem diminished in beauty but these features nonetheless command admiration especially on young plants. Its sweetly perfumed flowers attract beetles which transfer pollen deposited on their backs. After flowering has finished, the follicles are retained on the plant for long periods and the seeds cause the pods to rattle loudly in a breeze. Although stock are reported to browse the foliage during the flowering season, the species is not recommended for this purpose, unless grown expressly for it.

11A. *G. albiflora* Close-up of conflorescences (N.Marriott)

11B. *G. albiflora* Habit of plant, in cultivation near Geranium, S.A. (N.Marriott)

Grevillea alpina J.Lindley (1838)

Plate 12

T.L.Mitchell, *Three Expeditions into the interior of Eastern Australia* 2: 178 (England)

Mountain Grevillea, Cat's Claws

The specific epithet is derived from the Latin *alpinus* (alpine) in reference to the locality from which Lindley believed it was first collected. AL-PY-NA

Type: Mt William The Grampians, Vic., 1836, Major Mitchell's Third Expedition (lectotype: CGE) (McGillivray 1993).

Synonyms: *G. alpestris* C.F.Meisner (1856); *G. alpestris* var. *helianthemifolia* C.F.Meisner (1856); *G. alpina* var. *aurea* W.Guilfoyle (1909); *G. alpina* var. *dallachiana* G.Bentham ex W.Guilfoyle (1909); *G. dallachiana* F.Mueller ex Hazelwood (1972).

A **shrub**, decumbent to 30 cm, sometimes rounded to 1 m, or spindly and erect to 2 m. **Branchlets** terete, villous to sparsely silky. **Leaves** 0.5–3 cm long, 1.5–10 mm wide, often retrorse, sometimes ascending to spreading, sessile, simple, oblong to elliptic, obtuse to acute, mucronate; upper surface villous to glabrous and granulate, midvein evident; margin entire, recurved to revolute; lower surface villous, the indumentum often appressed, rarely absent, midvein prominent. **Conflorescence** 1–3.5 cm long, 2–3 cm wide, erect or decurved, simple, rarely few-branched, terminal, often on short branchlet or axillary, sometimes below first set of leaves, umbel-like to globose, sometimes 2–flowered, apical flowers opening first; peduncle and rachis tomentose to villous; bracts 0.5–4 mm long, narrowly triangular, variably persistent. **Flower colour:** either a single colour; cream, green, yellow, orange, pink or dull red, but more often a combination of these. **Flowers** adaxially acroscopic; pedicels 3–8 mm long, slender, tomentose; torus 1.5–2 mm across, oblique; nectary patent, tongue-like; **perianth** 3–10 mm long, 3–6 mm wide, oblong to ovoid, often markedly saccate, sometimes strongly curved from base, sometimes oblong–saccate below curve, ribbed, detaching quickly after anthesis, sparsely to densely tomentose outside, densely bearded inside near level of ovary, otherwise hairs sparse, cohering except along dorsal suture; limb revolute, ovoid, densely villous, sometimes keeled or crested, often with 2 prominent horns protruding from dorsal side; **pistil** 8.5–20.5 mm long; ovary sessile, densely white-villous; style not exserted before anthesis,

GREVILLEA ALPINA

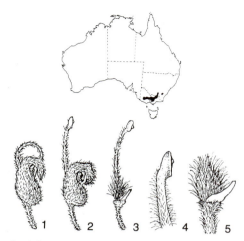

G. alpina

afterwards straight, villous especially in lower half; style end disc-like, glabrous; pollen presenter lateral, oblong–elliptic, flat. **Fruit** 8.5–12 mm long, 5–6 mm wide, erect on the stipe, ovoid, pubescent; style persistent; pericarp c. 0.5 mm thick. **Seed** 8–9.5 mm long, 2–3 mm wide, elliptic with recurved margin bordered by a waxy rim; surface minutely hairy.

Distribution Vic., N.S.W. and A.C.T. Widespread in Vic. in most areas from Melbourne northeast to the N.S.W. border and west to The Grampians. It occurs in N.S.W. at Albury and in the A.C.T. on Black Mountain. *Climate* Warm to hot, dry summer; cold to cool, wet winter. Rainfall 400–900 mm.

Ecology Generally occurs in well-drained, often rocky sites in dry sclerophyll forest, open woodland, scrub woodland, heathland or mallee in stony loam, clay, sand, sandy-loam or granite-loam. Flowers winter–summer, also autumn in favourable seasons. Pollinated by birds.

Major distinguishing features Leaves simple, entire, often retrorse; nectary conspicuous, tongue-like, extending laterally 1–3 mm beyond torus; perianth limb sometimes with prominent horn-like appendage; ovary sessile, densely hairy; pollen presenter lateral.

Related or confusing species Group 25, especially *G. chrysophaea*, which generally has ascending leaves more widely distributed along the branchlet, mostly longer than 2 cm, and with pominent secondary venation; its flowers have a less conspicuous nectary, extending laterally beyond the torus < 1 mm.

Variation An extremely variable species requiring further research into the many distinct populations. Throughout most of its range, each district has its own particular form with slightly different habit, leaves, flowers, flower colour, etc. Root-suckering populations are known from near Benalla and at Mhyree, near Wangaratta, Vic. We have grouped the forms superficially into five major groupings, some of whose features overlap.

Grampians form (type form) This form is most distinct and has the largest flowers, borne in short, few-flowered, erect conflorescences. Flowers are normally very bright as they lack the dense indumentum of many other forms and are a rich orange-red and yellow. Foliage is open with large, linear–elliptic leaves which are thicker and of quite a different texture and appearance from all other forms. Plants are normally small and decumbent but can range to 2 m at Mt Zero. They occur throughout The Grampians as well as the surrounding plains and hills.

Small-flowered form This includes populations from the Albury area, Canberra, Chiltern, Beechworth, Tooboorac. It is a quite distinct form, having a cluster of flowers at the end of the branch with 2-flowered conflorescences extending down the branchlet and massed together to give the impression of one large, pendent conflorescence with a number of small leaves between. Individual flowers have a squat perianth tube and very short styles (8.5–9.5 mm long). All populations are most attractive and showy in full flower. We believe that a detailed analysis of *G. alpina* will show that this taxon warrants formal recognition.

Northern Victorian form Plants in this group are closely related to the Small-flowered form, differing in their larger, less crowded leaves, an erect, open habit and slightly larger flowers which are normally 2-coloured, and displayed in graceful, pendent clusters. Isolated populations often occur on dry, stony hills in such places as the Warby Range, Strathbogie Range and several other areas. Plants are usually sold under locality names in the nursery trade.

Goldfields form Populations in this group typically have crowded foliage with small, elliptic, hairy, grey leaves and showy pendent conflorescences containing medium-sized flowers usually of only one main colour. Plants are dense and mounded to c. 1 m. They occur in the Central and Western Goldfield regions of Vic. in areas such as Bendigo, Castlemaine, Whroo Forest and St Arnaud. Plants are often sold under these locality names in the horticultural trade or under such trade names as Whroo Forest (white and green), Goldfields Apricot, Goldfields Pink and Goldfields Red.

Southern Hills forms These populations have elliptic, oval or occasionally almost round leaves and fairly small racemes of dull, woolly, orange-red and cream to dull red and cream flowers. They form open, rounded or occasionally erect shrubs to 2 m. Some well-known forms which take their name from their locality are Mt Slide, Mt Evelyn, Lerderderg Gorge, Cardinia, Kinglake, and the Dandenong Ranges. In cultivation, this form is the hardiest.

Conservation status Not presently endangered.

Cultivation Unfortunately, most forms of *G. alpina* have not proved to be either adaptable or long-lived in cultivation, despite a wide natural distribution and a long period of cultivation in Australia and overseas. Over 12 selections are in cultivation in the U.S.A. All forms grow best in cool-wet to hot-dry climates. Most are intolerant of summer rainfall with its attendant humidity but withstand frosts to -11°C. The species flourishes when planted in a very well-drained site in full sun to semi-shade in acidic, sandy or gravelly loam or in dry, stony soils. Plants in the wild usually grow in very dry, inhospitable sites with considerable competition from the surrounding vegetation. Re-creation of these conditions in the garden produce slower-growing but far longer-lived plants. Once established, plants do not require fertiliser. Summer watering is generally resented. Regular pruning may be needed to maintain shape and density. Most forms make delightful pot specimens which should be kept on

12A. *G. alpina* Goldfields form ex Morrl Morrl Forrest, Vic. (N.Marriott)

12B. *G. alpina* Typical Grampians form (N.Marriott)

12C. *G. alpina* Conflorescence Small-flowered form, Tooboorac Forest, Vic. (M.Hodge)

12D. *G. alpina* Goldfields form ex Whroo Forest, Vic. (N.Marriott)
12E. *G. alpina* Southern Hills Form ex Mt Dandenong, Vic. (N.Marriott)

the dry side during the summer months. Most forms are readily available in native plant nurseries in Vic.

Propagation *Seed* Sets copious quantities of seed that germinate well when pretreated. Care must be taken to ensure that seed is pure, as *G. alpina* hybridises readily. Most seed from garden plants is impure. *Cutting* Firm, young growth taken at most seasons normally strikes readily. *Grafting* Has been grafted successfully onto G. 'Canberra Gem' (8 forms), G. 'Poorinda Royal Mantle', G. 'Ned Kelly' and G. 'Moonlight'. 15 forms have been grafted onto *G. robusta*.

Horticultural features *G. alpina* is a desirable species with a host of quite spectacular forms and a profusion of variety in flower colour and foliage. When conditions are to its liking, it can be a long-lived species although it usually becomes very open and woody. This problem can be largely solved by regular pruning from an early age. Due to its reputation as a 'drop dead' plant, it cannot be recommended in the commercial landscape but it makes an ideal rockery plant or short-lived specimen plant in the home garden. All forms are most attractive to honey-eating birds.

Hybrids In the wild, hybrid populations have been recorded with a number of species, including *G. dryophylla* around Bendigo and *G. lavandulacea* in the Black Range of The Grampians. Chance hybrids have also been found with *G. obtecta*. Plants in McDonald Park, western Vic., have formed a hybrid swarm with a cultivated form of *G. rosmarinifolia* and many beautiful cultivars have arisen from this area, including G. 'Australflora McDonald Park'. It is the known parent of a number of garden hybrids including G. 'Bonnie Prince Charlie', and numerous Poorinda hybrids such as G. 'Poorinda Refrain', 'Beauty', 'Annette', 'Belinda', 'Elegance', 'Jeanie', 'Rachel', 'Splendour', 'Golden Lyre' and 'Tranquility'.

Early history *G. alpina* was first grown in England (as *G. alpestris*) in 1856 as a conservatory plant and is still grown there by a small circle of

12F. *G. alpina* Northern Victorian form with yellow flowers ex Warby Ranges (N.Marriott)

12G. *G. alpina* Goldfields form, Castlemaine/Maldon, Vic. (C.Woolcock)

12H. *G. alpina* Typical flowers on plants from the Warby Ranges, Vic. (N.Marriott)

enthusiasts. The earliest recorded cultivation in Australia was in 1858 at Melbourne Botanic Garden as *G. dallachiana*.

General comments Further research of the various populations of this exceedingly diverse species is needed to resolve its taxonomy.

Grevillea alpivaga M.Gandoger (1919)
Plate 13

Bulletin de la Société Botanique de France 66: 231 (France)

The specific epithet is derived from the Latin *alpinus* (alpine) and *vagus* (distributed), an allusion to the distribution given on the type specimen. AL-PEE-VAR-GA.

Type: Victorian Alps, Vic., Nov. 1903, C.Walter (holotype: LY). The type is thought to have been collected on Mount Buffalo.

A prostrate to erect, suckering **shrub** 0.3–1 m high, 0.3–1 m wide. **Branchlets** sharply angular, stout, silky with prominent granular, glabrous ribs decurrent from leaf bases. **Leaves** 1–2.5 (–3) cm long, 0.8–1.2 mm wide, sessile, simple, entire, ascending to spreading, very crowded, leathery, narrowly elliptical to linear; acute, pungent; upper surface flat, sparsely silky, 5-ribbed, 3 ribs on the upper side granular, usually prominent and parallel; margin vertically refracted; lower surface with a very short secondary refraction about an intramarginal rib, grooves beside midvein usually broad, silky, rarely enclosed, the attachment of leaf base not wider than lamina. **Conflorescence** 1 cm long, 1 cm wide, globose in bud, erect, sessile, simple, terminal, regular, umbel-like, often enclosed in foliage or about level with tips of leaves; rachis 2 mm long, brown-pubescent; bracts 1.5 mm long, not exceeding the buds, narrowly elliptic, incurved, brown-silky outside, falling before anthesis. **Flower colour:** perianth pale green to white with brownish hairs on limb; style white, becoming reddish after anthesis. **Flowers** acroscopic; pedicels 5–8 mm long, silky; torus 0.7 mm across, square, straight to slightly oblique; nectary obscure, arcuate to U-shaped; **perianth** 2.5–3 mm long, 0.7 mm wide, undilated, oblong-cylindrical, white-silky outside, scantily bearded inside adjacent to ovary, the hairs c. 0.2 mm long, slightly wavy, elsewhere glabrous; tepals separating to the base before anthesis, curling back to beard at anthesis; limb revolute, subglobose, silky; **pistil** 6.5–7 mm long; stipe 0.5–0.7 mm long, glabrous; ovary narrowly ovoid, style scarcely exserted from lower half on dorsal side, incurved, geniculate 1–2 mm below style end; style end abruptly divergent, undulate, usually with minute erect hairs; pollen presenter 0.5–0.8 mm long, 0.8 mm wide, round to squarish, oblique, flat with prominent stigma. **Fruit** 9 mm long, 4 mm wide, erect to slightly oblique on swollen, slightly incurved stipe, narrowly ellipsoidal, faintly warty; pericarp 0.2–0.3 mm thick. **Seed** 6 mm long, 2 mm wide, oblong with subapical pulvinus; outer face convex, slightly wrinkled, minutely pubescent; margins revolute with a waxy ridge along one side extending into a short basal and longer apical elaiosome to 1mm long; inner face flat, glabrous.

Distribution Vic., endemic on Mount Buffalo. *Climate* Summer hot, dry; winter cold, wet with several months snow. Rainfall c. 1200–1600 mm.

Ecology Occurs in granite sand pockets and shaly loam. Flowers spring–summer. Fire response unknown. Almost certainly pollinated by insects.

Variation This species varies in leaf length and habit but is otherwise morphologically uniform.

Major distinguishing features Leaves open to crowded, simple, entire, mostly < 2.5 cm long, narrowly linear; upper surface flat to slightly convex with prominent scabrid midvein and two similar, parallel intramarginal ribs; margin vertically refracted twice about intramarginal ribs; undersurface usually exposed in broad channels; inflorescence terminal, simple, sessile, regular, umbel-like, enclosed in foliage; rachis 2 mm long, brown-pubescent; bracts ovate, not visible beyond line of subglobose buds; perianth zygomorphic, hairy outside, bearded within; pistil 6.5–7 mm long, glabrous except a granular or minutely hairy style end; fruits faintly warted.

Related or confusing species Group 21, especially *G. micrantha*, *G. neurophylla* and *G. parviflora* with which it has formerly been included. *G. parviflora* differs in its subsecund inflorescence, looser, non-scabrid leaf margins, longer ovarian stipe (except Maroota–Berrilee specimens); it also has more slender branchlets. *G. micrantha* is closely related but is distinguished by its inner perianth surface being glabrous or almost so in the beard position. *G. neurophylla* differs in its longer leaves slightly convex on the upper surface, with lower surface enclosed by angularly refracted margins, less prominent, faintly granular intramarginal ribs and in its shorter pedicels (3–5 mm long). *Grevillea* sp. Temora–Barmedman is also related but differs in its more conspicuous beard, and its longer, narrower leaves with undersurface enclosed and its slightly longer pistil.

Conservation status 2RC recommended. Uncommon but not thought to be at risk. Distribution requires assessment.

Cultivation To our knowledge *G. alpivaga* has not been cultivated, but it should grow readily in an open position in well-drained, acidic to neutral sand, sandy loam or gravelly loam. Tolerance of very cold conditions is indicated from its natural habitat where temperatures as low as -8°C have been recorded. Extended dry periods may be tolerated but subsoil moisture should never be too low. It has a naturally compact habit which would be improved by occasional tip pruning. Once established, it should

G. alpivaga

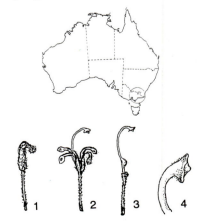

13A. *G. alpivaga* Flowers and foliage, Mt Buffalo NP, Vic. (I.McCann)

13B. *G. alpivaga* Flowering habit, Mt Buffalo NP, Vic. (P.Olde)

0.5–1 m wide with flexuose, ascending to spreading branches, dense to the ground. **Branchlets** terete, scabrous to sparsely hirsute. **Leaves** 3–7.5 cm long, 1–5 cm wide, concolorous, tangled, secund, ascending to erect, persistent, usually twice-divided with primary division pinnatisect; primary leaf lobes 3–7 per leaf, distant, cuspidate, obovate–cuneate, usually with secondary division; ultimate lobes 2–2.5 cm long, 1–3 cm wide, the ultimate secondary lobes broadly triangular, pungent; leaves subtending floral branches simple, linear; upper and lower surfaces similar, scabrous to sparsely hirsute; venation prominent, more conspicuous on undersurface; margin undulate, otherwise flat, coincident with a conspicuous, rounded, scabrous vein; texture firmly papery. **Conflorescence** 2–5 cm long, 1.5 cm wide, erect, sessile, terminal, usually simple, rarely 1–3 branched; unit conflorescence cylindrical, few-flowered, lax, scarcely to not exceeding foliage; rachis 1.5 mm wide at base, arising from a leaf-opposed rosette of bracts, villous; floral bracts 6–7 mm long, 1.5 mm wide, narrowly triangular, falling before anthesis. **Flower colour:** perianth reddish when young becoming dull yellow; style creamy yellow. **Flowers** regular, ascending; pedicels 2–3 mm long, villous, patent; torus c. 1 mm across, straight; nectary not evident; **perianth** 5–6 mm long, 1.5–1.8 mm wide, oblong, villous outside with both glandular and non-glandular hairs; tepals cohering to anthesis, becoming free to base and strongly rolled down after anthesis, exposing an inner surface either densely papillose or bearing short papilloid trichomes; limb densely villous, erect, subglobose; **pistil** 6–6.5 mm long; ovary sessile, densely villous with spreading to erect, straight hairs; style sharply folded above ovary, glandular-pubescent on lower filiform section, papillose and gradually dilated in apical half; style end broadly expanded;

not need fertiliser or watering. Not commercially available.

Propagation *Seed* Sets reasonable quantities in the wild. These should germinate if pretreated by nicking or peeling the testa. *Cutting* Should strike readily from cuttings taken in early spring using half-hardened new growth. *Grafting* Untested.

Horticultural features *G. alpivaga* is a bushy shrub with dull green leaves and conspicuous bright white flowers that are reminiscent of small dollops of snow attached to the ends of the branches. In the home garden this species would be best planted as a contrast to more colourful plants but would be suited to public and commercial landscapes, especially in areas with very cold winters. These suggestions lack horticultural experience with the species and are based on experience with closely related taxa.

General comments McGillivray (1993) referred *G. alpivaga* to synonymy as Form 'h' under a broadly circumscribed *G. linearifolia*. While taxonomy and species boundaries in this group are at present being reconsidered, the small population encompassed by plants here referred to as *G. alpivaga* is quite distinct and deserves formal recognition.

Grevillea althoferi P.M.Olde & N.R. Marriott (1993)   Plate 14

Nuytsia 9: 295 (Australia)

The specific epithet honours the naturalist and former curator of Burrendong Arboretum, near Wellington, N.S.W., Peter McDowell Althofer and his wife Hazel J. OAM, who was head propagator of the Arboretum. AL-THOFF-ER-EYE

Type: near Eneabba, W.A., 15 Sept. 1991, P.M.Olde 91/102 (holotype: NSW).

Compact, rounded, possibly suckering **shrub** 0.3–0.5 m high,

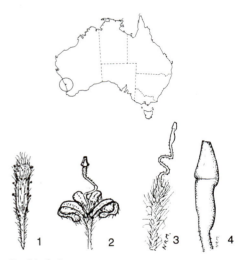

G. althoferi

14A. *G. althoferi* Flowering habit, near Eneabba, W.A. (P.Olde)

14B. *G. althoferi* Flowers and foliage (P.Olde)

pollen presenter c. 0.8 mm long, straight, conico-cylindrical with slightly cup-shaped apex. **Fruit** and **seed** not seen.

Distribution W.A., restricted to one known population south of Eneabba. Two other specimens have been seen from sites now cleared by sand-mining. *Climate* Hot, dry summer; cool, wet winter. Rainfall 475–525 mm.

Ecology Occurs in low heath dominated by *Grevillea integrifolia*, *G. shuttleworthiana* and *Eucalyptus* sp. aff. *tetragona* in grey sand with laterite. Flowers Sept.–Oct. Reproduction is probably by sucker.

Variation A stable, uniform species.

Major distinguishing features Leaves twice-divided with primary lobes cleft to midvein, distant, obovate-cuneate, with secondary triangular-toothed lobes; conflorescence scarcely exceeding leaves, cylindrical, open to loose; torus straight; nectary absent; perianth regular, inner surface papillose or minutely hairy; ovary hairy, sessile; style folded or kinked above ovary, dilated in apical half.

Related or confusing species Group 41, especially *G. rudis* which differs in its leaves with mostly primary division; its denser, many-flowered conflorescences, and its perianth with the inner surface smooth.

Conservation status 2E recommended. One extant population is known. Identification of new sites at which this species grows is urgently needed and a search of the general area of its occurrence is recommended.

Cultivation *G. althoferi* has not been cultivated to date but a position in full sun in well-drained, acidic to neutral, sandy loam is recommended. The species should tolerate cold weather but frosts may kill it, especially if severe or if the plant is very young. Not recommended for summer-humid climates.

Propagation Seed No seed has been seen in the wild. *Cutting* Attempts to propagate wild material have failed. Once established, however, material from cultivated plants will be easier to propagate, judging by experience with its closest relative, *G. rudis*. *Grafting* Initial attempts at grafting wild-source material onto *G. robusta* have failed.

Horticultural features *G. althoferi* is not a spectacular plant but is, nonetheless, quite attractive. Its dense, compact habit, and delicate yellow racemes make it suitable for rockery plantings or for feature planting. The blue-green foliage with its unusual leaf shape is its greatest horticultural asset and serves to make it a potentially useful, all-round garden plant for the enthusiast.

General comments Under the *International Code of Botanical Nomenclature* (Berlin Code), when two people of the same name are commemorated in a specific epithet, a plural Latin form should be used. In *Nuytsia* 9: 438 (1994), an editorial note proposed that the epithet *althoferi* be 'corrected' to *althoferorum*; we do not accept its proposal. The form *althoferi* used in the protologue was deliberately chosen, in the sense that although Peter and Hazel were married Hazel was born Johnston and therefore the singular family name was considered more appropriate.

Grevillea amplexans F.Mueller ex G.Bentham (1870)   Plate 15

Flora Australiensis 5: 488 (England)

The specific name is derived from the Latin *amplexare* (to encircle, surround), in reference to the fact that the basal lobes of the leaves sometimes overlap and encircle the stem. AM-PLEX-ANS

15A. *G. amplexans* in natural habitat, W of Arrino, W.A. (P.Olde)

15B. *G. amplexans* W of Arrino, W.A. (P.Olde)

Type: south-western W.A., date and collector not given, but possibly J.Drummond, Herb. F.Mueller (holotype: MEL).

A spreading, prickly **shrub** 1–2.5 m tall. **Branchlets** terete, glabrous, sometimes glaucous, rarely silky. **Leaves** 0.8–2.6 cm long, 1–3.5 cm wide, glabrous and often glaucous, spreading to patent, stiff to rigid, sessile, simple, rhombic to ovate with palmate venation, tipped by long, pungent spines; basal lobes retrorse to stem-clasping, pungent; margins flat with conspicuous edge-veins; venation prominent on both surfaces; upper surface and lower surface glabrous; texture firm, papery. **Conflorescence** ± erect, terminal or axillary, simple or branched, pedunculate; unit conflorescence 1–2 cm long, 1–2.5 cm wide, open, umbel like to subglobose; peduncle glabrous or silky; rachis glabrous; floral bracts 5–20 mm long, ovate, glabrous with ciliate margins, sometimes present at anthesis. **Flower colour:** perianth white or creamy white with pinkish tinges; style white. **Flowers** regular, glabrous: pedicels 8–12 mm long, filamentous; torus 0.5–1 mm across, straight to slightly oblique; nectary semi-circular, entire; **perianth** 3.5–4 mm long, 0.7–0.8 mm wide, oblong-cylindrical to ellipsoidal below curve; tepals separating below limb before anthesis and bowed out, splitting to base at anthesis, afterwards rolling back independently; limb 1.5–2 mm wide, erect, ovoid to subglobose, medially ribbed; **pistil** 2.5–5.8 mm long; 1–2.5 mm long, flexuose; ovary globose; style straight, constricted just above ovary, with a cylindrical to ovoid dilation above; pollen presenter straight, conical, the base broader than the style. **Fruit** 9–12 mm long, 5 mm wide, 5 mm deep, perpendicular to stipe

G. amplexans

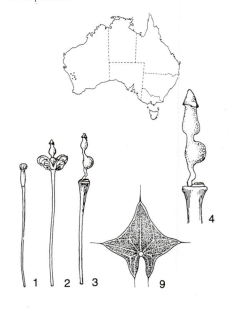

on curved pedicel, oblong, glabrous, smooth; pericarp 0.5–1 mm thick. **Seed** 5–6.5 mm long, 2.5 mm wide, ellipsoidal, biconvex with a membranous border on inner face, otherwise smooth.

Distribution W.A., from east of Geraldton south to near Arrino. ***Climate*** Hot dry summer; mild, wet winter. Rainfall 330–460 mm.

Ecology Grows on undulating plains in low mallee shrubland in yellow or white sand or sandy loam, usually over laterite, often on disturbed road verges. Flowers winter to spring.

Variation There is relatively little variation in this species.

Major distinguishing features Leaves stem clasping, glabrous; conflorescence ovoid; pedicels filiform; flowers glabrous; perianth regular; style dilated above a zone of constriction; pollen presenter conical, its base broader than the style; fruit oblong–ellipsoidal, smooth.

Related or confusing species Group 1, especially *G. adpressa*, *G. uniformis* and *G. phanerophlebia*, all of which have hairs on the leaf undersurface.

Conservation status Fairly widespread and common, but further clearing of its natural habitat for agriculture could see it come under threat.

Cultivation For successful cultivation, this demanding species requires an extremely well-drained sand or sandy, gravelly loam in an open, sunny situation. In cold, wet climates, it demonstrates little vigour and gradually dies out, with young plants notably susceptible to even light frosts. In these conditions, it must be regarded as unreliable. Best results have been achieved in areas with a hot, dry summer and mild, wet winter but, when conditions are to its liking, *G. amplexans* has proved to be extremely hardy. In fact, it has naturalised at Dave Gordon's arboretum at Glenmorgan, Qld, to such an extent that it has become a weed, hybridising with similar white-flowered species and wreaking havoc among the genetic purity of related species. It has been successfully cultivated in Vic., and at Burrendong Arboretum, N.S.W., as well as at Kings Park, Perth, W.A. In Sydney, it has been successfully held in pots using a well-drained, composted pine bark mix with slow-release fertiliser, but it is rarely available from nurseries.

Propagation *Seed* As with most grevilleas related to this species, it sets large quantities of seed that can be germinated readily following the standard peeling treatment. Without pre-treatment, the seeds average 27 days between sowing and germination (Kullman). Care should be taken to ensure that seed is taken from wild plants. *Cutting* Best results are achieved with firm, young growth taken from late spring to early autumn. Soft material or material taken in colder months gives poor results. *Grafting* Has been grafted onto *G. robusta*, G. 'Poorinda Anticipation' and G. 'Poorinda Royal Mantle' using the top-wedge and mummy methods.

Horticultural features The major feature of this species is its curious stem-clasping leaves that are often an attractive blue-green colour. When back-lit by the sun, the venation stands out quite beautifully. The rich bronze to reddish bronze young growth is also striking. The white flowers are not particularly showy, being borne in open, delicate clusters and are often somewhat hidden among the rigid, prickly leaves. This unusual plant is suitable as a low, screen plant or 'people-stopper' but is grown mainly by enthusiasts.

Hybrids In ideal conditions, *G. amplexans* readily self-sows in the garden and will hybridise with related species.

Grevillea anethifolia R.Brown (1830)
Plate 16

Supplementum Primum Prodromi Florae Novae Hollandiae exhibens Proteaceas Novas: 21 (England)

Anise-leaf Grevillea; Spiny Cream Spider Flower

The specific epithet alludes to a resemblance in foliage to the genus *Anethum* (dill) (Latin *folium* -a leaf). AN-EE-THI-FOLE-EE-A

Type: interior of N.S.W., 1817, A.Cunningham and C.Fraser on Oxley's First Expedition (lectotype: BM) (McGillivray 1993).

Erect, suckering, multi-stemmed **shrub** 0.5–2 m high. **Branchlets** terete, with a dense, appressed indumentum, brittle. **Leaves** 1.5–5 cm long, ascending, petiolate, crowded, leathery, divaricately divided into 3–5 subterete, pungent lobes, the lobes themselves often 2- or 3-forked; margins smoothly revolute to midvein; upper surface glabrous; lower surface obscured, usually glabrous, sometimes silky in grooves. **Conflorescence** 0.5–2 cm long, 1.5–2.5 cm wide, erect, dense, terminal or axillary, usually unbranched, shortly cylindrical to umbel-like, mostly enclosed within foliage; peduncle tomentose; rachis sparsely hairy to glabrous; bracts 0.6–1.2 (–5) mm long, ovate, falling before anthesis. **Flower colour:** perianth cream with the limb green in bud becoming cream or yellow; style white. **Flowers** glabrous, regular; pedicels 6.5–12 mm long; torus 0.5–0.8 mm across, ± straight; nectary conspicuous, U-shaped to oblong, toothed; **perianth** 3 mm long, 0.5–1 mm wide, oblong–obovoid to ellipsoidal below limb, glabrous outside, sparsely hairy inside especially near base; tepals separating below limb and bowed out before anthesis, afterwards free to base and rolling back independently; limb erect, globular; **pistil** 4–6 mm long; stipe 1.5–2.5 mm long, flexuose; ovary globose; style straight, constricted above ovary, with a globose to ovoid dilation above; pollen presenter slightly oblique, cylindrical to truncate-conic, the base broader than style. **Fruit** 5–8 mm long, 4–5 mm wide, 4 mm deep, horizontal to stipe, glabrous, rugose, oblong–ellipsoidal; pericarp c. 0.5 mm thick. **Seed** 6 mm long, 3 mm wide, ovoid; outer face smooth, convex; inner face convex, smooth except a waxy, annular rim.

Distribution Widespread across southern W.A., S.A. and N.S.W. In W.A., widespread between Cape Arid and the Stirling Ra. extending inland up to 100 km. In S.A., on Eyre Peninsula in the Gawler Ranges and Cunyarie Hills. In N.S.W., inland around Griffith, Rankin Springs and Hillston. *Climate* Hot, dry summer; cool, wet winter. Rainfall 250–400 mm.

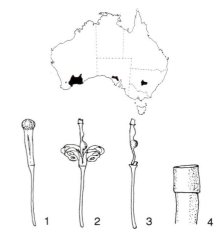

G. anethifolia

Ecology Grows on open, flat plains or rocky hills, in shrubland or mallee, usually in calcareous sand, sandy loam or granitic sand. Flowers late winter to early summer. Insect-pollinated; native bees have been seen in attendance.

Variation Populations with small fruits are sometimes scattered throughout the distribution of this species and deserve further field study. Plants from Mt Arid, W.A., have very conspicuous floral bracts (3–5 mm long, 2–4 mm wide).

Major distinguishing features Branchlets with a dense, appressed indumentum; flowers regular, glabrous; style slightly constricted above the ovary, dilated above; pollen presenter subcylindrical to truncate-conic; fruits rugose, oblong–ellipsoidal.

Related or confusing species Group 1, especially *G. biternata* and *G. paniculata*. *G. paniculata* differs in its conical pollen presenter and very strongly wrinkled fruits and usually has glabrous or sparsely silky branchlets. *G. biternata* differs in its conical pollen presenter.

Conservation status Not presently endangered in W.A. and S.A., but in N.S.W. much of its habitat has been cleared.

16A. *G. anethifolia* Close-up of flowers, Gawler Ra., S.A. (R.Bates)

16B. *G. anethifolia* Form with large bracts from Mount Arid, W.A., cultivated at Mt Annan Botanic Garden (P.Olde)

16C. *G. anethifolia* Plant in natural habitat, Jimberlana Hill, W.A. (N.Marriott)

Cultivation *G. anethifolia*, which is always found in dry situations, usually inland but sometimes along the coast in southern W.A., tolerates both drought and frost, but plants from some areas have not proved easy to grow. Plants propagated from some populations in N.S.W. and S.A. and coastal W.A. have been successfully established at Burrendong Arboretum, N.S.W., and in a few private gardens in Vic., and at Adelaide Botanic Gardens, S.A. Many people report success in cultivating it, but as there is widespread confusion with *G. paniculata*, these reports may be erroneous and we are restricting our comments to results from plants of known source. Plants from inland W.A. and S.A. have shown less willingness to adapt to cultivation, perhaps because of specific soil preference or lower climatic adaptability. The species prefers an open, sunny aspect in well-drained sand or gravelly loam in acidic to neutral pH range. Once established, it does not require further maintenance by way of watering or pruning. Plants sold in nurseries as *G. anethifolia* are often forms of *G. paniculata*.

Propagation *Seed* Fresh seed germinates readily given the standard peeling treatment. Young seedlings require an open, dry situation. *Cutting* Strikes well from firm, young growth at most times of the year. *Grafting* Untested to date.

Horticultural features The sweetly scented flowers that sometimes smother *G. anethifolia* in a mantle of creamy white are its main ornamental feature, although we have observed that the flowers are more prominently displayed on W.A. populations than on other populations. While the prickly foliage is an obvious disadvantage, if planted in the right situation, the deterrent effects can be useful. By regular trimming, it could be trained into a low hedge or screen plant. Well-grown plants are both compact and relatively dense, highlighting their suitability as a background plant. Although when not in flower it is a rather sombre-looking plant with dull green foliage, this can be used as a foil for brighter and more showy plants in the foreground. In full flower, especially if planted in a drift, it makes a bright contrast with the more colourful species.

Grevillea aneura D.J.McGillivray (1986)
Plate 17

New Names in Grevillea *(Proteaceae)* 1 (Australia)

Red Lake Grevillea

The specific name is derived from the Greek *neuron* (a sinew or nerve) and the prefix *a-* (without), in reference to the lack of evident venation on the leaves. AY-NEW-RA

Type: 16 miles (c. 25–26 km) E. of Red Lake, W.A., 2 Nov. 1967, J.S.Beard 5422 (holotype: PERTH).

A dense, bushy **shrub** 0.5–1.5 m tall. **Branchlets** terete to angular, silky. **Leaves** 4–7.5 cm long, ascending to spreading, crowded and tangled, shortly petiolate, divaricately tripartite to biternate; ultimate lobes 0.5–5 cm long, 0.8–1.1 mm wide, subterete with a longitudinal, hair-filled channel down each side, rigid, pungent; upper surface smooth, glabrous to silky-pubescent, the venation obscure; margins smoothly recurved; lower surface consisting mainly of the prominent, rounded midvein. **Conflorescence** 3.5–7 cm long, erect, often on pendulous branchlet, pedunculate, terminal, unbranched, oblong-secund; rachis often slightly incurved; peduncle and rachis silky to tomentose; bracts 0.5–0.8 mm long, ovate, silky outside, sometimes persistent to anthesis. **Flower colour:** perianth and limb red, orange-red or pale yellow; style orange or red with green to yellow tip. **Flowers** acroscopic: pedicels 2.5–3.5 mm long, silky; torus 1.5–2 mm across, oblique, cup-shaped; nectary V-shaped, entire; **perianth** 8–10 mm long, 2–3 mm wide, ovoid–S-shaped, silky outside, glabrous inside, cohering except along dorsal suture; limb revolute, silky, spheroidal, enclosing style end before anthesis; **pistil** 27–32 mm long; stipe 3.3–5 mm long, glabrous to sparsely villous; ovary villous; style at first exserted just below curve on dorsal side and looping upwards, refracted about ovary, after anthesis straight to undulate or slightly incurved, glabrous, sometimes with basal hairs, dilating evenly into a scarcely enlarged style end; pollen presenter very oblique to lateral, oblong–elliptic, convex. **Fruit** 10–13 mm long, 6 mm wide, erect, oblong–ellipsoidal, silky-pubescent with brown striping; pericarp c. 0.5 mm thick. **Seed** not seen.

Distribution W.A., from east of Lake King to east of Red Lake. *Climate* Hot, dry summer; mild, moist winter. Rainfall 250–350 mm.

Ecology Grows in heath or mallee scrub in yellow sand or sandy loam over laterite, usually on rises. Flowers winter to late spring. Probably pollinated by birds. After fire regenerates from seed.

Variation A stable species with little variation except habit which ranges from semi-prostrate to erect and pyramidal.

Major distinguishing features Leaves deeply and divaricately divided, the lobes subterete, with a groove along each side, rigid, pungent, the upper surface smooth and lacking venation; conflorescence secund with concave rachis; perianth zygomorphic, silky outside, glabrous within; ovary densely hairy, long-stipitate; pollen presenter very oblique to lateral; fruits striped with red hairs.

Related or confusing species Group 35, especially *G. treueriana*, the most closely related but differing in its shorter, angular leaf lobes, its shorter pistil and stipe and its slightly oblique pollen presenter; *G. beardiana* which has mostly simple leaves, and *G. secunda* which has a sessile ovary.

Conservation status Not currently under threat.

Cultivation *G. aneura* has been grown in Vic. for many years as Grevillea 'Red Lake' and was grown successfully for a few years in the Australian National Botanic Gardens, Canberra. Potted specimens also survived at Bulli, N.S.W., in the Grevillea Study Group collection for many years. It appears to be an adaptable species, tolerating both heavy frosts and extended dry periods and thriving in deep, well-drained sand or gravelly loam in either full sun or semi-shade. Soil should be acidic to

G. aneura

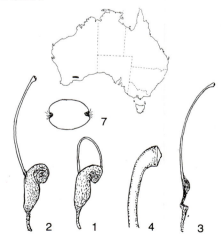

17A. *G. aneura* Close-up of conflorescences (P.Olde)

17B. *G. aneura* Habit and habitat, E of Lake King, W.A. (P.Olde)

17C. *G. aneura* Flowering habit (N.Marriott)

neutral. To date, success has been achieved only in a hot to cool, dry climate and it is doubtful if it would do well in summer-rainfall climates. Pruning may be necessary to improve the shape although it is a naturally dense shrub. It grows well in pots using a well-drained, open, composted pine bark mix with fertiliser, placed in full sun. Attempts are being made by members of the Grevillea Study Group to introduce it to cultivation.

Propagation *Seed* Untested. Should respond to standard peeling or nicking pretreatments. *Cutting* To date this species has proven difficult to strike from cutting, but material from cultivated plants may prove easier. Firm, young growth should give best results if taken in spring or autumn. *Grafting* Has been grafted onto G. 'Poorinda Royal Mantle' using the top wedge technique but died within a year. More research needed.

Horticultural features Although prostrate to decumbent for many years, *G. aneura* eventually forms a compact shrub usually to about 1 m tall. During spring, this decorative plant is resplendent with many clusters of pendant, orange to red incurved racemes, often weighing the branchlets down. The gold-tipped styles prominently exserted and contrasting with the bright orange-red perianth are both bright and attractive. The dense foliage is blue-green when young but mature plants have dull green foliage that provides an attractive, albeit prickly, backdrop to the flowers. Well-grown plants would be ideal as a feature plant or low screen plant in the landscape and the species could be considered for rockeries or as a ground cover. The dense, prickly foliage would provide an ideal shelter for small birds while the nectar-filled flowers would attract honeyeaters.

Grevillea angulata R.Brown (1830)
Plate 18

Supplementum Primum Prodromi Florae Novae Hollandiae exhibens Proteaceas Novas: 24 (England)

Aboriginal: anbirrim (Mayali), a name given for plants with anmirrk (needle), or prickly leaves. The Mayali noted it as a source of nectar for bees.

The specific epithet is derived from the Latin *angulatus* (angled), in reference to the angular sinuses between spines on the leaf margin of the Type specimen. AN-GEW-LAR-TA

Type: Sims Island., near Goulburn Gulf, [N.T.], 1818, A.Cunningham no. 163, on the 1st Voyage of the *Mermaid*, (lectotype: BM) (McGillivray 1975).

A single-stemmed, open to dense **shrub** usually 1–2.5 m tall, sometimes prostrate. **Branchlets** angular to terete, silky. **Leaves** 2–12 cm long, 1.5–3.5 (–5) cm wide, ascending to spreading, concolorous, usually oblong with 2–24 pungent tipped lobes separated by shallow or deeply arcuate sinuses regularly spaced around margin, sometimes simple and entire, elliptic; margins flat; upper and lower surfaces similar, minutely silky on one or both surfaces; venation conspicuous, the midvein prominent below; petiole up to 12 mm long. **Conflorescence** decurved to patent, pedunculate, terminal on short branchlets or axillary, usually simple, occasionally branched; unit conflorescence 1–5 cm long, conico-cylindrical, dense; peduncle silky; rachis glabrous to sparsely silky; floral bracts 0.5–1 mm long, ovate with ciliate margins, falling before anthesis. **Flower colour:** perianth green in bud, turning white to creamy white; style white with green style end. **Flowers** adaxially acroscopic, glabrous except inner perianth surface: pedicels 3.5–5 mm long; torus 2 mm across, oblique; nectary conspicuous, U-shaped, toothed or entire; **perianth** 5–8 mm long, 2–2.5 mm wide, oblong–ovoid below curve, slightly dilated at base, conspicuously bearded inside; dorsal tepals separating from ventral tepals, reflexing and exposing inner surface before or at anthesis; limb revolute, spheroidal, the style end partially exposed before anthesis; pistil 18–24 mm long; ovary glabrous; stipe 3–5 mm long; style at first exserted below curve on dorsal side and looped upwards, strongly incurved in upper half after anthesis, glabrous except papillae or short, erect hairs on style end, extending 2–3 mm from tip; pollen presenter oblique, ± obovate, convex. **Fruit** 10–14 mm long, 8.5–10.5 mm wide, oblique or transverse to stipe, ellipsoidal, verrucose; pericarp c. 1 mm thick. **Seed** 6–7 mm long, 3.5 mm wide, obovoid, membranously shallow-winged all round.

Distribution N.T., between Darwin and Gningarg Point on the mainland, extending from Sims Is. S to Oenpelli. *Climate* Monsoonal; summer hot, humid and wet; winter warm and dry. Rainfall 800–1000 mm.

Ecology Grows near creek beds in sand, on stony slopes, or sometimes on low sandstone platforms, in mixed shrubland or heath or with tall eucalypts. Coastal forms also grow in laterite. Flowers autumn to spring, although a few flowers may be found at any time except mid-summer. Regenerates from seed or epicormic buds on branchlets after low

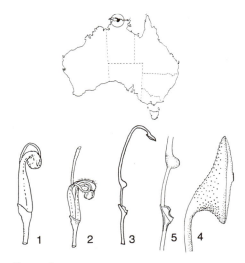

G. angulata

intensity fires. Bird- or insect-pollinated.

Variation *G. angulata* is a relatively uniform species with some variation in leaf shape.

Type form At the northern end of its range, *G. angulata* has shorter, broader leaves with 2–10 spines and noticeably angular, quite shallow sinuses between the lobes. It occurs on coastal islands and on the coastal strip between Murgenella and Gningarg Point where it suffers much wind-buffeting. Occasional plants with entire leaves also occur in this area. Plants in coastal situations are often completely prostrate.

Narrow-leaved form Plants from the Oenpelli–Nabarlek area grow to 1.5–2.5 m tall and have narrowly oblong leaves with 3–24 teeth per side, deeper arcuate sinuses and longer spines.

Major distinguishing features Leaves usually oblong, concolorous with flat, toothed margins, minutely silky on one or both surfaces; conflorescence < 5 cm long, decurved; outer perianth glabrous; inner perianth surface densely bearded; tepals separating below limb before anthesis and reflexing to reveal inner surface; pistil far exceeding perianth; ovary glabrous; stipe 3–4.5 mm long; style end minutely hairy or granular; pericarp c. 1 mm thick.

Related or confusing species Group 11, especially *G. prasina*, *G. glabrescens*, *G. aurea* and *G. brevis*, all of which differ in their glabrous leaves. *G. prasina* also has a smooth style end.

Conservation status Not presently endangered.

Cultivation Plants of *G. angulata* have generally flourished in Darwin and Brisbane. Grafted plants have even grown reliably in coastal Sydney. A warm position, perhaps near a north-facing wall, is advisable in cool-temperate areas because it succumbs to extended cold periods and frost. In its natural habitat, it occurs in sandy or laterite soils in open, well-drained, sunny positions where it can survive quite long dry periods. A similar position is suggested in cultivation. The open growth habit can be corrected by regular tip-pruning but allowing the

18A. *G. angulata* S of Murgenella, N.T. (P.Olde)

18B. *G. angulata* E of Murgenella, N.T. (P.Olde)

18C. *G. angulata* Prostrate form. Lateritic cliffs E of Murgenella, N.T. (P.Olde)

18D. *G. angulata* Southern form, near Oenpelli, N.T. (P.Olde)

soil to dry out will cause leaf drop and result in a lanky appearance. Over-wintering in a dry, warm glasshouse to prevent damage from cold is advised in very cold climates.

Propagation *Seed* Fresh seed should germinate well if pre-treated by peeling or soaking for 24 hours. *Cutting* Grows well from cuttings of firm, new growth taken in summer. *Grafting* Has been grafted successfully onto *G. robusta* and G. 'Poorinda Royal Mantle' using the top wedge and whip-and-tongue techniques.

Horticultural features *G. angulata* is admired for its symmetrically-toothed leaves and scented, creamy white flowers in short conico-cylindrical conflorescences. Flowering is restricted at southerly latitudes, but in tropical climates it may flower continuously in cultivation. Flowering is very prolific and will attract a variety of birds and insects to the garden. It is reported to be fast growing and would make an ideal screen or feature plant. It would have value as an ornamental in landscaping for its interesting leaves.

Grevillea annulifera F.Mueller (1864)
Plate 19

Fragmenta Phytographiae Australiae 4: 85 (Australia)

Prickly Plume Grevillea

The specific name is derived from the Latin *annulus* (a ring) and *ferre* (to bear or carry), in reference to the flowers having an annular (circular) nectary. AN-YOU-LIFF-ER-A

Type: Murchison River, W.A., Mar. 1859, A.Oldfield (lectotype: MEL) (McGillivray 1993).

An erect, irregular, glabrous **shrub** 2–4 m tall with dense, vegetative branches at the base, usually with one or more dominant floral branches. **Branchlets** terete, glaucous. **Leaves** 4.5–7 cm long, spreading, sessile but appearing long-petiolate, rigid, tripartite to subpinnatisect; leaf rachis decurved to divaricately refracted; leaf lobes 1.5–3.5 cm long, 5–9 per leaf, linear, rigid, pungent; upper surface glabrous, midvein a groove; margin angularly revolute; lower surface obscured, tomentose in grooves, the midvein prominent, glabrous. **Conflorescence** erect or decurved on emergent floral branches, terminal, usually paniculate; unit conflorescence 10–15 cm long, 4 cm wide, conical to cylindrical, dense; peduncle and rachis glabrous; floral bracts 3–5 mm long, imbricate, membranous, ovate, glabrous, falling before anthesis. **Flower colour:** perianth green in bud, becoming creamy white; style creamy white turning pink to red. **Flowers** adaxially acroscopic in bud, soon twisting irregularly with most directed ventrally away from rachis, glabrous outside; pedicels 5–9 mm long, expanded and ribbed at apex; torus 1.5–2.5 mm across, ± straight, square; nectary annular, sometimes irregularly so, entire; **perianth** 7–8 mm long, 2–3 mm wide, oblong-cylindrical with a slight annular dilation at base, bearded inside at base; tepals at first separating near base, then along dorsal suture, ultimately all free, reflexing laterally to expose inner surface at or before anthesis; limb revolute, spheroidal, firmly enclosing style end before anthesis; **pistil** 28–35 mm long, glabrous; stipe 4.5–7 mm long, inserted centrally in torus; ovary lateral, compressed-ellipsoidal; style minutely granular, elongate, at first exserted at curve and looped upwards, strongly incurved to semicircular after anthesis; style end dilated; pollen presenter almost lateral, obovate, conico-convex with a reflexed flange all round. **Fruit** 27–29 mm across, erect but transverse to stipe, lens-shaped or globose, smooth; pericarp 1.5–2 mm thick at suture. **Seed** 20 mm long, 18 mm wide, 7–8 mm deep, hemispherical with basal rim, unwinged, smooth.

Distribution W.A., from Shark Bay south to near Yuna, northeast of Geraldton. *Climate* Hot dry summer; mild, moist winter. Rainfall 250–440 mm.

Ecology Dominant in medium to deep yellow sand over loam in low heath and open mallee scrub. Flowers late winter to spring. Strong, foetid perfume emitted at night. Pollinators probably many, including beetles and marsupial mice; probable strong predisposition to self-fertilisation. Regenerates from seed.

Variation There is some variation habit. Some populations are robust shrubs with dense basal foliage and numerous emergent floral branches. Others are spindly with relatively few floral branches. The differences may be habitat-related.

Major distinguishing features Branchlets glabrous; leaves rigid, pungent, divaricately pinnatipartite, glabrous; conflorescence usually much-branched on emergent floral branches, glabrous, the units cylindrical; floral bracts membranous, quite enlarged; nectary annular; perianth zygomorphic, glabrous outside, hairy at base within, undilated except slightly at base; ovary glabrous; style semicircular, very conspicuous; pollen presenter convex to broadly conical; fruit globose, dehiscent; seed with spongy testa.

Related or confusing species Group 3, especially *G. candicans* which differs in its hard, nut-like indehiscent fruit, hairy branchlets, and shorter pedicels (< 4 mm long).

Conservation status 2RC. Although rare and endangered, this species is conserved in Kalbarri NP.

Cultivation *G. annulifera* is generally regarded as a 'difficult' species and does not adapt well to most garden conditions. Nonetheless, it has, at times, been successfully grown with spectacular results. Plants have been flowered at Glenmorgan, Qld, at Swan Hill, N.S.W., and at Coleraine, western Vic. Seeds sent to England by Mueller in 1880 were germinated and the plants flowered there in 1882 in the Royal Botanic Gardens, Kew, almost certainly under glass. In climates or soils to which it is not suited, it struggles to grow past the juvenile stage, and although it will survive cold, wet winters, it demands shelter from cold winds. In the wild, it thrives in a warm, sunny situation in deep, neutral to calcareous sand. Summer watering should be avoided. Seed catalogues occasionally list this species but it is rarely available in nurseries.

Propagation *Seed* Sets reasonable quantities of seed even in cultivation. Due to the thick corky material surrounding the embryo, the seeds take many years to germinate naturally. The purpose of this unusual seed coat, present in only a few grevilleas, is uncertain; it may contain inhibitors or act as an insulant. To overcome its effect, the corky layer must be carefully prised off to expose the white embryo which is then sown in the standard method.

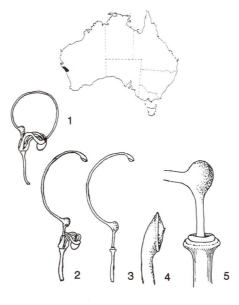

G. annulifera

19A. *G. annulifera* Flowering raceme (M.Hodge)

Germination time varies from 20 to 60 days, most seeds germinating within 30 days. Seedlings are prone to damping off and must be treated with a fungicide if they show any sign of collapse. *Cutting* Not generally grown by this technique as it has proved almost impossible to root even using fresh, firm, new growth. *Grafting* Several attempts on different rootstocks such as *G. robusta* and G. 'Poorinda Royal Mantle' have failed. Young seedlings have been grafted successfully onto both *G. robusta* and G. 'Poorinda Anticipation' using the top wedge method. This has resulted in rapid-growing plants, although long-term compatibility is untested. More research is needed.

Horticultural features In full flower, *G. annulifera* is a dazzling species with spectacular plumes of cream flowers ageing pink to red carried high above the foliage for several months of the year. These tend to diminish other, less favourable features such as the prickly foliage and the unpleasant odour of the flowers which emanates into the night air. Large, round fruits, known to some as 'native almonds', follow flowering. The hemispherical seeds are extremely large and most are eaten by small mammals or insects with the consequence that they never germinate in great numbers. This magnificent plant, despite appreciation of its floral attributes, cannot be recommended for general cultivation at this stage and continues to remain an enigma to growers of the genus.

General comments One grower (D.Gordon, Glenmorgan) reports that the fruits open explosively. The seeds have an almond-like taste and were eaten by Aborigines. Although they contain high levels of essential nutrients and would be very nutritious if used for human consumption, experience (P.J.Hocking, 1981) has shown that commercial cultivation would be limited by low seed set. Research could possibly overcome this problem.

19B. *G. annulifera* Plant in natural habitat, Pot Alley Gorge, Kalbarri NP, W.A. (P.Olde)

19C. *G. annulifera* Plant in natural habitat, S of Murchison River, W.A. (P.Olde)

Grevillea apiciloba F.Mueller (1876)

Fragmenta Phytographiae Australiae 10: 45 (Australia)

Specific epithet derived from the Latin *apex* (a tip, end) and *lobus* (a lobe), in reference to the terminal leaf lobes. A-PICK-I-LOBE-A

Type: E of Ularing, W.A., 10–15 Oct. 1875, J.Young (holotype: MEL)

Synonyms: *G. flabellifolia* S.Moore (1920); *G. hookeriana* race 'd' of McGillivray (1993).

Low, suckering **shrub** to 1 m or bushy single-stemmed **shrub** to 1.5 m. **Branchlets** angular to terete, pubescent to villous, brown-striate when young. **Leaves** 4–9 cm long, ascending to erect, ± sessile, usually obovate to cuneate with 3–5 linear to subtriangular lobes, sometimes simple, linear, entire, pungent to uncinate, leathery; lobes 0.5–3 cm long, c. 2 mm wide; juvenile leaves flabelliform, toothed across apex (up to 23 teeth); upper surface usually glabrous, sometimes pubescent when young, the venation obscure except sometimes midvein an impressed groove; margins smoothly rounded; lower surface usually partly exposed, sometimes the exposure restricted to the sinuses, pubescent, midvein and lateral veins prominent. **Conflorescence** 4–6.5 cm long, erect, terminal, simple, shortly pedunculate, oblong- to conico-secund; peduncle and rachis villous; bracts ovate, spreading, villous, usually falling before anthesis, sometimes persistent. **Flowers** acroscopic; pedicels 1–2 mm long, villous; torus c. 1 mm across, oblique; nectary tongue-like, conspicuous; **perianth** 6–8 mm long, 2 mm wide, ovoid to S-shaped, villous to silky outside, glabrous inside; limb revolute, subglobose, villous; **pistil** 18–21.5 mm long; ovary sessile, villous; style ± glabrous, exserted first near curve and looped strongly upwards, after anthesis straight or slightly wavy, erect, dilated slightly at style end; style end hoof-like; pollen presenter oblique, conical. **Fruit** 12–15 mm long, 7–9 mm wide, erect on curved pedicel, sometimes persistent, oblong-ellipsoidal, tomentose, with black or brownish stripes, some hairs glandular; pericarp 0.5–1 mm thick. **Seed** 9–11 mm long, 3.5–4.5 mm wide, compressed-oblong-ellipsoidal; outer face convex, smooth; inner face broadly gill-like around margin with central smooth flat area.

Major distinguishing features Leaves yellowish green, obovate to cuneate with some of undersurface exposed, usually 3–5-lobed, the margins smoothly rounded, the upper surface obscurely veined; conflorescence simple, secund; pedicels very short; nectary tongue-like, patent; perianth zygomorphic, hairy outside, glabrous within; ovary sessile, villous; pistil 18–21.5 mm long; style glabrous, usually black, straight; pollen presenter conical; fruit with reddish brown stripes and blotches; seed with gill-like waxy structures overlying outer margin of inner face.

Related or confusing species Group 35, especially *G. hookeriana* which differs in its simple or subpinnatisect leaves with linear lobes, the undersurface quite enclosed by margins. *G. apiciloba* is a relatively distinct species over a wide area and is usually easily recognised by its leaf shape. Some collections from plants with simple leaves may be difficult to separate.

Two subspecies are recognised based on foliar characters; there is some intergrading.

Key to subspecies

Most leaves > 3.5 cm long; conflorescence usually partly enclosed within foliage **subsp. apiciloba**

Most leaves < 2 cm long; conflorescence usually clearly exceeding foliage **subsp. digitata**

Grevillea apiciloba F.Mueller subsp. apiciloba Plate 20

Leaves 3.5–10 cm long, usually obovate–cuneate with prominent apical lobing and narrowly cuneate base, rarely some leaves simple, entire and linear; upper surface glabrous to pubescent. **Flower colour**: perianth usually silver-grey, pale green or brownish red; style red, maroon or black with green tip.

Distribution W.A., widespread in low rainfall areas, extending from north of Wubin to near Mt Churchman south to Lake Cronin and Narembeen, and east to Queen Victoria Rock, near Coolgardie. *Climate* Summer hot, dry; winter cool to mild, wet. Rainfall 250–300 mm.

Ecology Grows in open eucalypt forest and in heath, in yellow sand or in sand over laterite, sometimes in pure laterite. Flowers winter–late spring, sometimes in response to unseasonal rain. Regenerates from seed after fire. Pollinated by birds.

Variation Subsp. *apiciloba* varies in habit, in leaf length and indumentum and in flower colour. The most significant variation is in leaf size. Populations of subsp. *apiciloba* occur over a very wide area. Sometimes in successive sand heaths, leaf size varies markedly, although it tends to be consistent within the population. Occasional plants have mostly simple, linear leaves. Style colour varies from dull red to maroon to black. There is some variation in foliage and habit. Near Lake Cronin, we have seen plants to 0.5 m high with quite stout main stems and flowers with jet black styles. In this area, too, we collected from a plant with predominantly simple, linear leaves among a population with more typical leaves.

Conservation status Not presently endangered.

Cultivation Subsp. *apiciloba* has not been widely cultivated except by enthusiasts, mostly in Vic. Successful cultivation requires a dry climate and a full sun position in well-drained, neutral to slightly acidic sandy or loamy soil. It tolerates frosts to at least -6°C and endures extended dry periods without ill effect. Accordingly, it is long-lived and hardy

G. apiciloba subsp. *apiciloba*

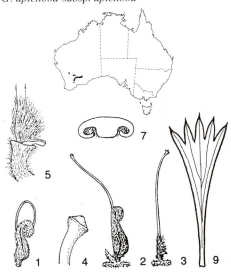

once established in harsh climates. Performance in summer-rainfall climates has not been fully assessed although potted plants flourished for many years in a well-drained mix at Bulli, N.S.W. Pruning may be needed in some conditions as it tends to become spindly, especially in harsh conditions or in shade. Being well adapted to poor soils, it does not require fertilising under normal conditions and summer watering could be harmful.

Propagation *Seed* Pretreated seed germinates well. Average germination time 35 days (Kullman). *Cutting* Firm, young growth from cultivated plants strikes rather erratically. Some forms strike more readily than others. Early spring and autumn appear to be the best times. *Grafting* It has been grafted successfully onto *G. robusta* and G. 'Poorinda Royal Mantle' using top wedge and whip grafts.

Horticultural features Subsp. *apiciloba* is an interesting shrub with attractive foliage and masses of usually black toothbrush conflorescences that attract comment, being of such an unusual colour. In the wild it may be seen as a robust, dense, spreading shrub to 1.5 m or as a spindly, low shrub to 50 cm. Flowers may have red, maroon-red, black-red or pure black styles. In cultivation, it tends to be an open shrub but this is possibly a reflection on the few places in which it has been grown. It would be useful for inland gardens as a screen or structure plant and is very popular with nectar-seeking birds.

General comments Subsp. *apiciloba* is conspecific with the Flabellate form of *G. hookeriana* of McGillivray (1993). It has no basis in the field. The two collections cited are from juvenile plants. Both *G. apiciloba* and *G. hookeriana* generally flower very precociously in the wild and have flabellate juvenile foliage.

Grevillea apiciloba subsp. **digitata**
(F.Mueller) P.M.Olde & N.R.Marriott (1994)

Plate 21

The Grevillea Book 1: 185 (Australia)

Based on *G. apiciloba* var. *digitata* F.Mueller, *Fragmenta Phytographiae Australiae* 10: 45 (1876, Australia).

Subspecific epithet derived from the Latin *digitatus* (finger-like), in reference to the leaf lobes DIDGE-IT-ARE-TA

Type: near Mt Churchman, W.A., 1875, J.Young (holotype: MEL)

Leaves 1–2 cm long, obovate–cuneate to digitate with ray-like veins from point of leaf attachment to pungent lobe tips, usually crowded together. **Flower colour:** perianth creamy white; style black to maroon-black with green tip.

Distribution W.A., confined to the south-west of the species' range, from near Merredin and Tandagin in the south to Burakin in the north. *Climate* Summer hot and dry; winter cool and wet. Rainfall 300–500 mm.

Ecology Grows in yellow sand in heath. Regenerates from seed. Flowers all year with a spring peak. Pollinated by nectarivorous birds.

Variation There is some variation in leaf size but this is otherwise a relatively uniform subspecies.

20A. *G. apiciloba* subsp. *apiciloba* Red flower form, near Southern Cross, W.A. (N.Marriott)

20B. *G. apiciloba* subsp. *apiciloba* A form with mainly simple leaves, near Mt Holland, W.A. (P.Olde)

20C. *G. apiciloba* subsp. *apiciloba* Flowers and foliage, near Southern Cross, W.A. (N.Marriott)

20D. *G. apiciloba* subsp. *apiciloba* Habit, near Mt Holland, W.A. (P.Olde)

G. apiciloba subsp. *digitata*

21. *G. apiciloba* subsp. *digitata* Flowers and foliage, near Wongan Hills, W.A. (N.Marriott)

Conservation status 3V recommended.

Cultivation Subsp. *digitata* has been cultivated successfully mostly in eastern Australia where it tolerates extended dry periods and frosts to c. -3°C once established. It likes well-drained, neutral to slightly acidic deep sand or gravelly loam in full sun. Tip pruning is recommended to maintain shape as cultivated specimens often become leggy. Supplementary watering or fertilising is not necessary. For many years it has been grown successfully by a few enthusiasts in a summer-humid climate in and near Sydney.

Propagation As for subsp. *apiciloba*.

Horticultural features Subsp. *digitata* is robust, spreading shrub with ascending branches angled at c. 45° to the plant. Its dull, yellowish green leaves are crowded along the branchlets which in turn are tipped with black-flowered conflorescences in abundance. Birds are readily attracted to the nectar-laden flowers and seek protection among its dense foliage. It has some landscape potential for its density and growth habit, especially in large-scale projects where special effects are sought. Home gardeners might also be tempted to grow it but allowance should be made for its pungent leaves and robust habit.

General comments Recognition of subsp. *digitata* is based on its occurrence is large populations over a considerable area. Intermediate populations that may be difficult to place in either subspecies are few and the separation seems worthwhile from conservation and horticultural aspects. The change from variety to subspecies is given to provide conformity with infraspecific usage in the genus and to acknowledge the extensive population base of this taxon.

Grevillea aquifolium J.Lindley (1838)

Plate 22

T.L.Mitchell, *Three Expeditions into the interior of Eastern Australia* 2: 178 (England)

Holly Grevillea

The specific epithet is a reference to the resemblance of the leaves to those of the holly, *Ilex aquifolium*. Orthographically, the epithet is a noun in apposition and not adjectival. Hence, *aquifolium* rather than *aquifolia* is the correct spelling. AK-WIFF-OLE EE-YUM

Type: Mt William, The Grampians, Vic., 1836, Major Mitchell's Third Expedition 232 (lectotype: CGE) (McGillivray 1993).

Synonyms: *G. variabilis* J.Lindley (1838); *G. aquifolium* var. *truncata* C.F.Meisner (1856); *G. aquifolium* var. *attenuata* C.F.Meisner (1856).

A prostrate, decumbent, clumping or suckering **shrub** to 30 cm or a dome-shaped, dense plant to c. 1.5 m or sometimes an open, small tree-like shrub to 4 m. **Branchlets** terete to slightly angular, tomentose. **Leaves** 2–10.5 cm long, 1–4.5 cm wide, ascending to spreading, petiolate, ovate to oblong, flat to sinuate or wavy, pungently toothed or pinnatifid with triangular lobes, rarely simple and entire; upper surface glabrous to pubescent with acutely angled venation visible; margins shortly recurved to strongly revolute; lower surface pubescent, the midvein and lateral venation conspicuous. **Conflorescence** 1.5–5 cm long, 2–2.5 cm wide, decurved, sometimes erect, pedunculate, terminal, secund, unbranched; peduncle and rachis densely villous; floral bracts 1–2.5 mm long, imbricate in bud, broadly ovate, villous outside, usually some persistent at anthesis. **Flower colour:** perianth cream, dull orange, grey or dull pink; style dull yellow, orange, pink or bright red. **Flowers** acroscopic; pedicels 1–2.5 mm long, tomentose; torus 1–1.5 mm across, slightly oblique; nectary U-shaped, conspicuous, toothed or wavy; **perianth** 8–10 mm long, 2–2.5 mm wide, ovoid, pubescent to villous outside, glabrous inside, cohering except along dorsal suture at anthesis; limb revolute, globular, densely villous, firmly enclosing style end before anthesis; **pistil** 21–26 mm long; stipe 1.8–3.8 mm long, villous; ovary densely villous; style glabrous, at first exserted from dorsal suture below curve and looped outwards, refracted from line of stipe, incurved after anthesis; style end hoof-like; pollen presenter oblique, oval, convex. **Fruit** 10–13 mm long, 6 mm wide, oblique on incurved stipe, ovoid–ellipsoidal, tomentose with purplish stripes; pericarp < 0.5 mm thick. **Seed** 7–9 mm long, 3 mm wide, oblong–ellipsoidal; outer face convex, smooth, basally rimmed; inner face flat or convex encircled by a waxy or papery scar or channel with a short, oblique wing at one or both ends.

Distribution Vic. and S.A. Widespread in western Vic. where it is centred on The Grampians and surrounding country, extending north into the Little Desert and south and west to the coast where infrequent. In S.A. it is confined to several small areas in the lower southeast of the State. *Climate* Hot, dry summer and cool to mild, wet winter. Rainfall 375 mm in the Little Desert to over 1000 mm in the wettest areas of The Grampians.

Ecology Grows in a wide range of habitats including wet to dry sclerophyll forest, open woodland, low to tall heathland, swampland, and mallee woodland. Soils include deep sand, sandy and gravelly loam, sandy clay, peaty sand and rarely, limestone outcrops. Flowers winter to summer. Pollinated by nectarivorous birds.

Variation One of the most variable of all *Grevillea* species with nearly every population varying slightly from the next. At this stage, until further taxonomic research is conducted, none of the forms listed below can be formally ranked. However, a number of horticultural forms can be separated.

Halls Gap form Growing c. 1–2 m tall and wide, with large, undulate, lightly hairy, shallow-toothed, prickly leaves, this form could be regarded as 'typical'. Many populations from elsewhere in The Grampians including Pomonal, Dadwell's Bridge and Balmoral fit this description showing only minor differences from each other. Style colour ranges from buff to bright red.

Deep-lobed form Normally a low and scrambling shrub, possibly due to its preference for higher sites on mountain tops and slopes such as Mt William, Mirranatwa Gap and Watgania Gap. It has deeply toothed, rigid, prickly, dark green leaves which are sparsely hairy on the upper surface. Flowers are usually bright red. Intermediates between this and the Halls Gap Form can often be found at the foot of the mountains or in the valleys between. The type collection of var. *attenuata* is of this form.

Fyans Valley form A most distinct, erect form confined to a small area in the Fyans Valley of The Grampians in an area that is permanently wet and semi-shaded. It has narrow, lightly lobed to prickly toothed leaves and very long, arching to upright branches. Plants are tall and open, often reaching 3–4 m in height with prominent pink to red flowers.

G. aquifolium

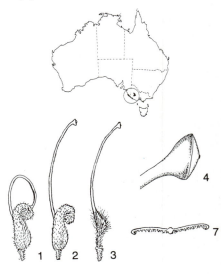

Mt Zero form Generally of upright habit to 2 m with very large, broad, green leaves and prominent, rich red flowers. It is confined to the northern end of The Grampians around Mt Zero, Mt Stapylton and Flat Rock.

Serra Rd form Plants in this area of The Grampians are extremely variable with forms either to 1 m and spreading low or mounded or completely prostrate. The forms in cultivation under this name are the prostrate forms which develop into extremely attractive, dense mats up to 2 m across. Leaves are usually small and lightly toothed but collections of all shapes and sizes can be made. Flowers are not usually very showy, ranging from dull orange to pale red. This form grows in deep sand with constant subsoil moisture and is sometimes short-lived even in the wild. Cultivated plants require summer watering.

Lake Wartook form Another variable population in The Grampians, ranging in habit from ground hugging mats to sprawling shrubs 1 m tall. Leaves are usually distinctly broad and wedge-shaped, often with very few teeth. Flowers vary from dull orange to bright red. This form grows in clay soils in winter-wet often almost swampy areas, although it prefers a well-drained situation in cultivation.

Mt Arapiles form An upright, often pyramidal shrub to 2 m with grey-green, velvety, lightly toothed leaves and a profusion of showy, red flowers. It is confined to Mt Arapiles and its surrounds, the best forms being found around the base of the mountain.

Stawell form A very distinctive population with medium to large, attractive, grey to silver, woolly leaves that are most variable in shape; some forms have long wedge-shaped leaves very similar to *G. ilicifolia* while others have entire, oval leaves. A feature of this population is the way the flowers change colour as they age. As a result, bi-coloured and even occasional tri-coloured forms can be found, often opening as a clear yellow and ageing to orange or red or both. In some plants, however, the colour remains the same throughout. This very hardy form is confined to dry, stony areas around Stawell and the Ironbark Ranges State Forest to the north.

Little Desert suckering form Most populations of *G. aquifolium* that have encroached into the Little Desert have adapted to the drier conditions by having tightly revolute leaf margins to reduce the leaf surface exposed to the heat and sun. This gives the leaves a curious tubular effect. This form appears to reproduce only by suckers and makes an attractive grey shrub to 1 m with bright red flowers. It requires a well-drained site in the garden and is quite drought hardy.

Cooack form This is a very interesting, erect form from the southern edge of the Little Desert. It is a large, open shrub to 2.5 m with similar leaves to the suckering form. Flowers are bright red or occasionally orange or yellow, and reproduction is from seed. This free-flowering form is rapidly being destroyed in its natural habitat by clearing for agriculture.

Portland, Mt Richmond, Kentbruck Heath forms These forms from the lower southwest of Vic. all have distinctive green leaves with lightly to deeply scalloped margins and a glabrous upper surface. They all grow in flat, swampy sites in black, acidic sand. Flowers range from creamy green to pale red, some forms including the beautiful Mt Richmond population changing from yellow to red as they age. All but the Kentbruck Heath population are close to extinction in the wild.

South Australian forms (Carpenters Rocks etc.) Confined to several small areas in south-eastern S.A. including Carpenters Rocks, Dairy Range and Lake Bonney. These populations generally have small to medium, lightly toothed or almost entire leaves which are almost glabrous on the upper surface. Habit ranges from prostrate mats to sprawling shrubs to 1.5 m. Flowering is profuse, ranging in colour from pink to pale red, with green and yellow-flowered forms also recorded. A unique feature of these populations is that most occur on alkaline soils with several populations growing in soil pockets on limestone outcrops.

Major distinguishing features Leaves usually toothed, pubescent on undersurface; conflorescence secund; bracts 2–4 mm long; inner perianth surface glabrous; style end hoof-like; ovary shortly stipitate, densely villous; fruit red-striped on incurved stipe.

Related or confusing species Group 35, especially *G. bedggoodiana*, *G. ilicifolia*, *G. infecunda* and *G. repens*. *G. ilicifolia* may be distinguished by its fruit, which is held erect on a straight stipe, the (normally) appressed indumentum on its ovary and lower leaf surfaces, and its smaller floral bracts (< 1 mm long) that fall very early. The style end is convex on the dorsal side. Some populations have spreading hairs on the ovary and this feature should not be relied on solely for identification. *G. bedggoodiana* has a pistil 12–16.5 mm long, the style strongly curved and hooked at the end. *G. infecunda* produces almost no pollen from poorly-developed anthers. The leaf undersurface is silky. *G. repens*, a prostrate, trailing shrub, has leaves almost glabrous below.

Conservation status Over most of its range *G. aquifolium* is widespread and extremely common, but several populations, notably at Cooack and Portland, have been reduced by clearing for agriculture and are close to extinction.

Cultivation A hardy and adaptable species growing well in cold-wet, hot-dry and even summer-rainfall climates from Brisbane to Adelaide and all points in between. Most forms withstand extended dry summers and frost to at least -7°C. At least five forms are cultivated in the U.S.A. today. *G. aquifolium*

22A. *G. aquifolium* Cooack form, in cultivation (N.Marriott)

22B. *G. aquifolium* Typical flowers and foliage (W.Blackburn)

22C. *G. aquifolium* Yellow-flowered form, Ironbark Forest, Vic. (N.Marriott)

22D. *G. aquifolium* Distinct form growing in limestone, Carpenters Rocks, S.A. (N.Marriott)

22E. *G. aquifolium* Note change in flower colour with age (N.Marriott)

22F. *G. aquifolium* Prostrate form from Serra Rd, The Grampians, Vic. (N.Marriott)

22G. *G. aquifolium* Stawell form (N.Marriott)

can be grown in most situations but, for greatest success, it should be planted in a sunny to semi-shaded site in well-drained, sandy or gravelly loam. All forms except the S.A. populations require an acidic soil while the latter will grow in either acidic or alkaline soils. Additional fertilisers are not required but shape and density is improved by regular tip-pruning. Several forms respond to summer watering. Prostrate forms are suitable for tub culture. A number of forms are usually available from specialist native plant nurseries especially in Vic.

Propagation *Seed* Sets copious quantities of seed which germinate well particularly if pretreated. *Cutting* Firm, young growth taken at most times of the year strikes readily. *Grafting* Unnecessary for most situations. Makes a hardy rootstock for some species. Has been grafted onto *G. robusta* using an approach graft.

Horticultural features *G aquifolium* is a very rewarding plant with a multitude of forms of all shapes and sizes. The foliage of all forms is very attractive while the variably coloured, toothbrush flowers are showy, and are popular with honey-eating birds, providing copious nectar for many months of the year. With so many forms to choose from, this long-lived species is highly regarded by landscapers as it can be used as a dense groundcover, a small or large dense screen, or as a feature plant.

Hybrids Natural hybrids occur between *G. aquifolium* and *G. microstegia* in several areas on and below Mt Cassell in The Grampians. There are also two collections of a hybrid between *G. aquifolium* and ?*G. montis-cole* on the eastern side of The Grampians. *G. aquifolium* is also the known parent of a number of garden hybrids including G. 'Australflora Copper Crest'.

History The first recorded cultivation of this species was at Melbourne Botanic Gardens in 1858 (as *G. variabilis*). References to introduction of this species by Cunningham into England around 1820 are clearly confused, since the species was not known at that time.

General comments The varieties based on leaf characters, var. *attenuata* and var. *truncata*, described by Meisner (1856) are encompassed by the normal variation in this species and have, accordingly, been reduced to synonymy.

Grevillea arenaria R.Brown (1810)

Transactions of the Linnean Society of London 10: 172 (England)

Sand Grevillea

The specific epithet is derived from the Latin *arenarius* (an inhabitant of sand), in reference to the locality of the type specimen, collected on a sandy island. ARREN-ARE-EE-A

Type: island of the Nepean R., a little above its junction with the Grose R., N.S.W., 1803–04, R.Brown 3325 (lectotype: BM) (McGillivray 1993).

Synonyms: *G. ferruginea* F.W.Sieber ex C.P.Sprengel (1827); *G. ferruginea* F.W.Sieber ex J.A.Schultes & J.H.Schultes (1827); *Embothrium arenarium* (R.Brown) G.L.M.Dumont de Courset (1814); ?*Lysanthe cana* J.Knight (1809).

A much-branched **shrub** 0.3–4 m high. **Branchlets** terete. **Leaves** 1–7.5 cm long, 0.3–1.5 cm wide, ascending, shortly petiolate, simple, entire; apex obtusely mucronate to acute; upper surface convex, usually inconspicuously veined, sometimes the venation obscure. **Conflorescence** 1–2.5 cm long, usually terminal on short branchlet, 2–10-flowered, occasionally branched; rachis tomentose to pubescent; bracts 1–2.5 mm long, narrowly triangular, sometimes persistent at anthesis. **Flowers** adaxially acroscopic, the buds spreading; pedicels 2–8 mm long, hairy, often retrorse; torus 1.5–2.5 mm across, oblique; nectary collar-like, spreading with an undulate and sometimes recurved margin, sometimes with inconspicuous hairs on upper surface; **perianth** 6–15 mm long, 3–5 mm wide, usually oblong with a small basal dilation, rarely ovoid-saccate, ridged, curved, sometimes strongly so, villous or sparsely so outside, white-bearded inside above level of ovary with a zone of dense, spreading hairs above beard almost to curve developed on ventral tepals, cohering except along dorsal suture before anthesis; limb ovoid to pyramidal, acute or tapered to a finely-drawn point, erect in bud, revolute at anthesis, the dorsal segments sometimes laterally reflexed after anthesis; **pistil** 24–32 mm long; ovary sessile, villous; style exserted below curve on dorsal side and looped upwards, straight after anthesis, pubescent, almost glabrous near enlarged, flattened style end; pollen presenter lateral, obovate, usually flat, umbonate. **Fruit** 12–22 mm long, 6–8.5 mm wide, often persistent, erect on pedicel, oblong/ellipsoidal, apex attenuated, usually longitudinally ridged, tomentose; pericarp c. 0.5 mm thick. **Seed** 10 mm long, 3.5 mm wide, ellipsoidal with strongly revolute margins bordered by a waxy rim, winged at apex; outer face wrinkled, minutely hairy; inner face obscured.

Major distinguishing features Leaves simple, entire; perianth limb acute or drawn into a long, finely-pointed tip; perianth zygomorphic, hairy on both surfaces, inner perianth surface with a dense, spreading indumentum above beard; nectary with hairs on upper surface (difficult to observe at 20× magnification); ovary sessile, hairy; style green, pubescent to loosely villous; pollen presenter lateral; fruit ridged on dorsal side with style erect.

Related or confusing species Group 25, especially *G. banyabba*, *G. masonii*, *G. montana*, *G. quadricauda* and *G. rhizomatosa*, all of which can be separated by their lack of a spreading indumentum above the beard on the inner perianth surface.

Variation Two subspecies are currently recognised.

Key to subspecies

Leaf undersurface silky, tomentose, villous or loosely so; pistil 22–27 mm long .... subsp. **arenaria**

Leaf undersurface velvety; pistil 26–32 mm long .... subsp. **canescens**

Grevillea arenaria R.Brown subsp. arenaria   Plate 23

Single-stemmed **shrub** 1–4 m high. **Branchlets** silky, villous, tomentose or pubescent. **Leaves** 1–7.5 cm long, 3–10(–15) mm wide, mostly linear to oblong–elliptic; upper surface silky to glabrous, granular; lower surface silky, tomentose or villous, midvein prominent; margins revolute to recurved. **Conflorescence** erect to decurved, sessile or pedunculate; unit conflorescence subcylindrical to few-flowered; peduncle and rachis silky to villous; bracts 1–2 mm long. **Flower colour:** perianth green to yellow with reddish tinges; limb pink to red; style green. **Flowers:** pedicels 2–8 mm long, silky to villous; **perianth** 6–15 mm long, oblong to strongly recurved, sparsely to densely hairy outside; limb with appendages almost absent to 5 mm long; **pistil** 22–26 mm long; pollen presenter 2–3 mm long.

Distribution N.S.W., irregularly distributed from Deua NP in the south to Richmond in the north.
Climate Summer hot, wet to dry; winter mild to cold, dry or sometimes wet. Rainfall 400–1000 mm.
Ecology Found in acidic gravelly loam or reddish brown loam, in poor, stony, shallow shale or sand, occasionally on limestone, usually occurs in woodland or on river banks. Flowers in almost every month but more prolific in spring.
Variation The following informal forms are recognised.

Silky form An erect shrub to 2.5 m usually with pendulous branches and oblong–elliptic, somewhat undulate leaves 2–5 cm long. The upper surface of

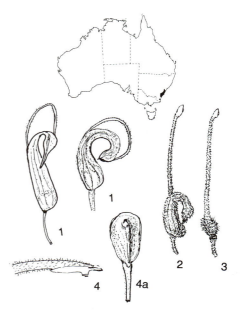

G. arenaria subsp. *arenaria*

mature leaves is dark green, often glabrous and scabrous, the undersurface loosely villous to silky. Conflorescences are pedunculate; flowers are green with pink to reddish limb, on pedicels 2–5 mm long, and have a sparsely pubescent indumentum. The perianth limb is very strongly revolute and tapers to a short point. The holotype of *G. arenaria*, which represents this form, was collected near Richmond, N.S.W., but has not been collected there in recent years. A small population is recorded nearby at Yarramundi. Other collections are mainly from limestone in the Goulburn–Braidwood district with a further collection from Deua on rhyolite. This form can sometimes be difficult to establish in cultivation and resents poor drainage.

Villous form This form occurs in N.S.W. from Bents Basin to Mittagong with occasional populations south to Mogo. It has a villous leaf undersurface, usually sessile, few-flowered conflorescences

23A. *G. arenaria* subsp. *arenaria* near Wingello. N.S.W. (P.Olde)

23B. *G. arenaria* subsp. *arenaria* near Goulburn, N.S.W. (N.Marriott)

(sometimes pedunculate) and pedicels 5–8 mm long. The perianth is not strongly recurved but rather oblong and the limb has appendages 3–5 mm long.

Conservation status Not presently endangered.

Cultivation This very adaptable, hardy plant is suitable for cultivation in most areas of Australia including Tas., as well as New Zealand. Southern forms are able to tolerate frosts to -11°C and extended dry periods. Robust plants are flourishing at the Australian National Botanic Gardens, Canberra, and at Burrendong Arboretum, N.S.W. Three forms are reportedly cultivated in U.S.A. It establishes well in summer-rainfall climates such as in Sydney where it succeeds in either partially shaded situations or full sun. Well-drained neutral to alkaline soils are important for successful cultivation, and once established it rarely requires much maintenance. Pruning is only required to maintain shape and neither fertiliser nor summer watering is usually required. Different forms are sometimes available at both specialist and general nurseries, mainly in the eastern States of Australia.

Propagation *Seed* Fresh seed should germinate well when given the standard peeling pretreatment. *Cutting* Firm, new growth taken at any time of the year strikes well. *Grafting G. robusta* has proved to be a suitable rootstock.

Horticultural features This highly variable subspecies has a number of interesting features depending on form but is used mainly as a foliage contrast or screen plant in large gardens. The greyish foliage can be quite attractive but flowers are usually obscured. The plant is constantly visited by birds seeking both nectar and shelter. It is suited to cool climates and is generally long-lived in cultivation.

24A. *G. arenaria* subsp. *canescens* Close-up of flowers (P.Olde)

Hybrids A cultivar known as G. 'Little Thicket', collected originally from the Braidwood area, may be a hybrid of *G. arenaria* and *G. lanigera*. Some plants from the Kowmung River are thought to have hybridised with *G. rosmarinifolia*. At Shaw's Creek, near Sydney, *G. arenaria* has hybridised with *G. mucronulata*.

Early history *G. arenaria* was introduced to England in 1803 by George Caley and was in cultivation at Kew in 1810. From here, it spread to the European continent and is recorded in cultivation in Amsterdam in 1857. The Sydney Botanic Gardens recorded it in cultivation in 1857.

General comments Further taxonomic study of the variation in subsp. *arenaria* is warranted as it still contains a considerable degree of population-based diversity.

Grevillea arenaria subsp. canescens (R.Brown) P.M.Olde & N.R.Marriott (1994) Plate 24

Telopea 5: 711 (Australia)

Based on *G. canescens* R.Brown, *Supplementum Primum Prodromi Florae Novae Hollandiae*: 18 (1830, England); *G. arenaria* var. *canescens* (R.Brown) G.Bentham (1870, England).

Hoary Grevillea

The subspecific epithet is derived from the Latin *canescens* (becoming grey), in reference to the greyish foliage. CAN-ESS-ENS

Type: banks of Cox R., Blue Mountains, N.S.W., 1817, A.Cunningham. (holotype: BM).

A single-stemmed **shrub** to 3 m, sometimes suckering and 0.3–1 m high. **Branchlets** velvety to villous. **Leaves** 1–4cm long, 0.4–1 cm wide, oblong, obovate to elliptic; upper surface pubescent; lower surface velvety, the midvein usually obscure, occasionally prominent; margin shortly recurved. **Conflorescence** decurved, pedunculate, subcylindrical to few-flowered; peduncle and rachis velvety; bracts 1.2–3 mm long, velvety. **Flower colour:** perianth green, yellow or red at the

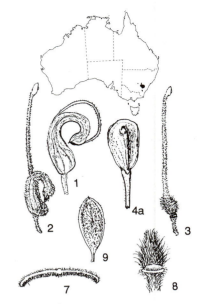

G. arenaria subsp. *canescens*

base with a pink to red limb; style green. **Flowers:** pedicels 3–5 mm long, velvety; **perianth** 6–10 mm long, strongly recurved, velvety; limb with appendages 0.2–3 mm long; **pistil** 26–32 mm long; pollen presenter 3–4 mm long.

Distribution N.S.W., from Gilgandra, Dubbo, Bathurst, Glen Davis, Kedumba and Megalong Valleys, Kiandra and Tumut. *Climate* Summer hot, dry or wet; winter cool to cold, wet or dry. Rainfall 400–800 mm.

Ecology Grows in deep yellow sand, granitic sand, brown gravelly loam or basalt-derived soils, usually in dry, sclerophyll forest. Flowers spring. We have observed nectarivorous birds probing the flowers on plants growing at The Rocks, near Bathurst.

Variation Subsp. *canescens* is readily distinguished

24B. *G. arenaria* subsp. *canescens* Red flowers on plants in Gilgandra Reserve, N.S.W. (P.Olde)

from other populations of *G. arenaria* by the velvety indumentum of its leaf undersurface, but there is a degree of floral diversity within the subspecies approaching that accommodated in subsp. *arenaria*.

Typical form This extremely beautiful form is a robust, dense shrub to 3 m tall, spreading 2–3 m wide and is distributed from the Megalong and Kanimbla valleys in the Blue Mountains south to Jenolan Caves and west through Bathurst to near Dubbo. It occurs mostly on granite. The mainly elliptic leaves with rounded apices are grey-green above and densely hairy below, rarely exceeding 2.5 cm long, and have a soft, velvety feel. The upper leaf surface bears a persistent appressed white indumentum and the flowers have spreading white hairs on the outer surface which dull the yellow to red flowers.

Gilgandra form Although here still regarded as part of subsp. *canescens*, plants from Gilgandra Nature Reserve differ in their lower, suckering habit and bright red flowers. Plants from Eumungerie and Goonoo Goonoo Forest near Dubbo have similar red flowers but also have mixed yellow or orange-yellow flowers. At Gilgandra, they rarely exceed 1 m and occur in yellow, sandy soil. This excellent horticultural selection should be more widely cultivated.

Kiandra form Collections from the Tumut–Kiandra area have a very obscure appendage 0.2–0.5 mm long on the perianth limb but otherwise agree with other populations.

Conservation status Not rare.

Cultivation Subsp. *canescens* has proved reliable and hardy in cultivation, growing easily in both moist, coastal and drier, inland situations. It tolerates frosts to at least -11°C and will grow in a variety of soil textures ranging from heavy loam to sand. A sunny position in well-drained soils is essential. Occasionally available from specialist nurseries.

Propagation *Seed* Grows readily from seed that has been pretreated. *Cutting* Strikes readily from cuttings using firm, new growth. *Grafting* Grafts easily onto *G. robusta* using the mummy technique, but grafting is rarely required.

Horticultural features An outstanding, ornamental foliage plant worthy of a position in any large garden or public landscape where floral brilliance is not a requirement. The velvety, grey leaves are both beautiful in appearance and soft to the touch and always attract admiration. Flowers of the Gilgandra form are very bright, however, and this form is worth growing in a rockery, small garden or as a tub plant. Well-drained mixes are essential. Nectar-feeding birds are strongly attracted.

Grevillea argyrophylla C.F.Meisner (1855)
Plate 25

Hooker's Journal of Botany & Kew Garden Miscellany 7: 75 (England)

Silvery-leaved Grevillea

The specific epithet is derived from Greek *argyros* (silvery) and *phyllon* (a leaf), in reference to the silvery undersurface of the leaves. AR-GYE-ROW-FILLER

Type: north of Swan R., W.A., 1850–51, J.Drummond, coll. 6 no.179 (lectotype: NY) (McGillivray 1993).

A spreading, compact **shrub** or small, broad-crowned **tree** 1.5–6 m high, sometimes almost prostrate in exposed conditions. **Branchlets** angular, pubescent to silky. **Leaves** 1.5–6 cm long, 1.5–15 mm wide, ascending, shortly petiolate, mucronate, obtuse or notched, simple, rarely bilobed, linear to obovate; upper surface sparsely silky, soon glabrous, midvein and sometimes lateral venation visible; margins revolute to shortly recurved; lower surface densely silky-white, midvein and acutely angled lateral veins ± glabrous and prominent. **Conflorescence** erect, pedunculate, axillary, dense, simple or few-branched; unit conflorescence c. 10 mm long, 20 mm wide, ovoid in bud, ovoid–umbel-like at anthesis; peduncle and rachis silky; bracts c. 0.5 mm long, linear, silky outside, falling before anthesis. **Flower colour:** perianth and style white or pinkish white. **Flowers** adaxially oriented; pedicels 1.5–4 mm long, silky; torus c. 0.5 mm across, ± straight; nectary prominent, half-moon shaped or cushion-like, entire; **perianth** 3–4 mm long, c. 0.5 mm wide, regular in late bud, slender, cylindrical to narrowly ovoid, silky outside, papillose inside about ovary; tepals at first separated along dorsal suture near curve, then all separated below limb, the dorsal tepals reflexed laterally and exposing inner surface, afterwards all free to base; limb revolute before anthesis, spherical–subcubic, the segments impressed at margins, sparsely villous with reddish hairs around tips; **pistil** 4–7 mm long, glabrous; stipe 0.3–0.7 mm long; ovary globose; style filiform, exserted first near curve and looped upwards, strongly incurved in apical half after anthesis, dilating smoothly into a slightly expanded style end; pollen presenter oblique, oblong–elliptic, convex to subconical. **Fruit** 6–10 mm long, 5 mm wide, erect on straight or curved pedicel, ovoid, rugose; pericarp c. 0.5 mm thick. **Seed** 7 mm long, 2 mm wide, ellipsoidal–ovoid, smooth, unwinged.

Distribution W.A., mainly in coastal situations from near Dandaragan and Jurien Bay to the Murchison R. *Climate* Hot, dry summer; mild, wet winter. Rainfall 300–550 mm.

Ecology Found in heath to tall shrubland, usually in white sand derived from limestone or among sandstone or limestone rocks. Flowers late winter to spring. After fire, regenerates from seed. Pollinated by insects.

Variation The width of the leaves varies considerably and both fine and broad leaf forms have been selected for cultivation. Semi-prostrate forms occur on the coast but it is not known if these forms retain their habit in cultivation.

Major distinguishing features Leaves simple, linear or obovate, dark green above, silky white below; conflorescence usually branched, umbel-like; pistil glabrous, 4–7 mm long; fruit rugose.

Related or confusing species Group 16, especially *G. commutata* and *G. hakeoides*. *G. commutata* has a longer pistil (12–14.5 mm long). Both subspecies of *G. hakeoides* have the leaf undersurface obscured by the margin. For many years *G. olivacea* (Group 14) has been sold as *G. argyrophylla*, particularly in S.A.; it has larger flowers (pistils 22–28 mm long).

Conservation status Not presently endangered.

Cultivation A tough, adaptable species suitable for a wide range of climatic conditions and soil types, with the possible exception of very acidic sands. It has been grown in S.A., Vic. and N.S.W. and is cultivated in the Western Australian Herbarium garden as well as in Kings Park, Perth. A small grove can be seen at Burrendong Arboretum, N.S.W., and at 'Myall Park', Glenmorgan, Qld. Frosts as low as -6°C have been tolerated and it is drought hardy. An open, sunny site is preferred and although it will

G. argyrophylla

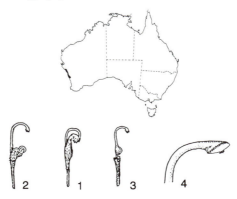

25A. *G. argyrophylla* Flowering branch (N.Gibb)

withstand semi-shade, it becomes more open in these situations. It flourishes in well-drained, alkaline to neutral, sandy soils but will happily adapt to acidic loams and even heavy clays. In warm, dry climates, it will even tolerate poor drainage for short periods. Although rarely necessary, pruning is readily accepted and new growth reshoots beautifully from the trunk if heavy trimming is required. *G. argyrophylla* is not widely available in nurseries at present.

Propagation *Seed* Tends to set few seeds in cultivation but germinates readily with standard peeling treatment. Usually germinates within 21 days. *Cutting* This species may strike with reluctance. Most success has been achieved using fresh, fairly firm, large cuttings to 15 cm, taken during early autumn. *Grafting* Due to its vigour and hardiness it is generally not necessary to graft *G. argyrophylla*. It may itself prove to be a suitable rootstock for use in alkaline soils. Has been grafted successfully onto *G. robusta*.

Horticultural features Massed displays of perfumed, white flowers and a compact, rounded habit are the main features of this species. In windy conditions the shimmering, silvery undersurface of the leaves adds an extra dimension to the foliage. In cultivation, flowering can extend from autumn to early summer. Makes an excellent, dense screen and is long-lived, which makes it valuable for the landscaper, although to date it is not used widely. It tolerates salt-laden winds to a limited degree and is suitable for coastal gardens. With a dense habit it is often used by birds for nesting and shelter. Generally pest- and disease-free.

Hybrids Natural hybrids with *G. commutata* have been recorded. These plants have larger, thicker and more compact flower heads. They are unknown in cultivation.

25B. *G. argyrophylla* Habit of plant S of Eneabba, W.A. (P.Olde)

25C. *G. argyrophylla* Plant in cultivation, Stawell, Vic. (N.Marriott)

Grevillea armigera C.F.Meisner (1856)
Plate 26

A.P.de Candolle, *Prodromus* 14: 373 Paris (France)

Prickly Toothbrushes

The specific epithet is derived from the Latin *armiger* (a bearer of weapons), alluding to the plant being 'armed' with pungent, leaf lobes. AR-MIDGE-ER-A

Type: near Swan R., W.A., early 1840s, J.Drummond coll. 4 no. 284 (lectotype: K) (McGillivray 1993).

A prickly, dense, irregular **shrub** to 3 m high, usually leaning at a slight angle and bearing spreading branches. **Branchlets** angular to terete, pubescent. **Leaves** 2.5–4.5 cm long, ascending, shortly petiolate, divaricately 2–3 times divided into pungent, rigid, narrowly linear lobes; leaf rachis curved, often angularly; upper surface granular, ± glabrous, midvein a faint groove; margins smoothly revolute to midvein; lower surface enclosed by margin, silky in grooves, midvein prominent. **Conflorescence** 5–9 cm long, erect, pedunculate, terminal, secund, dense, unbranched; peduncle appressed pubescent; rachis villous; bracts 0.5–1.5 mm long, ± ovate, falling before anthesis. **Flower colour:** perianth grey, greyish green or yellowish; style black or maroon-black. **Flowers** acroscopic: pedicels 1–2 mm long, silky, patent; torus c. 1 mm across, straight; nectary patent, tongue-like, wavy; **perianth** 6–7 mm long, 2–3 mm wide, narrowly ovoid, dilated at base, silky outside, glabrous inside, cohering except along dorsal suture; limb revolute, ovoid, enclosing style end before anthesis; **pistil** 18–23 mm long; ovary ± sessile, villous; style first exserted from dorsal suture below limb and looped upwards, slightly incurved after anthesis, later straight to undulate, glabrous except some basal hairs; style end hoof-like; pollen presenter oblique, elliptic, conical. **Fruit** 12–14 mm long, 7–9 mm wide, on sharply incurved pedicel, ellipsoidal, tomentose, reddish brown striped; pericarp c. 0.5 mm thick. **Seed** 9 mm long, 4 mm wide, ellipsoidal with a short apical wing; outer face smooth, convex, crimped around margin; inner face flat with an encircling, membranous ruff; margin scarcely recurved.

Distribution W.A., in the northern wheatbelt from Buntine to Dowerin. ***Climate*** Hot, dry summer; mild, wet winter. Rainfall 340–460 mm.

Ecology Found in sand heath or shrubland, usually in deep, yellow sand or gravelly loam. Flowers year-round with a peak in winter and spring. Pollinated by nectarivorous birds. Regenerates from seed.

Variation A relatively uniform species. A coarse-leaved, robust specimen was collected by Edwin Ashby early this Century near Botherling but no subsequent collection has been made of this form. Some populations west of Wongan Hills have silvery-grey rather than dull grey foliage.

Major distinguishing features Leaves divaricately 2–3 times divided into short pungent narrowly linear lobes; rachis angularly refracted; conflorescence erect, secund; pedicels 1–2 mm long; perianth zygomorphic, hairy outside, glabrous inside; ovary sessile, densely hairy; nectary patent, linguiform; style maroon–black; fruits with reddish striping.

Related or confusing species Group 35, especially *G. hookeriana* which has a ± straight leaf rachis with usually simple or once-divided leaves.

Conservation status Although relatively common, this species is often restricted to degraded road verges and could become endangered with further deterioration of its habitat.

Cultivation *G. armigera* is a tough plant, reliable for dry, inland conditions in sand, gravel, or well-drained, clay soils. Hardiness in humid, summer-rainfall climates has not been properly tested but the few so far attempted have usually been short-lived as it seems to resent summer watering. Not only is it a drought-hardy species but it will also withstand frosts to at least -6°C. It has been successfully grown in north-eastern and western Vic., at Glenmorgan, Qld and Kings Park, Perth, W.A. Successful growers report that it demands well-drained soils in full sun and, given these conditions, it will flower over an extended period. Judicious pruning may be necessary to modify the shape of the bush. Potted specimens do not thrive and do not respond to fertilisers. Plants are

G. armigera

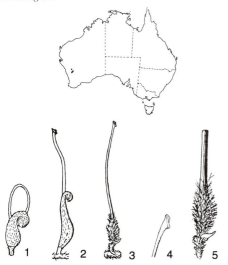

26A. *G. armigera* Conflorescence and fruit (P.Olde)

26B. *G. armigera* Plant habit, in cultivation, Stawell, Vic. (N.Marriott)

26C. *G. armigera* Flowering branch (N.Marriott)

occasionally available at specialist nurseries and seed can sometimes be purchased through native seed suppliers.

Propagation Seed Produces good quantities of viable seed which germinate readily when given the standard peeling pretreatment. Average germination is 34 days (Kullman). *Cutting* Strikes fairly readily from firm, fresh, new growth, especially taken from young, cultivated plants. Best strike rates are achieved during spring and early to mid-autumn. *Grafting* Has been grafted successfully onto *G. robusta* and G. 'Poorinda Royal Mantle' using the top wedge and mummy methods, but long-term compatibility is not known.

Horticultural features The main features of *G. armigera* are its asymmetrical habit, black styles and prickly blue-grey leaves contrasting with the white pubescent branchlets. The 'toothbrush' racemes, produced in great quantities over many months of the year, are regarded as attractive by many collectors for their unusual colour; flowers with black styles are somewhat unusual and these make a striking contrast with the yellowish perianth. The prickly foliage with its obvious disadvantages can be used effectively as a barrier or screen, and birds frequent it in search of both protection and nesting sites. The flowers also provide a source of nectar. Long-lived even under harsh, dry conditions, this species has been used by lovers of the genus as a curiosity specimen plant.

Grevillea asparagoides C.F.Meisner (1856)   Plate 27

A.P. de Candolle, *Prodromus* 14: 373, Paris (France)
The specific name refers to a supposed similarity to the genus *Asparagus*. ASS-PARR-AG-OY-DEES

Type: south-western W.A., 1848, J.Drummond coll. 4 no. 283 (lectotype: NY) (McGillivray 1993).

A bushy, prickly, lignotuberous **shrub** 0.5–2 m tall. **Branchlets**

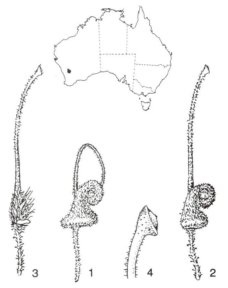
G. asparagoides

terete, appressed-pubescent, striate. **Leaves** 2–3.5 cm long, ascending, shortly petiolate, 2–3 times divaricately divided into rigid, narrowly linear, pungent lobes; leaf rachis decurved, sometimes angularly; margins angularly revolute to midvein; upper surface sparsely pubescent, soon glabrous and granulate, with conspicuous midvein and marginal vein; lower surface obscured, tomentose in grooves, midvein prominent. **Conflorescence** decurved, pendulous, pedunculate, terminal, simple or branched; unit conflorescence 2.5–6 cm long, 4–6 cm wide, hemispherical to subcylindrical, partially secund, loose; peduncle and rachis tomentose; floral bracts 1–2 mm long, triangular, tomentose outside, sometimes persistent at anthesis. **Flower colour:** perianth pinkish red to pinkish cream with a creamy pink limb; style red. **Flowers** acroscopic: pedicels 5–14 mm long, glandular-pubescent, patent to slightly retrorse; torus 1–2 mm across, ± straight; nectary patent, tongue-like, toothed or wavy; **perianth** 6–10 mm long, 3.5–5 mm wide, ovoid-saccate, narrowed markedly at the neck, tomentose to glandular-villous outside, glabrous inside, sometimes sparsely silky below limb, cohering except along dorsal suture; limb tightly revolute, globular, enclosing style end before anthesis; **pistil** 30–37 mm long; ovary sessile, villous; style elongate, exserted first at curve on dorsal side and looped upwards, incurved from c. halfway after anthesis, glandular-pubescent, dilating suddenly into a club-shaped style end; pollen presenter oblique, round, umbonate. **Fruit** 10–17 mm long, 7–8 mm wide, obliquely attached on reflexed pedicel, ovoid or oblong, glandular-pubescent; pericarp 0.5–1 mm thick. **Seed** 8.5 mm long, 3.2 mm wide, ellipsoidal with recurved margins; outer face convex, smooth with a pale border; inner face concave bordered with a papery, gill-like rim, otherwise smooth.

Distribution W.A., in scattered localities from near Perenjori to Wongan Hills. *Climate* Hot, dry summer; mild, moist winter. Rainfall 300–460 mm.

Ecology Grows in sandy loam and gravel in open sandheath and in low shrubland with scattered eucalypts. Flowers winter to spring. Bird-pollinated. The species attracts a procession of nectarivorous birds.

Variation Generally a uniform species. Some specimens have longer pedicels.

Major distinguishing features Leaves divaricately 2 or 3 times divided with pungent, narrowly linear lobes; indumentum of leaves and flowers glandular-pubescent; conflorescence secund-hemispherical, loose; pedicels elongate (5–14 mm long); perianth zygomorphic, hairy outside, glabrous or sometimes with scattered hairs behind the anthers inside; torus straight; nectary linguiform; ovary sessile, densely hairy; pollen presenter oblique; fruit with reddish markings.

Related or confusing species Group 35, especially *G. batrachioides*, which differs in its sessile leaves.

Conservation status Suggested 3V. *G. asparagoides* is confined mainly to road verges and is at risk due to diminishing and degrading natural habitat.

Cultivation G. asparagoides is a reasonably hardy species in a hot, dry climate with dry summer and wet winter, and will tolerate drought and frost to at

27A. *G. asparagoides* Close-up of conflorescence (N.Marriott)

27B. *G. asparagoides* Flowers and fruit (N.Marriott)

27C. *G. asparagoides* Plant in natural habitat (P Olde)

least −6°C. It is often cultivated in enthusiasts' gardens in Vic. and S.A. and is grown at Kings Park, Perth, W.A., as well as at Santa Cruz, U.S.A. Successful growers report that it flourishes only in well-drained sand or gravelly loam in full sun but, once established, requires almost no maintenance or attention except pruning to shape. It will adapt to dappled shade, although it is denser and more floriferous in an open site. It shows a decided distaste for poor drainage, cold, wet conditions and summer humidity and is short-lived unless grown in optimal climates. It is also intolerant of alkaline soils, preferring a neutral to acidic pH. Responds well to pot culture if grown in a large tub in a sheltered sunny situation but may succumb quickly if excess fertiliser is applied, preferring hungry mixes similar to the impoverished soil of its natural habitat. Has been sold in Vic. for many years but generally available only in specialist nurseries.

Propagation **Seed** Sets good quantities of seed in cultivation which germinate well with the standard peeling treatment. Average germination time is 42 days (Kullman). *Cutting* Strikes reasonably well from firm, new growth especially when taken in early autumn. Some trouble has been experienced with low root production. Cuttings placed under heavy mist propagation are troubled by fungal attack. *Grafting* Plants have been grafted successfully onto G. 'Poorinda Royal Mantle' using the top wedge graft. *G. robusta* and *G. rosmarinifolia* have also been used successfully.

Horticultural features *G. asparagoides* is a dense shrub with prickly grey-green, somewhat hoary foliage, white pubescent branchlets, and masses of showy, pendant racemes of flowers for many months. The smoky pink flowering racemes are quite large and hang loosely, enabling the individual flowers with their conspicuous styles and pretty colours to be well-displayed. It is a useful plant for landscapes and gardens, with its grey-green foliage providing an ideal foil to the flowers, at the same time making an effective screen plant. Large numbers of nectar-seeking birds are lured to the plant in search of the abundant nectar which its flowers produce, and it is also an excellent refuge for birds in search of shelter.

Grevillea aspera R.Brown (1810)

Plate 28

Transactions of the Linnean Society of London 10: 172 (England)

Rough-leaf Grevillea

The specific name is derived from the Latin *asper* (rough), in reference to the scabrous upper leaf surface. ASS-PER-A

Type: Bay X, [Port Lincoln, S.A.], 1802, R.Brown (lectotype: BM) (McGillivray 1993).

An erect, dense, sometimes suckering **shrub** 1–2.5 m tall. **Branchlets** terete, tomentose. **Leaves** 3–8 cm long, 3–12 mm wide, ascending, shortly petiolate, leathery, oblong to obovate, simple, obtuse and mucronate or acute; upper surface convex, penninerved, the venation longitudinal, villous when young, often becoming glabrous and scabrous; margin entire, recurved; lower surface densely silky or tomentose, midvein prominent. **Conflorescence** decurved, pedunculate, terminal or axillary, simple or rarely branched; unit conflorescence 2–3 cm long, 2.5 cm wide, open, globose to shortly cylindrical; peduncle and rachis tomentose; bracts 1–2.5 mm long, triangular, villous, falling well before anthesis. **Flower colour:** perianth pinkish red with a cream, green, yellow or white limb; style end green or yellow. **Flowers** acroscopic: pedicels 2.5–5 mm long, villous; torus 2–3 mm across, oblique to almost lateral; nectary thick, U-shaped, smooth; **perianth** 5–8 mm long, 2–3 mm wide, ovoid, villous outside, bearded inside; tepals faintly ribbed, separating along dorsal suture almost to base, exposing style end before anthesis, otherwise cohering after anthesis except

G. aspera

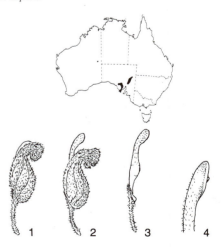

at limb; limb nodding, globular, villous; **pistil** 7.5–10.5 mm long, glabrous; stipe 1.6–3 mm long; ovary glabrous, ellipsoidal; style straight to slightly incurved, glabrous or sparsely silky, stout, barely exceeding perianth at anthesis; style end enlarged, spoon-like; pollen presenter lateral, obovate, ± flat or slightly concave. **Fruit** 13–17 mm long, 5–6 mm wide, erect, oblong–ellipsoidal or ovoid, glabrous, smooth, ribbed; pericarp c. 0.3 mm thick. **Seed** 8.5 mm long, 2.3 mm wide, ellipsoidal, smooth, double-grooved on inner face.

Distribution S.A. and W.A. Widespread in S.A. in the northern and central Flinders Ranges and on Eyre Peninsula. In W.A., known only from the Rawlinson Ra. ***Climate*** In S.A., hot, dry summer, mild, moist to wet winter; annual rainfall 200–550 mm. In the Rawlinson Ra. the rainfall is irregular and may occur at any season (200–220 mm).

Ecology Occurs on stony slopes and hillsides in open heath, thick scrub or Eucalypt woodland in sand, stony loam, clay-loam, or lateritic soils. Flowers winter to summer, the Gawler Ranges form usually flowering later then the Flinders Ranges form. Pollinated by birds.

Variation A variable species in leaf size, shape and colour as well as in flower size and colour.

Flinders Ranges form This form has sand-papery, oblong, olive-green leaves that are not as leathery as most other forms. Flowers are also smaller and less showy. It is widespread in the central and northern Flinders Ranges. It usually reaches 0.6–1 m tall and wide, both in the wild and in cultivation.

Eyre Peninsula form This form has variable, oblong to narrowly linear leaves that are sometimes less than 4 mm wide and generally stiffer than the Flinders Ranges form. A number of populations of this form have grey leaves. In addition, flowers tend to be larger and showier. It is widespread on hills and ranges on Eyre Peninsula.

Gawler Ranges form Plants from the Gawler Ranges on northern Eyre Peninsula can be quite variable but usually have broader leaves. This outstanding form with sleek, grey foliage grows into an exceptional plant of compelling beauty, flowering far more prolifically and for a longer period than other forms. Most populations grow on harsh, stony hillsides and are vigorously suckering. This is an ideal garden plant for dry, sunny sites. In cultivation this form usually reaches 1.5 m tall and wide.

Western Australian form Known from only one recent collection in the Rawlinson Ra., over 1000 km from the next nearest known population. The specimen has smoother leaves and branching conflorescences. Flowers are smaller than S.A. populations and there other morphological differences. Little is known of this form and more observations and collections are needed as it may be a new species.

Major distinguishing features Leaves simple, entire, the undersurface exposed, often scabrous above with evident lateral venation; conflorescence loose, subcylindrical to globose; perianth zygomorphic, hairy on both surfaces; ovary glabrous; pistil about the same length as the perianth; pollen presenter lateral, concave.

Related or confusing species Group 13, especially *G. parallelinervis* which can be distinguished by the smooth, ± parallel, longitudinal venation of its upper leaf surface. It has leaves less than 1.5 mm wide.

Conservation status In many areas plants are at risk from the grazing of feral goats, otherwise not endangered.

Cultivation *G. aspera* has been grown for many years in cool-wet or warm-dry climates where it is relatively resilient and adaptable, enduring both drought and frosts to at least -6°C. It is grown widely in Vic. and S.A. as well as at Burrendong Arboretum, N.S.W., and the Australian National Botanic Gardens, Canberra. It should be ideally suited to the calcareous sands of Perth. Best results are achieved when planted in a sunny or semi-shaded site in well-drained, acidic to alkaline sand or gravelly loam. Summer watering is not necessary, nor is additional fertilising. Young plants are usually compact and shapely but may become open with age; judicious pruning will rejuvenate the plant. It is an attractive plant for pot culture when planted in a medium to large tub. In a summer-humid climate, however, it flowers poorly and soon dies. The

28A. *G. aspera* Silver-leaf plant in natural habitat, Scrubby Peak, Gawler Ranges, S.A. (N.Marriott)

Flinders Range form is readily available at general and specialist nurseries in S.A. and Vic. Other superior forms are uncommon in cultivation.

Propagation *Seed* The Flinders Ranges form sets large quantities of seed which germinate well if pretreated before sowing. Often self sows in cultivation but no hybridisation has been noted. *Cutting* Firm, young growth of the Flinders Ranges form taken at most times of the year normally strikes well. Sucker growth of the Gawler Ranges form strikes fairly well when kept on the dry side, but material from mature plants gives a very poor strike.

28B. *G. aspera* Mt Stuart, Gawler Ranges, S.A. (R.Bates)

28C. *G. aspera* Close-up of flowers (N.Marriott)

Grafting Has been approach grafted successfully onto *G. robusta*. Mummy grafts have also proved successful.

Horticultural features *G. aspera* is an impressive, low-growing species with attractive, green to grey foliage and masses of showy, pendant racemes of red and cream flowers from late winter to summer. These are most popular with honey-eating birds. It is a very long-lived species that is ideally suited to commercial, public and private landscapes as a dense, low screen or feature plant. This is especially so in dry inland regions or in areas with dry, stony soils.

General comments Seed of this species was collected by William Baxter in 1823–24 and at about the same time it was introduced to England where it was grown in glasshouses for many years.

Grevillea aspleniifolia J.Knight (1809)
Plate 29

On the Cultivation of the Plants belonging ot the Natural Order of Proteeae: 120 (England)

Fern Leaf Grevillea

The specific name is taken from the name *Asplenium* (a genus of ferns) and the Latin *folium* (a leaf), in reference to the fern-like leaves. ASS-PLEN-IF-OLE-EE-AH

Type: originally cultivated in England in1806 by Mr Colville from seed collected at Port Jackson, but they were all lost. Replacement type: Port Jackson, N.S.W., in Joseph Banks' herbarium (neotype: BM) (McGillivray 1975).

Synonyms: *G. aspleniifolia* R.Brown (1810); *G. aspleniifolia* var. *shepherdiana* F.Mueller (1894); *G. shepherdii* J.H.Maiden & E.Betche (1916).

A sprawling, irregular **shrub** to 5 m with long, spreading branches. **Branchlets** terete, tomentose. **Leaves** 15–30 cm long, 5–15 mm wide, ascending, shortly petiolate, linear–lanceolate, acuminate, irregularly toothed or lobed, sometimes entire; upper surface sparsely tomentose, soon glabrous, midvein evident; margin shortly recurved to revolute; lower surface felted with curled hairs, midvein prominent, glabrous. **Conflorescence** 3–5 cm long, erect, terminal with subtending branchlets, sessile or shortly pedunculate, unbranched, secund; peduncle tomentose; rachis villous; floral bracts 0.5 mm long, ovate, villous, persistent. **Flower colour:** perianth purplish pink with a greyish white indumentum; style purplish pink with green tip. **Flowers** acroscopic; pedicels 1.5–3 mm long, tomentose; torus 1–1.5 mm across, ± straight; nectary U-shaped, entire; **perianth** 7–9 mm long, 2 mm wide, ± cylindrical to narrowly ovoid or sigmoid, appressed-villous outside, glabrous inside, cohering except along dorsal suture; limb revolute, obovoid, enclosing style end before anthesis; **pistil** 15–25 mm long; stipe 0.5–1.7 mm long, silky; ovary silky; style glabrous, silky at base, at first exserted on dorsal side below limb and looped upwards, reflexed above ovary, erect to slightly incurved just after anthesis, soon swept back; style end hoof-like; pollen presenter slightly oblique, broadly conical. **Fruit** 11–12 mm long, 5–6 mm wide, single-seeded, on incurved stipe, oblong or ellipsoidal with attenuated apex, flattened, ribbed, silky; pericarp 0.3 mm thick. **Seed** 14 mm long, 4 mm wide, ellipsoidal, both sides ± the same, convex, rugose, separated by a waxy border.

Distribution N.S.W., in the Blue Mountains, further northwest to the lower Cox R. and extending south to Bungonia Caves area. *Climate* Summer warm to hot, wet or dry; winter cool to cold, wet or dry. Rainfall 800–1200 mm.

Ecology Grows on slopes and ridges among exposed sandstone rocks in skeletal sandy soils, often high above rivers. Flowers principally winter to spring, but some flowers all year. Pollinated by birds.

Variation A species with little variation. Leaves range from entire to deeply and irregularly lobed.

Major distinguishing features Branchlets rounded, grey-tomentose; leaves long, lanceolate, irregularly toothed, often entire or almost so, undersurface densely covered with short, curled hairs; conflorescence secund; pedicels 1–2 mm long; perianth zygomorphic, hairy outside, glabrous within; ovary densely hairy, shortly stipitate; style purplish pink; fruits single-seeded, with reddish hairs; seed ellipsoidal.

Related or confusing species Group 35, especially *G. longifolia* which has reddish, angular branchlets, leaves with a silky, appressed indumentum of straight hairs on the undersurface, and with the margins usually more uniformly and deeply toothed; the conflorescence is also broader and more robust and the flowers have a pinkish red style. In seed characters, *G. caleyi* appears the most closely related species.

Conservation status Not presently endangered.

Cultivation *G. aspleniifolia* is a resilient plant in cultivation and has been grown successfully in all States of Australia except in tropical areas; it has succeeded in the U.S.A. and to a limited extent in England and Europe. It tolerates frost to at least -6°C and, in most areas including its natural habitat, survives extended, dry periods. It prefers a well-drained, acidic sand or loam in full sun for best results; plants grown in poor, sandy soil often become sparse and unattractive after a few years, while those grown in heavier loam or clay remain healthy and attractive for many years. It has somewhat fragile branchlets that are easily damaged by strong winds and, accordingly, should be grown in protected situations. It responds reasonably well to tip pruning and this may be needed to maintain shape. Applications of nitrogenous fertiliser produce vigorous new growth. Once established, it does not require watering unless in severe drought or planted in shallow soils. It is readily available in specialist nurseries on the east coast of Australia.

Propagation Seed Sets copious seeds that germinate readily following the usual peeling, nicking or soaking treatments. With fresh, ripe seed even this may not be necessary. *Cutting* Grows from cuttings of firm, new wood taken from late spring to autumn. Occasionally, some difficulty is experienced in obtaining a good strike. *Grafting* Untested.

Horticultural features *G. aspleniifolia* can be a very untidy, unattractive shrub if allowed to develop without attention although some growers like its long, lithe-branched habit. If pruned properly when young, it will develop into a shapely garden shrub, producing dense foliage and attractive pinkish new growth. The purplish pink flowers are prominent and prolific, attracting many honey-eating birds over

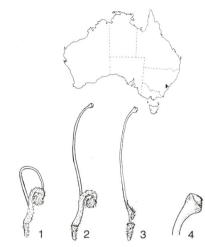

G. aspleniifolia

29B. *G. aspleniifolia* Plant in natural habitat, Lake Burragorang, N.S.W. (P.Olde)

29A. *G. aspleniifolia* Flowers and foliage, in cultivation, Brisbane, Qld (P.Olde)

a prolonged flowering season. It is relatively long-lived and could be used as an open screen or irregular specimen plant. Many growers report that it self-sows in the garden. The long, willowy leaves and the flowers are useful in floral arrangements.

Hybrids Hybridises with *G. caleyi* and *G. longifolia* in cultivation. G. 'Telopea Valley Frond' is a garden hybrid of *G. aspleniifolia* and *G. caleyi* and makes a beautiful, spreading, dense screen with attractive, ferny, soft leaves and a long flowering season.

History George Caley apparently supplied seed of this species in 1806 to Colvill's nursery in England where seedlings were raised. The type specimen is probably a flowering specimen in Joseph Banks' herbarium annotated by Dryander and available to R.A.Salisbury, the author of the botanical descriptions in Knight's book. The specimen was collected at Port Jackson.

General comments At a meeting of the Victorian Naturalists' Club in late 1893, Ferdinand Mueller presented a specimen of *G. aspleniifolia* which he called var. *shepherdiana*. It was collected along the banks of the Cole River, near Jervis Bay, N.S.W., by a son of Mr P.Shepherd. He described it as having leaves over 1 foot long but scarcely wider than ¼ inch and with its surface 'beset by minute hairlets'. This description fits the normal variation within *G. aspleniifolia* and the variety has accordingly been reduced to synonymy.

Grevillea asteriscosa L.Diels (1904)

Plate 30

Botanische Jahrbücher für Systematik 35: 151 (Germany)

Star-leaf Grevillea

The specific name is derived from the Latin *asteriscus* (a small star), in reference to the star-shaped leaves. ASS-TER-ISK-OH-SA

Type: 100 miles (c. 160 km) north of the Stirling Ra., W.A., 1879, T.Muir (holotype: B).

A dense **shrub** 1–2 m tall, 2–3 m wide. **Branchlets** terete, glandular-pubescent. **Leaves** 5–10 mm long, 8–18 mm wide, spreading, sessile, star-shaped, stem-clasping, palmately divided into 3–9 rigid, triangular, pungent lobes; upper surface pilose, sometimes mixed with glandular hairs, granulate, midvein obscure; margins recurved to revolute; lower surface glabrous or pilose, the veins raised prominently, glandular-pubescent to pilose. **Conflorescence** erect, unbranched, terminal, often on short branchlets, sessile, umbel-like, 4–6-flowered; rachis 1–3 mm long, pilose; floral bracts 2–4 mm long, triangular, pubescent outside, persistent at anthesis. **Flower colour:** red

G. asteriscosa

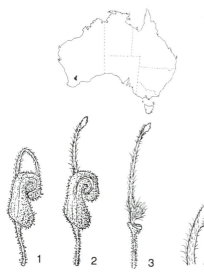

30A. *G. asteriscosa* Close-up (N.Marriott)

30B. *G. asteriscosa* Flowers and foliage. Note red new growth (N.Marriott)

30C. *G. asteriscosa* Habit, in cultivation, Stawell, Vic. (N.Marriott)

with a bluish red style. **Flowers** abaxially oriented: pedicels 5–10 mm long, glandular-pubescent; torus c. 3 mm across, very oblique; nectary broad, V-shaped, entire; **perianth** 8–10 mm long, 5 mm wide, oblong–ovoid with slight apical taper, ribbed, red-glandular-pubescent outside, bearded about ovary inside, otherwise glabrous; tepals separating along dorsal suture and between dorsal and ventral sutures at curve before anthesis, afterwards free to base but remaining cohering; limb tightly revolute, globular to spheroidal, the segments faintly ribbed, completely enclosing style end before anthesis; **pistil** 15.5–19.5 mm long; ovary appressed-villous; stipe 0.7–1 mm long, perpendicular to torus; style thick, hirsute, almost glabrous on ventral side, first exserted at curve and looped upwards, slightly curved to straight, merging evenly into a slightly dilated, clavate, glabrous style end; pollen presenter lateral, obovate, flat. **Fruit** 9–12 mm long, 6–7 mm wide, erect to slightly oblique, oblong–ellipsoidal, sparsely villous; pericarp c. 1 mm thick. **Seed** 7 mm long, 4 mm wide, ellipsoidal; outer face smooth and shiny with basal rim; inner face smooth and raised in centre surrounded by a broad waxy or papery border.

Distribution W.A., in the central wheatbelt; restricted to an area from Muntadgin south to near Pingaring and east of Bullaring. *Climate* Hot, dry summer; cool to mild, wet winter. Rainfall 350–450 mm.

Ecology Grows in heath or tall shrubland or near granite outcrops in heavy, gravelly loam or granitic soil, often on rises. Flowers winter to spring. Bird-pollinated.

Variation A stable species with little variation.

Major distinguishing features Leaves star-shaped; branchlets and flowers glandular-pubescent; conflorescence few-flowered; perianth oblong–ovoid, glandular-pubescent outside, bearded within; ovary densely hairy; stipe perpendicular to the very oblique torus; fruit oblong–ellipsoidal.

Related or confusing species Group 26. No confusing species.

Conservation status Suggested 2V. Restricted to a small area, often confined to road verges in mostly cleared country, and at risk of extinction in the wild.

Cultivation This hardy, reliable plant withstands hot, dry summers and winter frosts to at least -6°C. It does well in Vic. and has been grown since 1970 at Burrendong Arboretum, N.S.W., where it spreads up to 6 m wide. Growers at Rylstone, N.S.W., and Glenmorgan, Qld, as well as Sydney have succeeded with this species and report that, while it adapts satisfactorily to dappled shade, in full sun it is transformed into a dense, floriferous shrub. A variety of soils from sandy loam to well-drained clay are tolerated but it succumbs quickly if drainage is poor. It will grow well in hot, dry inland conditions as well as humid, coastal and cool southern climates but does not appear to adapt well to pot culture. It will grow vigorously without fertiliser but may require pruning to control its wide-spreading habit. Dry conditions will curtail flowering but it blooms almost continuously when subsoil is moist. Ready availability of this species has been restricted by its difficulty in propagation, and is only occasionally available from specialist nurseries.

Propagation Seed Fresh seed given the standard peeling treatment germinates readily. Hybridises in cultivation. *Cutting* Propagators using mist and bottom heat or extremely humid conditions have considerable difficulty in propagating this species by cutting. Others, however, with drier, less humid conditions, have had reasonable success, using firm, young wood during early autumn or spring. Liquid hormones 2000 ppm indol-butyric acidic (IBA), and 2000 ppm naphthalene acetic acidic (NAA) have been shown to significantly improve rooting percentages if applied to the base of the plant before setting in medium. Removal of leaves should be undertaken with care both for reasons of self-preservation and to prevent excessive stripping of the bark. Once rooted, plants grow readily to maturity. *Grafting* Has been grafted successfully onto *G. robusta* using the mummy and the top wedge techniques.

Horticultural features Unique leaves, glowing red flowers, attractive, velvet-red new growth and a dense, wide-spreading, growth habit are the main features of *G. asteriscosa*. Very floriferous and long-flowering under good garden conditions, it has been limited by its prickly foliage from gaining wide public acceptance especially for small gardens, but its other attributes and its desirability as a bird-attracting plant far outweigh this disadvantage. Plants over 25 years old are in cultivation and still performing strongly, and it would be suitable as a screen plant or feature plant.

Hybrids G. 'Merinda Gordon', a garden hybrid between *G. asteriscosa* and *G. insignis*, from Myall Park, Glenmorgan, Qld, has recently been registered with the Australian Cultivar Registration Authority.

Grevillea aurea P.M.Olde & N.R.Marriott (1993)
Plate 31

Telopea 5: 407 (Australia)

Deaf Adder Gorge Grevillea, Golden Grevillea

The specific epithet is derived from the Latin *aureus* (golden), in reference to the flower colour. AWE-REE-A

Type: Deaf Adder Gorge, N.T., 17 July 1978, D.J.McGillivray 3934 & C.Dunlop (holotype: NSW).

Synonym: *Grevillea angulata* Golden-flowered form, McGillivray (1993).

A tall, open **shrub** with erect branches 2–6 m tall. **Branchlets** angular, terete with age, silky or sparsely so. **Leaves** 7–16 cm long, 1.5–4.5 cm wide, spreading, petiolate, concolorous, oblong-dentate with 4–12 pungent lobes per side; upper and lower surfaces similar, sparsely silky when young but soon glabrous, minutely pitted; margins flat, sometimes sinuate; venation conspicuous, midvein prominent beneath. **Conflorescence** terminal, axillary or cauline, simple or few-branched, pedunculate, decurved; unit conflorescences 4–16 cm long, 2–3 cm wide, conico-cylindrical, lax; peduncle silky; rachis glabrous; bracts 0.6–1.5 mm long, ovate-acuminate, ciliate, falling before anthesis. **Flower colour**: young buds coppery brown; perianth yellow, orange to red; style yellow to orange. **Flowers** abaxially acroscopic: pedicels 4.5–8 mm long, glabrous, patent to retrorse; torus 1.5–2mm across, oblique; nectary U-shaped, smooth or toothed; **perianth** 5–7 mm long, 2–3 mm wide, oblong with slight basal dilation, strongly curved from base, glabrous outside, shortly bearded inside; tepals cohering after anthesis except along dorsal suture; limb revolute, spheroidal, flared open and exposing style end before anthesis; **pistil** 17–23 mm long; stipe 1.5–2.8 mm long; ovary glabrous; style at first exserted through dorsal suture below curve and looped upwards, usually sparsely granulate or shortly hairy from c. halfway, sometimes glabrous; style end slightly dilated, hairy or granulate on dorsal side; pollen presenter oblique, obovate, convex. **Fruit** 10–17 mm long, 7–8 mm wide, erect, ellipsoidal, smooth; pericarp 1–1.3 mm thick. **Seed** 7–11 mm long, 3–4 mm wide, oblong-ellipsoidal, mem-branously winged all round.

Distribution N.T., known from three localities in Kakadu NP. *Climate* Summer hot and wet; winter mild and dry. Rainfall 800–1000 mm.

Ecology Grows in rocky talus, on sandstone escarpment with spinifex or on sandy flats beside the upper reaches of creeks. Flowers autumn to winter, sometimes all year in cultivation. Regenerates from seed after fire. Probably pollinated by nectarivorous birds.

Variation There is some variation in conflorescence length, flower colour and leaf size but otherwise a stable taxon.

Major distinguishing features A shrub or small tree; leaves glabrous, concolorous, oblong, toothed; conflorescence terminal, axillary or cauline, 7–16 cm long, conico-cylindrical; perianth strongly curved, 2–3 mm wide, maintaining its shape after anthesis, glabrous outside, bearded inside; ovary

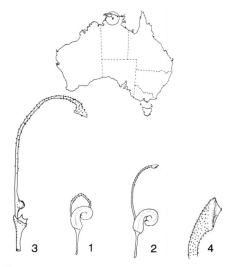

G. aurea

31A. *G. aurea* Yellow-flowered form, Kakadu NP, N.T. (D.Forster)

31B. *G. aurea* Conflorescence and foliage (M.Hodge)

31C. *G. aurea* Habit, near Barramundi Gorge, Kakadu NP, N.T. (D.Forster)

glabrous; style granular or shortly hairy in upper half; style end hairy or granular; seed winged all round.

Related or confusing species Group 11, especially *G. glabrescens* which, like *G. angulata*, differs in its shorter conflorescence, its perianth 1.5–2 mm wide, the tepals of which separate below the limb and reflex to reveal their inner surface before anthesis.

Conservation status Not presently endangered.

Cultivation *G. aurea* has been widely cultivated in Darwin for a number of years and has been successfully grown in Brisbane. Grafted plants have flourished even in Sydney, performing well in well-drained, sandy, acidic soils in full sun. It is, however, subject to wind-throw (C.R.Dunlop pers. comm.) and needs to be planted in a protected position or staked. Summer watering improves the appearance and judicious pruning will make it more compact. It does not tolerate frost and can be set back by long cold periods.

Propagation *Seed* Germinates readily, especially if pre-treated by peeling, nicking or soaking. *Cutting* Strikes reasonably well from cuttings taken off firm, new growth in spring or autumn. *Grafting* Has been grafted successfully onto *G. robusta* and *G.* 'Poorinda Royal Mantle' using top wedge, whip-and-tongue and mummy grafts.

Horticultural features This stunning, free-flowering plant is a small, shrub-like, open tree in the wild. It features oblong, blue-green leaves with symmetrically arranged teeth, while red-tinged new growth provides a noteworthy contrast. However, it is the flowers that cluster over the plant for months on end that attract most comment. The coppery brown buds pass through reddish brown to bright golden yellow or deep reddish orange as the flowers mature. The long, curved inflorescences hang like clusters of yellow grapes from the leaf axils or the ends of the branches, sometimes even appearing directly from the stem. So prolific and long-flowering is *G. aurea* that it is highly sought after in nurseries and gardens everywhere that it becomes known. In full flower, it is a spectacular plant destined to be much more widely cultivated.

Grevillea australis R.Brown (1810)

Plate 32

Transactions of the Linnean Society of London 10: 171 (England)

Alpine Grevillea, Southern Grevillea

The specific epithet is derived from the Latin *australis* (southern). The species was described from collections made in Tasmania, the most southerly distribution of the genus. OSS-TRAR-LISS

Type: along the banks of the Derwent R. [near Launceston, Tas.], 1804, R.Brown (lectotype: BM) (McGillivray 1993).

Synonyms: *G. tenuifolia* R.Brown (1810); *G. australis* var. *linearifolia* J.D.Hooker (1847); var. *erecta* J.D.Hooker (1847); var. *montana* J.D.Hooker (1847); var. *subulata* J.D.Hooker (1847); *G. stuartii* C.F.Meisner (1854).

An open to dense **shrub**, prostrate to 2 m tall. **Branchlets** angular to terete, silky to tomentose. **Leaves** 0.5–5 cm long, 0.5–5.5 mm wide, ascending to spreading, simple, entire, leathery to pliable, either subterete with tightly rolled margins obscuring undersurface, or narrowly linear, oblong to obovate or elliptic with undersurface exposed and margins evenly and smoothly recurved, sometimes unevenly refracted; leaf apex acute or obtuse, pungent; upper surface convex or flat, silky but soon glabrous, usually shiny, midvein visible, sometimes longitudinally veined; lower surface obscured either by the margin or by a dense, silky indumentum; midvein on lower surface (when visible) densely silky, raised for about half its length, then disappearing into lamina in apical half. **Conflorescence** erect, sessile or very shortly pedunculate, terminal or axillary in upper branches, simple, rarely 2–3-branched; unit conflorescence umbel-like, usually < 1 cm long, 1 cm wide; peduncle and rachis tomentose; floral bracts 0.5–1.5 mm long, ovate to oblong, tomentose, usually persistent. **Flower colour:** perianth white or creamish white; style white sometimes turning pink to red. **Flowers** adaxially acroscopic: pedicels 1.2–4 mm long, silky to tomentose, slender; torus 0.5–1 mm across, straight, square; nectary inconspicuous, cushion-like, smooth; **perianth** 3–6 mm long, 1 mm wide, oblong, silky outside, glabrous inside with a small, glabrous cushion inside at base, the cushion sometimes bearing short hairs; tepals ribbed,

G. australis

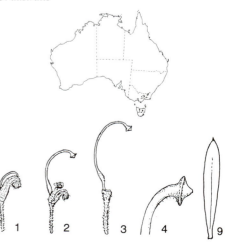

32A. *G. australis* var. *tenuifolia* Brindabella Range, A.C.T (N.Marriott)

32B. *G. australis* Ground-hugging plant, Ben Lomond NP, Tas. (N.Marriott)

separating along dorsal suture before anthesis, afterwards separating to base; limb revolute, ovoid, enclosing style end before anthesis; **pistil** 6–7.5 mm long, reflexed from line of pedicel; stipe 0.4–1.1 mm long; ovary glabrous; style exserted below curve from dorsal suture, reflexed from line of pedicel, smoothly to angularly incurved after anthesis, geniculate c. 1mm from the flat, plate-like style end, short hairs or papillae extending 0.5mm from the end; pollen presenter very oblique to straight, oblong–elliptic, broadly conical. **Fruit** 8.5–9.5 mm long, 4–5 mm wide, narrowly ovoid, glabrous, granulose, longitudinally lined; pericarp c. 0.2 mm thick. **Seed** 6.5–7.5 mm long, 2–2.7 mm wide, oblong–ellipsoidal; outer face smooth; inner face flat, rimmed by a waxy border.

Distribution Eastern Australia except Qld. Widespread in Tas. but mainly in the north-east and central area from sea level to c. 1200 m, the only *Grevillea* occuring naturally in this State. Vic., in the alpine or high country areas in the eastern portion of the State. N.S.W. and A.C.T., in the alpine or high country areas of the Snowy Mountains. **Climate** Summer cool and mild; winter cold and wet with frequent snowfalls. Rainfall 600–2400 mm.

Ecology Grows in heath, herbfield and subalpine woodland, usually around rocks, beside streams and in swamps, in peaty sand or granitic loam; often covered by snow for long periods during winter. Flowers spring to summer. Insect-pollinated. After fire regenerates from seed.

Variation There is little variation in the flowers of this species, but foliage and habit vary markedly. Prostrate forms tend to have broader, more leathery leaves. Leaves vary in their shape, thickness, the shape of their apex, the venation, the degree of revolution of the margin, and their orientation on the plant. J.D.Hooker described seven varieties based on leaf characters and a number are still applicable in horticulture although their botanical status is doubtful owing to the number of intermediates. Hooker's descriptions relate only to Tasmanian collections but there is some referability to mainland populations. D.J.McGillivray (1993) has noted differences in leaf shape and venation on plants from the mainland. Some specimens, notably broad-leaved ones, have conspicuous longitudinal venation that is not evident on Tasmanian specimens. We acknowledge here four varieties.

var. *australis* [syn. var. *linearifolia*] An erect, shrub 0.5–2 m tall with spreading, linear to lanceolate leaves 0.8–4 cm long, 1–3 mm wide with shortly recurved margins and the undersurface clearly visible. The leaves are either pliable or stiff and sometimes pungent. Leaves that fit this description are also found on the mainland but often have faintly visible longitudinal venation. A form which occurs around Corra Linn, Tas., has been cultivated for many years and has considerable horticultural value in cold-winter climates. Hooker's var. *erecta* also appears to be part of this variety, mainly having leaves at the shorter end of the spectrum.

var. *planifolia* J.D.Hooker (1847) An erect shrub with thick, flat, leathery, elliptic to oblanceolate leaves up to 4mm wide, 1.5–3 cm long with a narrowed, sharp tip and shortly recurved margins. The leaves are further distinguished by the raised midvein on the upper surface. This form is found in the Tasmanian lowlands between Launceston and Devonport.

var. *brevifolia* J.D.Hooker (1847) (syn. var. *montana* J.D. Hooker) A prostrate to low, much-branched shrub spreading to c. 1 m with linear–oblong to narrowly lanceolate leaves 0.5–1 cm long, 2–3 mm wide with a narrowed, pungent tip. The type specimen was collected by R.C.Gunn on Mt Barrow at Great Lake, Tas., at an elevation of 3500 feet. Some plants on the mainland also fit this description, differing only in the longitudinal venation sometimes visible on the upper surface. One form has been cultivated for many years. Prostrate forms occur at higher elevations.

var. *tenuifolia* (R.Brown) C.F.Meisner (1856) (syn. var. *subulata* J.D.Hooker) This variety is found both in Tas. and on the mainland and has spreading to erect, subterete leaves, usually < 1 mm wide, with margins so tightly rolled as to obscure the undersurface completely, leaving a longitudinal groove visible at the junction of their edges.

Broad-leaf mainland form In leaf shape and size, this form is very similar to var. *planifolia* but differs in the 3 conspicuous, longitudinal nerves. It is generally prostrate to shrubby and occurs at elevated sites in Vic. and N.S.W.

Major distinguishing features Branchlets terete; leaves simple, entire, the midvein on the undersurface obscured by the margins or, if exposed, becoming obscured in upper half of lamina; midvein of leaf undersurface with same indumentum as adjacent lamina; margin smoothly revolute; perianth zygomorphic, hairy outside, usually glabrous inside; ovary glabrous; pedicel < 4 mm long; pistil < 8 mm long; style with obscure hairs in apical 0.5 mm; fruit narrowly ellipsoidal.

Related or confusing species Group 21, especially *G. linearifolia* which has leaves with the midvein on the undersurface ± glabrous and always visible for its full length; its leaf margins are sharply and angularly refracted about the marginal vein.

Conservation status Not presently endangered.

Cultivation *G. australis* will tolerate a wide range of conditions and is hardy to frost, ice and snow while, at the same time, capable of withstanding hot, dry summers provided moisture is adequate. It has been grown successfully in Qld and Sydney but does not flower well at these latitudes. It flowers well in Tas., A.C.T. and Vic. Two forms are in cultivation in the U.S.A. In humid, coastal situations that do not receive very cold winters it may be short-lived. It appears to need cold conditions for good bud initiation and flower development and grows best in an open or partially shaded situation in well-drained, acidic, sandy or heavy soils. Its shape is improved with pruning. Summer watering enhances its appearance during extended, dry spells especially if grown in light soils. Prostrate forms in tubs will cascade attractively. In the wild, these forms tend to closely follow the contours of rocks which radiate warmth and assist survival. It is widely available in native plant nurseries where several select forms have been introduced into cultivation.

Propagation *Seed* Sets reasonable quantities of seed that should germinate well when given the standard peeling treatment. *Cutting* Firm, young growth taken at most times of the year strikes readily. *Grafting* Untested.

Horticultural features *G. australis* has small, white, terminal flowers and dark green leaves, often with a silvery underside. Prostrate forms make excellent ground covers. Erect forms make interesting garden shrubs, spectacular in full flower and sufficiently dense to provide low screening. In suitable conditions flowering extends over several months and is accompanied by a pervasive, honey-like scent through the garden. The aroma in confined conditions can produce nausea and it is advisable not to leave vases of this species inside overnight.

General comments The taxonomy of *G. australis* remains in need of further detailed work.

Grevillea baileyana D.J.McGillivray (1986)   Plate 33

New Names in Grevillea *(Proteaceae)* 2 (Australia)

Scrub Beefwood, White Oak, Brown Silky Oak

Named in memory of F.M.Bailey (1827–1915), and his descendants J.F.Bailey (1866–1938) and C.T.White (1890–1950). BAY-LEE-ARE-NA

Type: Johnstone R., Qld, collected prior to May 1886 by Qld Woods and Forests (neotype: BRI).

Replaced synonym: *Kermadecia pinnatifida* F.M.Bailey (1886) in *Queensland Woods*: 69, n. 332a.

Note: When transferred to *Grevillea* (May 1886), this taxon took the specific epithet from its basionym *Kermadecia pinnatifida*, but an earlier use of the name *Grevillea pinnatifida* by Jacques (1843) for a different species rendered Bailey's name illegitimate under the rules of botanical nomenclature and the latter had to be replaced.

A medium to large **tree** 6–30 m tall with hard, scaly, grey bark. **Branchlets** slightly angular, grooved, ferruginous-silky. **Leaves** 6–30 cm long, 1–6(–10) cm wide, ascending, petiolate, leathery, glabrous above, silky below, with mixed grey and rusty hairs, the latter falling with maturity; adult leaves simple, entire, ± elliptic or obovate-oblong, acute to obtuse; juvenile leaves pinnatifid, cleft into 5–9 broad, oblong to lanceolate lobes; margins flat; venation brochidodromous, midvein broad and conspicuous above, prominent below with faint penninervation. **Conflorescence** erect, usually terminal, sometimes axillary, usually 5–10-branched; unit conflorescence 6–14 cm long, 1.5–2.5 cm wide, dense, cylindrical; peduncle and rachis silky; bracts 0.5 mm long, ovate, silky, falling before anthesis. **Flower colour:** perianth greenish in late bud, creamy-white in flower; style white. **Flowers** adaxially acroscopic: pedicels 1.6–2.4 mm long, silky; torus c. 1 mm across, straight to slightly oblique; nectary prominent, U- or V-shaped, entire; **perianth** 3–5 mm long, 0.8 mm wide, oblong, silky outside, bearded inside below ovary, otherwise glabrous or a few scattered hairs; tepals first separating to base along dorsal suture, then all separating below limb and reflexing to expose inner surface before anthesis, afterwards free to base; limb nodding to revolute, spherical, enclosing style end before anthesis; **pistil** 9–13.5 mm long, glabrous; stipe 1.1–2 mm long; ovary globose; style exserted through lower part of dorsal suture, strongly incurved to C-shaped after anthesis; style end expanded, flanged with wavy margin, its base with a prominent stay attached to ventral side of style; pollen presenter oblique, oblong-elliptic, convex to subconical. **Fruit** 13–19 mm long, 7.5–11 mm wide, persistent, slightly to very oblique to pedicels, compressed, oblong or obovoid with decurved apiculum, smooth to faintly granulose; pericarp c. 0.5 mm thick. **Seed** 15 mm long, 7 mm wide, oblong-elliptic, flat, dull and slightly mottled both sides, prominently ridged near margin on outer face, channelled on reverse, winged all round.

Distribution Occurs in Qld, along the coast and up to c. 50 km inland, from Cape York Peninsula south to Ingham. Also in Papua-New Guinea, along the south coast. Climate Tropical with very wet, hot, humid summer and mild, drier winter. Rainfall 1200–3000 mm.

Ecology Grows in and on margin of primary and secondary rainforest, sometimes on dry, flat terrain, sometimes on ridges, slopes, creek banks or gullies. Often in granite-derived soil. Common as regrowth in disturbed rainforest. Flowers late spring to summer. Insect-pollinated.

Variation A widespread species with minor variation.

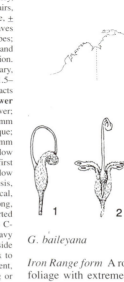

G. baileyana

Iron Range form A robust form which has juvenile foliage with extremely broad leaf lobes, although its leaves revert to simple-elliptic at maturity. Local reports (P.Radke pers. comm.) suggest that it is more floriferous and vigorous than forms further south.

Papua-New Guinea form Similar in most respects to Australian plants except that adult leaves often remain divided.

Major distinguishing features Tree; juvenile leaves broadly pinnatifid, with rusty hairs on lower surface; leaves bifacial with looping venation; conflorescence paniculate, much-branched with cylindrical units; torus straight; perianth zygomorphic, oblong, hairy outside, hairy at base inside; pistil glabrous; pollen presenter very oblique, conico-convex; fruit compressed; pericarp c. 0.5 mm thick, with recurved apiculum; seed winged all round.

Related or confusing species Group 3, especially

33A. *G. baileyana* Flowering branch (G.Norris)

33B. *G. baileyana* Habit, in cultivation, Brisbane, Qld (G.Norris)

G. hilliana which differs in the grey hairs on its leaf undersurface, its mostly simple conflorescences, its flowers directed at right angles to the rachis, with a longer pistil (13.5–16 mm) and thicker pericarp (1.6–2.2 mm thick). *G. hilliana* generally has a more southerly distribution but overlaps with *G. baileyana* at the northern end of its range.

Conservation status Not presently endangered.

Cultivation *G. baileyana* is long-lived and establishes readily in a wide range of coastal and near-coastal climates as far south as Melbourne where it will even tolerate mild frosts when planted in a protected situation. To date it has been cultivated mainly by enthusiasts in Qld who report that, apart from summer watering in dry conditions, it requires little maintenance, although wind and exposure can damage the branches. In situations where it is protected by other plants, it grows rapidly in rich, well-drained but moisture-retentive, acidic soils which should contain a high level of composted material for best results. It is invigorated by applications of low-phosphorus fertiliser, sprouting lush new growth in response. The more interesting juvenile foliage can be retained for many years by pruning the branches lightly and it can be held for quite a long time in pots, making it suitable as an indoor plant in bright, warm and preferably humid conditions. It is fairly readily obtainable from specialist native plant nurseries, especially in Qld where it may still be sold under its superseded name, *G. pinnatifida*.

Propagation *Seed* Usually grown by this method. Germinates readily without pretreatment but nicking the long edge of the seed coat or soaking in warm water for 24–36 hours may aid germination. *Cutting* Firm young growth strikes readily if taken in summer and housed in hot, moist conditions. Excess foliage should be removed rather than trimmed as this can cause leaf fungal problems. *Grafting* Untested to date. Recently has been used as a rootstock with some success.

Horticultural features *G. baileyana*, along with a number of other Australian rainforest plants, has great potential for horticulture. The luxuriant, attractively lobed, juvenile leaves which persist on the plant for a number of years and the coppery leaf undersurface are its most admired features. While it grows to a medium to large tree in the wild, it rarely exceeds 8–10 m in cultivation where it is best suited to a rainforest plantation. It flowers prolifically, bearing bunched racemes of white curled flowers which swathe the tree-top in a fragrant, honey-scented cloud of white. In more temperate regions such as Sydney, flowering is inhibited and may not occur for many years. Nonetheless, it is a very attractive, long-lived foliage plant. Leaf miners and chewing insects occasionally damage the leaves and can at times reduce their attractiveness, but appropriate insect sprays and the vigour of the species generally manage to quickly conceal any damage.

Uses *G. baileyana* has a fine, white or pinkish oak-grained timber useful in furniture manufacturing and wood turning.

Grevillea banksii R.Brown (1810)

Plate 34

Transactions of the Linnean Society of London, Botany 10: 176 (England)

Red Silky Oak, Dwarf Silky Oak, Banks' Grevillea, Byfield Waratah, Kahili Flower (Hawaii)

The specific name honours Joseph Banks (1743–1820), the 'Father of Australian Botany'. BANKS-EE-EYE

Type: Facing Is., Keppel Bay, [near Gladstone, Qld], 1802, R.Brown 3344 (lectotype: BM) (McGillivray 1993).

Synonyms: *G. forsteri* T.Moore (1874); *G. robusta* var. *forsteri* L.H.Bailey (1927); *Grevillea banksii* forma *albiflora* (Degener) Degener & Degener (1959); *Stylurus banksii* forma *albiflora* Degener (1932).

A dense, leafy **shrub** to 3 m or open, slender **tree** to 10 m, sometimes a prostrate or decumbent shrub. **Branchlets** terete, silky. **Leaves** 14–30 cm long, ascending, petiolate, usually deeply pinnatifid to pinnatipartite with 4–12 pliable, lanceolate or linear acute lobes 0.5–1.5 cm wide, rarely simple; margins recurved; upper surface concave or flat, silky to glabrous; lower surface densely silky to villous, midvein prominent. **Conflorescence** erect, pedunculate, simple or few-branched, terminal or in upper leaf axils; unit conflorescences 5–12 cm long, 5–10 cm wide, cylindrical, dense, opening irregularly; peduncle tomentose; rachis glandular-villous; bracts 1–2.5 mm long, ovate, glandular-villous outside, falling before anthesis. **Flower colour:** perianth grey when young, then creamy-white or red, sometimes pink or apricot with a yellow limb; style red to creamy-white with yellow tip. **Flowers** basiscopic; pedicels 3–10 mm long, glandular-villous, patent to retrorse; torus 2–3 mm across, straight to slightly oblique; nectary U-shaped, shallowly lobed or toothed; **perianth** 10–15 mm long, 3–5 mm wide, ± oblong, finely ridged, detaching quickly, glandular-pubescent outside, glabrous inside, cohering except along dorsal suture before anthesis; limb revolute, densely glandular-pubescent, globular, not immediately relaxed at anthesis, flared open along dorsal suture and exposing style end before anthesis; **pistil** 32–50 mm long, sometimes digynous; ovary sessile,

34B. *G. banksii* Typical tree form (M.Hodge)

34A. *G. banksii* Common tree form, in cultivation, Brisbane, Qld (P.Olde)

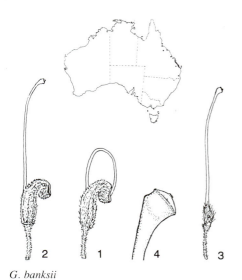

G. banksii

villous; style glabrous, sparsely hairy at base, exserted first along dorsal suture at curve and looped upwards, gently incurved after anthesis; style end evenly dilated, club-shaped; pollen presenter oblique, oblong, convex. **Fruit** 15–25 mm long, 10 mm wide, sessile or oblique on short pedicel, compressed-ovoid, tomentose with reddish patches; pericarp c. 0.5 mm thick. **Seed** 8–11 mm long, oblong-elliptic, winged all round.

Distribution Qld, on the coast and surrounding islands from about Ipswich to Yeppoon, extending W to the hills W of Eidsvold, with a disjunct population between Ingham and Townsville and a few southern occurrences near Esk and Coominya. Allan Cunningham collected the species on the upper Brisbane R. on open barren hills. *Climate* Subtropical to tropical; humid, wet summer; warm, sometimes wet winter. Rainfall 1000–1600 mm.

Ecology Grows on rocky granitic ridges in gravelly clay, in shallow sand on open windswept headlands and in open forested areas. Flowers late winter to spring. Bird-pollinated.

Variation There is considerable variation in habit, indumentum, leaf size and lobing, and flower colour and size. Individual flowers sometimes have 2 pistils but this feature may vary from season to season. Sometimes the style will lift the perianth from the torus, and, with the style end firmly clasped by the perianth limb, dangle the perianth like slippers on a peg. This feature is likewise inconsistent and may vary from season to season and from plant to plant, being usually more common when the plant is under some sort of stress (P.Althofer, pers. comm.).

Tree form This form is widespread in south-eastern Qld between Maryborough and Bundaberg and has broad leaf lobes and either cream or red flowers. It has an open habit, growing to 6-10 m, and occurs mainly in forests. Both colour forms grow together, although pure stands of one or other form are known, while intermediate colour forms such as pink and apricot often occur in mixed populations. It has a distinct flowering season (spring). After opening of the forest canopy or other disturbance, it will often appear in great numbers but tends to die out as the forest matures.

var. *forsteri* This widely cultivated variety has no known wild origin, although it was almost certainly introduced to cultivation by Charles Moore in 1853 after a visit to Qld where he met Bidwell from Wide Bay. The name first appeared as *G. fosteri* in 1855–56 as a manuscript name, in the Report of the Director of the Botanic Gardens, Sydney, in a list of plants sent to D.Moore, Royal Botanic Gardens, Dublin, and to Veitch's Nursery, Kings Road, London, but was not formally published until 1874. The name has gained popular currency because of the horticultural superiority of this plant. *G. fosteri* and *G. banksii* var. *forsteri* are presumably the same plant. This long-lived and adaptable variety is a silvery-leaved shrub to 3 m that flowers continuously with a spring peak. Both red and white flower forms are known.

Townsville form Conspecific with the tree form, this form occurs in a disjunct population between Ingham and Townsville. Foliage is similar to the typical tree form but both the inflorescence and the flowers are smaller, the racemes rarely exceeding c. 6 cm and the pistil 32–34.5 mm long.

34C. *G. banksii* Shiny-leaf form ex Byfield, Qld (P.Olde)

34D. *G. banksii* Colour variant, near Byfield, Qld (A.Don)

34E. *G. banksii* Colour variant, Agnes Waters Headland, Qld (P.&H.Shaw)

34F. *G. banksii* Prostrate coastal form ex Round Hill Head, Qld (N.Marriott)

Prostrate coastal form A vigorous, semi-prostrate to decumbent form, spreading in cultivation about 4 m across. It originates from windswept coastal headlands such as Round Hill Head, Stockyard Point and other exposed areas. The main stem tends to grow erect for about 15 cm after which the branches turn down and run flat along the ground. The upper surface of the leaves has persistent appressed hairs, giving a silvery-grey colour. Flowers are creamy-white or red, sometimes a mixture.

'Ruby Red' This delightful plant of uncertain source was probably selected from an occasional variant among other prostrate plants, probably on a headland such as Emu Park (D.Hockings) or Agnes Water (P.Shaw). It has a completely prostrate habit, small silvery leaves with narrow lobes (c. 4 mm wide), a short, bright red inflorescence (5–6 cm long) and a pistil c. 35 mm long. In cultivation, it does not tolerate cold climates. This plant should be sought in the wild to verify its source.

Shiny-leaf form Originating in the Five Rocks area near Byfield, this usually prostrate form has a dark green, ± glabrous upper surface to the leaves and broad leaf lobes (c. 12 mm wide). Flowers are deep red or white or a mixture of both, the perianth densely covered with sticky, glandular hairs. This form may attain a height of c. 60 cm and is reasonably compact.

Major distinguishing features Leaves deeply once-divided with the undersurface exposed; conflorescence cylindrical; flowers basiscopic; floral bracts ≤ 2.5 mm long, perianth zygomorphic, glandular-hairy outside, glabrous inside; ovary sessile, villous; fruits compressed, with red-brown patches; seed winged all round.

Related or confusing species Group 35, especially *G. sessilis* and *G. whiteana*. *G. sessilis* has almost sessile flowers. *G. whiteana* differs in its narrower leaf lobes (< 4 mm wide) with the undersurface usually enclosed by the margin.

Conservation status Not presently endangered.

Cultivation The typical form of *G. banksii* is not widely cultivated outside the Brisbane area, although plants are sometimes seen in cultivation as far south as Sydney. Most forms are frost-sensitive and prefer a temperate to tropical climate. Var. *forsteri* is widely grown in all areas of Australia as well as in temperate and tropical climates in other parts of the world. It has proved a most tolerant and attractive plant for the garden with a preference for frost-free conditions, although it will tolerate light frosts that may burn the foliage. It has been established in inland gardens such as Burrendong Arboretum, N.S.W., and Glenmorgan, Qld. *G. banksii* prospers in an open, sunny position in well-drained, acidic to neutral, sandy to clay soil and likes plenty of water, particularly in summer, to which it responds by more prolific flowering. Pruning makes the tree form more dense and compact, but is generally not needed for the prostrate forms unless they are growing too vigorously. Do not prune into old wood too heavily as this can kill the plant. It responds to light applications of fertiliser low in phosphorus.

Propagation *Seed* Most forms are readily propagated from fresh seed that does not need peeling or nicking. Plants will self-sow and seedlings are common in gardens. Soaking in warm water for 24–36 hours will improve germination of unpeeled seed. *Cutting* Most forms except *G. banksii* 'Ruby Red' grow readily from cuttings of firm new growth at almost any season except winter. Cuttings strike most readily in summer. Best results have been achieved with G. 'Ruby Red' by setting the cuttings in pure perlite. *Grafting* This species has not been widely grafted but has been used as a rootstock in Qld and N.S.W. where it grows easily. G. 'Ruby Red' grows very well when grafted onto *G. robusta*.

Horticultural features *G. banksii* is a hardy, rewarding long-lived plant suitable as a small open tree, screen, specimen or ground cover. Its attractive, silvery, divided leaves and striking red or cream flowers that are full of nectar and strongly bird-attracting are widely admired. May become a little woody with age but its long flowering period, particularly in var. *forsteri*, provides excellent value. Leaves and stems are sometimes attacked by sooty mould, a fungus associated with nectar secretions. Although a little unsightly, it does little harm to the plant and can be removed by spraying with a mixture of water and 10% bleach during cooler weather. Sometimes attacked by a psyllid which causes bud drop and by a mite which disfigures the leaves. These can be treated with a systemic insecticide or miticide.

History Earliest recorded Australian cultivation of *G. banksii* appears to date from c. 1853 but probably occurred much earlier, possibly as early as 1830 or soon after settlement of the Brisbane area. Seed and young plants were sent to the Royal Gardens, Kew, in 1802 by Peter Good. It was again introduced to England via Veitch's Nursery in 1856, possibly as seed but, apart from a few plant catalogues (e.g. Osborn 1870), is not mentioned in the horticultural literature until 1874. From England it was introduced to the Continent, probably around 1868, and was well established in Europe and elsewhere prior to 1900, especially in favourable areas such as Hawaii and South Africa where it was mistakenly called *G. caleyi* (T.R.Sim). Seed was listed on Van Geert's catalogue of Stove and Glasshouse Plants in Belgium in 1883.

Hybrids Var. *forsteri* is the parent of a number of well-known garden hybrids, including G. 'Robyn Gordon', well-established in horticulture as one of the best of the *Grevillea* hybrids to date. Others include G. 'Superb', G. 'Mason's Hybrid' (syn. G. 'Ned Kelly'), G. 'Pink Surprise', G. 'Misty Pink', G. 'Honey Gem', G. 'Moonlight' and G. 'Sylvia'. We do not regard the registered variety 'Kingaroy Slippers' as valid since it describes a feature common within the species (see Variation, above).

General comments The hairs on leaves of *G. banksii* cause dermatitis in sensitive people. Cyanophoric glucosides have been found in the flowers and fruit.

Grevillea banyabba P.M.Olde & N.R.Marriott (1994) Plate 35

Telopea 5: 719 (Australia)

The specific epithet is an allusion to the Banyabba Nature Reserve, where the type was collected. The epithet is used as a noun in apposition, not adjectivally. BAN-YABB-A

Type: Fortis Creek, 3.5 km along track from Coaldale–Grafton Rd, N.S.W., 21 Sept. 1992, P.M.Olde 92/100 & D.Mason (holotype: NSW).

Open, little-branched **shrub** 1–1.5 m high. **Branchlets** slender, terete, loosely villous. **Leaves** 2.5–3.8 cm long, 0.5–1 cm wide, narrowly obovate to oblong-elliptic, shortly petiolate, simple, usually acute, occasionally obtuse and mucronate, the mucro c. 0.5 mm long; upper surface coarsely granular to scabrous, glabrous or with a few scattered, appressed hairs, the midvein, lateral veins and reticulum slightly raised; margins entire, shortly recurved to almost flat; lower surface openly silky with prominently raised midvein; texture papery. **Conflorescence** erect, terminal or axillary, pedunculate to subsessile, simple or few-branched; unit conflorescence loose, subsecund, 6–14-flowered with basipetal development; peduncle (0–)10–25 mm long, loosely brown-villous; rachis 10–25 mm long, loosely brown-villous; bracts 2–2.5 mm long, narrowly triangular with attenuate apex, loosely brown-villous outside, usually persistent to anthesis. **Flower colour:** perianth at first green, soon turning red with the basal dilation green turning yellow; style green. **Flowers** in bud aligned parallel to rachis, ultimately acroscopic; pedicels 7.5–10 mm long, loosely brown-villous; torus 1.5–1.8 mm across, oblique c. 40°; nectary glabrous, arcuate, thin, rising 0.2–0.5 mm above toral rim, the margin smooth to undulate; **perianth** 12 mm long, 4.5 mm wide, erect till late bud, oblong with slight basal dilation, sparsely silky to loosely villous outside; tepals with prominent midrib, detaching soon after anthesis, bearded inside c. 3 mm from base, the hairs reflexed and extending over a glabrous basal cavity, sparsely silky above beard to limb; limb revolute at anthesis, angularly ovoid–acuminate, loosely tomentose, the segments prominently ribbed to apex, the apex drawn into an appendage 0.7–1.5 mm long; **pistil** 25–27 mm long; stipe c. 0.3 mm long, villous; ovary subsessile, white-villous, style straight to slightly incurved, openly tomentose in proximal third, bearing a mixed indumentum of short, erect trichomes c. 0.05 mm long and longer biramous trichomes in distal two-thirds, the indumentum extending onto style end; pollen presenter 2.2.5 mm long, 1.5 wide, lateral, obovate, ± flat; stigma distally off-centre. **Fruit** (immature) erect, villous with erect style. **Seed** not seen.

Distribution N.S.W., in the NE in the Fortis Creek–Coaldale area, confined mostly to Banyabba Nature Reserve. *Climate* Hot, wet summer; mild, dry or occasionally moist winter. Rainfall 900–1200 mm.

Ecology Grows in well-drained sandy soil in open, eucalypt forest, near or at the top of ridges. Flowers Aug.–Oct., possibly much longer.

Conservation status Suggested 2EC.

Variation There is some variation in the amount of indumentum on the perianth.

Major distinguishing features Leaves simple, entire, usually acute, sparsely hairy beneath; conflorescence pedunculate; perianth hairy on both surfaces, bearded at base inside, the limb acute; ovary sessile, villous; pollen presenter lateral.

Related or confusing species Group 25, especially *G. arenaria* which differs in the dense indumentum of its leaf undersurface and the dense spreading indumentum above the beard on the inner perianth surface.

Cultivation *G. banyabba* has only recently been introduced to cultivation through the endeavours of Mr D.Mason of Coraki. Although all introductions have been as grafted plants, it is anticipated that plants grown on their own roots in well-drained, sandy soil will prosper in the manner of related species. A semi-shaded to full sun position would be suitable. Frost- and drought-tolerance are untested but related species tolerate quite extreme cold. Grafted plants have flourished in cultivation from Brisbane to western Vic. and flowered in their

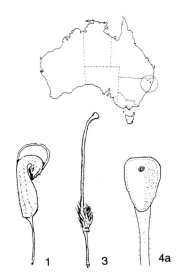

G. banyabba

35A. *G. banyabba* In cultivation, Coraki, N.S.W. ex Fortis Creek, N.S.W. (P.Olde)

first season. The species is not yet available commercially.

Propagation *Seed* Untested. *Cutting* Grows reasonably well from cuttings using firm, young growth. *Grafting* Grafts successfully onto *G. robusta* using the mummy graft.

Horticultural features *G. banyabba* is an impressive plant with an open habit and beautiful, relatively large flowers. The slender branchlets are often slightly weighed down by the predominantly red and green flowers that bunch at their ends. Nectar-feeding birds are attracted. Has potential as a feature or shrubbery plant. Potted specimens should do well as they flower well for some time.

Grevillea barklyana F.Mueller ex G.Bentham (1870)   Plate 36

Flora Australiensis 5: 436 (England)

Gully Grevillea

This species was named in honour of Henry Barkly (1815–1898), Governor of Vic. 1856–1863. BAR-KLEE-ARE-NA

Type: on the upper Tarwin and Bunyip Rivers, Vic., date unknown, F.Mueller (lectotype: K) (McGillivray 1993).

A tall **shrub** or small **tree** to 8 m. **Branchlets** angular, silky. **Leaves** 5–27 cm long, 2.5–12 cm wide, ascending, petiolate, leathery, ovate to obovate in outline, entire or irregularly cleft with 2–7(–11) broad, acute or obtuse ± triangular lobes; upper surface ± glabrous, flat, conspicuously penninerved; margins flat to slightly recurved; lower surface white-pubescent with short, curly hairs. **Conflorescence** 5–10 cm long, erect, pedunculate, terminal, unbranched, secund; peduncle silky; rachis tomentose; bracts 2.5–3 mm long, oblong-ovate, silky outside, falling before anthesis. **Flower colour:** perianth brownish pink or pale pink; style pale pink to reddish pink. **Flowers** acroscopic; pedicels 1.5–2.5 mm long, pubescent; torus 1–1.5 mm across, straight or slightly oblique; nectary U-shaped to arc-like, lightly toothed; **perianth** 6–9 mm long, 1.5–2.5 mm wide, oblong-ovoid to S-shaped, tomentose outside, glabrous inside, cohering except along dorsal suture; limb revolute, ovoid to globular, completely enclosing style end before anthesis; **pistil** 18.5–25 mm long; stipe 0.6–2 mm long, silky; ovary silky; style ± glabrous except at base, first exserted on dorsal side in lower half, looped outwards and upwards, slightly refracted above ovary, straight to undulate after anthesis, later swept back; style end hoof-like; pollen presenter very oblique, oblong-elliptic, conical. **Fruit** 13.5–17 mm long, 7.5–8 mm wide, erect on incurved stipe, oblong or ellipsoidal but attenuate, flattened, silky, purple striped or blotched, dorsally ridged; pericarp c. 0.5 mm thick. **Seed** 12 mm long, 5 mm wide, elliptic with marginal rim on inner face, rugose on outer face.

Distribution Vic., confined to several small populations along the tributaries of the upper Bunyip R., N of Labertouche, in western Gippsland. *Climate* Mild, dry summer; cold, wet winter. Rainfall 1000–1100 mm.

Ecology Grows on steep gully slopes in sheltered, wet sclerophyll eucalypt forest as a tall understory plant with mintbushes (*Prostanthera*) and other shade-loving species, in well-drained, rocky clay-loam that is moist for most of the year. Flowers spring to early summer. Pollinated by birds.

Variation A reasonably uniform species with minor variation in leaf lobing and flower colour.

Major distinguishing features Arborescent habit; leaves with 2–7 triangular lobes; conflorescences secund; floral bracts 2.5–3.5 mm long, falling before anthesis; perianth zygomorphic, hairy outside, glabrous inside; ovary and stipe silky; fruit incurved, with purplish stripes.

Related or confusing species Group 35, especially *G. macleayana* which is distinguished by its simple or rarely 1-lobed leaves (lobes not triangular), persistent bracts and spreading-villous ovary.

Conservation status 2VC. Restricted to an extremely small population in a limited area of reserved State Forest and could be seriously endangered if bushfires were to pass through the area in quick succession.

G. barklyana

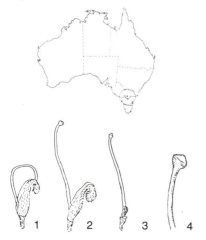

35B. *G. banyabba* Flowers and foliage (P.Olde)

Cultivation *G. barklyana* does not adapt well to environments dissimilar to its natural habitat, and rarely flourishes except in a cool–wet climate. Although it has been established successfully in warm-dry climates, such as at Rylstone and Burrendong Arboretum, N.S.W., as well as in numerous Victorian gardens, attention must be given to providing a sheltered position in semi-shade. While it does withstand full sun, its large leaves become easily wind-damaged in exposed sites and it is subject to wind-throw because of its relatively weak, shallow root system. It accommodates frosts to at least -3°C, but it does not tolerate extended, dry conditions. Soil should be well-drained, acidic sandy-loam or clay loam but must retain subsoil moisture during the summer months. It does not require fertiliser and rarely requires pruning except sometimes to control its often wide-spreading lower branches. It responds to summer watering. The true species is seldom available from native plant nurseries where hybrids are often sold incorrectly as *G. barklyana*. These are generally readily identifiable by their regularly-lobed leaves, the true species having few lobes except at the juvenile stage.

Propagation *Seed* Sets numerous large seed that germinate readily, especially when pretreated or sown before the testa turns brown. When conditions are to its liking, it often self-sows throughout the garden. Care must be taken that garden-grown seedlings are pure, however, as numerous inferior hybrids have been observed in cultivation. *Cutting* Fairly large cuttings, with leaves trimmed considerably to reduce evaporation, generally strike well at most times of the year. Tips should not be removed or a strong lead trunk will not develop and a misshapen specimen will be the result. Young rooted cuttings often need light staking to promote a straight trunk and to give the generally weak root system a chance to strengthen. *Grafting* Untested.

Horticultural features *G. barklyana* is an attractive, small upright tree with a dense canopy of large, irregularly-lobed, dark green leaves with a white

Grevillea batrachioides

36A. *G. barklyana* Large plants in natural habitat, Ryson's Creek, Vic. (N.Marriott)

36B. *G. barklyana* Flowers and foliage (N.Marriott)

underside. When growing in favourable conditions, it is fast-growing, developing into a luxuriant, leafy specimen which might be thought more at home in a tropical rainforest. Unfortunately, the large 'toothbrush' racemes, which are produced in profusion, are usually very pale pink, often going unnoticed except by honey-eating birds which feast on the copious nectar. Occasionally, plants with brighter pink or reddish flowers can be found and cuttings should be taken from these showier forms. It is a very long-lived species in cultivation, making a superb shade plant when grown as overhead cover in a fernery. When planted in a moist, shady gully it often self-seeds to create a small forest. Due to its requirements for shelter and moisture, it is not suited to commercial landscapes but is often a wonderful addition to private or public gardens.

Hybrids Known parent of the garden hybrids G. 'Jessie Cadwell' and G. 'Copper Rocket'.

General comments *G. barklyana* subsp. *macleayana* is treated as a distinct species in this work.

Grevillea batrachioides F.Mueller ex D.J.McGillivray (1986) Plate 37

New Names in Grevillea *(Proteaceae)* 2 (Australia)

The specific epithet is derived from the Greek *batrachion* (a ranunculus), alluding to the similarity of the leaves of this species to those of some species of *Ranunculus*. The name was suggested by Mueller but not published. BAT-RACK-EE-OY-DEES

Type: south-western W.A., probably 1850–51, J.Drummond (holotype: MEL).

A bushy **shrub** to 2 m. **Branchlets** angular to terete, tomentose. **Leaves** 1–4 cm long, ascending to spreading, ± sessile, crowded, deeply and divaricately 1–3 times divided; rachis refracted; leaf lobes 3–20 mm long, 1 mm wide, narrowly linear, rigid, pungent, secondary lobes occasionally of unequal length; upper surface sparsely pubescent, soon glabrous, midvein conspicuous; margins angularly refracted about an intramarginal vein to midvein; lower surface obscured, silky in grooves, midvein prominent. **Conflorescence** 2–5 cm long, erect to decurved, unbranched, terminal, hemispherical and partially secund, loose; peduncle and rachis pubescent; bracts 1 mm long, broadly ovate, pubescent, persistent to anthesis. **Flower colour**: perianth pale pink to creamy pink with richer pink overtones; style deep red. **Flowers** acroscopic: pedicels 7–15 mm long, glandular-villous; torus 1.5 mm across, straight or slightly oblique; nectary prominent, V-shaped, lipped, cup-like; **perianth** 7–11 mm long, 3.5–5 mm wide, ovoid, grossly saccate, sharply narrowed at neck, openly tomentose outside, glabrous inside except a few trichomes on attenuated segment of tepals, cohering except along dorsal suture, partially exposing style end before anthesis; limb revolute, globular, villous; **pistil** 30–38 mm long; ovary sessile, villous; style elongate, first exserted near curve on dorsal side and looped upwards, glabrous at base and apex, otherwise sparsely glandular-pubescent, dilating into style end; pollen presenter oblique, elliptic, flat, umbonate. **Fruit** 17–23 mm long, 9–10 mm wide, 6 mm deep, erect, ellipsoidal but attenuate, flattened, dorsally ribbed, tomentose with reddish striping; pericarp 0.5 mm thick at suture. **Seed** 11.5 mm long, 3.5 mm wide, ellipsoidal; outer face smooth; margin revolute with a narrow waxy or papery border; inner face smooth, channelled near rim.

Distribution W.A., near Mt Lesueur. *Climate* Hot, dry summer; mild, moist winter. Rainfall 500–550 mm.

Ecology Occurs on sandstone pavement in shallow, sandy loam soils or in rock crevices. Flowers late spring. Pollinator unknown but almost certainly birds.

Variation A species with little variation. Flowers of some plants are slightly paler than others.

Major distinguishing features Leaves sessile, deeply and divaricately 1–3 times divided, the undersurface obscured, secondary lobes sometimes of unequal length; conflorescence loose, secund to hemispherical; perianth zygomorphic, grossly saccate, hairy outside, glabrous inside; ovary sessile, densely hairy; fruit red-striped.

Related or confusing species Group 35, especially *G. asparagoides* and *G. maxwellii*, both of which have petiolate leaves.

Conservation status 2EC-t. Presumed extinct until 1991 when one of the authors located a population of 10 plants near Mt Lesueur. The lead was given from a specimen collected on a vegetation survey in 1982 by E.A.Griffin.

Cultivation Recently introduced to cultivation and being tested by growers from Qld to Vic. The natural habitat indicates a well-drained, sunny position. Soil should be neutral to possibly slightly alkaline sandy or light loam. The area receives low rainfall, most of which occurs in winter. Grafted plants have been established in Qld and N.S.W. using *G. robusta* as the rootstock; but a number of these have died, indicating a possible compatibility problem.

Propagation Seed Sets copious quantities of seed in the wild and should respond to standard methods of germination. Closely related species germinate within 40 days. *Cutting* Firm, young growth strikes reasonably well, particularly if taken in spring or early autumn. *Grafting* Has been grafted onto *G. robusta* using the mummy technique. Grafted wild material reached flowering within four months of despatch, but as mentioned above there may be a compatibility problem. Testing with other rootstocks is needed.

Horticultural features *G. batrachioides* is an extremely beautiful plant with conspicuous, terminal pink or creamy pink flowers contrasting with crowded grey-green leaves. It has potential as a very showy specimen plant. An excellent 'bird' plant for its nectar-laden flowers and its protective, prickly foliage, this species can be recommended for horticulturists everywhere. Members of the Grevillea Study Group, especially in Qld where early success was achieved, are working to increase numbers of this species and aim to spread material as widely as possible.

G. batrachioides

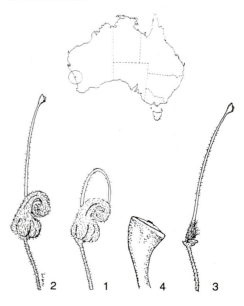

37A. *G. batrachioides* Flowers and foliage (P.Olde)

37B. *G. batrachioides* Plant in natural habitat near Mt Lesueur, W.A. (P.Olde)

Grevillea baueri R.Brown (1810)

Transactions of the Linnean Society of London 10: 173 (England)

Bauer's Grevillea

Named after Ferdinand Bauer (1760–1826), a botanical artist who accompanied Robert Brown on Flinders' *Investigator* voyage of 1801–03. BOWER-EYE

Type: near the Cowpastures, Nepean R., N.S.W., 19 Oct. 1803, R.Brown 3330 (lectotype: BM) (McGillivray 1993).

A bushy, much-branched **shrub** 0.5–1 m high, up to 2 m wide. **Branchlets** angular, tomentose to pubescent. **Leaves** 1–3 cm long, 3–10 (15) mm wide, ascending to spreading, crowded, simple, sessile, oblong to elliptic, obtuse to mucronate; upper surface flat or convex, smooth to finely granular, usually glabrous, sometimes sparsely silky, midvein evident; margins entire, flat or shortly recurved; lower surface glabrous, the prominent midvein sometimes sprinkled with hairs, lateral veins prominent. **Conflorescence** 1–4 cm long, 2 cm wide, erect, sessile or shortly pedunculate, simple or branched, terminal or axillary, secund-subglobose, apical flowers opening first; peduncle pubescent; rachis glabrous or almost so; bracts 1.5–2.9 mm long, triangular, sparsely silky, usually persistent at anthesis. **Flowers** adaxially acroscopic: pedicels 3.5–7 mm long, slender, ± glabrous; torus 1.5–2 mm wide, oblique; nectary conspicuous, kidney-shaped, smooth; **perianth** 6–8 mm long, 3.5–5.5 mm wide, strongly curved, oblong with strong basal dilation, glabrous outside, bearded inside, cohering except along dorsal suture; limb revolute, glabrous or with sprinkled hairs, globular to spherical; style end partially exposed before anthesis; **pistil** 16–23 mm long; stipe 0.5–1.4 mm long, villous; ovary villous; style villous, inflexed above ovary, stout, at first exserted at curve and looped upwards, afterwards straight to gently incurved, at length straight; style end glabrous, flattened, disc-like; pollen presenter lateral, obovate, flat, umbonate. **Fruit** 13–14 mm long, 6–7 mm wide, erect or oblique on pedicel, narrowly ovoid, ridged, sparsely hairy, rugose; pericarp c. 0.5 mm thick. **Seed** 8 mm long, 4 mm wide, elliptic; outer face convex; inner face channelled; margin revolute.

Major distinguishing features Leaves simple, entire, oblong to ovate, ± glabrous, smooth or granulate, sometimes a few hairs on either surface; conflorescence secund-globose, perianth zygomorphic, glabrous outside or sometimes a few hairs on limb, hairy within; ovary villous, sessile or shortly stipitate; style hairy; pollen presenter lateral.

Related or confusing species Group 20, especially *G. iaspicula* which differs in its glabrous ovary.

Variation Two subspecies are currently recognised.

Key to subspecies

Leaves oblong-elliptic to narrowly oblong-ovate, 3–7 mm wide, smooth on upper surface or with a few granules; conflorescence simple, erect; bushy shrub 0.5–1 m high
subsp. **baueri**

Leaves oblong-ovate to ovate, usually 5–10 mm wide, conspicuously granular and sometimes sparsely hairy on upper surface; conflorescence usually branched, decurved; bushy shrub 0.3–0.5 m high or erect spindly shrub to 1.2 m
subsp. **asperula**

Grevillea baueri R.Brown subsp. baueri
Plate 38

Bushy **shrub** 0.5–1 m high. **Leaves** oblong-elliptic to narrowly oblong-ovate, 3–7 mm wide, smooth on upper surface or with a few granules. **Conflorescence** simple, erect. **Flower colour**: perianth pink or reddish pink and cream, becoming paler at anthesis, rapidly turning black after anthesis; style red; occasionally all cream or pinkish cream.

Synonyms: *G. pubescens* W.J.Hooker (1826) non R.Graham; *G. baueri* var. *pubescens* (W.J.Hooker) C.F.Meisner (1856); *G. baueri* var. *glabra* C.F.Meisner (1856).

Distribution N.S.W., from Camden to Bundanoon.
Climate Summer hot, dry or wet; winter cold, wet or dry. Rainfall 600–800 mm.

Ecology Occurs in woodland or heath, in sand or sandy loam, generally with sandstone. Flowers principally winter and spring. Pollinated by birds.

Variation Leaf width and flower colour vary a little.

Conservation status Becoming less common but not presently endangered.

Cultivation Subsp. *baueri* has been cultivated successfully over a wide area ranging from central Qld to Tas. and across to S.A. and Perth. It is also cultivated in the U.S.A. and England where it survives heavy frosts and icy conditions. It tolerates extended dry periods provided adequate subsoil moisture is present. Summer-humid climates are tolerated, but growers report that the grevillea bud-drop psyllid affects flowering in these areas and it seems to prefer a drier climate. A position in full sun or partial shade in well-drained, acidic sandy soil suits it best and in these conditions it flourishes into a long-lived, bushy shrub requiring little maintenance. It responds to summer watering and light applications of nitrogenous fertiliser such as blood and bone. Pruning is generally unnecessary, although older plants should be pruned to maintain shape and density. Older plants are sometimes affected by white scale and sooty mould. The species is ideally suited to tubs or pots and thrives in an open, sandy mix with fertiliser.

Propagation *Seed* Fresh seed germinates well without any treatment but should be collected only from wild sources as this species hybridises readily in cultivation. *Cutting* Propagates readily from cuttings taken all year using firm, young growth. *Grafting* Generally unnecessary.

Horticultural features Subsp. *baueri* grows naturally into a compact, rounded shrub with decorative coppery-bronze or reddish new growth complementing the adult leaves for long periods. The shapely dark green leaves are shiny and often wavy, and attractive red and cream flowers cluster in their axils in profusion over an extended period. After anthesis, flowers quickly blacken and this can detract from the floral impact, especially when most flowers have still to open. This strongly bird-attracting plant is hardy and long-lived in cultivation

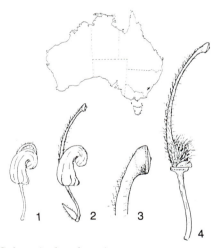

G. baueri subsp. *baueri*

38A. *G. baueri* subsp. *baueri* Unusually-coloured flowers, near Berrima, N.S.W. (P.Olde)

38B. *G. baueri* subsp. *baueri* Habit, near Berrima, N.S.W. (P.Olde)

under suitable conditions and is widely used in landscaping as a feature plant or low screen. Easy to grow, it can be recommended to beginners or to those seeking low-maintenance gardens.

Hybrids It hybridises readily, especially in cultivation. Narrow-leaved plants in a churchyard at Berrima have hybridised, possibly with *G. juniperina*, and grow in a prolific flowering colony. Flower colours range from creamy-white to pure red. It is also the known parent in a number of popular garden hybrids such as G. 'Austraflora Pendant Clusters', G. 'Pink Star' and G. 'Poorinda Rondeau'. It is known to hybridise readily with *G. lanigera* and *G. rosmarinifolia*. The cultivar, G. 'John Evans', arose in the Sydney garden of the bearer of its name. It has narrow leaves but is otherwise typical.

Early history Subsp. *baueri* was introduced to English glasshouses in 1823–24 by Allan Cunningham and is still cultivated there among a limited circle of enthusiasts. It was listed among the plants growing at Hügel's garden in Vienna in 1831. The first recorded cultivation in Australia was c. 1853 when it was listed among plants despatched from the Royal Botanic Gardens, Sydney, to Hobart.

General comments McGillivray (1993: 279) stated that some collections from the Bundanoon–Kangaroo Valley area approach subsp. *asperula* and cannot be placed with certainty in either subspecies.

Grevillea baueri subsp. asperula
D.J.McGillivray (1986) **Plate 39**

New Names in Grevillea *(Proteaceae)* 2 (Australia)

The subspecific name is derived from the Latin diminutive *asperulus* (a little roughened), in reference to the upper surface of the leaves. ASS-PER-RULE-A

Type: Round Hill, S of Sassafras, N.S.W., 20 Sept. 1961, E.F.C.Constable (holotype:NSW).

Open, spindly, erect **shrub** to 1.2 m or a compact, domed shrub to 50 cm. **Leaves** oblong-ovate to ovate, usually 5–10 mm wide, conspicuously granular and sometimes sparsely hairy on upper surface. **Conflorescence** usually branched, decurved. **Flower colour**: perianth deep pink and cream, quickly turning black; style pinkish-red.

Distribution N.S.W., from near Morton NP south to the Budawang Ra. *Climate* Summer mild to hot, wet or dry; winter mild to cold, wet or dry. Rainfall 600–800 mm.

38C. *G. baueri* subsp. *baueri* Close-up (N.Marriott)

Ecology Grows in open, dry, sclerophyll forest or heath, in skeletal, sandy soil on sandstone. Flowers principally in winter and spring. Pollinated by birds.

Variation In exposed areas, subsp. *asperula* forms a compact, low shrub. In shrubland and protected areas, it ages to an open, spindly shrub.

Conservation status Not presently endangered.

Cultivation This hardy plant has been cultivated widely for many years and thrives in most well-drained situations. An acidic soil in either full sun or partial shade should support well-grown attractive specimens needing little maintenance. It has been cultivated successfully in New Zealand as well as in most Australian States in both coastal and inland areas. Frosts to at least -10°C are accommodated, as well as extended dry conditions provided there is some subsoil moisture. Erect forms are more open and should be tip-pruned when young. Light applications of low phosphorus fertiliser and

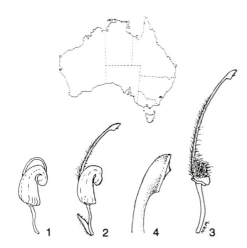

G. baueri subsp. *asperula*

summer watering will improve appearance, inducing vigorous new growth and inhibiting leaf drop. Often available from native plant nurseries, especially in the eastern States, under the name '*G. baueri* Hairy-leaf form'.

Propagation *Seed* Sets reasonable quantities of seed that germinate well when given the standard peeling treatment. *Cutting* Readily propagated from cuttings of firm, new growth taken at most times of the year. *Grafting* Untested. Not usually necessary.

Horticultural features Subsp. *asperula* is a rewarding plant valued for both foliage and flowers. The broad, heart-shaped leaves are undulate and have an intriguing, sand-papery texture while bronze new growth provides a pleasing contrast for long periods. Red and cream flowers cluster in the outer foliage from late autumn to spring with odd flowers present all year, making this plant valuable in gardens designed to attract birds. It is relatively long-lived and useful in landscaping, as either a low screen or background plant. Unfortunately, it tends to become leggy with age and should not be used in a foreground. Old flowers ageing black also detract from its horticultural potential. Attention should be paid to introducing compact forms.

Grevillea baxteri R.Brown (1830)

Plate 40

Supplementum Primum Prodromi Florae Novae Hollandiae exhibens Proteaceas Novas: 22 (England)
Named after William Baxter, who collected plants in southern Western Australia from 1823–29 for Henchman's Nursery, London, as well as the Botanic Gardens, Sydney. BAKS-TER-EYE

Type: between Cape Arid and Lucky Bay, [W.A.], 1825, W.Baxter (holotype: BM).

A spreading, dense, sometimes suckering **shrub** 1–3.5 m tall. **Branchlets** angular becoming terete, silky to pubescent. **Leaves** 8–12 cm long, ascending, shortly petiolate, pliable, pinnatipartite into narrowly linear lobes 2–6 cm long with obtuse, scarcely pungent apices; young growth rusty; upper surface tomentose, soon glabrous, midvein impressed; margins angularly revolute to midvein below; lower surface obscured, silky in the two grooves, otherwise glabrous; midvein prominent. **Conflorescence** 4.5–9.5 cm long, erect, pedunculate, terminal, unbranched, secund; peduncle and rachis silky to tomentose; bracts 1.3–2 mm long, ovate, imbricate, rusty-tomentose outside, falling before anthesis. **Flower colour:** perianth green to brownish fawn; style orange, usually orange-red, occasionally yellow or red. **Flowers** acroscopic: pedicels 1.5–2 mm long, silky; torus 1–2 mm across, ± straight; nectary thick, kidney-shaped, cupped, the margin wavy or smooth; **perianth** 8–11 mm long, 2.5 mm wide, ovoid to S-shaped, rusty, silky to tomentose outside, glabrous inside, coherent except along dorsal suture; limb strongly revolute, globular, densely brown tomentose, enclosing style end before anthesis; **pistil** 22–25 mm long; stipe 0.9–1.3 mm long, silky; ovary silky; style silky for 4–8 mm, otherwise ± glabrous, sometimes pubescent on ventral side or near apex, at first exserted from just below curve on dorsal side, retrorse to S-shaped just after anthesis; style end conspicuous, hoof-like; pollen presenter straight to slightly oblique, elliptic to round, broadly convex, umbonate. **Fruit** 14–20 mm long, 6–8 mm wide, on incurved pedicels, flattened, ellipsoidal but attenuated, prominently ridged, sparsely silky; pericarp c. 0.5 mm thick. **Seed** 10 mm long, 3.5 mm wide, oblong-elliptic.

Distribution W.A., near the south coast from the Truslove–Scaddan area east to Israelite Bay. *Climate* Hot, dry summer and cool to mild, wet winter. Rainfall 350–400 mm.

Ecology Grows in sand, sandy loam and granitic loam in low heath to tall open shrubland and open mallee. Flowers winter to summer.

Variation Fairly uniform in morphology but variable in flower colour. McGillivray (1993) also discussed a collection from the Alexander R. with finer, smaller floral parts.

Major distinguishing features Leaves pinnatipartite, the midvein evident on upper surface; conflorescence secund; perianth zygomorphic, hairy outside, glabrous inside; ovary densely hairy, shortly stipitate; style robust with an indumentum extending up to 8 mm.

Related or confusing species Group 35, especially *G. cagiana*, which has shorter conflorescences (2–6 cm), an ovarian stipe 1.9–5 mm long, the style glabrous except basal hairs restricted to 1.5 mm from the ovary, and the style end narrower.

Conservation status 3RC. Widespread in the region but usually restricted to degraded road verges. The species occurs in Cape Arid NP.

Cultivation *G. baxteri* is seldom cultivated and has so far proved difficult to propagate and establish. Its natural habitat suggests a need for extremely well-drained soils as it occurs in neutral to alkaline, white sand over laterite gravel in a dry climate. Frosts to at least -4°C have been recorded with no ill effect. It usually grows in full sun in dense heath or scrub, protected by the surrounding vegetation. It is too large to consider for long-term pot culture, although it could be successfully cultivated in large tubs for many years and made attractive by judicious

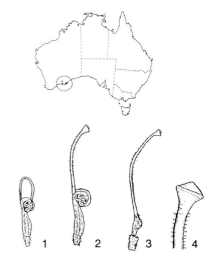

G. baxteri

39A. *G. baueri* subsp. *asperula* Habit, in cultivation, Stawell, Vic. (N.Marriott)

39B. *G. baueri* subsp. *asperula* (M.Hodge)

40A. *G. baxteri* Close-up (P.Olde)

40B. *G. baxteri* Plant in natural habitat, N of Esperance, W.A. (P.Olde)

pruning and annual fertilising. Members of the Grevillea Study Group are attempting to introduce this species to cultivation but plants are not available in nurseries. Grafted specimens have been established near Sydney (P. & N. Abell) and north of Brisbane (E. & P. Burt). *G. concinna* subsp. *lemanniana*, mistakenly identified as *G. baxteri* in the past, has been sold under this name in some Victorian nurseries.

Propagation *Seed* Sets few seeds but these should germinate readily given the standard peeling pretreatment. *Cutting* Has proved difficult to strike from cuttings. Limited success has been achieved using firm, young growth taken in spring or early autumn. *Grafting* Has been grafted successfully onto *G. robusta* using top wedge and mummy grafts.

Horticultural features Conspicuous, orange-red 'toothbrush' conflorescences, large shrubby habit and attractive, divided leaves are the main features of *G. baxteri*. The dull-green leaves with a white undersurface complement the white branchlets and silky bronze new growth to add interest to the general appearance of the plant. Strongly bird-attracting conflorescences are boldly presented at the ends of the branches over a long period, and interesting bird-like fruits adorn the plant after flowering. For reasons of both conservation and horticultural value, serious attempts must be made to introduce this species to cultivation. Evidence from wild populations shows that it is a long-lived species and would make a fine landscape plant in dry areas with sandy soil. It should thrive in most coastal, sandy areas on the south coast of Australia.

Grevillea beadleana D.J.McGillivray (1986)
Plate 41

New Names in Grevillea *(Proteaceae) 2* (Australia)

Named in honour of Professor Noel C. Beadle, Emeritus Professor of Botany, University of New England, co-author of *Flora of the Sydney Region*, for a long time the only comprehensive guide to the flora of the Sydney area. BEEDLE-ARE-NA

Type: SSW of Chaelundi Falls, Guy Fawkes NP, N.S.W., 13 June 1982, J.B.Williams (holotype: NSW).

A dense, single-stemmed **shrub** to c. 2.5 m. **Branchlets** terete with some ridging, tomentose. **Leaves** 8–16 cm long, 5–10 cm wide, spreading, petiolate, ± ovate in outline, bipinnatifid; lobes linear or triangular 0.2–1 cm wide, acute, not pungent; upper surface sparsely pubescent, midvein an impressed groove; margins shortly recurved; lower surface white pubescent, midvein and lateral veins prominent. **Conflorescence** 3.5–5 cm long, erect, pedunculate, terminal, simple, secund; peduncle and rachis ± villous; bracts 5–6 mm long, imbricate in bud, narrowly triangular, villous, persistent at anthesis. **Flower colour:** perianth grey; style purple-mauve to burgundy with green tip. **Flowers** acroscopic; pedicels 1.7–2.5 mm long, villous; torus 1–1.5 mm across, ± straight; nectary semi-circular, faintly toothed; **perianth** 5–6 mm long, 1–1.5 mm wide, oblong-ovoid, pubescent outside, glabrous inside, cohering except along dorsal suture; limb revolute, globular, densely villous, enclosing style end before anthesis; pistil 15–18.5 mm long; stipe 0.5–1 mm long, silky; ovary appressed-villous, purple blotched; style glabrous, slightly refracted above the ovary, at first exserted below curve on dorsal side, straight to gently incurved after anthesis, retrorse at length; style end hoof-like; pollen presenter straight to slightly oblique, oblong-elliptic, conical. **Fruit** 9–10.5 mm long, 5 mm wide, borne on incurved stipes, ovoid, apically attenuate, dorsally concave, tomentose, conspicuously purple-striped; pericarp c. 0.3 mm thick. **Seed** 9 mm long, 4 mm wide, ellipsoidal, outer face rugulose, inner face channeled around the rim with a flat central portion.

40C. *G. baxteri* Yellow-flowered form, Cape Arid NP, W.A. (P.Olde)

G. beadleana

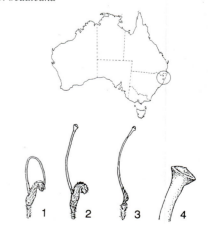

41A. *G. beadleana* Close-up of flowers and foliage (N.Marriott)

41B. *G. beadleana* Habit in cultivation, Tynong North, Vic. (N.Marriott)

Distribution N.S.W., near Walcha (last collected in 1887), Ebor and Torrington. ***Climate*** Hot, dry summer and cold, wet or dry winter. Rainfall 800–1250 mm.

Ecology Grows in gritty loam among granite outcrops in open scrub or woodland at the top of gorges and near creek-lines. Also in undulating terrain in skeletal, granite soil. Flowers spring. Killed by fire and regenerates from seed. Bird-pollinated.

Variation A fairly uniform species. The only specimen from near Walcha (where it is now apparently extinct) has coarser leaf lobes than those from other areas.

Major distinguishing features Leaves bipinnatifid with oblong, acute lobes; conflorescence secund; floral bracts 5–6 mm long, narrowly triangular, persistent at anthesis; perianth zygomorphic, hairy outside, glabrous inside; ovary densely hairy, shortly stipitate; fruit incurved with reddish stripes.

Related or confusing species Group 35, especially *G. caleyi* which has pinnatipartite leaves, and *G. aspleniifolia* which has elongated, broadly linear, entire or toothed leaves; both have much shorter floral bracts.

Conservation status 3RC. Apart from the original collection at Walcha, the species is confined to a small population of fewer than 50 plants in Guy Fawkes NP and a recently discovered one of c. 4000 plants at Binghi, north of Torrington, on private land.

Cultivation Limited experience has shown this species is reasonably adaptable to cultivation, performing well in a wide range of conditions. Frosts to at least -4°C are endured and it is unaffected by dry conditions provided adequate sub-soil moisture is present. Since its introduction to cultivation in 1983, it has been established by enthusiasts in Sydney and Vic. where reports indicate that it prospers in well-drained, acidic sand or sandy loam in full sun and responds to good growing conditions by forming a compact habit and dense, healthy foliage. Occasionally, in both summer-humid and dry-inland conditions, plants die for unknown reasons, possibly because of poor drainage. Pruning may be necessary to retain shape but otherwise maintenance is seldom required, especially if a position protected from strong winds is selected. It makes a good specimen in a tub for a number of years but requires a well-drained potting mix and annual fertilising. Although only recently re-discovered, it has already become widely available through specialist nurseries.

Propagation *Seed* Sets numerous seeds that germinate particularly well when given the standard peeling treatment. Self-sows in the garden. *Cutting* Strikes reasonably well from cuttings of firm, young growth taken in spring and autumn. Resents excessive moisture or misting. *Grafting* Has been grafted successfully onto G. 'Poorinda Royal Mantle' and *G. robusta* using the top wedge and mummy techniques. Indications are that it is hardy enough to grow on its own roots at least in Sydney and Melbourne.

Horticultural features *G. beadleana* is a fascinating shrub, distinguished by its outstanding, fern-like foliage. In cultivation, it is usually a spreading, dense shrub with dull-green leaves invested with a pubescent white and brown undersurface and brownish branches in graceful contrast. New growth is a pale fawn to brown and very floppy, creating a gold-like fleck in the shrub in late summer. Prominent, long bracts are noticeable on the young conflorescence which matures into an attractive 'toothbrush' type with nectar-filled flowers crowded along the axis. The flowers have striking burgundy styles that contrast oddly with the grey perianth, and flowering continues from spring to autumn, luring continuous visits from nectar-feeding birds. Attractively striped seed pods adorn the rachis when flowering has finished. Being reasonably long-lived, it has potential as a screen or specimen plant and would be suitable for both coastal and inland plantings.

General comments This species was rediscovered in 1982 in Guy Fawkes NP, being previously known from one collection made in 1887 by Captain Crawford near Walcha. The latter form, which has more coarsely lobed leaves, should be sought in the gorge system of the Walcha area.

Grevillea beardiana D.J.McGillivray (1986) Plate 42

New Names in Grevillea *(Proteaceae)* 2 (Australia)
Named in recognition of John Beard who, among other achievements, surveyed and mapped the vegetation of W.A., collecting many species new to science. BEER-DEE-ARE-NA

Type: c. 45 km E of Newdegate, W.A., 30 Oct. 1962, J.S.Beard 2179 (holotype: PERTH).

Spreading **shrub** 0.3–0.6(–1.5) m tall. **Branchlets** terete, tomentose. **Leaves** 1–5 cm long, 1–2 mm wide, erect to ascending, ± sessile, crowded, slightly curved, rigid, pungent, simple, linear, rarely bipartite; upper surface glabrous or sparsely silky, midvein obscure; margins smoothly revolute to midvein below; lower surface enclosed, silky in grooves, midvein prominent. **Conflorescence** 2–4 cm long, erect or deflexed, pedunculate, terminal, simple (rarely 2-branched), secund; peduncle and rachis tomentose; bracts 0.5–0.8 mm long, ± ovate, tomentose outside, often persistent to fruiting. **Flower colour:** perianth and style bright red or orange with green style end. **Flowers** acroscopic: pedicels 1.5–3.5 mm long, silky; torus 1–1.5 mm across, straight; nectary arc-like to kidney-shaped, smooth; perianth 7–9 mm long, 2–3 mm wide, ovoid to slightly S-shaped, silky outside, glabrous inside; cohering except along dorsal suture; limb strongly revolute, globular, enclosing style end before anthesis; **pistil** 28–32.5 mm long; stipe 2.5–5.1 mm long, silky, glabrous on ventral side; ovary appressed-villous; style glabrous, first exserted just below curve and looped upwards and outwards, after anthesis gently incurved, strongly so in apical few mm; style end scarcely dilated; pollen presenter oblique, obovate, convex. **Fruit** 7.5–11.5 mm long, 5–6 mm wide, erect, ellipsoidal or ovoid, tomentose with faint striping, fragile; pericarp c. 0.5 mm thick. **Seed** 7 mm long, 4 mm wide, ellipsoidal, outer face smooth, crimped submarginally, inner face smooth, concave.

Distribution W.A.; generally found from Newdegate E to c. 50 km W of Kumarl. *Climate*

G. beardiana

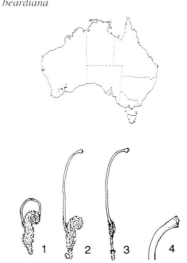

42A. *G. beardiana* Massed flowering, near Lake King, W.A. (C.Guthrie)

42B. *G. beardiana* Close-up of red flowers and foliage (N.Marriott)

42C. *G. beardiana* Plant in natural habitat near Lake King, W.A. (C.Guthrie)

Hot, dry summer; cool, wet winter. Rainfall 250–400 mm.

Ecology Usually found in heath in sand over laterite or on heavily lateritic rises. It also occurs in mallee communities and occasionally in granitic loam and sandy clay. Flowers spring to summer. Pollinated by birds.

Variation There is variation in habit and leaf width as well as minor variation in flower colour. Plants range from almost prostrate to 1.5 m high in habit.

Narrow-leaf form At the eastern edge of its range, *G. beardiana* has consistently narrower, more rigid leaves (c. 1 mm wide), often with a very prominent midvein on the lower surface. Leaves are usually a distinctive bright green, especially new growth. Flower colour is usually bright red.

Broad-leaf (typical) form In the western part of its range plants have broader (1–2 mm) sometimes divided leaves that are usually dark to dull-green, more pliable and not mucronate. Flowers are brilliant red or orange-red but occasional plants have clear orange or yellow flowers.

Major distinguishing features Leaves mostly simple, linear, stiff, pungent, < 2 mm wide with venation obscure on upper surface, the margins smoothly rounded; conflorescence secund; floral bracts < 1 mm long; perianth zygomorphic, hairy outside, glabrous inside; ovary densely hairy; stipe 2.5–5 mm long; fruit with reddish hairs.

Related or confusing species Group 35, especially *G. coccinea* and *G. concinna*, both of which have a prominent midvein visible on the upper surface. *G. concinna* also differs in its ovarian stipe < 2.5 mm long and its narrowly triangular floral bracts > 1 mm long. *G. coccinea* also has a stipe < 2.5 mm long, a shorter pistil (19–23.5 mm long), a ± straight pollen presenter and angularly revolute leaf margins.

Conservation status Fairly common over its range, especially east of Lake King where it is conserved in Frank Hann NP and other reserves.

Cultivation *G. beardiana* has been introduced to cultivation only recently, in W.A. and Vic. In these areas it has proved hardy to extended, dry summers and frosts to at least -6°C. Potted plants have succeeded in summer-rainfall climates near Sydney but performance in the ground there is untested. It has shown a preference for a warm, sunny position in extremely well-drained, acidic to neutral sand or open gravelly loam. Neither cold, wet climates nor poorly drained conditions are tolerated at all. Plants rarely require pruning and should not be watered once established. In pot culture, the species can be maintained using well-drained mixes but foliage often drops or remains sparse and more experimentation is required to improve presentation. Occasionally available in specialist nurseries in Perth, W.A., *G. beardiana* is also being introduced to cultivation elsewhere by members of the Grevillea Study Group. Indications are that the Broad-leaf form is easier to grow than the Narrow-leaf form.

Propagation *Seed* Sets seed profusely. Germinates quite readily when given the standard peeling treatment. Seedlings require good drainage and warmth to become established. *Cutting* Firm, young material strikes well when taken during the warmer months, especially during early autumn. *Grafting* Has been grafted successfully onto *G. robusta* and *G.* 'Poorinda Royal Mantle' using the top wedge technique.

Horticultural features *G. beardiana* is a tidy, compact shrub of compelling beauty when in full flower. The neat habit and dense green foliage are highlighted by the bright green new growth, but it is the magnificent floral display that impresses and which makes it possibly the most attractive of the small western grevilleas. On some plants, the bright orange to red flowers smother the bush in spring. On less floriferous plants, the flowers still make a stunning contrast with the bright foliage and provide plenty of nectar for birds. In cultivation, flowering extends over many months and finishes much later than in the wild. This plant could be an ideal choice for small gardens or rockeries although it should be remembered that hardiness is not fully tested. Older plants both in the garden and pot culture tend to become open and woody, especially if they receive too much water which causes the leaves to drop. This is particularly so with the Narrow-leaf form.

Grevillea bedggoodiana J.H.Willis ex D.J.McGillivray (1986)   Plate 43

New Names in Grevillea *(Proteaceae)* 3 (Australia)

Enfield Grevillea

The specific name honours Mrs Stella Bedggood (1916–1978), a foundation member of the Ballarat Field Naturalists' Club, whose special interest was the Enfield State Forest to which this species is confined. BEDJ-GOOD-EE-ARE-NA

Type: Little Hard Hills, Enfield Forest Park, SSW of Ballarat Post Office, Vic., 23 Oct. 1978, A.C.Beauglehole 60988(A) (holotype: NSW).

Prostrate or scrambling, decumbent **shrub** to c. 0.5 m tall. **Branchlets** terete, tomentose. **Leaves** 2–7 cm long, 1–3.5 cm wide, ascending, ± sessile or shortly petiolate, simple, pungent-toothed, ovate to oblong; upper surface faintly pubescent, soon ± glabrous, the midvein, lateral veins and reticulum evident; margins shortly recurved; lower surface openly pubescent with prominent penninervation. **Conflorescence** 2–6.5 cm long, erect to decurved, pedunculate, terminal, unbranched, secund, dense; peduncle bracteate, tomentose to villous; rachis villous; bracts 2–3 mm long, ovate to obovate, tomentose outside, persistent to anthesis. **Flower colour:** perianth green turning pale pink; style greenish pink turning dark pink. **Flowers** acroscopic: pedicels 1.5–2.5 mm long, tomentose; torus 1–1.5 mm across, ± straight; nectary U-shaped, toothed; **perianth** 5–6 mm long, 2 mm wide, narrowly ovoid, tomentose outside, glabrous inside, cohering except along dorsal suture before anthesis; limb revolute, villous, globular to spheroidal, enclosing style end before anthesis; **pistil** 12–16.5 mm long; stipe 1–1.6 mm long, silky; ovary appressed-villous; style glabrous except basal hairs, at first exserted below curve on dorsal side, slightly refracted from line of stipe, incurved after anthesis; style end scarcely dilated, hooked, hoof-like; pollen presenter oblique, oblong to round, convex. **Fruit** 8.5–10 mm long, 5–6 mm wide, oblique on incurved stipe, oblong/ellipsoidal, tomentose with reddish stripes or blotches, longitudinally ridged; pericarp c. 0.5 mm thick. **Seed** not seen.

Distribution Vic., confined to the Enfield–Smythesdale area in the central-west. *Climate* Warm, dry summer; cold, wet winter. Rainfall 650–750 mm.

Ecology Grows in eucalypt woodland on undulating hills in rocky to sandy clay. Flowers late spring. Probably pollinated by birds.

Variation A fairly uniform species with only minor variations in habit and foliage.

Major distinguishing features Leaves toothed with the undersurface openly pubescent; conflorescences secund, the rachis with a spreading indumentum, the flowers changing from green to pink at anthesis such that the conflorescence appears two-coloured;

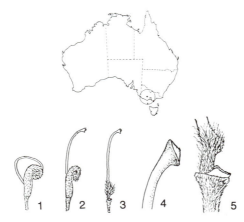

G. bedggoodiana

floral bracts 2–3 mm long, often persistent; perianth zygomorphic, glabrous inside; ovary densely hairy, shortly stipitate; pistil 12–16.5 mm long; fruit on incurved stipe with reddish blotches.

Related or confusing species Group 35, especially *G. aquifolium*, *G. obtecta* and *G. infecunda*. *G. aquifolium*, of which *G. bedggoodiana* was once regarded as a form, has a straighter, longer pistil (21–26 mm long). *G. obtecta* has conspicuous bracts 5–10 mm long. *G. infecunda* has a pistil 18–26 mm long, poorly developed anthers and a silky indumentum on the undersurface of the leaves.

Conservation status 2RC. Extremely limited in its distribution and found mostly within the Enfield State Forest. This area was recently proposed for clear-felling and planting with pine but fortunately this was rejected. The area is auriferous and is presently covered by a gold-mining lease. It should be given permanent protection as a conservation reserve.

Cultivation G. bedggoodiana has been in cultivation in Vic. for many years where it has proved fairly hardy and reasonably adaptable. It can be grown in cool to cold-wet and cool to warm-dry climates and has also succeeded in summer-rainfall climates such as Sydney in areas where humidity is not excessive. Dry summers occur in the wild but, in cultivation, it succumbs in extended dry conditions unless there is plentiful sub-soil moisture. It is hardy to frosts to at least -6°C. When planted in a sunny to semi-shaded site in well-drained, acidic, sandy or gravelly clay or loam, it performs well but at times is very slow-growing. It responds to annual top dressings of slow-release fertiliser as well as additional watering during hot, dry weather. When grown in a medium to large tub, it makes an attractive trailing specimen especially when planted in association with an upright, contrasting shrub. Specialist nurseries in Vic. sometimes stock this species as Grevillea 'Enfield'.

Propagation *Seed* Sets few seed but these germinate quite well if peeled before sowing. *Cutting* Firm, young growth taken at most times of the year normally strikes readily. *Grafting* Untested.

Horticultural features In the wild, *G. bedggoodiana* can be a very sparse, poor-looking plant. When grown in good garden conditions, it is transformed into an attractive, dense ground-cover, often with distinctive, ascending branches around the edge. The small, toothed leaves are attractive and the bright green foliage provides a nice contrast to the unusual bi-coloured conflorescences which are very popular with nectar-feeding birds. In the right conditions, it is a long-lived species in cultivation and makes an excellent ground-cover or rockery plant especially in cooler climates. More testing is needed to give a better guide to performance in a wide range of conditions.

Grevillea benthamiana D.J.McGillivray (1986)   Plate 44

New Names in Grevillea *(Proteaceae)* 3 (Australia)

Fergusson River Grevillea

Named in memory of George Bentham (1800–1884), brilliant British botanist who, although he never visited Australia, was the author of the first comprehensive flora of Australia, *Flora Australiensis* (1863-1878). BEN-THAM-EE-ARE-NA

Type: 1 km N of Stuart Hwy bridge over Fergusson R., N.T., 15 July 1978, D.J.McGillivray 3922 (holotype: NSW).

Erect lignotuberous **shrub** 2–4 m high, usually with 1 dominant stem. **Branchlets** angular, villous. **Leaves** 2.5–5 cm long, 3–4.5 cm wide, spreading, petiolate, ± ovate in outline, usually twice-, sometimes three-times pinnatipartite into many, short, fine, pungent, ± linear lobes; leaf rachis recurved; upper surface usually tomentose, ultimately glabrous with faint midvein; margins revolute to midvein below; lower surface obscured, villous in the two grooves, midvein prominent. **Conflorescence** erect, pedunculate, terminal on leafless floral branches beyond

43A. *G. bedggoodiana* Enfield State Forest, Vic. (N.Marriott)

43B. *G. bedggoodiana* Typical bi-colour conflorescences (N.Marriott)

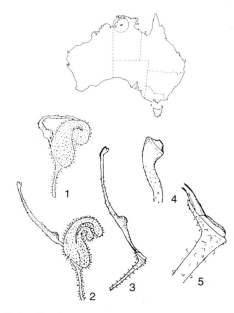

G. benthamiana

44A. *G. benthamiana* Close-up of conflorescence (W.Payne)

foliage, rarely in upper axils, simple or branched; unit conflorescence 8–10 cm long, 3.5 cm wide, conico-cylindrical, dense; peduncle and rachis glandular-pubescent; bracts 2.3–3.8 mm long, triangular, glandular-pubescent, falling before anthesis. **Flower colour:** perianth dark red or dark purple, rapidly turning black; style pinkish red. **Flowers** acroscopic, glandular-pubescent outside; pedicels 6.5–7 mm long; torus 2–3 mm wide, straight to slightly oblique, extending markedly beyond dorsal side of pedicel; nectary long-U shaped, smooth, cup-shaped; **perianth** 6–7 mm long, 3–4 mm wide, broadly ovoid, slightly depressed just above dilation, strongly curved, narrowing at neck, inner surface sparsely villous, cohering except along dorsal suture; limb revolute, globular to spherical, completely enclosing style end before anthesis; **pistil** 14.5–15.5 mm long; stipe 5.1–5.8 mm long, lateral and adnate to torus and ± perpendicular to pedicel; ovary glabrous; style minutely glandular-pubescent, glabrous near style end, at first exserted near base on dorsal side, straight after anthesis, very oblique to perianth; pollen presenter oblique, elliptic, convex with basal collar. **Fruit** 16–19 mm long, 13–15 mm wide, oblique on pedicel, round or obovoid, glabrous, wrinkled; pericarp 1.5–2 mm thick. **Seed** 8–10 mm long, 7–8 mm wide, ovate to round, flat, winged all round.

Distribution N.T., known from 10 populations between Pine Creek and Daly River. *Climate* Hot, wet summer; warm, dry winter. Rainfall 800–1200 mm.

Ecology Grows in sandy, rocky soil or stony ridges in low, open eucalypt woodland. Flowers winter. Nectar-feeding birds have been seen in attendance.

Variation A uniform species. Yellow-flowered plants have been collected.

Major distinguishing features Leaves prickly, bipinnatipartite with pungent, narrowly linear lobes; conflorescence branched, the units conico-cylindrical borne on long, leafless peduncles beyond leaves; torus ± straight and extending beyond dorsal side of pedicel; perianth zygomorphic, glandular-pubescent outside, bearded inside; ovary glabrous, stipitate; stipe of ovary adnate to torus and ± at right angles or very oblique to line of pedicel; seed winged all round.

Related or confusing species This distinctive species is rarely confused or misidentified. It appears to be most closely related to species in Group 10.

Conservation status 3R. Does not occur in any reserve.

Cultivation *G. benthamiana* is grown mainly in Darwin but attempts have begun at cultivating it in and around Brisbane where grafted plants have been successfully established. It should be suitable for tropical or subtropical climates in sandy soil where it receives summer rain and relatively dry, warm to hot winters. Although it has proved difficult to establish in subtropical and temperate climates, more testing is required over a wider range of conditions. A few nurseries in Darwin sometimes list this species in their catalogues.

Propagation *Seed* Plants have been successfully germinated from seed using standard peeling treatment but have damped off in cool weather. *Cutting* May be more amenable in warmer climates but has so far shown some reluctance to strike from cutting in cooler climates where leaf drop is a problem. *Grafting* Has been grafted successfully onto *G. robusta* using the mummy method.

Horticultural features In the wild, *G. benthamiana*

44B. *G. benthamiana* Rare pink and yellowish-cream flower form, N.T. (G.Wightman)

44C. *G. benthamiana* Close-up of flowers and foliage (A.Fairley)

44D. *G. benthamiana* Plants in natural habitat, Fergusson River, N.T. (A. & L. Murray)

45. *G. berryana* Close-up of conflorescences (K.Atkins)

is a spindly, open small tree with lignotuber from which new growth springs following fires that may occur annually, in the dry season. It has considerable potential for cultivation, being valued for both its hairy, grey leaves and its large, showy conflorescences. The uniquely coloured flowers are clustered on long, stalked racemes which arise at odd angles from the upper branchlets. After flowering, large, round, pale green fruits adorn the rachis. The spectacular and unusual features which typify this plant suggest it as a specimen plant for tropical and possibly sub-tropical gardens. Appropriate garden conditions would undoubtedly improve the appearance of the plant.

Grevillea berryana A.J.Ewart & Jean White (1909) **Plate 45**

Proceedings of the Royal Society of Victoria 22(1): 14–15 (Australia)

Named after Professor R. Berry (1867–1962), Professor of Anatomy at Melbourne University. BERRY-ARE-NA

Type: Malcolm, W.A., Dec. 1907, F.A.Rodway 321 (holotype: MEL)

A small, rough-barked, suckering **tree** or silvery **shrub** 2–7 m tall. **Branchlets** terete, silky. **Leaves** 6–27 cm long, spreading to ascending, sometimes weeping down, sinuate, petiolate, either simple and entire, tripartite or pinnatipartite with broad sinuses; leaves or leaf lobes narrow-linear; upper surface silky, soon glabrous, midvein inconspicuous or ingrooved; margins revolute to the midvein below; lower surface obscured except at the sinuses of divided leaves, silky in the two grooves, broad midvein prominent. **Conflorescence** erect, terminal, 2–5 branched; unit conflorescence 2–6 cm long, 1 cm wide, cylindrical, dense; peduncle and rachis silky with occasional glandular hairs; bracts 1–3 mm long, ovate to triangular, glandular tomentose outside, falling before anthesis. **Flower colour**: perianth pale cream to yellow; style creamy white. **Flowers** horizontal with the ventral suture directed away from the rachis; pedicels 1.2–2 mm long, openly villous; torus 1–1.5 mm across, straight to slightly oblique; nectary U-shaped, entire; **perianth** 3–5 mm long, 1–1.5 mm wide, oblong with slight basal dilation, strongly curved, sparsely villous to almost glabrous outside, glabrous inside, the ventral suture directed at right angles to rachis, the tepals separating below limb before anthesis, reflexing and everting at or after anthesis; limb nodding to revolute, globular, densely silky, enclosing style end before anthesis; **pistil** 10.5–13 mm long; stipe 1.3–2.8 mm long, glabrous; ovary sprinkled with spreading, easily detached hairs; style glabrous except a few hairs at the base, strongly curved; style end scarcely dilated; pollen presenter oblique, oblong-elliptic, conical. **Fruit** 10–17 mm long, 9–12 mm wide, sometimes persistent, oblique, flattened, ovoid or round, apiculate, glabrous, rugulose; pericarp 1–1.5 mm thick. **Seed** 5–6.5 mm long, 4 mm wide, ovate-apiculate, membranously winged all round with raphe conspicuous, the surfaces flat to slightly convex, smooth but dull.

Distribution W.A., widespread from near Mt Magnet N to the lower Fortescue R. and almost to the coast, E to the Gibson Desert and S to Menzies. *Climate* Hot, dry summer; warm to mild, winter. Rainfall irregular, 180–300 mm.

Ecology Found in a wide variety of habitats including rocky hills, open spinifex plains and *Eremophila* association and in *Acacia aneura* (Mulga) shrubland, usually in lateritic red clays. Flowers summer. Insect-pollinated.

Variation A uniform species.

Major distinguishing features Leaves dorsiventral, double-grooved below, the margins revolute but undersurface exposed at sinuses; perianth zygomorphic, the limb densely hairy, glabrous inside; ovary glabrous or sparsely hairy, stipitate; pollen presenter an oblique cone; fruit compressed; seed winged all round.

Related or confusing species Group 3, especially *G. nematophylla* which differs in its smooth, silvery, adult bark and leaves without apparent midvein (upper and lower surface appear the same). The outer surface of the perianth (usually) and the ovary (always) are glabrous. *G. berryana* never has terete leaves.

Conservation status Not presently endangered.

Cultivation *G. berryana* has been cultivated rarely, although its wide distribution in a dry region suggests a certain toughness. In these areas, it tolerates heavy frosts and extended drought. Its reaction to summer humidity is unknown, but potted specimens have survived for many years in Sydney. It is unlikely to grow well in extended cold or wet conditions. The

G. berryana

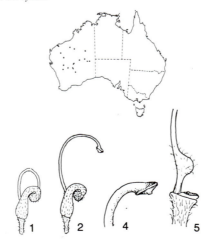

natural habitat suggests that it is suited to a well-drained, gravelly loam or clay soil in full sun. In pots it has been grown in an open, free-draining potting mix with slow-release, low-phosphorus fertiliser. Although unavailable from nurseries, seed merchants sometimes list it.

Propagation *Seed* Sets good quantities of seed that germinate well using standard treatments such as peeling or soaking. *Cuttings* Untested to date. *Grafting* Has been grafted successfully onto *G. robusta* using cotyledon grafts, and G. 'Poorinda Royal Mantle' using top wedge grafts.

Horticultural features *G. berryana* has a number of attractive features that make it worth consideration for cultivation, especially in inhospitable areas of low rainfall. An unusual shrub or small tree with dark, rough, fissured bark, it has roots from which suckers arise and form new plants, often at quite a distance from the parent plant. This results in a small community of scattered trees. The leaves of young plants are finely divided and silvery, those of adults dark green. In summer, the tree sports conspicuous panicles of creamy-white, scented flowers, attracting many insects.

Grevillea biformis C.F.Meisner (1845)

in J.G.C.Lehmann (ed.) *Plantae Preissianae* 2: 258 (Germany)

The specific epithet refers to the two leaf shapes present on the holotype, linear leaves on the floral branches and obovate leaves on the vegetative material. It is derived from the Latin bi (two) and forma (a shape). BY-FOR-MISS

Type: south-western W.A., 1844, J.Drummond coll. 3 (lectotype: NY) (McGillivray 1993).

Erect **shrub** 1.5–2.5 m high. **Branchlets** silky, terete. **Leaves** (juvenile) broadly (to 22 mm wide) to narrowly obovate; upper surface glabrous to silky, lower surface silky with prominent midvein and reticulum; juvenile silky leaves sometimes persisting on adult growth; adult leaves usually unifacial, occasionally bifacial (subsp. *cymbiformis*) 5–16 cm long, 1–2.5(–12) mm wide, ascending to erect, sessile, simple, linear, subterete, or obovate, straight to strongly incurved, acute to obtuse with short, blunt point; margins entire, flat; venation parallel-ribbed, the ribs silky and obscure or glabrous and prominent with silky grooves, ribs on one surface 3–5 and unequal in width, on the other up to 10 and equal in width. **Conflorescence** erect, pedunculate, terminal or usually in upper axil, simple or branched; unit conflorescence (2.5–)4–13 cm long, 0.8–1.5 cm wide, cylindrical; flowers opening ± synchronously, sometimes from apex first; peduncle silky; rachis silver-silky, usually brown (sometimes yellow) beneath indumentum, occasionally sparsely silky or glabrous; bracts 1.2–1.5 mm long, ovate-acuminate to caudate, silky-villous with white and red hairs, falling early. **Flowers** regular, glabrous, adaxially oriented; pedicel 1.2–4 mm long, the apex 4-lobed; torus c. 1 mm across, straight; nectary absent, sometimes present; **perianth** (below limb) 2–3 mm long, 0.5–1 mm wide, oblong; tepals channelled inside, with obscure to prominently exserted V-shaped labia below anthers, before anthesis separating to base and bowed out below the cohering limb, afterwards all free; limb 2 mm long, ellipsoidal, the segments flanged at margins, ribbed at base; distance from base of anthers to narrowed section of tepal 0.6–1 mm; **pistil** 5.5–8.5 mm long, glabrous; stipe 0.5–1 mm long; ovary 1 mm long, lateral, obovoid; style 4–6 mm long, 0.25–0.4 mm thick, smooth or granular, deeply grooved, before anthesis exserted near base and folded outwards, afterwards noticeably kinked at middle, becoming straight to undulate; style end 0.7–1 mm across, straight; pollen presenter 1.2–1.8 mm long, narrowly conico-cylindrical with cupped stigma. **Fruit** 7–13 mm long, 2.5–4 mm wide, oblique, narrowly obovoidal to obovoid, smooth, often persistent; pericarp 0.2–0.5 mm thick. **Seed** 4–8.5 mm long, 1.2–2 mm wide, obovoid or narrowly so, rugose to smooth; apex and base with a membranous band 0.5–1 mm wide; outer face convex; inner face flat.

Major distinguishing features Leaves simple, entire, usually unifacial, narrowly linear to terete, ribbed, rarely when juvenile or in subsp. *cymbiformis*, bifacial with upper surface glabrous, lower surface silky; conflorescence cylindrical; nectary absent; rachis usually silky, brown; bracts 1–1.5 mm long; flowers glabrous, regular, ovary shortly stipitate; pollen presenter conico-cylindrical; fruit narrowly obovoid to obovoid.

Related or confusing species Group 2, especially *G. ceratocarpa*, *G. eremophila* and *G. incurva*. *G. ceratocarpa* differs in its longer, narrower fruit, spreading foliar and branchlet indumentum, longer floral bracts and lobed ovary. *G. eremophila* differs similarly except that it shares an appressed foliar and leaf indumentum with *G. biformis*. *G. incurva* differs in its shorter, subterete leaves.

Variation Two subspecies are recognised.

Key to subspecies

Leaves obovate with glabrous, smooth upper surface and silky lower surface; fruit 3–4 mm wide .... subsp. **cymbiformis**

Leaves narrowly linear, both sides silky with glabrous to silky ribs; fruit 2.5–3 mm wide .... subsp. **biformis**

Grevillea biformis C.F.Meisner subsp. biformis   Plate 46

Leaves (juvenile) broadly (up to 22 mm wide) to narrowly obovate, the upper surface glabrous to sparsely silky, the lower surface silky and with prominent midvein and reticulum, juvenile, silky leaves sometimes persisting on adult growth; adult leaves usually unifacial, occasionally bifacial, 5–16 cm long, 1–2.5(–12) mm wide, linear, straight or slightly incurved; margins flat; venation longitudinally parallel-ribbed, the ribs silky and obscure or glabrous and prominent with silky grooves, ribs on one surface 3–5 and unequal in width, on the other up to 10 and equal in width. Unit **conflorescence** (2.5–)4–13 cm long; rachis silver-silky, usually brown (sometimes yellow)

46A. *G. biformis* subsp. *biformis* Pink-flowered form, N of Gingin, W.A. (F. & N. Johnston)

46B. *G. biformis* subsp. *biformis* Northern form, near Kalbarri NP, W.A. (N.Marriott)

G. biformis subsp. *biformis*

beneath indumentum, occasionally sparsely silky or glabrous; pedicels 1.2–4 mm long. **Flower colour**: perianth and style creamy white, rarely the perianth pink. **Flowers**: **perianth** (below limb) 2–3 mm long, 0.5–1 mm wide; distance from base of anthers to narrowed section of tepal 0.6–1 mm; limb 2 mm long, ellipsoidal, the segments ribbed at base; **pistil** 5.5–8.5 mm long; stipe 0.5–1 mm long; ovary 1 mm long; style 4–6 mm long, 0.25–0.4 mm thick, scarcely longer than perianth; style end 0.7–1 mm across; pollen presenter 1.2–1.8 mm long, narrowly conico-cylindrical. **Fruit** 2.5–3(–4) mm wide.

Synonyms: *G. integrifolia* subsp. *biformis* (C.F.Meisner) D.J.McGillivray (1986); *G. stenocarpa* G.Bentham (1870) nom. illeg.

Distribution W.A. (south-west), widespread in the northern, central and southern wheatbelt regions. **Climate** Hot, dry summer; mild, wet winter. Rainfall 300–500 mm.

Ecology Occurs in open sandplain to scrub and mallee country, often in sandy loam, also in gravelly sand, sand or clay loam, occasionally in poorly drained sites. Flowers spring–summer.

Variation Two forms are recognised. These warrant further investigation to establish whether sufficient discontinuities are present to provide a basis for formal recognition.

Northern (typical) form This form has long, linear leaves (usually 7–13 cm long) yellowish green in dried specimens, which are multi-ribbed (up to 10 on one surface), the ribs glabrous. Conflorescence is usually longer (7–13 cm long), wider (1–1.5 cm wide); pedicels 2.5–4; style granular. It is found from the Murchison R. south to Wongan Hills.

Southern form Very similar to the Northern form except that leaves are silky all over, on average slightly narrower, and less noticeably ribbed. In addition, the unit conflorescence is usually shorter (3–6 cm long), narrower (0.8–1 cm wide) and more enclosed in the foliage; pedicels 1.2–2 mm long; style smooth to granular. It is found in the Lake Grace area, but is also known from Bendering, Hyden, and Kulin.

Conservation status Not presently endangered.

Cultivation Subsp. *biformis* has not been cultivated widely to date although it has been grown successfully at Kings Park, W.A., in the National Botanic Gardens, Canberra, and in a few private gardens in inland Vic. It is hardy to extended, dry conditions and frosts to at least -5°C, depending on provenance. It prefers a mediterranean-type climate and has not adapted to summer rainfall climates. Some populations grow in poorly drained sites and if these forms could be selected there should be a far greater chance of success in cultivating this reliably. It performs best in full sun but sheltered, in deep, well-drained acidic to neutral sand, sandy loam or gravelly loam. Fertiliser and summer watering are resented once established and it rarely requires pruning. Although not usually available at nurseries, seed can sometimes be purchased from specialist suppliers, usually as *G. biformis*.

Propagation *Seed* Germinates readily if pre-treated by nicking. Seedling leaves are broad, gradually progressing to narrow, adult leaves. Average germination time 37 days (Kullman). *Cutting* Firm, young growth taken during the warmer months generally strikes reasonably well. *Grafting* Has been grafted successfully on to a number of rootstocks including *G. robusta*, G. 'Poorinda Royal Mantle' and G. 'Poorinda Anticipation' using both top wedge and side-graft techniques.

Horticultural features Subsp. *biformis* is a medium to large, generally compact shrub that produces a spectacular display of cream or yellow flower spikes. Pink-flowered selections from near Moora are also very desirable and attempts should continue to introduce it to cultivation. In the wild, it is a long-lived species and the dual foliage creates interest. Flowers are attractively perfumed. The erect habit indicates use as a screen, feature or foliage contrast plant in the home landscape.

Grevillea biformis subsp. cymbiformis
P.M.Olde & N.R.Marriott (1994)   **Plate 47**

The Grevillea Book 1: 176 (Australia)

The subspecific epithet is derived from the Latin *cymba* (a boat) and *-formis*, (-shaped), in reference to the leaf shape. SIM-BIF-FOR-MISS

Type: Erindoon Rd, 11.4 km S of Eneabba–Coolimba Rd, S of Eneabba, W.A., 15 Sept. 1991, P.M.Olde 91/103 (holotype: NSW).

Juvenile **leaves** not seen; adult leaves bifacial, 3–9 cm long, 1–12 mm wide, usually obovate with concave, glabrous, smooth upper surface and silky lower surface with prominent midvein, sometimes narrowly linear with a distinct narrow deeply cleft line of separation down middle of adaxial surface, otherwise ribbed as for subsp. *biformis* (Northern form); margins flat. Unit **conflorescence** 8–13 cm long; rachis silver-silky, usually brown beneath indumentum, occasionally sparsely silky; pedicels 3–4.5 mm long. **Flower colour**: perianth and style creamy white. **Flowers**: nectary obscure, erect, horn-like; **perianth** (below limb) 3 mm long, 0.5–1 mm wide, on the inside lamina divided below anthers into an outer flange-like section and an inner, prominently raised labia-like section c. 1 mm long; limb 2.2 mm long, ellipsoidal, the segments not ribbed at base; **pistil** 7.5–8 mm long; stipe 0.5–1 mm long; ovary 1 mm long; style 5.5 mm long, 0.4 mm thick, granular on inner surface; style end 0.7 mm across; pollen presenter 1 mm long, narrowly conico-cylindrical. **Fruit** 10–12 mm long, 3–4 mm wide, obovoid, rugulose.

Distribution W.A., confined to a small area SSW of Eneabba. **Climate** Summer hot, dry; winter cool, wet. Rainfall 500 mm.

Ecology Grows in grey or white sand over laterite in low heath, often dominant. Flowers spring–summer. Pollinated by insects. Regenerates from seed.

Variation The presence of linear-leaved plants in some populations suggests the possibility that subsp. *cymbiformis* may be a clinal variant of subsp. *biformis* with abnormal retention of juvenile foliage in adult plants. They otherwise differ in the presence of a small nectary and the larger fruit of subsp. *cymbiformis*.

Conservation status Recommended 2E.

Cultivation As for subsp. *biformis*

Propagation As for subsp. *biformis*

Horticultural features The unusual, boat-shaped, green leaves make this an interesting acquisition for the horticulturist. The conflorescences are long and conspicuous and the white flowers are sweetly scented. It would make an excellent specimen or low screen plant but further trials are warranted before recommendations can be made.

G. biformis subsp. *cymbiformis*

47. *G. biformis* subsp. *cymbiformis* Flowers and foliage (P.Olde)

Grevillea bipinnatifida R.Brown (1830)
Plate 48

Supplementum Primum Prodomi Florae Novae Hollandiae exhibens Proteaceas Novas: 23 (England)

Fuchsia Grevillea, Grape Grevillea

Epithet derived from the Latin *bi-* (twice-), *pinnatus* (feathered) and *-fidus* (divided, usually in the outer third), in reference to the bipinnatifid leaves. BY-PINNA-TIFF-ID-A

Type: The original collection (near the Swan R., [W.A.], 1827, C.Fraser) is missing. Neotype: Swan View, W.A., C.A.Gardner, Dec. 1926 (PERTH). (McGillivray, 1993).

Synonyms: *G. bipinnatifida* var. *glabrata* C.F.Meisner (1845); *G. bipinnatifida* var. *vulgaris* C.F.Meisner (1845).

A spreading, dome-shaped **shrub**, 30 cm to 2 m high, sometimes prostrate, often lignotuberous. **Branchlets** angular, ridged, sometimes terete, tomentose to glabrous. **Leaves** 4–15 cm long, 2.5–11 cm wide, ascending, sessile or petiolate, rigid, often strongly wrinkled, ovate in outline, pinnatifid or bipinnatifid; leaf lobes 0.3–1 cm wide, oblong or triangular, pungent; upper surface glabrous, sometimes glaucous, sometimes sparsely pubescent with pubescent midvein; margins flat or shortly recurved; lower surface glabrous and glaucous or appressed-pubescent, venation with midvein and lateral veins conspicuous on undersurface, inconspicuous above. **Conflorescence** decurved to erect, pedunculate, terminal, simple (sometimes few-branched); unit conflorescence 4–20 cm long, 4.5–5 cm wide, conspicuous, secund, lax, often pendulous; peduncle and rachis tomentose; bracts 4–5.5 mm long, ovate to triangular, villous outside, falling before anthesis. **Flower colour**: perianth dull red, pink, pale green, pale orange or grey with red stripes; limb orange or red; style red. **Flowers** acroscopic; pedicels 5–17 mm long, tomentose to villous; torus 1.5–2 mm wide, ± straight; nectary conspicuous, linguiform; **perianth** 10–12 mm long, 3.5–6 mm wide, oblong-ovoid, dilated at base, strongly curved, ribbed, tomentose outside, glabrous inside, cohering except along dorsal suture and sometimes just below curve; limb very conspicuous, densely pubescent, globular to spheroidal, revolute and often displaced laterally, enclosing style end before anthesis; **pistil** 34–42 mm long; ovary sessile, glandular pubescent; style elongate, exserted just below curve and looped upwards and inwards, slightly infracted above ovary, gently incurved after anthesis, pubescent to silky or sparsely so in lower half, dilating suddenly and evenly into the glabrous style end; pollen presenter very oblique, oblong-obovate, convex. **Fruit** 17–21 mm long, 9–10 mm wide, very oblique, oblong, glandular-pubescent, with reddish stripes and blotches; pericarp c. 1 mm thick. **Seed** 13–15 mm long, 4.5 mm wide, pubescent, ellipsoidal but lower side straight, narrowly winged all round except broader at apex; outer face convex, slightly compressed and flange-like at margin; inner face flat with a central, black elliptic area; margin revolute on straight side, otherwise shortly recurved.

Distribution W.A.; from near Cataby to Collie, usually near or on the scarp of the Darling Ra. *Climate* Hot, dry summer; cool, wet winter. Rainfall 600–1000 mm.

Ecology Grows in moist sand, gravel, sandy- or clay-loam and granitic loam, in open eucalypt forest, open damp heath, and open woodland. Flowers winter to summer. Bird-pollinated. Regenerates mainly from lignotuber after fire.

Variation There is considerable variation in this species in habit, leaf division and lobe width, indumentum and flower colour. Some of the variation is listed below.

Green-leaved (type) form The typical form is found widely over the whole range of the species. It has broad, bright green leaves usually with broad lobes. In some plants the leaves are stiff and quite crinkly. Flowers are either orange-red or dull red. Some populations have densely red-pubescent new growth and very hairy, dark red flowers.

Prostrate green-leaved form This form is found in low-lying, swampy terrain both north (around Bullsbrook) and south (around Harvey) of Perth. It is almost prostrate with leaves deeply dissected and lobes both widely spaced and narrow (to 0.5 cm). It grows in mats up to 1 m wide. Flowering is sparse and usually quite dull.

Glaucous form This variety typically grows to 2 m high and 3 m wide with large, glabrous and glaucous, blue-grey leaves, striate or angular branchlets and long, showy, red to dull-orange conflorescences. The outermost leaf segments are often pinnatifid rather than twice-divided and the lobes are quite large. Sometimes flowers are rather insipid, having a grey perianth with red striping and pale red style. It usually occurs in scattered populations in the hills of the Darling Ra., growing in heavy, gravelly clay. Although usually found in pure populations, plants are also seen scattered among populations of the Green-leaved form without exhibiting any particular habitat preference.

Prostrate glaucous form This form usually occurs in sandy, gravelly areas in the northern part of the range between Bullsbrook and Northam. Leaves are usually smaller and flowers a dull orange-red. In this form, habit varies from prostrate, with branches running close to the ground, to decumbent and mounding to c. 50 cm.

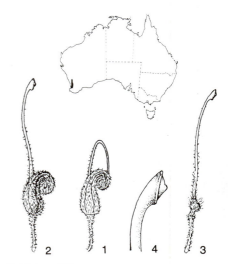

G. bipinnatifida

Major distinguishing features Leaves dorsi-ventral, usually bipinnatifid or sometimes the primary lobes cleft almost to the midvein; conflorescences 4–15 cm long, secund, lax, usually pendulous; pedicels > 5 mm long; perianth zygomorphic, glandular-hairy outside, glabrous inside; ovary sessile, densely hairy; fruits with reddish markings.

Related or confusing species Group 35. This distinct species is not usually confused with other species. Its closest relatives appear to be *G. asparagoides*, *G. batrachioides* and *G. maxwellii* but none of these has bipinnatifid leaves.

Conservation status Not presently endangered.

Cultivation *G. bipinnatifida* is a resilient, adaptable species that will grow in a wide range of conditions from cool-wet to warm-dry and even summer-rainfall climates. It is hardy to extended dry conditions in heavy soils provided there is a ready supply of subsoil moisture, but not in sandy soils that dry out too deeply. Frosts to at least -6°C have been recorded without effect on its foliage. Today,

48A. *G. bipinnatifida* Glaucous form, Chittering Valley, W.A. (N.Marriott)

48B. *G. bipinnatifida* Decumbent habit of a glaucous form, near New Norcia, W.A. (M.Hodge)

48C. *G. bipinnatifida* Prostrate green leaf form, Bullsbrook, W.A. (N.Marriott)

48D.*G. bipinnatifida* Conflorescence (A.Cavanagh)

it is widely grown in Australian gardens and is also grown in the U.S.A. Warm, sunny or semi-shaded situations in well-drained loam, sandy loam or heavy clay, in acidic to neutral soils produce long-lived, healthy plants. Summer watering and low-phosphorus fertiliser applied sparingly also enhance its appearance. Plants occasionally become leggy and open with age but respond vigorously to very heavy pruning with new growth arising from the lignotuber as well as old branches. It is an attractive pot plant when grown in larger tubs using a well-drained, nutritious potting mix and is obtainable at most native plant nurseries, particularly in W.A. where several forms are sold.

Propagation Seed Sets copious quantities of seed that germinate readily. However, as it can hybridise readily with other species, resultant seedlings may not be pure. *Cuttings* Strikes well from cuttings using firm, young growth taken in spring or autumn. Some propagators prefer to use soft tip growth which also gives good results in the right conditions. Due to their large size, the leaves often require trimming down to one or two lobes in order to reduce the evaporative surface and take up less space in the propagating frame. *Grafting* It has been successfully grafted onto *G. robusta*, G. 'Poorinda Royal Mantle', G. 'Ivanhoe', and G. 'Poorinda Anticipation' mainly using top wedge or approach grafts.

Horticultural features A first-rate grevillea in every sense. The compact, dense habit with shapely, robust foliage resplendent with long, pendulous racemes combine to make a superior plant for horticulture. The strongly bird-attractive flowers are produced abundantly over a long period and complement the differing foliage colours. The large fruits are attractive to parrots. Some plants have striking red new growth and interesting buds. There are a number of foliage forms and flower colours, all of which have merit and are useful in landscaping both as feature plants and for foliage contrast. This reliable, long-lived species is an important cog in the hybrid industry and the differing forms should be tested with compatible species to extend the range of plants available to gardeners.

Hybrids No natural hybrids have been recorded but it is one parent of a number of garden hybrids including G. 'Robyn Gordon', G. 'Superb', G. 'Boongala Spinebill', G. 'Mason's Hybrid' (syn. G. 'Ned Kelly') and G. 'Clare Dee'.

Early history *G. bipinnatifida* was grown in England as early as 1837, probably from the collections of Baron Karl von Hügel who visited Australia from November 1833 to October 1834. It was grown in his garden in Vienna, and later in the famous garden of Prince de Demidoff at San Donato, Italy, to which Hügel's collection was transferred.

Grevillea biternata C.F.Meisner (1845)
Plate 49

in J.G.C.Lehmann [ed.] *Plantae Preissianae* 1: 549 (Germany)

Specific epithet derived from the Latin *bi-* (twice-) and *ternatus* (in threes), in reference to the foliage of the type specimen which is twice three-forked. BY-TER-NAR-TA

Type: Swan R. colony, W.A., 184-, J.Drummond coll. 1 no. 624 (lectotype: K) (McGillivray 1993).

Synonym: *G. biternata* var. *leptostachya* G.Bentham (1870).

Suckering **shrub** 1.5–2.5 m high. **Branchlets** softly tomentose-pubescent, sometimes sparsely so to almost glabrous. **Leaves** 1–8.5 cm long, sessile or shortly petiolate but appearing prominently stalked, ascending to spreading, divaricately divided, tripartite to biternate, sometimes simple and entire; simple leaves and lobes 0.8–7 cm long, 1–2.8 mm wide, narrowly linear, pungent; upper surface tomentose to glabrous, smooth or granulate, often channelled beside midvein; lower surface bisulcate, midvein prominent; margins revolute, enclosing undersurface. **Conflorescence** erect, sessile or shortly pedunculate, usually axillary, sometimes terminal, simple or few-branched; young conflorescence with imbricate bracts usually evident for long period; unit conflorescence globose to subcylindrical, open; peduncle tomentose; rachis tomentose to villous; bracts 1–2.5 mm long, linear to narrowly ovate, tomentose, villous or woolly, persistent usually to anthesis. **Flower colour**: perianth and style white. **Flowers** glabrous, regular; pedicels 3.5–6 mm long, filamentous; torus c. 0.5 mm wide, straight; nectary U-shaped, scarcely evident; **perianth** 5 mm long, 0.5 mm wide, actinomorphic, oblong-obovoid below limb; tepals first separating below limb, free to base before anthesis, afterwards rolling down independently; limb erect, globose; **pistil** 3–4 mm long, glabrous; stipe c. 1 mm long; ovary globose; style constricted above ovary, dilated above, again contracted below style end; pollen presenter conical, erect. **Fruit** 7–10 mm long, 4–6 mm wide, transverse to stipe, oblong-ellipsoidal, glabrous, warty; pericarp c. 0.5 mm thick. **Seed** 5–7 mm long, 2.8–3.5 mm wide, oblong-ellipsoidal, convex on both sides, smooth to faintly rugose.

Distribution W.A., from New Norcia to North- ampton. *Climate* Hot, dry summer; cool, wet winter. Rainfall 400–600 mm.

Ecology Grows in sandy or gravelly loam in open heath and low mallee woodland. Flowers late winter–spring. Regenerates from seed or sucker after fire. Pollinated by insects, possibly native bees.

Major distinguishing features Branchlets usually tomentose-pubescent or sparsely so, rarely almost glabrous; leaves divaricately divided or simple; conflorescence usually axillary, sessile, scarcely exceeding leaf stalk; floral rachis tomentose to woolly; floral bracts tomentose to villous, usually evident on undeveloped conflorescence for a long period; flowers glabrous, regular; pistil glabrous; ovary stipitate; style contracted above ovary, then dilated and again contracted; pollen presenter conical; fruit warty.

Related or confusing species Group 1, especially *G. paniculata* which differs in its strongly wrinkled fruit, appressed-hairy or glabrous branchlets and subulate leaf lobes, and *G. triloba* which has broader leaves with prominent secondary venation on the upper surface.

Variation There is some variation between populations, most notably a form with simple leaves.

G. biternata

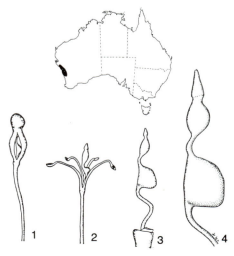

49A. *G. biternata* Flowers and foliage (P.Olde)

49B. *G. biternata* Close-up (N.Marriott)

49C. *G. biternata* Simple-leaf form from near Nanson, W.A. (P.Olde)

Southern populations (i.e. from New Norcia to Watheroo) have hairier branchlets than those further north.

Simple-leaf form This occurs near Watheroo; it usually has some divaricately divided leaves.

Conservation status Not rare, although no specimen has been collected that quite matches the type.

Cultivation *G. biternata* has been grown in western Vic. for many years. It tolerates frost to at least -4°C and is drought resistant. A well-drained, neutral to acidic loam or sandy loam in full sun is ideal. Does not need pruning except for containment in a limited space; prune after flowering. It is reliable and long-lived. Rarely stocked by nurseries under its correct name.

Propagation *Seed* Untested but sets prolific seed that should germinate in suitable conditions. *Cutting* Strikes readily from firm, young growth in spring. Wild material can be difficult to strike. *Grafting* Untested.

Horticultural features This dense, dull grey-green shrub is valuable as a contrast and screen plant in the landscape. The prickly foliage may suggest planting away from paths, but it is an ideal barrier plant. Small birds find the foliage suitable for shelter and nesting. Spring transforms it for a long period into a mass of white, sweetly scented flowers that attract many insects. An excellent feature plant.

General comments McGillivray (1993) included *G. biternata* in *G. paniculata* but we consider it to be distinct. His populations a, b, c, d, e, f, h and i are here provisionally included in *G. biternata* but further revision of this group is in hand. *G. biternata* has also long been confused with *G. curviloba* subsp. *incurva*.

Grevillea brachystachya C.F.Meisner (1848)   Plate 50

in J.G.C.Lehmann (ed.), *Plantae Preissianae* 2: 254 (Germany)

Short-spiked Grevillea

The specific name is derived from the Greek *brachys* (short) and *stachys* (a spike of flowers), in reference to the short, dome-shaped conflorescence. BRACK-ISS-TACK-EE-A

G. brachystachya

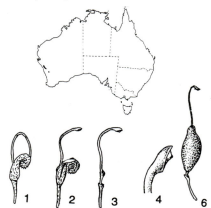

Type: south-western W.A., c. 1844, J.Drummond, coll. 2, no. 319 (holotype: NY).

A dense **shrub** to 2.5 m. **Branchlets** terete, silky, sometimes striate. **Leaves** 1.5–12 cm long, 1–2 mm wide, ascending to erect, shortly petiolate, rigid, simple, linear, pungent; upper surface usually glabrous, convex, the venation obscure or with faint longitudinal grooves; margins entire, smoothly revolute to midvein below; lower surface obscured, bisulcate, silky in the two grooves, midvein prominent and c. level with margin. **Conflorescence** erect, pedunculate, terminal, simple or few-branched; unit conflorescence 0.5–2.2 cm long, c. 2 cm wide, umbel-like; peduncle silky; rachis tomentose; bracts 1–3.5 mm long, linear-elliptic to ovate, villous, falling before anthesis. **Flower colour**: perianth cream or greenish cream; style pink to cream or green. **Flowers** adaxially oriented: pedicels 3–8.5 mm long, slender, sparsely silky, often with intermixed glandular hairs; torus c. 1.5 mm across, cup-shaped, slightly oblique; nectary not evident above toral rim, semi-circular; **perianth** 4–6 mm long, 1.5 mm wide, narrowly triangular to oblong, undilated, glabrous or sparsely glandular-pubescent outside, papillose inside below and about ovary, sometimes with an inconspicuous beard; tepals keeled, at first separated along dorsal suture, all tepals soon free below limb, the dorsal tepals reflexed and everted before anthesis, afterwards free to base; limb erect in late bud, revolute at anthesis, subcubic-globular, the apex depressed, the segments conspicuously keeled, glandular-pubescent to silky; style end visible before anthesis; **pistil** 12–14(–17) mm long, glabrous; stipe 1.5–2.5 mm long, inserted just within upper toral margin, adnate at base within torus; ovary ovoid to globose; style at first exserted near curve and looped upwards, after anthesis straight above ovary, strongly incurved in upper half; style end flanged; pollen presenter lateral or very oblique, oblong-obovoid, flat to convex, umbonate. **Fruit** 17–21 mm long, 8–11 mm wide, appearing erect, but very oblique to pedicel, granulose, ovoid with a secondary, obtuse apical attenuation 2–8 mm long, 2–3 mm wide; pericarp 0.5–1 mm thick. **Seed** (not seen, from McGillivray 1993) 14–15 mm long, 7 mm wide, oblong to obovate, marginate.

Distribution W.A., from Wongan Hills in scattered localities north beyond the Murchison R. where it is more frequent. *Climate* Hot, dry summer; cool to mild wet winter. Rainfall 200–400 mm.

Ecology In red or yellow sand or gravelly sandy loam, in open eucalypt forest, tall shrubland and occasionally in sandplain. Flowers winter to spring. Pollinator unknown, probably insects.

Variation A relatively uniform species.

Major distinguishing features Leaves simple, entire, narrowly linear, rigid, pungent, double-grooved below, the venation obscure above; conflorescence umbel-like; torus slightly oblique, cup-shaped; nectary obscure; perianth zygomorphic, hairy or papillose on both surfaces; pistil glabrous;

50A. *G. brachystachya* Habit in degraded road verge, E of Wongan Hills (N.Marriott)

50B. *G. brachystachya* Unusual fruits (P.Olde)

50C. *G. brachystachya* Flowering branch (P.Olde)

style attached at base of pollen presenter; fruit with an obtuse apical attenuation.

Related or confusing species Group 16, especially *G. hakeoides* subsp. *stenophylla* which differs in its smaller, rugose fruits lacking an obtuse apical attenuation, its more conspicuously ridged leaves, its densely silky outer perianth surface and its more conspicuous nectary.

Conservation status 3VC. Restricted to a few, small, widespread localities, often in degraded road verges.

Cultivation For many years typical *G. brachystachya* has been cultivated successfully at Glenmorgan, Qld, but rarely elsewhere. There it has grown into a robust, hardy shrub in well-drained, gravelly loam in an open sunny position. It seems to be an adaptable species and tolerates a surprising degree of cold as quite heavy frosts sometimes occur where it grows naturally. Once established, no supplementary watering or other maintenance is required unless it needs trimming to maintain shape. Able to withstand extended dry periods, it flowers prolifically and sets abundant seed in cultivation. Nurseries do not sell this species, as plants sold under this name have been misidentified in horticulture.

Propagation *Seed* Grows readily from seed if given the standard peeling treatment. *Cutting* Difficult to strike from cutting. Firm, young growth taken from cultivated plants should strike reasonably well. *Grafting* Early trials using G. 'Poorinda Royal Mantle' have failed. *G. diversifolia* might be a good rootstock for this species.

Horticultural features *G. brachystachya* is a robust shrub with ascending to erect branches and thin, dull-green leaves of medium length. Creamy-green heads of flower, redolent with a sweetly-scented perfume in the middle of the day and attended by numerous insects, moths and butterflies, cluster at the ends of the branches in spring. An additional feature is the uniquely shaped, dark fruits which dot the plant after flowering and sometimes persist all year. It is suitable as a screen plant in dry, inland areas in well-drained soils where it appears to be both long-lived and disease-free.

Hybrids For many years the name *G. brachystachya* has been used incorrectly in the horticultural trade for a hybrid grevillea of uncertain origin; this pink-flowered hybrid is a probable cross between *G. commutata* and *G. pinaster* and has been recently registered by a Perth nurseryman as G. 'Frosty Pink'.

Grevillea brachystylis C.F.Meisner (1845)

in J.G.C.Lehmann (ed.), *Plantae Preissianae* 1: 538 (Germany)

Short-styled Grevillea

The specific epithet is derived from the Greek *brachys* (short) and *stylos* (a wooden pole, in botany a style), in reference to the short style which is only partially exposed at anthesis. BRAK-EE-STILE-ISS

Type: Molloy Plain, W.A., 20 Dec. 1839, L.Preiss (lectotype: LD) (McGillivray 1993).

A much-branched, spreading or erect **shrub** 0.3–1 m tall. **Branchlets** up to 60 cm long, somewhat wiry, angular to terete, ridged, villous to glabrous. **Leaves** 1–14 cm long, 2–10 mm wide, spreading to erect, sessile, simple, linear to obovate, obtuse, mucronate; upper surface convex, finely granulate, pubescent, soon ± glabrous, midvein an impressed groove; margins entire, recurved to revolute; lower surface silky to villous, midvein prominent; texture chartaceous. **Conflorescence** deflexed, subsessile or pedunculate, terminal or axillary, simple, occasionally branched, arising in sequence from same peduncle; unit conflorescence 1–1.5 cm long, 2–2.5 cm wide, umbel- to wheel-like, 4–7-flowered; peduncle bracteate, villous; rachis villous; bracts 3–4 mm long, imbricate in bud, ovate, tomentose outside, persistent to anthesis. **Flowers** acroscopic: pedicels 2.5–5 mm long, tomentose to villous; torus 2–4 mm across, very oblique to lateral; nectary V-shaped, entire; **perianth** 6–12 mm long, 3–4 mm wide, cohering at anthesis, oblong, geniculate from c. halfway, dilated ventrally at base, ridged, sparsely tomentose outside, pubescent inside from ovary to limb, sometimes persistent to fruiting, cohering except along dorsal suture from curve to limb; limb revolute, ovoid with a point, dilated on dorsal side, densely brown tomentose; segments separating at anthesis; **pistil** 7–15 mm long; stipe 2–5 mm long, pilose; ovary villous; style stout (1–1.5 mm wide), pubescent, exposed at curve before anthesis but not exserted, afterwards straight and scarcely exceeding perianth; style end beaked; pollen presenter lateral, oblong-elliptic, concave. **Fruit** 12–17 mm long, 4–6 mm wide, erect, ovoid, tomentose; pericarp c. 0.5 mm thick. **Seed** 7 mm long, elliptic, with recurved margins, smooth both sides with an apical eliasome 2–4 mm long.

Major distinguishing features Leaves simple, entire; conflorescence umbel- to wheel-like; torus very oblique; perianth often sharply kinked from c. halfway, dilated at base, hairy inside and out; ovary hairy, stipitate; style end prominently beaked, scarcely to not exceeding perianth; pollen presenter lateral, concave.

Related or confusing species Group 23, especially *G. bronwenae* which differs principally in its undilated perianth and in the absence of an appendage or beak on the style end. *G. bronwenae* has a blue style end. See also *G. brachystylis* subsp. *australis*.

Variation Two geographically disjunct subspecies are recognised.

Key to subspecies

Leaves very thin; style end red to orange-red; perianth sparsely or densely hairy   subsp. **brachystylis**

Leaves leathery; style end blue; perianth densely hairy
      subsp. **australis**

Grevillea brachystylis C.F.Meisner subsp. brachystylis   Plate 51

Leaves very thin. **Flower colour**: perianth scarlet or orange-red with a brownish red limb; style end red to orange-red. **Perianth** sparsely or densely hairy.

Distribution W.A., on the coastal plain east of Busselton. *Climate* Mild, dry summer; cool to cold, wet winter. Rainfall 800–1100 mm.

Ecology Grows in damp to wet, grey sand and sandy or gravelly loam in heath or woodland. Flowers winter to spring. Regenerates from seed or lignotuber after fire. Pollinated by birds.

Variation There is some variation in habit, indumentum, the shape of the flowers and the length of the peduncle.

Prostrate (typical) form A much-branched, prostrate plant up to 1 m wide, becoming decumbent with age. Conflorescence usually sessile. Common, usually in low, moist heath.

Erect form Some robust plants are up to 1 m tall and have markedly pedunculate conflorescences. These are extremely showy, with prominent flowers about twice the usual size. Usually in forest.

Conservation status 2E suggested. Restricted to small populations, most in degraded road verges and disturbed sites where it is being choked by weeds.

Cultivation Subsp. *brachystylis* has been grown with limited success and has not proved hardy or adaptable except in cool, wet climates similar to its natural habitat. Frosts at -6°C will damage the foliage, although light frosts are tolerated. It has been successfully grown in gardens in western Vic. and at Burrendong Arboretum, N.S.W., and flourished for many years at Brookvale Park, Qld. Both forms are growing well at Mount Annan Botanic Garden, N.S.W. As would be expected from its natural habitat, an assured water supply is required for success, most specimens succumbing in extended, dry conditions. This is especially so in summer when hot, dry conditions can be fatal. With summer watering, this species makes a glorious display in either full shade or full sun. A cool, moist but well-drained, acidic soil and partial shade provide the optimal conditions. Annual pruning is rarely necessary. The soft, hairy foliage is prone to insect attack at times, though rarely heavily enough to threaten the plant. It can be grown well in pots, in a well-drained, acidic peaty mix with low phosphorus, slow-release fertiliser. Specialist nurseries occasionally stock this species.

Propagation *Seed* Sets good quantities of seed that germinate readily especially if given the standard peeling treatment. Fresh green seed broken out from unripened, green pods and sown immediately germinates readily. *Cutting* Strikes readily from cuttings of firm to semi-hard wood at most times of the year. *Grafting* Grafts readily onto G. 'Poorinda Royal Mantle', G. 'Poorinda Anticipation' and *G. robusta* using both top wedge and mummy grafts. Appears to be incompatible with *G. robusta*, usually dying after 12–18 months.

Horticultural features Subsp. *brachystylis* in its typical form has an interesting, mat-forming or decumbent habit with dense, dark green foliage and bright, prominent red or orange-red flowers. The erect form is by far the more showy

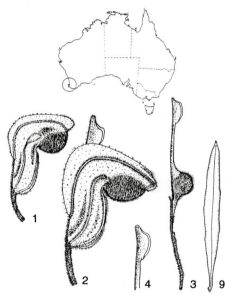

G. brachystylis subsp. *brachystylis*

51B. *G. brachystylis* subsp. *brachystylis* Flowering habit (N.Marriott)

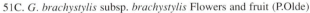

51A. *G. brachystylis* subsp. *brachystylis* in natural habitat, Vasse, W.A. (C.Woolcock)

51C. *G. brachystylis* subsp. *brachystylis* Flowers and fruit (P.Olde)

and should be more widely cultivated. Flowering commences in late winter and, in ideal conditions, will continue into late summer. Another attractive feature is its egg-shaped follicles which at first are a pretty orange-red colour similar to though not as bright as the flowers, before fading to the more usual grey-brown. The retention of dead flowers on the plant during the flowering season detracts from its appearance. If grafted onto a suitable rootstock, *G. brachystylis* would make an outstanding, spectacular ground cover or rockery plant, useful in general landscaping. On its own roots, it is short-lived and remains a plant strictly for the collector or enthusiast.

Grevillea brachystylis subsp. australis
G.J.Keighery (1990)   **Plate 52**

Nuytsia 7: 125 (Australia)

The subspecific epithet is derived from the Latin *australis* (southern), an allusion to the more southerly distribution of this taxon. OSS-TRARL-ISS

Type: Scott River Rd, Scott River NP, W.A., 29 Jan. 1988, G.J.Keighery 9711 (holotype: PERTH).

Leaves leathery. **Flower colour**: perianth dull red; style end blue-grey. **Perianth** densely hairy.

Distribution W.A., restricted to the Scott R. area in the far south-west. *Climate* Cool, wet winter; mild, dry summer. Rainfall 900–1100 mm.

Ecology Grows in winter-wet heath in sandy loam often over laterite. Flowers spring. Regenerates profusely from seed and from lignotuber. Pollinator unknown, probably birds.

Variation A subspecies with minor variation in habit.

Conservation status 2EC suggested. Known from six sites, mostly road verges, but two occur in a NP.

Cultivation Subsp. *australis* has adapted quite well to cultivation in good garden conditions. Plants have succeeded for many years in sandy loam at Mount Annan Botanic Garden, N.S.W. The species is not affected by frost or persistent cold and rain and tolerates a higher watering regime than most grevilleas, but persistent humidity is not to its liking. It does not need pruning, and once established will persist for many years in an open, full-sun situation

G. brachystylis subsp. *australis*

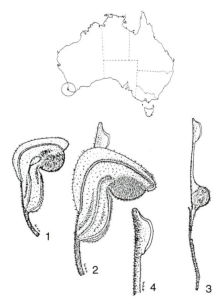

52A. *G. brachystylis* subsp. *australis* Close-up of flowers (N.Marriott)

52B. *G. brachystylis* subsp. *australis* Flowering habit (P.Olde)

but declines in vigour if crowded by other plants. In good conditions it self-sows in the garden. It is unlikely to be drought-hardy and subsoil moisture should be monitored in extended, dry conditions. Flowering may be affected in subtropical climates, although beautiful specimens were seen in cultivation near Oakey, Qld. Potted specimens need a well-drained mix with appropriate slow-release fertiliser. It is occasionally stocked in specialist native plant nurseries, especially in Vic.

Propagation Seed Germinates readily from seed, apparently with little treatment. *Cutting* Firm, young growth strikes readily. *Grafting* Both G. 'Poorinda Royal Mantle' and G. 'Poorinda Anticipation' have proven hardy rootstocks. Plants grafted onto *G. robusta* have lacked vigour.

Horticultural features Subsp. *australis* is an excellent specimen plant for the garden and is extremely eye-catching when mass-planted along the borders of large beds. Its suitability in a wide range of conditions makes it ideal in landscapes where some care is provided. The showy red flowers with their unusual blue style end always attract attention and appear over many months. It is well-suited for rockeries and small gardens and can also be used as a groundcover or grown in pots for many years.

Grevillea bracteosa C.F.Meisner (1848)
Plate 53

in J.G.C.Lehmann (ed.) *Plantae Preissianae* 2: 254 (Germany)

Bracted Grevillea

The specific epithet is derived from the Latin *bractea* (in botany, a floral bract), a reference to the prominent scaly bracts of the conflorescence. BRAK-TEE-OWE-SA

Type: south-western W.A., 1844–1845, J.Drummond coll. 3 no. 269 (holotype: NY).

An erect, rounded, spreading **shrub** 1–2 m tall. **Branchlets** almost terete, silky. **Leaves** 5–25 cm long, 1–3 mm wide, ascending to spreading, petiolate, simple, usually linear, rarely divided at the base into 2 or 3 linear segments, leathery, obtuse, mucronate to pungent; upper surface ± glabrous, or sprinkled with silky hairs, midvein broad and conspicuous; margins entire, angularly refracted, usually revolute to midvein; lower surface bisulcate, glabrous, midvein prominent. **Conflorescence** erect, pedunculate, simple or branched, terminal or axillary; unit conflorescence 1–3 cm long, 1.2–3.5 cm wide, dome-shaped to ± globular, apical flowers opening first; peduncle silky; rachis glabrous; bracts 7–14 mm long, broadly elliptic to obovate, conspicuous and overlapping in bud, papery, sparsely silky outside, glabrous inside, with densely ciliate margins, persistent at anthesis but soon falling. **Flower colour:** perianth pale green to greenish pink; style deep rose-pink, pale pink or white, often turning pale pink. **Flowers** adaxially acroscopic: pedicels 2.5–

5 mm long, glabrous; torus 1–1.5 mm across, square, straight; nectary semi-circular, not conspicuous; **perianth** 4–7 mm long, 1.5–3 mm wide, oblong, ribbed, glabrous outside, pubescent inside to level of ovary; limb revolute, scarcely dilated, spheroidal, not relaxing at anthesis; tepals ± cohering to level of ovary, the upper half reflexing laterally at anthesis to form a circular, undulate platform below level of ovary; **pistil** 16–23 mm long; stipe 5–7 mm long, glabrous, flattened, channelled; ovary globular, glabrous; style exposed to base before anthesis, refracted above ovary, strongly incurved to C-shaped, minutely papillose above ovary, swollen 2–4 mm before enlarged, evenly dilated style end; pollen presenter lateral, circular, flat. **Fruit** 12–20 mm long, 5–6 mm wide, erect, narrowly obovoid with a long apical dilation, smooth, glabrous; pericarp c. 1 mm thick. **Seed** 8–13 mm long, 2.5 mm wide, narrowly ovoid, acute; outer face convex, smooth; inner face flat with a central area and outer channelled area; margin slightly winged, shortly recurved.

Distribution W.A., from northeast of Geraldton south to near Mogumber. *Climate* Hot, dry summer; cool to mild, wet winter. Rainfall 300–600 mm.

Ecology *G. bracteosa* occurs in granitic loam in open to tall shrubland and occasionally on sandplain. Flowers winter to spring. It regenerates from seed but is declining rapidly due to clearance of habitat and consequent decrease of pollinators. It is now mostly reduced to weedy road verges and semi-cleared farmland.

Variation There are two forms but the differences are insufficient to warrant formal infraspecific recognition. More southerly collections appear to have smaller conflorescences but the differences are inconsistent.

Small-conflorescence form This occurs in the southern part of the species' range, between Mogumber and Pithara. Although it is more common than the larger form, it occurs mostly in cleared areas and is reduced to weedy road verges and small reserves. It has a greater range of flower colour from white to deep rose-pink and the conflorescences are noticeably smaller. It forms a shrub to c. 1.5 m and grows in shallow, gravelly loam, often near outcropping granite.

Major distinguishing features Leaves simple, linear, bisulcate on undersurface; conflorescence globose; floral bracts large, papery, conspicuous to anthesis; perianth markedly zygomorphic, the tepals reflexing and forming a rounded, undulate platform below the ovary; ovary glabrous, very long-stipitate; style minutely papillose above ovary, tumid in apical few mm; fruits elongate with apical attenuation of the persistent, swollen style.

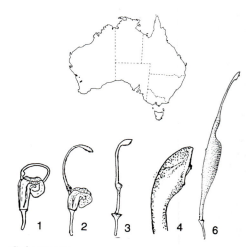

G. bracteosa

Related or confusing species Group 36. This is a distinct species that appears to have no close relatives.

Conservation status Suggested 3E. Restricted to small, isolated populations mostly on private land and road verges.

Cultivation This hardy, adaptable species succeeds in a wide range of soils and climates and tolerates both extended dry conditions and frost to at least -6°C without ill effects. Reliable results can be expected in all except tropical climates and it grows well in both coastal and inland gardens. It has succeeded in several Sydney gardens and is also grown by enthusiasts in Vic. and Qld. For optimal flowering and growth, *G. bracteosa* needs a well-drained, acidic to neutral, sandy loam in full sun. These conditions produce a fast-growing, long-lived, hardy shrub. Heavy soils, although tolerated, should be opened up with gravel or gypsum. Once established, it requires little maintenance except pruning to maintain shape. It is not generally suited to pot culture due to its large, open habit but is becoming more frequently available at specialist native plant nurseries.

Propagation Seed Sets many seeds that germinate well given the standard peeling treatment. No hybrids are known. Cutting Strikes well from firm

53A. *G. bracteosa* Close-up of unusual white-flowered form, near Miling, W.A. (P.Olde)

53B. *G. bracteosa* Flowers and developing fruit (P.Olde)

53C. *G. bracteosa* Plant on gravelly hillside with *G. petrophiloides*, E of Miling, W.A. (N.Marriott)

young growth taken in spring or autumn. *Grafting* Has been grafted onto a number of rootstocks including G. 'Poorinda Royal Mantle' and *G. robusta*. Resultant plants have proved extremely vigorous and hardy.

Horticultural features *G. bracteosa* has an open, rounded habit and delightful, unique flowers valued in both gardens and floral arrangements. The pale green, long, narrow leaves widely spaced along the stem contrast nicely with the pale brown branchlets; but it is the terminal, globular heads of pink flowers, enveloped in bud by large, papery bracts, that stand out in a free-flowering display over many months. These flowers are useful in cut-flower arrangements and do not drop their perianth as they age in the vase. Unusual, bean-like follicles also add interest after flowering. Landscapers should consider wider usage of this species which is suitable as a feature plant or as a screen plant where dense growth is not required, such as near windows. The small-conflorescence form tends to be a smaller, more compact shrub, more floriferous and with more intensely coloured flowers than the typical form, and select clones would have a wide application. It is vital that more growers and nurseries appreciate this plant because of its rarity in the wild.

Grevillea brevis P.M.Olde & N.R.Marriott (1993) Plate 54

Telopea 5: 410 (Australia)

Named from the Latin *brevis* (short), in reference to the short pistil by comparison with related species.
BREV-ISS

Type: 18.5 km S of Gimbat homestead (below edge of Marawal Plateau), Kakadu NP, N.T., 13°44'S 132°36'E, 21 April 1990, A.V.Slee 2689 & L.Craven, (holotype: CANB).

Synonym: *G. angulata* var. ?*lancifolia* F.Mueller ex G.Bentham (1870).

Single-stemmed **shrub** 1–2.5 m high. **Branchlets** angular to terete, ridged, glabrous to sparsely silky. **Leaves** green, sometimes with a yellowish tinge in dried specimens, flat, 7–15 cm long, 0.6–1.5 cm wide, petiolate, glabrous, concolorous, narrowly elliptic to elliptic to oblong-elliptic, either entire or some leaves with 1–3 triangular pungent lobes; upper and lower surfaces minutely pitted; apex acute to acuminate, non-pungent; base attenuate; margins flat with rounded edge-veins; venation conspicuous, more prominent on upper surface. **Conflorescence** decurved, terminal, shortly pedunculate, simple or few-branched; unit conflorescence 1–2 cm long, conico-cylindrical, open to dense; peduncle glabrous to sparsely silky; rachis glabrous; bracts 0.7–1.3 mm long, ovate, glabrous or sparsely ciliate, falling before anthesis. **Flower colour**: perianth white to yellow or creamy green; style green to pale cream. **Flowers** acroscopic, glabrous; pedicels 3–3.5 mm long; torus c. 1 mm across, oblique; **perianth** 4.5 mm long, 1.5 mm wide, narrowly ovoid-attenuate, bearded inside; tepals at first separated along dorsal suture, then all free below limb, the dorsal tepals reflexed and everted to expose inner surface, detaching or shrivelling soon after anthesis; limb cohering before anthesis, spheroidal, revolute in upper half of perianth, the style end partially exposed before anthesis; pistil 8–12.5 mm long; stipe 1.5–2 mm long, glabrous; ovary glabrous; style at first exserted below curve on dorsal side and looped upwards, refracted above ovary, strongly incurved after anthesis, smooth, glabrous except minute trichomes or tubercles in apical 2-3 mm, sometimes extending down style; style end scarcely dilated; pollen presenter oblong to obovate, convex; stigma central to slightly off-centre. **Fruit** 13–15 mm long, 12 mm wide, 10 mm deep, ellipsoidal, oblique to adaxially transverse to stipe with suture directed outwards; pericarp c. 1 mm thick at centre of suture. **Seed** 7–8 mm long, 3.5–4 mm wide, obovoid to ellipsoidal, winged all round, flat on inner face with marginal curvature and rim evident.

Distribution *G. brevis* is confined to a few localities on the eastern and western sides of the Marawal Plateau, Kakadu NP, usually near the edge of the escarpment. It occurs in the Douglas Springs-Bloomfield Springs area and in an area due south of Big Sunday (Niljanjurrung). The area is within the 'sickness' country of the Jawoyn Aboriginal people. *Climate* Hot, wet summer; warm, dry winter. Rainfall 800–1000 mm.

Ecology Found on top or just below the top of the sandstone plateau in lateritised, rocky sand in *Asteromyrtus* heathland or in brown, kaolinised clay mixed with laterite in broad, shallow valleys, sometimes in deep grey sand. Flowers autumn–winter. Response to fire and pollinator unknown.

Variation Most plants from the eastern side of the Marawal plateau have some toothed leaves and rarely grow taller than 1.5 m whereas those from the eastern side have almost all leaves entire and often exceed 2 m in height. The presence of much-narrowed (usually pale green to yellow) leaves on some specimens appears to be either a virally or environmentally induced deformity. The deformity is more noticeable in dried specimens than in living plants. We have also noted it in some plants of *G. glabrescens*. A biological and ontogenetic study of its leaves would be worthwhile.

Major distinguishing features Leaves simple to 3-toothed, concolorous, glabrous with flat margins; conflorescence conico-cylindrical; perianth zygomorphic, glabrous outside, bearded inside; pistil glabrous except the granular style end, the style clearly exceeding the perianth.

Related or confusing species Group 11, especially *G. glabrescens* which has wider leaves (> 2 cm) that are always toothed and a pistil more than 15 mm long.

Conservation status The species has a limited distribution but is relatively common at each site. A code of 2RC-t is recommended.

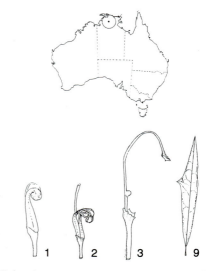

G. brevis

54A. *G. brevis* Habit and habitat, Marawal Plateau, Kakadu NP, N.T. (P.Olde) 54B. *G. brevis* Marawal Plateau, Kakadu NP, N.T. (P.Olde)

Cultivation *G. brevis* is a relatively compact shrub suited to tropical and subtropical climates. Its adaptability and hardiness in cultivation are untested, but given the relative success of related species, it should do well in gardens as far south as Brisbane and Geraldton and in other mild, frost-free climates. Sandy, acidic soils should best suit it, although heavier soils with good drainage should be satisfactory. A position in full sun is recommended. The species grows in exposed heaths and should develop a strong root system.

Propagation *Seed* Sets many seeds in the wild that should germinate satisfactorily with the usual peeling treatment. Long soaking in warm water should be beneficial. *Cutting* Should strike well using firm, new growth. *Grafting* Related species have been grafted successfully onto *G. robusta*.

Horticultural features *G. brevis* is a neat, interesting plant with yellowish white flowers set against the pale green leaves. It is not particularly showy but has potential as a low, screen or foil in large gardens and in massed roadside plantings. Perhaps suited more to the collector.

Grevillea bronwenae G.J.Keighery (1990)
Plate 55

Nuytsia 7: 128 (Australia)

The specific epithet honours Bronwen Keighery, wife of the author of the name. BRON-WENN-EYE

Type: Sabina Rd, Whicher Ra., 15 km S of Busselton, W.A., 1986/87, B.J. & G.J.Keighery s.n. (holotype: PERTH).

An erect, few-branched, non-lignotuberous **shrub** 1–1.8 m tall.

Branchlets terete, sparsely tomentose. **Leaves** 1–16 cm long, 2–14 mm wide, erect to ascending, sessile, papery, simple, oblong to obovate, obtuse, mucronate to acuminate; upper surface flat to convex, granulate, pubescent but soon ± glabrous, midvein evident; margins entire, recurved to revolute; lower surface sparsely villous to glabrous, midvein prominent. **Conflorescence** c. 2 cm across, erect to decurved, sessile or shortly pedunculate, terminal and axillary, umbel-like or wheel-like, 6–9-flowered; peduncle and rachis villous; bracts 3–4.5 mm long, narrowly triangular, villous outside, usually falling before anthesis. **Flower colour**: perianth scarlet; style end blue. **Flowers** adaxially acroscopic: pedicels 3–4.5 mm long, tomentose; torus 4–5 mm across, very oblique; nectary V-shaped; **perianth** 10–14 mm long, 2–4 mm wide, erect, oblong with cuneate base, ribbed, sparsely tomentose outside, bearded inside, cohering except along dorsal suture from curve to limb; limb nodding, ovoid-apiculate, villous; **pistil** 8–12 mm long; stipe 2–4.5 mm long; ovary villous; style thin, 0.6–1 mm thick, pubescent, exposed at curve, not exserted before anthesis, straight; style end thickened, scarcely exceeding perianth at anthesis; pollen presenter lateral, oblong-elliptic, concave. **Fruit** 15 mm long, 5 mm wide, erect, ovoid, tomentose; pericarp c. 0.5 mm thick. **Seed** 6 mm long, elliptic; margins strongly revolute, apex with elaiosome 2 mm long.

Distribution W.A., in the Whicher Ra.–Jarrahwood area. *Climate* Mild, dry summer; cool to cold, wet winter. Rainfall 800–1100 mm.

Ecology Grows in well-drained sandy soil over laterite, sometimes along drainage lines, usually in eucalypt woodland. Flowers winter to spring. Pollinated by nectarivorous birds. Regenerates from seed.

Variation In some collections, the conflorescence has a short peduncle.

Major distinguishing features Habit erect, usually 1–1.5 m tall; leaves relatively thin, the undersurface sparsely hairy; floral bracts linear to narrowly triangular; perianth zygomorphic, oblong-cuneate, undilated at base, cohering at anthesis, limb nodding; style 0.6–1 mm thick, the style end blue, without appendage, scarcely to not exceeding perianth.

Related or confusing species Group 23, especially *G. brachystylis* which is distinguished by its beaked style end, ventrally dilated perianth, ovate floral bracts, perianth usually densely hairy and geniculate in upper half, and broader style.

Conservation status Suggested 2R. Although reasonably secure in the Whicher Ra.–Jarrahwood area, its limited distribution would place it at risk if land management policies were to change for this area.

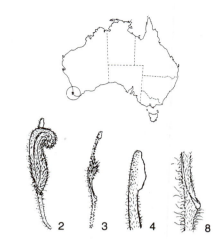

G. bronwenae

55A. *G. bronwenae* Flowering shrub in cultivation, Stawell, Vic. (N.Marriott)

55B. *G. bronwenae* Flowering branches (P.Olde)

Cultivation G. bronwenae has been grown with limited success but has not proved hardy. Plants have survived for a number of years in locations as widely spaced as Perth, Sydney, Burrendong Arboretum, Melbourne and western Vic., but more often it has been short-lived. Although it will tolerate light frosts to c. -3°C, heavier frosts are likely to kill it. Successful growers have used a well-drained, acidic to neutral loam or gravelly loam in full sun or partial shade. Occasional pruning is necessary as it has a strong tendency to become lanky, while an assured water supply, especially in hot, extended dry conditions, is required. In summer-rainfall climates such as Sydney, it appears to be prone to attack by white louse scale which severely debilitates the plant and ultimately kills it if not treated. It grows extremely well in pots in a well-drained, peaty mix with appropriate slow-release fertiliser and where attention can be given to controlling infestations of scale. Native plant nurseries specialising in grevilleas sell this species, usually as *G. brachystylis*, the species in which it was formerly included.

Propagation Seed Sets reasonable quantities of seed that germinate readily, particularly if peeled. *Cutting* Strikes readily from cuttings taken at most times of the year. Firm to semi-hardwood material strikes best. *Grafting* Grafts readily onto G. 'Poorinda Royal Mantle' and G. 'Poorinda Anticipation' using both top wedge and approach grafts. Plants grafted onto *G. robusta* succumb after 6–9 months, but when an interstock such as G. 'Poorinda Anticipation' is used they grow on vigorously.

Horticultural features This plant, with its brilliant scarlet flowers and blue styles, is widely regarded by growers as one of the choicest grevilleas. Apart from the striking show created by its free-flowering habit and eye-catching flowers, flowering in cultivation is continuous with clusters arising one after another from the same point on the axis. This glorious display continues at full peak from autumn to spring with odd flowers appearing spasmodically over the rest of the year. Late in the season, dying flowers tend to spoil the effect, a minor blemish on a plant of such compelling beauty. An additional feature that often attracts comment is the bright red colour of newly formed fruits which develop soon after flowering. If grafted onto a suitable rootstock, it makes an exceptional feature plant, but on its own roots is usually short-lived. As it is easy to propagate, many growers are happy to treat it as an annual, balancing the need to continually replace it against its incomparable floral display.

Grevillea buxifolia (J.Smith) R.Brown (1810)

Transactions of the Linnean Society of London 10: 174 (England)

Basionym: *Embothrium buxifolium* J.Smith (1794) in *Specimen of the Botany of New Holland* 3: 29 t. 10. Type not cited.

Grey Spider Flower, Box Leaf Grevillea

The specific epithet is derived from the generic name *Buxus* (a box-tree) with the Latin *folium* (a leaf), in reference to the similar shape of their leaves. BUCKS-I-FOLE-EE-A

Type: N.S.W., early 1790s, J.White, (lectotype: LINN) (McGillivray 1993).

An erect, symmetrical to spreading **shrub** 1.5–2.5 m tall, 1.5–2.5 m wide. **Branchlets** slightly angular to terete, villous. **Leaves** 1–1.5 cm long, 5–8.5 mm wide, ascending, ± sessile, simple, ovate to elliptic; upper surface pubescent when young, soon glabrous and granulose, rarely glabrous and shiny, midvein an impressed groove; margins entire, shortly recurved; lower surface villous, midvein prominent. **Conflorescence** 3 cm long, 3–4 cm wide, erect, sessile, terminal, unbranched, umbel-like, the apical flowers opening first; peduncle and rachis villous; bracts 2–3 mm long, linear, villous, falling before anthesis. **Flowers** abaxially oriented: pedicels 8–12 mm long, brown-villous; torus 1.5–2 mm across, square, ± straight; nectary conspicuous, semi-circular to square, lightly toothed; **perianth** 5 mm long, 2–2.3 mm wide, cylindrical, strongly curved from c. halfway, villous outside, woolly inside; tepals splitting to base on dorsal side before anthesis, ventral tepals remaining joined, reflexed and everted from c. half-way forming a silvery pubescent platform extending out from ovary before anthesis; limb ± pyramidal, densely villous, revolute, cohering after anthesis, the style end fully exposed before anthesis; **pistil** 19–21 mm long, villous; stipe 2 mm long, glabrous on ventral side; ovary villous; style exposed before anthesis except at base, afterwards strongly incurved in upper half; style end conspicuous, thickened, c. 2 mm wide, bearing a prominent reflexed horn 2–3.5 mm long on dorsal side; pollen presenter circular, lateral, surrounded by a conspicuous, wrinkled rim, slightly conical. **Fruit** 18–21.5 mm long, 8–9 mm wide, erect, ovoid, sparsely villous; pericarp c. 0.5 mm thick. **Seed** 12 mm long, 2.5–4 mm wide, oblong with a subapical cushion-like swelling, minutely pubescent on both surfaces; outer face convex, smooth with slight wrinkling; inner face concave, obscured, the margin strongly but loosely revolute, narrowly winged along one side, the wing drawn to a terminal point 2.5–4 mm beyond testa.

Major distinguishing features Branchlets villous; leaves simple, entire; conflorescence umbel-like; flowers abaxially oriented, greyish; perianth zygomorphic, densely villous on both surfaces, everting to form a round platform above ovary before anthesis; ovary villous, sessile; style with a spreading indumentum; style end corniculate; pollen presenter orbicular or elliptic, c. 2 mm wide.

Related or confusing species Group 22. *G. sphacelata* is most closely related but differs in the mostly appressed indumentum on flowers and branchlets.

Variation Three subspecies are recognised.

Key to subspecies

1 Pollen presenter orbicular, c. 2 mm wide
  2 Stylar appendage 2–4 mm long; pistil > 18 mm long
      subsp. buxifolia
  2* Style end either lacking an appendage or the appendage < 1 mm long; pistil < 14 mm long   **subsp. ecorniculata**
1* Pollen presenter elliptic, c. 1 mm wide
      subsp. phylicoides

56. *G. buxifolia* subsp. *buxifolia* Close-up (M.Hodge)

Grevillea buxifolia (J.Smith) R.Brown subsp. buxifolia   Plate 56

Leaves 1–1.5 cm long, 5–8.5 mm wide, ovate to elliptic. **Conflorescence** 3 cm long, 3–4 cm wide; pedicels 8–12 mm long. **Flower colour**: perianth brown and grey, sometimes with pinkish tinges; style grey with a chocolate-brown tip; pedicels brown. **Perianth** 5–6 mm long, 2–2.3 mm wide; **pistil** 19–21 mm long; stipe 2–2.3 mm long; style end appendage 2–3.5 mm long; pollen presenter c. 2 mm wide, orbicular. **Fruit** 18–21.5 mm long, 8–9 mm wide.

Synonyms: *Embothrium genianthum* A.J.Cavanilles (1798); *Stylurus buxifolia* (J.Smith) J.Knight (1809).

Distribution N.S.W., mostly in coastal situations between Waterfall and the coast S of Newcastle. *Climate* Cool to mild, wet or dry winter; warm to hot, wet summer. Rainfall 800–1200 mm.

Ecology Usually occurs in heath or dry sclerophyll forest in skeletal, sandy soil on sandstone. Flowers late winter to spring, sometimes with a few flowers at other seasons. Killed by fire and regenerates from seed. Pollinator unknown, most probably native bees or larger insects. Strongly perfumed, especially on warm, still days.

G. buxifolia subsp. *buxifolia*

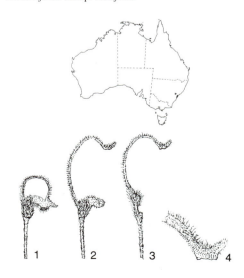

Variation A widespread, relatively uniform species. An unusual plant with densely clustered, large conflorescence has been photographed in Muogamarra Reserve (J.Steenson pers. comm.) but is not in cultivation. Occasional flowers with two styles have been observed.

Conservation status Not presently endangered.

Cultivation Subsp. *buxifolia* has been cultivated in all eastern States of Australia as far north as Brisbane with varying degrees of success. Under most conditions, it is relatively hardy and will tolerate frosts to at least -6°C as well as extended dry conditions. It appears to be somewhat particular about soil texture, demanding a sandy, open soil for long-term success, but it has been successfully grown for short periods in heavy, well-drained loam. If planted in a sunny, open position or partial shade in acidic, sandy soil it will thrive but requires summer watering for best results. Light tip-pruning may be necessary to maintain shape as it tends to drop its lower leaves with age and has a natural tendency to legginess. Fertiliser is usually unnecessary but can be applied lightly to improve vigour. White scale is sometimes seen on plants in the wild and in cultivation but is easily treated. Potted plants do well in a sand and peat mix with slow-release fertiliser. Specialist native plant nurseries on the east coast often stock this species.

Propagation *Seed* Easy to propagate from fresh seed sown in neutral or acidic sandy media, preferably given the standard nicking or peeling treatment. *Cutting* Somewhat difficult from cutting, especially from plants in the wild. Young, semi-mature tip growth can be used usually in late spring or early autumn. Soft material will die readily. Conditions in the propagation house should not be too wet or the hairy branchlets will hold water and rot. *Grafting* First trials have shown this to be difficult to graft.

Horticultural features Subsp. *buxifolia* is a curious, extremely desirable plant with attractive foliage and charming, woolly flower-heads. The heart-shaped leaves are quite small, shapely and crowded along the woolly grey branchlets. New growth, which in selected forms is quite conspicuous, is an attractive bronze or red-brown and is visible on vigorous, young plants for many months of the year. The grey flowers, which always create interest, are borne at the ends of the branches in prominent heads over a long period and are also lightly scented, especially noticeable about the middle of the day. The pollen carried up by the newly released styles of young flowers is an attractive caramel colour. The species is not long-lived in cultivation and plants may need replacing after about 5 years. Grown mainly as an oddity or feature plant, this species makes an interesting contrast among other plants of the genus.

Hybrids Known parent in a number of popular hybrids such as G. 'Evelyn's Coronet' which arose in a home garden. G. 'Lyn Parry' is a natural hybrid collected from the wild near Kariong and is a presumed hybrid with *G. sericea*.

Early history Along with many other plants, *G. buxifolia* was introduced to England by Colonel William Paterson in about 1791 when seeds were sent to Lee & Kennedy, nurserymen of Hammersmith. It is reputed to have been the first Australian *Grevillea* to flower in cultivation, in 1795. It was listed in Loddiges' Nursery Catalogue in 1804 and cultivated at Kew in 1810. Hügel listed it among his collection in 1831 at Vienna. In Australia, the first mention of it occurs among lists of plants despatched to Hobart from the Sydney Botanic Gardens in 1851.

Grevillea buxifolia subsp. ecorniculata
P.M.Olde & N.R.Marriott (1994)   **Plate 57**

Telopea 5: 709 (Australia)

The subspecific epithet is derived from the Latin *e* (without), and *corniculatus* (horned), in reference to the stylar appendage being more or less absent by contrast with subsp. *buxifolia*. EE-CORN-ICK-YOU-LAR-TA

Type: Staircase Hill, Putty Rd., 82.8 km N of Windsor, N.S.W., 24 Sept. 1989, R.O.Makinson 384 (holotype: NSW).

Synonym: *Grevillea buxifolia* subsp. *buxifolia* Race 'a' of D.J.McGillivray (1993).

Leaves 1–2 cm long, 3–6 mm wide, elliptic or narrowly so. **Conflorescence** 1–2 cm long, 3 cm wide. **Flower colour**: perianth brown and grey, sometimes with pinkish overtones outside, whitish grey inside; style grey with a chocolate-brown tip; pedicels brown. **Flowers**: pedicels 6–9 mm long; **perianth** 5–6 mm long, 1.5–2 mm wide; **pistil** 11–13 mm long; stipe 1 mm long; appendage of style end to 1 mm long or absent; pollen presenter orbicular, c. 2 mm wide. **Fruit & seed** not seen.

Distribution N.S.W., confined to a small area between Putty, Gospers Mountain and Wollombi. *Climate* Cool to mild, wet or dry winter. Warm to hot, wet summer. Rainfall 800–1200 mm.

Ecology Usually found in heath or in dry sclerophyll forest in skeletal, sandy soil on sandstone. Flowers usually all year but principally winter and spring. Pollinator unknown.

Variation There is minor variation in the style end but otherwise this subspecies is morphologically uniform.

Conservation status Not presently endangered.

G. buxifolia subsp. *ecorniculata*

Cultivation Subsp. *ecorniculata* is not known in cultivation, but it occurs in a habitat similar to those of the other subspecies. It should do well in a sunny to partially shaded situation in well-drained, acidic sandy soil especially one derived from sandstone. Regular tip-pruning would produce a compact plant, as it tends to legginess in the wild. Fertiliser should be kept to a minimum, although applications of well-rotted animal manure should induce vigorous growth.

Propagation *Seed* Untested, but fresh nicked or peeled seed should germinate readily. *Cutting* Untested. Other subspecies have proved difficult. *Grafting* Untested.

Horticultural features An interesting plant with very prominent lightly scented grey flowers on most branchlets. More trials are needed before recommendations on use can be made.

57. *G. buxifolia* subsp. *ecorniculata* Conflorescence (P.Olde)

Grevillea buxifolia subsp. phylicoides
(R.Brown) D.J.McGillivray (1986) **Plate 58**

New Names in Grevillea *(Proteaceae)* 3 (Australia)
Based on *Grevillea phylicoides* R.Brown, *Transactions of the Linnean Society Botany* 10: 174 (1810, England).

Grey Spider Flower
Named for its similarity in habit and foliage to the South African genus *Phylica*. FYE-LICK-OY-DEES

Type: hills near the banks of the Grose R., N.S.W., probably between 1803 and 1804, R.Brown (holotype: BM).

Synonyms: *G. scabrifolia* M.Gandoger (1919); possibly *G. collina* (J.Knight) R.Sweet.

Leaves 2–3 cm long, 2–6 mm wide, linear to oblong-lanceolate. **Conflorescence** 1.5–2 cm long, 2 cm wide. **Flower colour:** Grey or pinkish grey. **Flowers:** pedicels usually 2–6 mm long; **perianth** 2–3 mm long, 1 mm wide; **pistil** 10.5–13 mm long; style end appendage 1–4 mm long; pollen presenter c. 1 mm wide, elliptic (more evident on drying). **Fruit** c. 15 mm long, 5 mm wide.

Distribution N.S.W., from the Culoul Ra. near the Colo R. to the Bilpin area and the Blue Mountains generally, extending to the Dural–Richmond–Wisemans Ferry area with disjunct populations around Pigeon House S of Ulladulla and near Nowra. *Climate* Summer hot, wet, sometimes dry; winter cold, wet or dry. Rainfall 800–1200 mm.

Ecology Grows in open heathland and dry, sclerophyll forest in sandy, skeletal sandstone derived soils. Flowers all year but principally spring and summer. Killed by fire and regenerates from seed only. Pollinator unknown but bees are strongly linked.

Variation McGillivray (1993: 312–314) extensively discussed variation. The following two forms are recognised, distinguished mainly on the size of the appendage on the style end. There is some apparent intergradation between subsp. *phylicoides* and subsp. *buxifolia* in the Wisemans Ferry area. Plants from the Bilpin–Kurrajong areas share features of both forms listed below.

Typical form (Blue Mountains form) This form is found widely in the lower Blue Mountains and on Kings Tableland near Wentworth Falls, extending SW to near Camden. It is a shrub to c. 1.5 m with

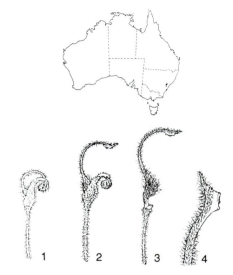

G. buxifolia subsp. *phylicoides*

conspicuous flowers whose style has an appendage c. 2 mm long. The appendage is not as sharply reflexed and erect as in subsp. *buxifolia*. Leaves are narrowly elliptic to oblong, sometimes ovate. (Race 'd' *sensu* McGillivray.)

Colo form This form from the Culoul Range area and extending to the Wiseman's Ferry-Dural area differs in its more conspicuous appendage (3–4 mm long) (Race 'b' *sensu* McGillivray).

Conservation status Not presently endangered.

Cultivation Not widely cultivated and little information can be given authoritatively, but guidance can be sought from the conditions given for subsp. *buxifolia*. Natural habitat indicates that it will tolerate heavy frosts and icy winter conditions for a short period. Extended dry periods may also be experienced in the natural habitat. A full sun or partially shaded site in a well-drained, sandy, acidic soil would be suitable. Not generally available in nurseries.

Propagation *Seed* Germinates reasonably well from seed if peeled or nicked. *Cutting* The hairy branchlets can cause difficulty in striking especially using material taken from wild plants. Fresh, firm young growth taken in late spring to early autumn can be struck successfully. *Grafting* Untested.

Horticultural features An attractive plant similar to subsp. *buxifolia* and with similar horticultural features. Flowering is prolific and can almost hide the foliage. Flowers have a light, spicy scent. Not a top-rating species but one for the enthusiast. Useful in landscaping as a low screen or fill-in.

Early history *G. buxifolia* subsp. *phylicoides* was cultivated in England from seed sent by Allan Cunningham in 1823–24.

Grevillea byrnesii D.J.McGillivray (1986)
Plate 59

New Names in Grevillea *(Proteaceae)* 3 (Australia)
Named in honour of Norman Byrnes for his contribution to the botany of tropical Australia. BURNS-EE-EYE

Type: between Gibbie Creek and Mt Sanford Station, N.T., 7 July 1978, D.J.McGillivray 3902 (holotype: NSW).

Single-stemmed **shrub** or small **tree** with elongated, erect branches 4–5.5 m high. **Branchlets** angular to terete, silky. **Leaves** 7.5–12 cm long, 3.5–7 cm wide, ascending, petiolate, concolorous, leathery, ± ovate to obovate, simple, the margins entire or lightly toothed, the lobes inconspicuous, pungent;

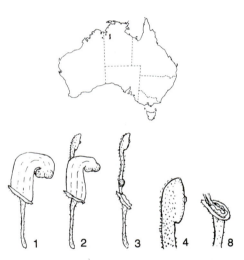

G. byrnesii

58A. *G. buxifolia* subsp. *phylicoides* Flowering habit, near Windsor, N.S.W. (P.Olde)

58B. *G. buxifolia* subsp. *phylicoides* Conflorescences, Blue Mountains form, Blue Mountains, N.S.W. (P.Olde)

59A. *G. byrnesii* Massed flowering, Keep River NP, N.T. (T.Blake)

59B. *G. byrnesii* Plant in natural habitat, Jasper Gorge, N.T. (D.McGillivray)

59C. *G. byrnesii* Close-up of foliage and flowers (D.McGillivray)

upper and lower surfaces similar, tomentose, with conspicuous venation and flat, undulate margins. **Conflorescence** decurved, pedunculate or rarely sessile, cauline or axillary, sometimes terminal on short branchlets, simple or branched; unit conflorescence 5–12 cm long, 2–3 cm across, conico-cylindrical, open; peduncle tomentose; rachis pubescent; bracts 2.5–5 mm long, ovate, glandular-pubescent outside with ciliate margins, falling before anthesis. **Flower colour**: perianth orange; style orange with yellow tip. **Flowers** acroscopic, sometimes abaxially oriented; pedicels 6–9 mm long, sparsely glandular-pubescent, retrorse; torus 3–5 mm across, very oblique; nectary long U-shaped, cupped, uneven or bi-lobed; **perianth** 5–6 mm long, 3–4 mm wide, oblong with slight basal annular dilation, strongly curved, coherent but flared open on the dorsal at the curve just before anthesis and exposing the style end, soon all free and deciduous, sparsely glandular-pubescent to almost glabrous outside, pilose inside; limb very conspicuous in bud, revolute with expansion of the perianth, spheroidal to subcubic, the segments thick-walled, glandular-pubescent; **pistil** 9–11 mm long; stipe 2.2–3.5 mm long, glabrous, adnate to the torus over most of its length; ovary ovoid, glabrous; style sparsely glandular pubescent, not exserted before anthesis, afterwards straight and scarcely exceeding perianth, the broadly expanded, club-shaped style end slightly inflexed; pollen presenter lateral, obovate to round, almost flat, mammiform. **Fruit** 17–24 mm long, 15 mm wide, erect on reflexed pedicels, almost round, rugose; pericarp 1–2 mm thick. **Seed** 7–8.5 mm long, 3.5–5 mm wide, compressed-ellipsoidal, broadly and membranously winged all round; raphe conspicuous, biconvex, smooth or finely wrinkled.

Distribution N.T., mainly in the Victoria R. district. W.A., confined mainly to the east Kimberley region and extending to the Great Sandy Desert. *Climate* Summer hot, moist to wet; winter warm, dry. Rainfall 300–800 mm.

Ecology Grows on rocky clifftops, sandstone ridges and red sand dunes, on lateritic gravelly rises and in skeletal sandy soil, generally in open woodland. Flowers autumn–winter. Killed by fire and regenerates from seed. Native bees have been seen in attendance but it may be pollinated by birds.

Variation A relatively uniform species with minor differences in leaf and flower morphology.

Major distinguishing features Erect, single-stemmed habit; leaves concolorous, the lamina broad, flat, usually lightly toothed; conflorescence decurved, usually axillary, sparsely glandular-pubescent; pedicels 6–9 mm long; torus very oblique; perianth zygomorphic; ovary glabrous; style glandular-pubescent, sometimes sparsely so, scarcely exceeding perianth at anthesis; pollen presenter lateral; seed membranously winged all round.

Related or confusing species Group 9, especially *G. wickhamii* which differs in its usually shorter pedicels (mostly < 6 mm long) and its silky style, although there are some collections with a glandular-pubescent style. These may then be distinguished by their shorter pedicels (< 6 mm long). Most collections of *G. wickhamii* have two-armed hairs on the outer perianth surface or are glabrous. *G. wickhamii* also has stout lateral branches from the base and is a more compact shrub with spreading branches and red flowers, although orange, pink and yellow forms are known. Occasionally the perianth indumentum is either absent or sparse, as in *G. byrnesii*, but in such cases the pedicels do not exceed 6 mm long.

Conservation status 3VC suggested. Often found in small populations, but distributed over a wide area.

Cultivation *G. byrnesii* has been in cultivation since 1990. Most plants are grafted and still at the nursery stage, but at least one grower has flowered it. It should be suitable for tropical and subtropical climates with summer rainfall and a dry winter. Experience with similar species such as *G. wickhamii* could be a guide. It will probably tolerate light frosts but not continuous cold weather and should withstand extended dry periods. Judging by its natural habitat, it requires sandy or gravelly, well-drained soil in a sunny position. Allow 1–1.5 metres between plants.

Propagation *Seed* Sets large quantities of seed in the wild that should germinate readily if soaked or nicked. *Cutting* Untried, but firm, young growth taken in the warm months of the year should strike well. *Grafting* Has been grafted successfully onto *G. robusta* using the mummy graft.

Horticultural features *G. byrnesii* appears to be one of those horticultural gems that have escaped the attentions of growers, no doubt due to its isolated occurrence in the wild and a certain confusion with the closely related *G. wickhamii*. This delightful, winter-flowering plant has an open, conical habit with large, blue-grey, rounded adult leaves and attractive bronze new growth. Long, decurved, orange-flowered conflorescences arise in profusion from the leaf axils or on short side branches and even, sometimes, directly from the main stem, hanging like beads among the leaves. This combination highlights the genuine charms of this magnificent species which will undoubtedly take its place in tropical gardens.

Grevillea cagiana D.J.McGillivray (1986)
Plate 60

New Names in Grevillea *(Proteaceae)*: 3 (Australia)

Red Toothbrushes

The specific epithet is an acronym of the initials of Charles Austin Gardner (1896–1970), a dominant force in the botany of the Western Australian flora from 1920 to 1970. To his friends, Gardner was known as 'CAG'. KAG-EE-ARE-NA

Type: 3.5 km SSE of Kukerin, W.A., 26 June 1976, D.J.McGillivray 3534 & A.S.George (holotype: NSW).

Bushy, green to grey, single-stemmed **shrub** 1–4 m tall. **Branchlets** terete to angular, silky to tomentose, sometimes ridged. **Leaves** 5–16 cm long, ascending, shortly petiolate, sometimes simple and narrowly linear, usually pinnatipartite, with 2–11 linear, acute, occasionally pungent lobes 0.8–1.8 mm

G. cagiana

wide; upper surface tomentose or sparsely so, rarely glabrous, sometimes velvety, midvein and sometimes longitudinal veins conspicuous; margin angularly revolute to midvein; lower surface bisulcate, tomentose in grooves, midvein prominent. **Conflorescence** 2–6 cm long, erect, pedunculate, terminal, unbranched, oblong-secund; peduncle silky or tomentose; rachis usually tomentose; bracts 0.6–1 mm long, ovate to rhombic, tomentose, falling before anthesis. **Flower colour:** perianth green, yellow, pink or orange; style bright orange or red with yellowish style end. **Flowers** acroscopic; pedicels 1–2.5 mm long, tomentose to silky; torus 1–2 mm across, straight or slightly oblique; nectary arcuate, cup-shaped with smooth margin; **perianth** 6–10 mm long, 2–3 mm wide, ovoid to S-shaped, silky to villous outside, glabrous inside, cohering except along dorsal suture; limb revolute, globular to spheroidal, enclosing style end before anthesis; **pistil** 16.5–24 mm long; stipe 1.5–5 mm long, villous; ovary villous; style glabrous except a few basal hairs, sometimes papillose on ventral side, at first exserted just below curve on dorsal side, gently incurved to erect and undulate after anthesis, slightly reflexed above ovary; style end hoof-like; pollen presenter oblique, round, conical. **Fruit** 15–25 mm long, 10 mm wide, erect on curved pedicel, compressed-ellipsoidal but attenuate, ribbed and striped on dorsal side, silky; pericarp 0.5–1 mm thick. **Seed** 9–12 mm long, 2.5–4 mm wide, oblong-elliptic with a short, excurrent wing at both ends; outer face smooth with a few, longitudinal wrinkles, crimped near margin, convex; margin recurved, narrowly winged along one side; inner face slightly convex at centre, surrounded by a slight elliptic ridge and a submarginal, broad and shallow channel.

Distribution W.A., widespread in the inland south-west between Merredin, the Bremer Ra. and Coolgardie. *Climate* Hot, dry summer; mild, moist winter. Rainfall 250–350 mm.

Ecology Dominant in dense to open heath or tall shrubland in white sand over laterite, deep yellow sand, sand over clay or occasionally in gravelly clay. Flowers winter–summer. Regenerates from seed after fire. Pollinated by birds.

Variation There is some vegetative and floral variation in *G. cagiana*. The style may be either orange or red. Leaves may be all simple, all divided, or mixed, and either glabrous or hairy. There is also considerable variation between populations in stipe length (McGillivray 1993). The taxonomy of this species warrants further investigation.

Typical form Widespread in sandy heaths around Southern Cross, this is a greyish shrub to 3 m with pinnatipartite leaves with a persistent indumentum. The perianth limb is brownish and the ovarian stipe c. 3 mm long.

60A. *G. cagiana* Robust form (P.Olde)

60B. *G. cagiana* Habitat of Simple-leaved form, NE of Lake King, W.A. (N.Marriott)

60C. *G. cagiana* Flowers and foliage (N.Marriott)

Green-leaved form North of Lake King, plants often grow to c. 1.5 m tall, have finer, less hairy leaves and the style is usually bright orange. They dominate in low heath in yellow sand or in lateritised sandy loam. Some nurseries have sold this form as G. *hookeriana* 'Lake King'.

Robust form This dense, spreading form grows to 3–4 m with a spread of up to 4 m. It has long, flexible, silvery grey leaves, white, tomentose branchlets, a densely hairy perianth and red style; the stipe of the ovary is 4–5 mm long. Occurs around Jitarning in white sand over laterite.

Simple-leaved form A greyish compact form to 1.5 m, dominant in low, sandy heath near the Bremer Ra. and extending into Frank Hann NP. This form often has predominantly simple leaves mixed with a proportion of leaves with 3–5 lobes that are usually broader than in other forms. Leaves have a persistent grey indumentum. The stipe of the ovary is c. 2 mm long.

Major distinguishing features Leaves or leaf lobes narrowly linear; conflorescence secund, erect; perianth zygomorphic, hairy outside, glabrous inside; ovary densely hairy, conspicuously stipitate; style glabrous except at base; pollen presenter oblique, c. 1 mm wide; fruit compressed-ellipsoidal, attenuate, dorsally ribbed, and with reddish stripes.

Related or confusing species Group 35, especially *G. baxteri*, *G. beardiana*, *G. coccinea*, *G. concinna* and *G. hookeriana*, all of which, with the exception of *G. beardiana* and some collections of *G. concinna*, have the stipe of the ovary < 1.5 mm long. *G. baxteri* is very closely related to *G. cagiana* but is distinguished by its generally longer and more robust conflorescences, the stipe 0.9–1.3 mm long, a more hairy style, pollen presenter 1–1.5 mm wide and a thicker nectary. *G. beardiana* has shorter leaves (1–5 mm long) and a pistil 28–31.5 mm long. *G. concinna* has shorter, deflexed conflorescences. *G. coccinea* generally has simple, linear to narrowly obovate leaves, floral bracts 2.5–5 mm long. *G. hookeriana* differs in its sessile ovary and prominent, tongue-like nectary. It generally has black or maroon-black styles.

Conservation status Not presently endangered, but the Robust form occurs in a limited area that is mostly cleared and is locally threatened.

Cultivation *G. cagiana* has been widely grown for many years especially in Vic., although plants have also been successfully cultivated at Rylstone, N.S.W., and inland Qld as well as around Perth. It grows best in areas with a hot, dry summer and cool, moist winter. It will tolerate frosts to at least -6°C and extended dry periods, but does not adapt readily to humid conditions. Best results are achieved if it is given an area of 2–3 sq. m. to grow and planted out in a warm, sunny situation, in deep, well-drained, acidic sand or sandy, gravelly loam. In a cool, wet or humid climate, plants with pubescent leaves usually develop black spotting due to fungal attack. In a shaded situation it develops leaf blackening and few flowers. Less hairy forms may do better in these situations. As the branches are somewhat brittle, it should be given protection by other plants or kept heavily pruned. Fertiliser is not recommended as the species is adapted to low nutrient levels. It is generally too large to grow in pots but remains attractive for many years if grown in large tubs in a well-drained, sandy mix. Occasionally available from native plant nurseries in Vic.

Propagation *Seed* Seed germinates well if peeled. Seedlings are vigorous and establish rapidly; they require a warm, sunny situation and a well-drained potting mix. *Cutting* Extremely difficult to strike from cutting. The limited success so far has been with firm to semi-hardwood material taken during the warmer months. *Grafting* Grafts onto several rootstocks including G. 'Poorinda Royal Mantle', G. 'Poorinda Anticipation', G. 'Moonlight' and *G. shiressii* using top wedge, approach and mummy grafts. This is the preferred method of propagating this species because of the difficulties with cuttings. Has been almost impossible to graft onto *G. robusta*.

Horticultural features One of the most attractive of the shrubby grevilleas. Most forms develop into robust, dense shrubs with silvery-grey to bright green foliage borne on upswept branches. The spreading, dense habit makes it an ideal screen plant and the finely divided leaves add a touch of delicacy. The toothbrush flowers are either orange or red, and while never prolific usually stand out prominently against the foliage. Flowering is continuous for most of the year and this is, accordingly, a valuable plant for nectar-seeking birds. The Robust form is particularly attractive and has, as an additional feature, branchlets covered in a woolly white indumentum.

Grevillea calcicola A.S.George (1967)
Plate 61

Journal of the Royal Society of Western Australia 50: 97 (Australia)

The specific epithet alludes to the preference of this species for limestone soils and is derived from the Latin *calx* (lime) and *-cola* (a dweller or inhabitant). KAL-SEE-KOLA

Type: Charles Knife Rd, Cape Range, W.A., 30 Aug. 1960, A.S.George 1331 (holotype: PERTH).

Much-branched **shrub** 2–4 m high. **Branchlets** angular, ridged, pubescent. **Leaves** 12–20 cm long, ascending, shortly petiolate, pinnatipartite with c. 7 closely-aligned, pungent, linear lobes 1.3–2.5 mm wide; upper surface silky becoming ± glabrous, midvein an impressed groove; margins revolute to midvein; lower surface with 2 villous grooves, midvein prominent. **Conflorescence** erect, paniculate, rarely unbranched, terminal, usually exceeding foliage; unit conflorescence 6–7 cm long, 3–5 cm wide, dense, cylindrical; peduncle silky becoming glabrous, with a green to black triangular swelling at base of secondary peduncles; rachis usually glabrous; bracts 0.6–0.8 mm long, ovate, ciliate, villous outside, falling before anthesis. **Flower colour**: Perianth and style off-white or cream. **Flowers**

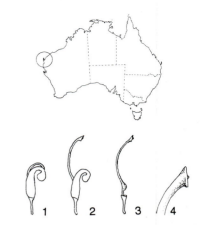

G. calcicola

acroscopic; pedicels 3–4 mm long, glabrous; torus slightly oblique, c.1 mm across; nectary U-shaped, smooth or broadly toothed; **perianth** 3.5–4 mm long, 1.5 mm wide, cylindrical to oblong, glabrous outside, sparsely bearded inside; tepals first separated along dorsal suture, then all separate, the dorsal tepals reflexed and everted to expose inner surface; limb revolute, spheroidal, enclosing style end before anthesis; **pistil** 11.5–12 mm long, glabrous; stipe 1.3–1.8 mm long; ovary subglobose, ± central on stipe; style lateral to ovary, first exserted along dorsal side and looped upwards, after anthesis strongly incurved to C-shaped, later almost straight; style end flattened; pollen presenter lateral, oblong-elliptic with prominent base, flat to concave; stigma distally off-centre. **Fruit** 21–27 mm long, 10–13 mm wide, oblique to stipe, oblong-ovoid with a basal heel and decurved apical beak, slightly compressed, viscid when immature, slightly rugose; pericarp c. 0.5 mm thick. **Seed** elliptic, winged.

Distribution W.A., confined to the Cape Ra. **Climate** Hot, dry summer with occasional storms; mild, short moist winter. Rainfall 180–280 mm but irregular.

Ecology Grows in spinifex and low, mallee-eucalypt shrubland in limestone, often on hilltops. Flowers winter. Response to fire unknown; some regeneration from seed. Probably pollinated by insects.

Variation A uniform species.

Major distinguishing features Leaves pinnatipartite, the lobe undersurface 2-grooved; conflorescence usually much-branched with

61A. *G. calcicola* Close-up of conflorescences (F. & N. Johnston)

61B. *G. calcicola* Flowering branch (F. & N. Johnston)

61C. *G. calcicola* Shrub in natural habitat, North-west Cape, W.A. (F. & N. Johnston)

cylindrical racemes, the secondary peduncles with triangular bases; nectary U- or V-shaped; perianth zygomorphic, glabrous outside, hairy inside; pistil glabrous, stipe < 2 mm long; pollen presenter flat to concave; fruit compressed; seed winged.

Related or confusing species Group 3. *G. leucadendron* and related species are closely related but differ in their unifacial leaves and erect, conical pollen-presenter.

Conservation status 2VC-i. Present status uncertain.

Cultivation *G. calcicola* is, perhaps surprisingly for a conspicuous species, almost unknown in cultivation, its isolated location restricting access to propagating material. The information provided here is speculative. It is likely, given its natural habitat, to need an alkaline sandy soil but, like most grevilleas, requires good drainage and full sun for best growth. While it is possibly best suited to subtropical gardens and may not tolerate frost, it would most probably resent humid, summer-rainfall conditions. It is likely to be well-adapted to extended dry conditions and should thrive in coastal gardens in Geraldton and Perth.

Propagation *Seed* Use standard treatments before sowing. *Cutting* Try firm, young growth for best results, ideally in late spring. *Grafting* Recently has been grafted successfully onto *G. robusta* using the mummy method.

Horticultural features This is a most attractive plant with its dense foliage and conspicuous panicles of creamy-white flowers. Flowering is mainly in winter but is likely to be extended in different climates. Bushes in the wild tend to become straggly, but in cultivation tip-pruning or selective removal of untidy branches should improve appearance. Uses such as screening and specimen planting are suggested. Every effort should be made to introduce this species to cultivation especially as it appears suited to alkaline soils.

Grevillea caleyi R.Brown (1830)

Plate 62

Supplementum Primum Prodromi Florae Novae Hollandiae exhibens Proteaceas Novas: 22 (England)

Caley's Grevillea; Fern Leaf Grevillea

The specific epithet honours George Caley, the enigmatic naturalist-explorer who collected for Joseph Banks in N.S.W. from 1800 to 1810. KAY-LEE-EYE

Type: near Sea Sight Hill, N.S.W., Feb. 1805, G.Caley (lectotype: BM) (McGillivray 1993). Caley probably collected the type in the Mona Vale area.

Synonym: *G. blechnifolia* A.Cunningham ex W.Hooker (1832).

A wide-spreading, long-branched **shrub** up to 3 m high, 4 m wide. **Branchlets** reddish, terete, villous. **Leaves** 5–18 cm long, 3–7 cm wide, spreading, petiolate, oblong to elliptic in outline, pinnatipartite; lobes oblong 1.5–3.5 cm long, 2–6 mm wide; upper surface sparsely pubescent becoming ± glabrous, hairs retained on rachis and impressed lateral veins; margins strongly recurved; lower surface loosely and softly brown-villous, midvein prominent. **Conflorescence** 4–8 cm long, erect, pedunculate, terminal or axillary in upper axil, unbranched, oblong-secund; peduncle and rachis villous; bracts 1–1.5 mm long, ovate, villous outside, falling before anthesis. **Flower colour**: perianth brownish grey; style maroon-red with green tip. **Flowers** acroscopic; pedicels 2–3 mm long, villous; torus c.1.5 mm across, ± straight; nectary semi-circular, wavy; **perianth** 8–10 mm long, 1.5 mm wide, ovoid to slightly S-shaped, rusty villous outside, glabrous inside, cohering except along dorsal suture; limb revolute, obovoid, enclosing style end before anthesis; **pistil** 25–28 mm long; stipe 1–1.5 mm long, silky; ovary appressed-villous; style glabrous except basal hairs near ovary, refracted above ovary almost at right angles, exserted first just below curve and looped upwards, straight to undulate after anthesis, ultimately strongly retrorse; style end hoof-like; pollen presenter oblique, oblong-elliptic, convex. **Fruit** 17–21 mm long, 8–11 mm wide, single-seeded, perpendicular to pedicel on incurved stipe, compressed oblong-ellipsoidal with an apical attenuation, longitudinally purple-striped, pubescent; pericarp c.1 mm thick. **Seed** 15–20 mm long, 4–5 mm wide, solitary, ellipsoidal, ± the same both sides, rugose with a central waxy border.

Distribution N.S.W., restricted to a small area in the northern Sydney suburbs, mainly in the Mona Vale area. *Climate* Hot, wet summer; cool, wet or dry winter. Rainfall 1000–1200 mm.

Ecology Restricted to eucalypt woodland on lateritised, sandstone ridges. Flowers late winter to summer. Killed by fire but regenerates readily from seed. Pollinated by birds.

Variation A uniform species showing little variation.

Major distinguishing features Branchlets villous; leaves pinnatipartite, with oblong lobes; conflorescence secund; perianth zygomorphic, glabrous inside; ovary shortly stipitate, densely hairy; fruits reddish-striped on strongly incurved stipe, single-seeded; seed ellipsoidal.

Related or confusing species Group 35. *G. caleyi* is a distinctive and easily recognisable species, most closely related to *G. aspleniifolia*.

Conservation status 2VC. Restricted and endangered. Most of the extant habitat of this species is on private land or road verge, but a small population is included within Ku-ring-gai Chase National Park.

Cultivation *G. caleyi* has been widely cultivated in gardens from Brisbane to Melbourne as well as inland, at least as far as Canberra and north-eastern and western Vic. Although sometimes difficult to establish, it is reasonably long-lived in most situations and tolerates frosts to at least -6°C, but it is not truly drought hardy, and even though it can tolerate extended dry periods, it appreciates summer watering. Like most grevilleas, it prefers a deep, gravelly, acid loam or sandy loam in full sun but will tolerate partial shade. Consistent and regular pruning may be necessary to maintain shape. Having brittle branches and a shallow root system, it is susceptible to strong winds and should be planted in a protected position. Its branches are sometimes ringbarked by a borer which feeds on the cambial layers, whereupon the shrub produces a gelatinous substance in response.

G. caleyi

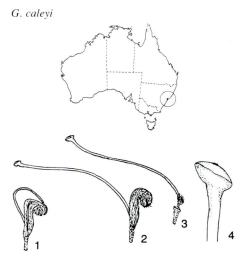

62A. *G. caleyi* Flowers, fruit and foliage (P.Olde)

62B. *G. caleyi* Shrub in natural habitat near Mona Vale, N.S.W. (N.Marriott)

Propagation *Seed* Easily propagated from fresh seed at any season and will self sow in the garden. Better germination is obtained using the standard peeling treatment. Hybridises readily in cultivation and, as a result, often does not breed true. *Cutting* Readily propagated from cuttings of half-hard wood in early spring or late autumn. Due to the hairiness of the branchlets, cuttings should be kept in a drier-than-usual propagating mix. *Grafting* Has been grafted successfully onto *G. robusta* using approach and mummy grafts.

Horticultural features *G. caleyi* is regarded as an outstanding foliage plant both for the form of its leaves and its hairy, purplish-red, new growth. The deeply and regularly divided leaves are quite fern-like and most attractive. They are sometimes used in flower arrangements. The purplish-red 'toothbrush' racemes are prolific, continuous for a long period and are very attractive to birds. The follicles, too, are eye-catching and resemble small colourful miniature birds. Allan Cunningham in 1825 reported that the seeds were a favourite food of black cockatoos. It makes an ideal screen plant or feature plant and, when planted among other shrubs, has a habit of producing branches that poke through other plants to the light beyond, creating a somewhat bizarre effect.

Hybrids In cultivation *G. caleyi* hybridises readily with related species such as *G. longifolia* and *G. aspleniifolia* and many beautiful, hardy hybrids have resulted, among them G. 'Telopea Valley Frond' and several Poorinda hybrids such as G. 'Poorinda Empress' and G. 'Poorinda Emblem'. Many plants sold in nurseries under the name *G. caleyi* are actually hybrids and care should be taken when purchasing this species that the plant is correctly named.

History *G. caleyi* was grown in England as early as 1830 from seed sent to William Aiton by A.Cunningham [Curtis' *Botanical Magazine* (1832)]. It was also grown at Woburn (1833) and Cambridge (1845) and was listed in the Botanic Gardens at Batavia (Indonesia) in 1851. In Australia, *G. caleyi* was introduced very early to cultivation and was first recorded in the Macarthur papers in 1843. Shepherd's Nursery of Sydney listed it in their 1851 catalogue.

General comments There is concern that wild populations of this species will hybridise with garden plants from nearby populated areas. The pure gene pool could thus be lost or considerably contaminated.

Grevillea calliantha R.O.Makinson & P.M.Olde (1991)   Plate 63

Telopea 4: 351 (Australia)

Foote's Grevillea, Black Magic Grevillea

The specific epithet is derived from the Greek *callos* (beauty) and *anthos* (a flower). KAL-EE-AN-THA

Type: near Cataby, W.A., 27 Sept. 1989, B.J.Conn 3283 & J.A.Scott (holotype: NSW).

A low-spreading, compact **shrub** c. 1 m tall, 2–3 m wide. **Branchlets** angular to terete, pubescent, brown-striped. **Leaves** 2–7.5 cm long, 3 cm wide, ascending, ± sessile, ovate to obovate in outline, pinnatipartite, the rachis ± straight; lobes 5–7, 1–4.5 cm long, 1–1.5 mm wide, linear, weakly pungent; upper surface pubescent becoming glabrous and faintly granulate with age, midvein scarcely evident; margins smoothly revolute to midvein; lower surface similar, grooves beside midvein hairy, midvein prominent. **Conflorescence** 5–7 cm long, deflexed to decurved, pedunculate, simple, terminal or axillary, many-flowered, somewhat lax, hemispherical-secund, sometimes on older stems; peduncle and rachis pubescent to villous; bracts 2.2–2.9 mm long, spreading, ovate, tomentose, persistent to anthesis. **Flower colour**: perianth pale yellow outside, apricot inside, sometimes reddish in throat; style maroon-black to reddish. **Flowers** acroscopic; pedicels 1–2 mm long, pubescent to villous; torus 1.3–2.1 mm across, oblique; nectary linguiform, spreading to patent; **perianth** 7–8 mm long, 2.5 mm wide, ovoid, villous outside, glabrous inside, cohering except along dorsal suture; limb revolute, densely villous, spheroidal, enclosing style end before anthesis; **pistil** (28.5-)30–40 mm long, sometimes digynous; ovary sessile, villous; style glabrous or minutely glandular-pubescent, villous at base, exserted first from near curve on dorsal side and looped upwards, incurved after anthesis, dilating evenly into style end; pollen presenter oblique, oblong-elliptic, convex. **Fruit** 13–18 mm long, 8–10 mm wide, oblong-ellipsoidal, compressed, oblique, pressed close to rachis, glandular-pubescent, sometimes red-striped; pericarp c. 1 mm thick. **Seed** 12.5 mm long, 5 mm wide, elliptic; outer face convex, smooth, ribbed at edge, border smooth; inner face with papery, radially oriented, gill-like platelets around margin.

Distribution W.A., confined to the Cataby area north of Perth. *Climate* Hot, dry summer; cool, wet winter. Rainfall 600–800 mm.

Ecology Occurs on rises in sand over laterite. Flowers winter–spring but some flowers all year. After fire regenerates from seed but has been noted sprouting from lower stems after a mild burn (P.Armstrong, pers. comm.). Probably pollinated by birds.

Variation A uniform species.

Major distinguishing features Leaves pinnatipartite, the margins smoothly rounded; venation of upper surface obscure; leaf rachis straight;

G. calliantha

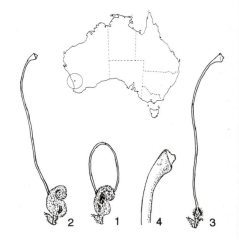

conflorescence deflexed, secund, lax; perianth zygomorphic, glabrous inside; ovary villous, sessile; pistil 30–40 mm long, blackish-maroon; fruit glandular-pubescent, sometimes with reddish striping; seed with gill-like platelets on inner face.

Related or confusing species Group 35, especially *G. crowleyae*, *G. armigera* and *G. hookeriana*, all of which have black or red-black styles. Both *G. armigera* and *G. hookeriana* have a pistil < 21.5 mm long. *G. crowleyae* has leaves with narrower lobes (0.8 mm wide) and hairs extending up to 10 mm along the style.

Conservation status Suggested 2EC-i. There are a few small populations scattered along a narrow, weed-infested road verge. Recently, about 120 plants have been found in bush on private land.

Cultivation Although *G. calliantha* has not been widely cultivated, it is rapidly becoming better known. It was introduced to cultivation by Zanthorrea Nursery near Perth in 1984 and has been successfully established in W.A. and Vic. Evaluation of adaptability to summer rainfall conditions is incomplete. In Vic., recent abnormal summer rains killed several plants, but potted specimens have remained healthy for many years at Bulli, N.S.W., and at the Royal Botanic Gardens, Sydney. Foliage is not damaged by frosts to around -3°C. A neutral to slightly alkaline, deep sand or loam in full sun or partial shade gives best results and once established it requires little maintenance. Mature plants should not be watered in summer. Pruning is unnecessary as the bush is naturally compact. Potted plants

63C. *G. calliantha* Conflorescence (N.Marriott)

require an open, free-draining mix with slow-release fertiliser. A few native plant nurseries in Perth and Vic. sell this species, usually as G. 'Black Magic'.

Propagation *Seed* Sets plentiful seed that germinates readily if pretreated. *Cutting* Firm, young growth strikes readily from cuttings taken at most times of the year. *Grafting* Successfully grafts onto G. 'Poorinda Royal Mantle' and *G. robusta* using several methods.

Horticultural features *G. calliantha* is a most desirable plant with attractive grey-green foliage and a compact, spreading habit. The branches tend to form layers bordered by a line of pendant conflorescences, presenting a somewhat exotic, oriental profile. Flowering is prolific. The flowers have unusual red-black styles highlighted by the apricot yellow perianth which in young flowers also has red tinges. Large quantities of nectar are produced within the perianth tube, inducing honeyeaters to regularly patronise the plant. It could be used as a low screen or feature plant as it appears to be both hardy and long-lived in most areas. As this species is endangered in the wild, only cultivated plants should be used for commercial propagation, whether by cutting or seed.

General comments The common name, given in a recent publication by the Department of Conservation and Land Management, W.A., acknowledges Nicholas Foote, who discovered the species.

Grevillea candelabroides C.A.Gardner (1964) Plate 64

Journal of the Royal Society of Western Australia 47: 56 (Australia)

The specific epithet, derived from the Latin *candelabrum* (a candelabrum) and the Greek *-oides* (like, resembling), refers to the conflorescence. KAN-DEL-AB-ROY-DEES

Type: near Ajana, W.A., 4 Jan. 1959, C.A.Gardner 12062 (holotype: PERTH).

A robust **shrub** to 5 m with spreading to erect branches. **Branchlets** terete, silky, soon glabrous. **Leaves** 13–22 cm long, ascending, petiolate, ovate in outline, pinnatipartite; lobes 7–14, 5–16 cm long, narrowly linear, closely aligned, pungent; upper surface silky, soon glabrous, glaucous, midvein and longitudinal veins evident in the lamina; margins

63A. *G. calliantha* Flowers and foliage (N.Marriott)

63B. *G. calliantha* Precarious habitat, near Cataby, W.A. (N.Marriott)

revolute to midvein, midvein flat or prominent; lower surface bisulcate, the grooves hairy, midvein prominent. **Conflorescence** erect, terminal, paniculate and candelabra-like, barely exceeding the foliage; unit conflorescence 10–25 cm long, 2–4 cm wide, dense, conico-cylindrical; peduncle and rachis glabrous; bracts 2.5 mm long, ovate with reddish tips, closely packed, silky outside, falling soon after expansion of the buds. **Flower colour**: perianth and style white to creamy white. **Flowers** acroscopic, irregular in bud; pedicels 3–8 mm long, glabrous; torus 1–1.5 mm across, oblique; nectary annular to V-shaped, cupped, entire; **perianth** 4–6 mm long, 1.5 mm wide, cylindrical, slightly dilated in middle, strongly curved, glabrous outside, shortly bearded inside at base, at first separate along dorsal suture, then all free below limb, the dorsal tepals reflexing and everting inner surface, all free to base and soon deciduous; limb revolute, globular, enclosing style end before anthesis; **pistil** 11–15 mm long, glabrous; stipe 2–3.5 mm long, terete, inserted centrally on torus and ovary; ovary globular; style lateral, at first exserted from below curve on dorsal side and looped upwards, strongly incurved after anthesis, suddenly dilated at style end; pollen presenter oblique, conical. **Fruit** 13–15 mm long, 8–9 mm wide, horizontal to stipe, flattened-ovoid with a decurved style base, viscid when young; pericarp 0.3 mm thick. **Seed** 7–8 mm long, 3–4 mm wide, elliptic, very thin, ± the same both sides, minutely and densely verruculose, faintly crimped around rim, broadly winged all round.

Distribution W.A., in the northern wheatbelt from the Murchison R. S to near Coorow. *Climate* Hot, dry summer; mild, wet winter. Rainfall 300–450 mm.

Ecology Grows in deep, yellow to grey, sandy soils often over laterite gravel in open heathland to tall shrubland. Flowers late spring to summer. Regenerates from seed after fire. Pollinated by insects.

Variation A uniform species.

Major distinguishing features Leaves pinnatipartite; conflorescence with many vertical branches; nectary annular to V-shaped, conspicuous; perianth zygomorphic, glabrous outside, bearded in basal 2 mm inside; pistil glabrous; stipe of ovary 2–3.5 mm long; pollen presenter conical; fruits, compressed-ovoid with a decurved style base, horizontal to stipe; pericarp wall very thin; seed winged.

Related or confusing species Group 3, especially *G. stenobotrya* which differs in its hairy outer perianth surface and its shorter stipe (< 2 mm long). Sometimes *G. stenobotrya* has an almost glabrous outer perianth surface.

Conservation status 3VC. Occurs in small populations over a wide range. Much of its former habitat has been cleared for agriculture.

Cultivation *G. candelabroides* has not been widely cultivated but it has been grown successfully in cool to warm, dry climates. Humid areas with summer rainfall are apparently not to its liking although grafted plants have succeeded. Like most grevilleas it is intolerant of poor drainage but is otherwise hardy. Once established, it tolerates frosts to at least -4°C and extreme drought. It has been grown in Melbourne, inland Vic., inland Qld, S.A. and Perth and it is also cultivated in the U.S.A. at the University of California, Santa Cruz. A full-sun situation in well-drained, acidic to neutral sand or sandy loam produces the most reliable results and it grows into a vigorous, rounded shrub which rarely needs pruning or watering. Fertilising is not recommended and could be quite harmful, encouraging short-term lushness while shortening the useful life of the plant. If grown in a large tub using a well-drained, gravelly mix, it can be induced to grow into an attractive flowering specimen. This method of cultivation is useful in areas where it is difficult to grow in the ground. It is occasionally available from specialist native plant nurseries in W.A., S.A. and Vic.

Propagation *Seed* Sets many seeds both in the wild and in cultivation. Even without special treatment these normally germinate readily in spring or autumn. Seedlings need a warm, sunny position or they can damp off rapidly. Average time for germination 27 days (Kullman). *Cutting* Seldom attempted from cutting since branches rapidly become woody, giving poor strike rates. The ease of propagation by seed makes that method more attractive. *Grafting* Has been grafted successfully onto a number of rootstocks including G. 'Poorinda Royal Mantle' and *G. robusta* using several techniques. Resultant plants have grown strongly and rapidly even with summer-rainfall.

Horticultural features For many reasons, *G. candelabroides* is widely regarded as one of the most spectacular grevilleas. It grows naturally into a well-shaped plant, an appealing character even if other features are ignored, but its finely divided, blue-green leaves and beautiful, large branching racemes of white flowers held erect just above the foliage are its strongest attributes. The floral arrangement is unusual and strongly reminiscent of a candelabra. Apart from their natural beauty, the showy flowers are sweetly scented and attract butterflies and other insects which assist in pollination. Although suitable for dense screening, it would be better planted as feature where its summer-flowering habit can be used to extend the flowering period of the garden. In addition, flowering may continue beyond the usual season when it is grown in suitable garden conditions.

Grevillea candicans C.A.Gardner (1942)
Plate 65

Journal of the Royal Society of Western Australia 27: 170 (Australia)

The specific epithet refers to the white colour of the flowers and is derived from the Latin *candescere* (to grow white). KAN-DEE-KANS

Type: 48 km N of Galena, W.A., Sept. 1940, W.E.Blackall 4718 (holotype: PERTH).

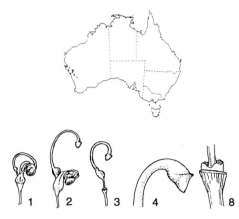

G. candelabroides

64A. *G. candelabroides* Candelabra-like conflorescence (F.Alley)

64B. *G. candelabroides* (N.Marriott)

Grevillea candicans

A dense, bushy **shrub** 2–3 m tall. **Branchlets** terete, tomentose. **Leaves** 8–24 cm long, ascending, petiolate, pinnatipartite; leaf lobes 5–7, 4–15 cm long, 1–2 mm wide, narrowly linear, pungent; upper surface convex, ± glabrous with scattered, appressed hairs, midvein and longitudinal marginal veins prominent; margins angularly refracted; lower surface bisulcate, woolly in grooves, midvein prominent. **Conflorescence** erect, pedunculate, terminal or axillary in upper axils, simple or few-branched, usually exceeding foliage; unit conflorescence 14–21 cm long, 5–8 cm wide, conico-cylindrical, dense to loose; peduncle and rachis glabrous or with sprinkled hairs; bracts 10–16 mm long, 8–13 mm wide, imbricate in bud, ovate, papery, sparsely tomentose, usually falling at or before anthesis. **Flower colour**: green in bud with bronze-pink bracts; perianth and style becoming creamy white or white; style end becoming yellow. **Flowers** acroscopic; pedicels 1.5–4 mm long, glandular-pubescent; torus ± 2 mm across, straight; nectary circular, cup-shaped; **perianth** 5 mm long, 2 mm wide, cylindrical, strongly curved from base, glabrous to sparsely glandular-pubescent outside, bearded inside near base, at first separating along dorsal suture, then all separate below limb, the dorsal tepals reflexed to expose inner surface; limb revolute, spheroidal, enclosing style end before anthesis; **pistil** 27–30 mm long, glabrous; stipe 3–4.5 mm long, terete, centrally inserted; ovary globose; style reflexed above ovary, at first exserted below curve on dorsal side, strongly incurved in a semi-circle after anthesis; style end conspicuous; pollen presenter oblique, obovate, slightly conical. **Fruit** c. 25 mm diam., a globular, pitted ball, falling but not opening when ripe; pericarp 8–9 mm thick. **Seed** 12–15 mm long, 10–12 mm wide, hemispherical; testa spongy.

Distribution W.A., inland from the central west coast from SE of Geraldton to the Murchison R. area, with a disjunct population E of Dalwallinu. *Climate* Hot, dry summer; mild, wet winter. Rainfall 300–430 mm.

Ecology Occurs in deep yellow or white sand in open shrubland and open woodland. Flowers spring to early summer. Regenerates from seed after fire. Pollinator unknown, possibly mammals such as honey possums or large insects.

Variation A uniform species with little variation except one population, noted in McGillivray (1993).

Xantippe form A population of smaller-flowered plants with more divided leaves occurs at this disjunct locality c. 200 km SE of the major distribution of the species.

Major distinguishing features Leaves pinnatipartite; lobes linear, pungent; conflorescence conical, conspicuous, simple or few-branched; floral bracts very large, papery; nectary annular; perianth zygomorphic, strongly curved, glandular-pubescent outside; pistil glabrous; stipe 3–4.5 mm long; fruit an indehiscent nut.

Related or confusing species Group 3. *G. annulifera*, a glabrous shrub, is closely related but differs in its paniculate conflorescence, its divaricately divided, pungent leaves and its dehiscent fruit.

Conservation status 3E. Mostly restricted to road verges. Much of its former habitat has been cleared for agriculture.

Cultivation Cultivation has been limited by difficulties in propagation but *G. candicans* has succeeded in warm, dry localities such as northern and western Vic. and inland Qld as well as summer-humid ones such as Brisbane. It has proven hardy to frosts to at least -4°C and tolerates extended dry conditions or drought, but hardiness and adaptability are not fully tested. Successful growers have planted it in a warm to hot, sunny situation in very well-drained, acid to neutral sand or gravelly loam. Shade should be avoided. Once established, it does not usually require any maintenance, growing naturally into a dense, compact shrub. It would make an outstanding potted specimen in a large tub using a

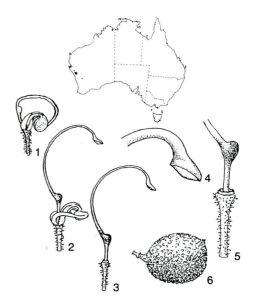

G. candicans

65A. *G. candicans* Markedly bracteose conflorescences (P.Olde)

65B. *G. candicans* Massed flowering (N.Marriott)

65C. *G. candicans* Plant in natural habitat, near the Murchison River, W.A. (P.Olde)

well-drained sand and peat mix with slow-release fertiliser. Native plant nurseries rarely stock this species, although it is occasionally available through seed suppliers.

Propagation **Seed** Due to the nature of the hard, woody fruits, seed is always readily available under large plants in the wild, but these require treatment prior to sowing. Firstly, the hard shell must be cracked and removed; the safest way is squeezing in a vice (use of a hammer usually results in damaged seed). Once this has been removed, the seed inside must have its tough skin removed, as is the case for most species. At this stage, viability can be readily assessed by checking colour of the embryo. Brown or discoloured embryos should be discarded in favour of creamy white viable ones. Average germination time is 39 days (Kullman). *Cutting* Several attempts using firm to hard tip growth have failed. More testing should be done using material from cultivated plants. *Grafting* Has been grafted successfully onto several rootstocks, including *G. robusta*, *G.* 'Poorinda Anticipation' and *G.* 'Poorinda Royal Mantle', mainly using the top wedge method, but most grafts, after initial vigorous growth, have collapsed and proved incompatible. Further testing is needed.

Horticultural features *G. candicans* is a curious but beautiful plant with green, pine-like foliage and massed racemes of large, white flowers with their large styles distinctively curled. Young buds have conspicuous, large, pinkish bronze bracts that fall as the flowers develop. Flowering is followed by large, round woody fruits that resemble quandong fruits. They are unique in the genus in that they do not split open at maturity but drop to the ground where they take many years to decompose and release the seed. Flowers are very strongly perfumed, especially at night when they attract night-flying moths and other insects and potential pollinators. This is a magnificent feature plant especially suited to gardens in warm, dry climates.

Grevillea candolleana C.F.Meisner (1845)
Plate 66

in J.G.C.Lehmann (ed.), *Plantae Preissianae* 1: 541 (Germany)

Toodyay Grevillea

Meisner did not explain the etymology but was probably honouring both Augustin de Candolle (1778–1841) and his son Alphonse (1806–1893), prominent Swiss botanists who published many works including a *Prodromus* intended to cover the flora of the world but never completed. KAN-DOLL-EE-ARE-NA

Type: near the Swan R., W.A., J.Drummond coll. 1, no. 628 (lectotype: NY) (McGillivray 1993).

Spreading to dome-shaped **shrub** c. 0.5 m high. **Branchlets** ± terete, tomentose. **Leaves** 1–3.5 cm long, 1–9 mm wide, ascending, sessile, simple, undulate, ovate, elliptic to linear, obtuse, mucronate, usually uncinate; upper surface convex, sometimes concave longitudinally or undulate, villous becoming glabrous and smooth, midvein and intramarginal veins prominent, lateral venation faintly evident; margin entire, strongly recurved; lower surface white-pubescent, midvein evident. **Conflorescence** 1–2 cm long, 2–3 cm wide, erect, sessile, terminal or axillary in upper axils, simple, umbel-like; rachis villous; bracts 1.6–3.2 mm long, narrowly ovate, villous outside, persistent at anthesis. **Flower colour**: perianth and style cream or creamy white; style end yellow, orange or red. **Flowers** abaxially oriented: pedicels 7–15 mm long, villous; torus c. 1 mm across, ± straight; nectary V-shaped or cushion-like, smooth; **perianth** 3–4 mm long, 1–2 mm wide, cylindrical, the tepals separating to base on dorsal side, cohering on ventral side but rolling down strongly at anthesis forming small platform in front, ± uniformly villous on both surfaces; limb conspicuous, tightly and strongly revolute, villous, ovoid, the style end exposed before anthesis, dorsal tepals auriculate adjacent to anthers; **pistil** 9.5–11.5 mm long, villous, completely exposed before anthesis through expansion of perianth; stipe 0.3–0.6 mm long; ovary oblong; style not bowed out before anthesis, straight after anthesis; style end incurved, glabrous; pollen presenter lateral, concave, oblong, exceeded and partly obscured by a tongue-like, curled appendage. **Fruit** 8.5–11 mm long, 4.5–5 mm wide, oblique on pedicel, ovoid, faintly ridged, pubescent; pericarp 0.5–1 mm thick. **Seed** 7.5 mm long, 2.5 mm wide, elliptic; outer face smooth, convex with an apical transverse rib beneath a short elaiosome, margins revolute; inner face flat to concave, smooth.

Distribution W.A., confined to a few small areas near Toodyay, NE of Perth. *Climate* Hot, dry summer; cool, wet winter. Rainfall 600–800 mm.

Ecology Grows on well-drained ridges in gravelly and stony clay-loam in open Wandoo forest. Flowers winter to spring. Fire response unknown. Pollinator unknown, although the form of flower opening suggests native bees.

Variation A relatively uniform species with little variation.

Major distinguishing features Leaves simple, entire, the upper surface smooth, the undersurface exposed; conflorescence sessile, umbel-like; perianth zygomorphic, villous inside and out; dorsal tepals auriculate at the anthers; pistil villous; style end glabrous, with an incurled appendage.

Related or confusing species Group 22, especially *G. scabra* which is distinguished by its scabrous upper leaf surface.

Conservation status 2E. Restricted to road verges and private property in a very small area. Immediate,

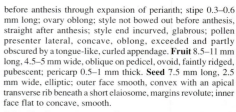

G. candolleana

66A. *G. candolleana* Massed flowering (P.Olde)

66B. *G. candolleana* Flowers and foliage (N.Marriott)

66C. *G. candolleana* Regenerating in gravel scrape, Toodyay, W.A. (P.Olde)

active conservation measures are urgently required.

Cultivation *G. candolleana* has been grown for short periods at Burrendong Arboretum, N.S.W., and in a few private gardens in Vic. but, while hardy to frost to at least -6°C, has proved unreliable in cultivation, most plants dying within a few years. Accordingly, it must be regarded as a somewhat 'touchy' species that does not thrive over a wide range of conditions. It seems to prefer a cool, dry climate but its tolerance of extended, dry periods and summer humidity remain untested. For optimal growth, a warm, sunny or semi-shaded site should be chosen, with well-drained, acidic loam or gravelly loam that retains some sub-soil moisture during summer, although it will not tolerate poor drainage. It responds vigorously to slow-release fertiliser but this may shorten its life. It makes an ideal pot plant and should be grown in a well-drained, nutritious mix. Rarely available commercially.

Propagation Seed Sets moderate amounts of seed both in the wild and in cultivation; these germinate well if peeled or soaked. *Cutting* Firm, young growth strikes well at most times of the year. *Grafting* Has been grafted successfully onto *G. robusta*.

Horticultural features *G. candolleana* is a delightful, small shrub producing a massed spring display of woolly, cream lightly perfumed flowers. The curious, brightly coloured yellow, orange or red style ends make a striking contrast with the flower colour and are often commented upon by growers. The species can be used to highlight other plants in the garden or as an eye-catching border plant. A massed planting would be very effective, or it would make an ideal rockery plant. The oval leaves, usually dark green above and white below, often turn an attractive bronze in the winter months. Its short lifespan on its own roots severely limits its usefulness for commercial landscapes.

Grevillea capitellata C.F.Meisner (1856)
Plate 67

A.P. de Candolle, *Prodromus Systematis Naturalis Regni Vegetabilis* 14: 356 (France)

The specific epithet is derived from the Latin diminutive *capitellatus* (bearing small heads), in reference to the conflorescences. CAP-IT-ELL-ARE-TA

Type: Illawarra district, near Port Jackson, N.S.W., Oct. 1818, A.Cunningham (lectotype: G-DC) (McGillivray 1993).

Compact, much-branched, sometimes lignotuberous **shrub** to 50 cm, sometimes prostrate. **Branchlets** usually arching, angular, prominently ribbed, pubescent to villous, rarely sparsely so to almost glabrous, usually secund; new growth rusty. **Leaves** 2–9 cm long, 2–8 mm wide, spreading to retrorse, shortly petiolate, simple, narrowly elliptic to oblong-lanceolate, leathery, acute to obtuse, mucronate; upper surface ± glabrous or with curled hairs at base, sparingly punctate, lateral and intramarginal veins usually raised with scattered granules on intramarginal veins; margins entire, refracted vertically about intramarginal vein; lower surface silky, overlain by many ascending, long black hairs, midvein more prominent. **Conflorescence** erect, terminal, pedunculate, simple, very condensed, subglobose to subsecund 1.5 cm long, 2 cm wide; peduncle short (c. 5 mm long) relatively stout (0.7–1 mm thick), villous, rachis villous; bracts 2–5.5 mm long, narrowly elliptic to linear, rusty-villous, falling usually just before anthesis. **Flower colour**: perianth dull, black-red or red; style maroon-red. **Flowers** acroscopic; pedicels 1.5–2 mm long, villous; torus 1–1.8 mm across, slightly oblique; nectary arcuate with a broad, smooth margin; **perianth** 6–7 mm long, 1.5 mm wide, oblong to ellipsoidal, villous outside, bearded inside c. level with ovary, sometimes with extensive papillae elsewhere, cohering except along dorsal suture before anthesis, afterwards all tepals free to beard and independently rolled down; limb revolute, spheroidal, villous, enclosing style end before anthesis; **pistil** 10–11.5 mm long; stipe 1.5–2.5 mm long, slightly flattened with 2 longitudinal grooves; ovary oblong-ovoid, not prominent; style glabrous except sparse hairs near pollen presenter, exserted first near base on dorsal side and bowed out, after anthesis angularly incurved, geniculate c. 1–2 mm from flattened, plate-like style end after anthesis; pollen presenter oblique, oblong, convex, umbonate. **Fruit** 18 mm long, 5.5 mm wide, erect, ± cylindrical or narrowly ovoid, smooth; style persistent; pericarp c. 0.3 mm thick. **Seed** 8.5–9 mm long, 2.5 mm wide, oblong with a subapical cushion-like swelling, minutely pubescent on both surfaces; outer face convex, smooth with slight wrinkling; inner face concave, obscured, the margin strongly but loosely revolute, narrowly winged along 1 side, the wing drawn to a terminal apical point 2.5–4 mm beyond testa, sometimes also with a short basal point.

Distribution N.S.W., confined to the Illawarra area between Bulli, Mt Ousley, Cordeaux Dam and Cataract Dam. *Climate* Hot, wet summer, cool to cold, wet or dry winters. Rainfall 1000–1200 mm.

Ecology Grows in or at the margins of swamp or low-lying, poorly drained depressions in moist, grey, sandy loam with laterite. Flowers winter to early summer. Regenerates from seed or lignotuber after fire. Pollinator unknown.

Major distinguishing features Branchlets angular, ribbed, villous; leaves leathery, simple, entire, the undersurface silky with many long ascending hairs, the upper surface conspicuously veined, the veins with many granules, the margins straight, refracted vertically about intramarginal vein; conflorescence subglobose, very condensed; peduncle stout; perianth zygomorphic, villous outside, bearded and papillose within; pistil glabrous except minute trichomes on style end; fruit smooth, ellipsoidal.

Related or confusing species Group 21, especially *G. diffusa* which differs in its thinner and mostly smaller leaves with a silky indumentum, the lateral venation obscure on the upper leaf surface, the leaf margins shortly and irregularly recurved; peduncle fine, and *G. oldei*, which differs in its wiry peduncle and its style hairy over most of its length.

Variation A relatively uniform species.

Conservation status 2V suggested. Much of the habitat of this species remains undisturbed, being part of the catchment land protected for the Sydney water supply.

G. capitellata

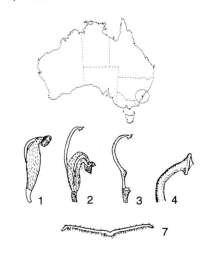

67A. *G. capitellata* Habit of growth in cultivation (P.Olde)   67B. *G. capitellata* Flowers and foliage (P.Olde)

Cultivation *G. capitellata* is a very adaptable, hardy plant that has been cultivated successfully in the eastern States of Australia for many years. Light frosts and extended dry periods are tolerated well and it is unaffected by summer humidity. It thrives in an open, well-drained, gravelly loam or sandy, acidic loam but may be more tolerant of poor drainage than other related species. When grown in full sun it is a compact, dense plant but in shade (which it survives tolerably well) it may need pruning to prevent legginess. Summer watering is appreciated and light fertiliser improves appearance and vigour. Rarely available from native plant nurseries.

Propagation **Seed** Untested but fresh seed pretreated by nicking or peeling should germinate readily. **Cutting** Readily propagated from cuttings of firm, new growth taken at any time of the year. **Grafting** Has been grafted successfully onto *G. robusta* using the top wedge technique.

Horticultural features: *G. capitellata* is a distinctive plant with arching branchlets and dense foliage. The leaves are relatively long, leathery and have a somewhat shaggy indumentum. The small flower heads are terminal but inconspicuous, being dull maroon, the colour further hidden by the dense indumentum and the foliage. In cultivation, it develops a dense spreading to prostrate habit which renders it useful as a ground cover or rockery plant. Generally a low-maintenance, pest-free plant with much to recommend it for broad-scale landscaping where brightly coloured flowers are not essential.

General comments McGillivray (1993) submerged *G. capitellata* in *G. diffusa*, as the 'robust form', but it is virtually sympatric with the typical silky form of *G. diffusa* and, given its distinctive morphology, should in our opinion be recognised as a distinct species.

Grevillea centristigma (D.J.McGillivray) G.J.Keighery (1992)   Plate 68

Nuytsia 8: 227 (Australia)

Based on *G. drummondii* subsp. *centristigma* D.J.McGillivray (1986)

The specific epithet alludes to the stigma which is located in the middle of the pollen presenter and is derived from the Latin *centralis* (central, in the middle) and *stigma* (a mark–in botany, the stigma). SENTRY-STIG-MA

Type: 24 km SE of Pemberton on road to Northcliffe, W.A., 1 Oct. 1967, P.G.Wilson 6289 (holotype: PERTH).

A low, compact or erect, rounded **shrub** 0.3–1 m tall. **Branchlets** angular to terete, tomentose, soon glabrous. **Leaves** 1.5–4 cm long, 0.2–0.9 cm wide, ascending, sessile, simple, elliptic to obovate, soft; upper surface sparsely villous, becoming ± glabrous, the midvein faint; margins entire, shortly recurved, ciliate; lower surface sparsely villous, midvein and lateral veins prominent. **Conflorescence** 0.3–1 cm long, 1 cm wide, erect to decurved, subsessile, unbranched, terminal on short lateral branchlets, 5–7-flowered, umbel-like; peduncle and rachis sparsely villous; bracts 1.2–3 mm long, narrowly triangular, sparsely villous outside, persistent to anthesis. **Flower colour**: perianth pale yellow-orange; style end red; flower colour deepens with age. **Flowers** adaxially oriented; pedicels 2.5–3 mm long, sparsely villous; torus 1.5–2 mm across, very oblique; nectary U-shaped, smooth; **perianth** 4–6 mm long, 2 mm wide, ellipsoidal, ribbed, pilose outside, glabrous inside except a beard at throat, cohering except along dorsal suture from curve to limb; limb nodding, globular, the segments keeled; **pistil** 5–6 mm long; stipe 1.5–2 mm long; ovary villous; style not exserted before anthesis, afterwards straight and barely exceeding perianth, villous; style end thick, spoon-like; pollen presenter lateral, round, flat to concave; stigma central. **Fruit** 8–12 mm long, 5 mm wide, erect, ovoid, tomentose; pericarp c. 0.5 mm thick. **Seed** 6–7 mm long, elliptic, acute, margins recurved, channelled on inner face around rim, otherwise smooth.

Distribution W.A., widespread in the southern Jarrah and Karri forests from Armadale to the Whicher Range and SE to at least Shannon. *Climate* Warm, dry summer; cool, wet winter. Rainfall 800–1200 mm.

Ecology Grows in shaded to semi-shaded sites in Jarrah and Karri forest in gravelly loam; sometimes around the edge of granite outcrops in wet sand. Flowers winter to spring. Some plants may be lignotuberous but most appear to be killed by fire and regeneration is from seed. Probably pollinated by birds.

Variation A relatively uniform species with minor variation in habit and indumentum.

Robust form One population near Jarrahdale in the Darling Range has a fairly compact, rounded habit to c. 1 m tall, a dense villous indumentum on leaves and branchlets and slightly larger leaves and flowers. Conflorescences are more conspicuous although still somewhat hidden.

Major distinguishing features Leaves simple, entire, narrowly elliptic to obovate, sparsely hairy, ciliate; conflorescence few-flowered; perianth zygomorphic, yellow, hairy outside, bearded at throat inside; ovary stipitate, villous; style scarcely to not exceeding perianth; pollen presenter round, flat to concave with the stigma central.

Related or confusing species Group 23, especially *G. drummondii*, *G. fistulosa*, and *G. pimeleoides*. In *G. drummondii* the perianth is glabrous outside. *G. fistulosa* differs in its rolled leaf margins, pedicels 5–6.5 mm long and all-red flowers. *G. pimeleoides* has leaves mostly broader than 1.5 cm., a conflorescence with 13–18 flowers, a pistil 6–8 mm long with an obovate pollen presenter and a sub-basal stigma.

Conservation status Not currently endangered.

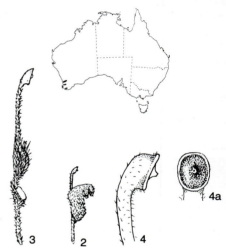

G. centristigma

Cultivation *G. centristigma* has been in cultivation for a number of years and is proving reasonably hardy and adaptable. It succeeds in cool-wet and warm-dry climates but will not take summer humidity and rain. It tolerates frost to at least -6°C as well as short dry periods, although sub-soil moisture is essential. It grows well in full sun to full shade with best results in partial shade. In a well-drained, acidic, sandy or heavy loam soil, it grows reliably into a neat, compact low shrub that rarely requires pruning or fertilising. In dry conditions, summer watering produces a vigorous burst of new growth. It makes an interesting pot plant but requires a well-drained peaty mix with fertiliser. Native plant nurseries and seed suppliers rarely stock this plant.

Propagation *Seed* Sets small quantities of seed that should respond to standard pre-sowing treatments. *Cutting* Strikes well from cuttings of firm, young growth at any time of the year. *Grafting* Has been grafted successfully onto *G. robusta* and G. 'Poorinda Royal Mantle' using top wedge and mummy grafts.

Horticultural features *G. centristigma* is valued mainly as a foliage plant because of its soft, hairy leaves that are somewhat silky to the touch. It is an attractive, low, mounded shrub or scrambling undershrub and would be useful in a semi-shaded landscape as a low rockery plant or small mounded specimen, but if not pruned may become open and sparse. The yellow flowers are interesting and attractively coloured but very small and usually hidden by the foliage.

Grevillea ceratocarpa L.Diels (1904)

Plate 69

Botanische Jahrbücher für Systematik 35: 157 (Germany)

The specific epithet is derived from the Greek *ceras* (a horn) and *carpos* (a fruit) a reference to the pointed, ear-like projections on the apex of the fruit.
SER-RAT-OWE-CAR-PA

Type: Bronti, near Southern Cross, W.A., 20 Nov. 1901, L.Diels 5980 (lectotype: B).

Synonym: *G. integrifolia* subsp. *ceratocarpa* (L.Diels) D.J.McGillivray (1986).

Erect **shrub** 0.5–1(–1.5) m high. **Branchlets** pubescent, terete. **Juvenile leaves** broadly elliptic (up to 22 mm wide), the upper surface loosely villous with a spreading indumentum, the midvein obscure, the lower surface similar with prominent midvein and reticulum, margins flat to angularly infracted; **adult leaves** usually bifacial, 5.5–8.5 cm long, 2–10 mm wide, ascending to erect, sessile, leathery, simple, narrowly elliptic or obovate, the rachis straight, acute to obtuse with short, blunt, frequently recurved point, sometimes the leaves unifacial, linear, ribbed both sides, deeply grooved on adaxial surface beside midvein; margins entire, flat to undulate, sometimes angularly infracted; upper surface tomentose with spreading hairs, venation obscure; lower surface tomentose, longitudinally parallel-ribbed, the ribs prominent and sometimes glabrous, usually slightly more obscure than midvein. **Conflorescence** erect, pedunculate, terminal or axillary usually in the upper axils, simple or branched; unit conflorescence 7–?15 cm long, 1–1.5 cm wide, cylindrical, the flowers opening ± synchronously, sometimes from apex first; peduncle glabrous; rachis yellow, glabrous; bracts 2–4 mm long, narrowly obovate-spathulate to linear, loosely villous to villous, (usually with long, spreading, wavy white hairs only), falling early. **Flower colour**: perianth and style creamy white. **Flowers** regular, glabrous, adaxially oriented; pedicel 2.2–3 mm long; torus c. 0.4 mm across, straight; nectary absent, sometimes present as a short, horn-like structure; **perianth** (below limb) 3–3.5 mm long, 0.5 mm wide, oblong; tepals channelled inside, with obscure to prominently exserted V-shaped labia below anthers, before anthesis separating to base and bowed out below coherent limb, afterwards all free; limb 2.2 mm long, ellipsoidal, the segments slightly flanged at margins, an obscure midrib at base of each; distance from base of anthers to narrowed section of tepal 0.5–1 mm; **pistil** 7.5–10 mm long, glabrous; stipe 0.5–1.2 mm long; ovary 1 mm long, lateral, obovoid with 2 apical, ventral lobes; style 6–8 mm long, 0.25 mm thick, minutely granular, deeply grooved, before anthesis exserted from near base and folded outwards, afterwards noticeably kinked about middle, becoming straight to undulate; style end 0.5–0.7 mm across, straight; pollen presenter 1.5–1.8 mm long, cylindrical-urceolate; stigma cup-shaped. **Fruit** 12–18 mm long, 1.5–2.5 mm wide, oblique, narrowly obovoid-cylindrical with slightly recurved apex, smooth, often persistent; pericarp 0.2–0.5 mm thick. **Seed** 4–8.5 mm long, 1.2–2 mm wide, obovoid or narrowly so, rugose to smooth; apex and base with a membranous band 0.5–1 mm wide; outer face convex; inner face flat.

Distribution W.A., restricted to an area centred on Southern Cross, bounded by Merredin, Tandegin, Parker Range and Woolgangie. *Climate* Summer hot, dry; winter wet, cold. Rainfall 250–300 mm.

Ecology Usually grows on open sandplain, occasionally in mixed, taller scrub in yellow sand, sometimes white sand with laterite. Flowers late spring–summer.

Major distinguishing features Branchlet and leaf indumentum spreading; leaves unifacial and ribbed both sides or bifacial and prominently veined on lower surface; margins flat to angularly involute; conflorescence cylindrical; rachis yellow, glabrous; bracts linear 2–4 mm long, loosely villous; nectary absent to obscure; flowers regular, glabrous; ovary ventrally bilobed at apex; pollen presenter cylindrical–urceolate; fruit narrowly obovoid–cylindrical with recurved apex.

Related or confusing species Group 2, especially *G. eremophila* and *G. biformis*, both of which differ in their appressed foliar and branchlet indumentum.

68B. *G. centristigma* Flowering branch (C.Woolcock)

68C. *G. centristigma* Habit, in cultivation, Stawell, Vic. (N.Marriott)

68A. *G. centristigma* Flowers and fruit (N.Marriott)

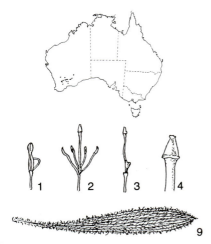

G. ceratocarpa

G. biformis also differs in its shorter fruit and usually hairy floral rachis. *G. eremophila* is a more robust shrub (to 3 m high) and usually has shorter, silky-tomentose floral bracts (< 2 mm long), longer leaves and conflorescences.

Variation A relatively uniform species.

Conservation status Not presently endangered.

Cultivation *G. ceratocarpa* has not adapted well to cultivation, being rarely found in gardens despite many attempts, and is widely regarded as 'touchy'. Its natural occurrence in a low-rainfall, low humidity inland climate and its poor adaptability (on present knowledge), suggests replication of its home environment will bring best success. It is hardy to extended dry conditions and frosts to at least -5°C. A well-drained, deep, acidic to neutral sand or sandy loam in full sun is recommended. Once established it dislikes fertiliser and summer watering but has succeeded in large tubs and open-ended pipes etc. placed in a warm, sunny site. Not usually available in nurseries or from seed suppliers.

Propagation *Seed* Sets numerous fruit but seed is rarely collected. The standard pretreatment of nicking the testa should produce good results. *Cutting* Difficult to strike. Some success has been achieved using firm, young growth taken during the warmer months of the year. Cuttings do best in a dry, unmisted cutting frame with hand-watering. *Grafting* Untested to date.

Horticultural features *G. ceratocarpa* is an attractive, showy plant with most interesting woolly white, lanceolate leaves. It has potential as a feature or contrast plant and its bright, white flowers, borne so conspicuously above the foliage, would contrast strongly with other garden foliage and provide a lift to the garden. The flowers are also delicately perfumed and attract many insects from late spring. Difficulties so far experienced in introducing it indicate that collectors and enthusiasts will gain more reward than commercial growers.

69. *G. ceratocarpa* Flowers and foliage (W.R.Elliot)

Grevillea christinae D.J.McGillivray (1986)   Plate 70

New Names in Grevillea *(Proteaceae)* 4 (Australia)

The specific epithet honours Mrs Christine Cornish who assisted Mr Don McGillivray with his revision of the genus *Grevillea*. KRISS-TEEN-EYE

Type: near Mortlock R., SW of Goomalling, W.A., 1 Sept. 1979, A.S.George 15738 (holotype: PERTH).

Open or bushy **shrub** to 1 m. **Branchlets** flexuose, angular, silky when young, soon glabrous. **Leaves** 2–6 cm long, 1–6.5 mm wide, patent, often solitary and widely spaced, sessile, simple, linear-lanceolate, acute, pungent; upper surface silky becoming glabrous, midvein and 2 longitudinal, intramarginal veins evident; margins entire, recurved to revolute; lower surface glabrous, sometimes sparsely silky, midvein prominent. **Conflorescence** c. 1 cm long, 1 cm wide, erect, pedunculate, axillary or terminal on short branchlets, unbranched, few-flowered, (usually c. 12 per head), umbel-like; peduncle and rachis tomentose; bracts 0.9–1.2 mm long, narrowly triangular, tomentose outside, usually falling before anthesis. **Flower colour**: perianth and pistil white; style end sometimes turning pink. **Flowers** adaxially oriented; pedicels 4–5.5 mm long, silky; torus c. 1 mm wide, square to round, ± straight; nectary arcuate, entire; **perianth** 3–4 mm long, 1.5 mm wide, oblong, silky outside, bearded inside near base, at first separating along dorsal suture, all tepals then separating below limb, the dorsal tepals reflexed and exposing inner surface before anthesis, afterwards all free to base and rolling down independently; limb tightly revolute, spheroidal, enclosing style end before anthesis; **pistil** 7–8.5 mm long; stipe 0.6–0.8 mm long; ovary ovoid, glabrous; style glabrous except inconspicuous hairs or papillae in apical few mm, grooved along ventral side, at first exserted along dorsal suture and looped upwards, incurved after anthesis, geniculate near apex; style end flat, plate-like; pollen presenter slightly oblique, conical, the base flanged. **Fruit** 10–15 mm long, 3–5 mm wide, erect, narrowly ovoid-ellipsoidal, glabrous, ribbed; follicle wall c. 0.3 mm thick. **Seed** 12 mm long 3 mm wide, elliptic; outer face smooth, margin revolute with a short apical elaiosome; inner face concave.

Distribution W.A., known from only 4 populations between Watheroo and Goomalling. *Climate* Hot, dry summer; cool, wet winter. Rainfall 350–400 mm.

Ecology Grows in *Acacia*-York Gum woodland and low shrubland in winter-moist, sandy clay-loam close to salt marsh vegetation. Flowers winter to mid-spring. Apparently killed by fire and regenerates from seed. Pollinator unknown, probably insects.

Variation A uniform species with minor variation.

Major distinguishing features Branchlets flexuose, angular; leaves simple, entire, glabrous or sprinkled with silky hairs, widely spaced, the undersurface exposed; conflorescence few-flowered, open, umbel-like; perianth zygomorphic, undilated, hairy on both surfaces; ovary glabrous; style end minutely hairy or papillose; fruit faintly ribbed.

Related or confusing species Group 21, especially *G. costata* and *G. inconspicua*, both of which differ in having the leaf undersurface concealed by the revolute margins and in the more prominently ribbed fruits. There is also a close relationship, in conflorescence and floral morphology, with some members of Group 16, which includes *G. hakeoides*, which differs in its smooth, glabrous style end.

Conservation status 3V. Known from four populations, two of which have a good number of plants.

Cultivation *G. christinae* has not been grown widely but where grown has so far proved to be an adaptable, hardy plant, tolerating frosts to at least -4°C and extended dry periods but not summer moisture and humidity. Closely related species are long-lived in cultivation. It likes a well-drained acidic to neutral or slightly alkaline sandy loam and may be salt-tolerant. It thrives in a warm, sunny to semi-shaded situation. Regular tip-pruning will assist to make the plant more compact. Plant at 1 m spacing for best effect. Occasionally available from native plant nurseries in Vic.

Propagation *Seed* Untested. *Cutting* Firm, young growth strikes readily at any time of the year. *Grafting* Has been grafted successfully onto *G. robusta*.

Horticultural features *G. christinae* is an intriguing plant with many rather wavy, red branchlets adorned with well-spaced, spreading leaves, in all looking something like arms of barbed wire. Apart from its curiosity value, this is a rather unremarkable shrub for most of the year, although with age it forms a dense, rounded shrub of excellent landscape value, with the bonus of an all-over cover of white flowers. This spectacular winter display provides a stark contrast to the surrounding dull green vegetation of most gardens at this time of the year, while the scented flowers have the added attraction of introducing a pleasant honey perfume to the air, especially noticeable around the middle of the day. Its rarity in the wild and the consequent danger of extinction provide additional incentives for cultivation. It would surely be of considerable value in large, urban landscapes, both for its hardiness, its unusual foliage and its delightful flowering habit. Testing should commence for its suitability in revegetating salt-affected soils.

G. christinae

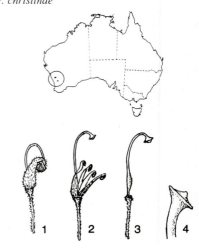

70A. *G. christinae* Flowering branch (N.Marriott)

70B. *G. christinae* Shrub in full flower, in cultivation, Stawell, Vic. (N.Marriott)

Grevillea chrysophaea F.Mueller ex C.F.Meisner (1854) Plate 71

Linnaea 26: 357 (Germany)

Golden Grevillea

The specific epithet is derived from the Greek *chrysos* (gold) and *phaios* (dusky, brown), a reference to the golden-brown flowers. KRY-SOW-FEE-A

Type: precise locality unknown, Vic., 1851–52, F.Mueller (holotype: NY). The ?isotype was collected by John Dallachy in August 1850 'about 20 miles this side of Geelong on a sandy ridge' (note on the herbarium sheet, MEL), probably referring to the Brisbane Ranges.

Synonym: *G. chrysophaea* var. *canescens* H.Williamson (1927).

Single-stemmed **shrub** 0.3–2 m high. **Branchlets** angular, ribbed, tomentose. **Leaves** 1–6 cm long, 5–15 mm wide, ascending to spreading, ± sessile, simple, oblong-elliptic to elliptic, obtuse, mucronate; upper surface convex, pubescent to glabrous, minutely granular (rarely almost smooth), midvein and lateral veins yellowish, evident; margins entire, shortly recurved or revolute; lower surface tomentose to velvety, midvein prominent. **Conflorescence** decurved, shortly pedunculate, terminal, simple or few-branched; unit conflorescence 1.5–2 cm long, 3 cm wide, 2–8-flowered, ± secund, umbel-like to globose, the outermost flowers opening first; peduncle and rachis tomentose to villous; bracts 1.2–2.7 mm long, ovate-acuminate, villous to tomentose outside, falling before anthesis. **Flower colour**: perianth golden tan, yellow or golden yellow; style reddish brown to orange-red. **Flowers** acroscopic; pedicels stout, 4–6 mm long, tomentose to villous; torus 1.5–2 mm across, oblique; nectary extending < 1 mm above toral rim, oblong-arcuate to U-shaped, collar-like, the margin sometimes recurved; **perianth** 5–10 mm long, 3–5 mm wide, ± oblong with ventral dilation at base, very strongly curved to geniculate midway, the limb usually adjacent to perianth base in bud, sparsely tomentose to villous outside, bearded in lower half inside and sparsely hairy above; tepals cohering except along dorsal suture, the style end exposed before anthesis; limb conspicuous, revolute, ± ovoid, emarginate or depressed, the tepal limbs prominently keeled, the keels sometimes attenuated into horn-like projections; **pistil** 15–21.5 mm long; ovary sessile, villous; style first exserted at curve on dorsal side and looped upwards, gently incurved becoming straight after anthesis, villous, the hairs usually coloured; style end flattened, disc-like; pollen presenter lateral, elliptic, convex, rimmed. **Fruit** 11.5–13 mm long, 5.5–6 mm wide, erect, ellipsoidal, villous, longitudinally ridged; pericarp c. 0.5 mm thick. **Seed** 10 mm long, 2.5 mm wide, ellipsoidal with apical elaiosome; outer face convex, wrinkled; inner face flat, minutely pitted; margin strongly recurved, bordered by a waxy edge.

Distribution Vic., widespread mainly in the south from the Brisbane Ranges, north of Geelong, and in Gippsland from Licola to Woodside. *Climate* Warm to hot, dry summer; cool to cold, wet winter. Rainfall 500 mm in the Brisbane Ranges, 1500 mm near Licola.

Ecology Grows in a variety of habitats, including open to closed eucalypt woodland, banksia woodland, mixed scrub on rocky hillsides and heathland, in sand, sandy clay, sandy loam or stony clay. Flowers winter to early summer, often greatly extended under cultivation. Regenerates from seed or sucker; sometimes killed by fire. Pollinated by birds.

Variation A species that shows some variation in habit (low, spreading or erect), leaf size and shape, flower colour and indumentum. The following forms are recognised.

Gippsland form This is usually lower and more rounded than the Brisbane Ranges form and occurs in widely disjunct areas from west and east Gippsland. Flower colour ranges from dull green-yellow to gold. In good forms, the flowers are spectacular although often hidden in the foliage. One collection from Spermwhale Head was described as a variety because of the white-villous flowers. Habitat ranges from stony clay on dry mountain ridges to deep sand in coastal scrub. A form from Providence Ponds has attractive grey-green leaves. Plants near Gormandale are at risk from clearing for pine planting.

Brisbane Ranges form This is generally an open to straggly shrub with rusty-golden, conspicuously hairy flowers that are slightly larger than other forms. It occurs in the Brisbane Ranges, where it

G. chrysophaea

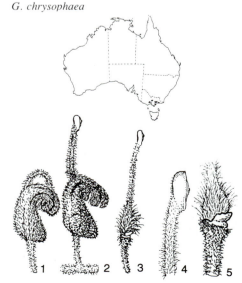

grows in poor sandy soil. This is an excellent plant to cultivate in dry conditions as well as beneath established trees. At the western end of its range near Steiglitz it tends to be a more compact, rounded shrub with rather dull flowers. It is sometimes misidentified as *G. floribunda*, a species to which it is very closely related.

Holey Plains form This form, found SW of Sale, has the narrowest leaves (c. 5 mm wide) and is either low and spreading or prostrate, to 1 m wide. A selected form from near Rosedale has massed flowers smothering the plant in winter and spring. The deep golden flowers are less hairy and smaller than most forms. This form may represent a distinct infraspecific taxon.

Major distinguishing features Single-stemmed habit; leaves simple, elliptic to obovate, entire, strongly convex with visible secondary veins; lower surface usually velvety; floral rachis 1–6.5 mm long; nectary oblong-arcuate, collar-like, not exceeding torus; perianth zygomorphic, strongly recurved, oblong with basal ventral dilation, the hairs outside mostly pale, inside crowded in a beard near base; limb prominently keeled or with apical horns; ovary sessile, villous; style with coloured hairs; pollen presenter lateral.

Related or confusing species Group 25, especially *G. alpina*, *G.* sp. aff. *chrysophaea*, and *G. floribunda*. *G. alpina* has usually smaller leaves (1–2 cm long, c. 3 mm wide) with obscure secondary veins, floral rachis usually longer than 10 mm, slender pedicels, usually smaller flowers revolute in upper half and a patent tongue-like nectary extending laterally 1–3 mm beyond the torus. *G.* sp. aff. *chrysophaea* differs in its suckering habit, red and yellow flowers and ascending, shovel-like nectary extending up to 1.5 mm beyond the torus. *G. floribunda* differs in its longer floral rachis (usually > 15 mm) and ellipsoidal perianth with the beard inside near the curve.

Conservation status Not presently endangered. A widespread species occurring in several National Parks.

Cultivation *G. chrysophaea* has been in cultivation for many years and has proven quite hardy and reasonably adaptable over a wide range of conditions. It is commonly grown in Vic. and occasionally in N.S.W., Canberra, Qld and elsewhere including the U.S.A. A thriving large clump of this species grows at Burrendong Arboretum, N.S.W. It grows readily in most climates as long as humidity is not excessive. The Brisbane Ranges form has proved the hardiest and will tolerate extended, dry conditions as well as frosts to at least -4°C. Best results are achieved when it is planted in a sunny to semi-shaded site in well-drained, acidic sandy or gravelly loam. Most forms do not require fertilising or summer watering, but all respond to regular tip-pruning to improve shape and density. Low, spreading forms make attractive pot plants. Victorian native plant nurseries generally stock this species.

Propagation *Seed* Sets small quantities of seed that germinate well if peeled. Hybridisation may occur in the garden, so if the true species is required care should be taken to ensure seed is from the wild. *Cutting* Firm, young growth taken at most times of the year normally strikes readily. *Grafting* Several forms have been grafted onto *G. robusta* using the mummy and whip methods. Early growth rates are encouraging.

Horticultural features *G. chrysophaea* is a pleasing shrub with soft, green leaves and very attractive, large dull yellow to golden flowers produced for many months. Sometimes these flowers are extremely bright and floriferous. The flowers are very popular with honey-eating birds. In cultivation, it makes an attractive shrub for planting beneath large trees where it will grow well, once established. Landscapers have successfully used it as a low screen plant or specimen plant but it is generally more suited to garden landscapes than commercial or public plantings.

Hybrids In disturbed areas in the Brisbane Ranges, hybrids have been recorded with *G. rosmarinifolia* and *G. steiglitziana*. It is also the known parent of a number of garden hybrids such as *G.* 'Australflora Pendant Clusters'.

History This species was first recorded in cultivation at the Melbourne Botanic Gardens in 1858.

General comments *G. chrysophaea* var. *canescens* was described by H.D.Williamson, differing in its 'young growth quite white with a fine woolly tomentum', but McGillivray (1993) placed it in synonymy. Nonetheless, the taxonomy of *G. chrysophaea* as a whole is unresolved and warrants further study.

71A. *G. chrysophaea* Close-up of the Gippsland form, in cultivation, Burrendong Arboretum, N.S.W. (P.Olde)

71B. *G. chrysophaea* Gippsland form. Flowering branch from a brightly flowered plant (N.Marriott)

Grevillea sp. aff. chrysophaea
Plate 72

Strongly suckering **shrub** 0.3–1.5 m high. **Branchlets** terete or almost so, sparsely pubescent. **Leaves** 1–3 cm long, 7–12 mm wide, ascending to spreading, sessile, simple, oblong-elliptic to elliptic, obtuse, mucronate; upper surface convex, becoming glabrous, wrinkled, sparsely granular, the midvein evident, lateral and intramarginal veins obscure to faint; margins entire, shortly recurved; lower surface loosely tomentose to velvety, midvein prominent. **Conflorescence** decurved, shortly pedunculate, terminal, usually simple; unit conflorescence 2–8-flowered, ± secund, umbel-like to globose, the outermost flowers opening first; peduncle and rachis tomentose to villous; bracts 1.5 mm long, ovate, acuminate, villous outside, falling before anthesis. **Flower colour**: perianth red, pink, sometimes the limb yellow; style red or orange–red. **Flowers** adaxially acroscopic; pedicels stout, 4 mm long, villous; torus 1.5–2 mm across, oblique; nectary ascending, shovel–like, entire or toothed, exceeding torus by c. 1.5 mm; **perianth** 5–10 mm long, 5 mm wide, ± oblong with ventral dilation at base, strongly curved, sparsely glandular-tomentose outside, bearded near base inside, elsewhere glabrous or with scattered hairs; tepals cohering except along dorsal suture, the style end exposed before anthesis; limb conspicuous, revolute, subcubical, emarginate in side view, the segments channelled; **pistil** 17–18 mm long; ovary sessile, villous; style at first exserted at curve on dorsal side and looped upwards, gently incurved to straight after anthesis, villous with coloured hairs; style end disc-like; pollen presenter lateral, elliptic, convex, rimmed. **Fruit** 14–16 mm long, 6.5–7.5 mm wide, erect, ellipsoidal, villous with many tan-coloured hairs along suture as well as occasional elsewhere; suture prominent, keel-like, a faint ridge along each side; pericarp c. 0.5 mm thick. **Seed** 10 mm long, 3 mm wide, ellipsoidal with apical elaiosome, the outer face convex, smooth, shiny; inner face flat, minutely pitted, the margins strongly recurved, with a waxy edge along one border.

Distribution Vic., confined to the lower Nowa Nowa–Buchan area, east Gippsland. *Climate* Warm to hot, dry summer; mild to cool wet winter. Rainfall 600–800 mm.

Ecology Grows in various habitats from open eucalypt woodland and sandy loam to granitic hillsides. Flowers winter to early summer but for longer in cultivation. After fire regenerates from sucker and seed. Bird-pollinated.

Major distnguishing features Habit strongly suckering; leaves simple, entire, elliptic to obovate, strongly convex; upper surface usually minutely granular; lower surface densely hairy with curled hairs; floral rachis 1–6.5 mm long; nectary ascending, shovel-like, sometimes toothed; perianth zygomorphic, oblong with ventral basal dilation, sparsely hairy outside, bearded inside in lower half; ovary densely hairy, sessile; style with coloured hairs; pollen presenter lateral.

Related or confusing species Group 25, especially *G. alpina* and *G. chrysophaea*. *G. alpina* has

G. sp. aff. *chrysophaea*

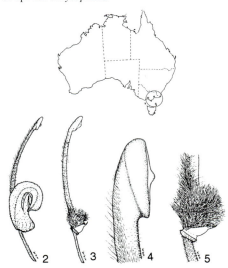

72. *G.* sp. aff. *chrysophaea* Cult. ex Nowa Nowa, Vic. at Stawell, Vic., (N.Marriott)

smaller leaves, usually 1–2 cm long, c. 3 mm wide and a patent tongue-like nectary extending laterally beyond the torus 1–3 mm. *G. chrysophaea* differs in its less conspicuous, collar-like nectary, its single-stemmed habit and its mainly yellow to gold flowers that are often villous.

Variation Varies in habit, from low and spreading to erect and few-branched. Flower colour also varies somewhat.

Conservation status Not presently endangered.

Cultivation This has been grown for many years in Vic. and N.S.W. where it has proved very successful. Foliage is not damaged by frosts or light snow and long dry periods are not harmful. It appears to thrive in temperate climates but its performance in subtropical, summer-humid conditions is unknown. A light, well-drained yet moist acidic to neutral sand or sandy loam best suits its suckering habit. A position in full sun is recommended. Occasionally available, particularly at specialist native plant nurseries in Vic., usually as *G. chrysophaea* 'bicolor'.

Propagation *Seed* Untested. *Cutting* Firm, young growth taken at most times of the year usually strikes readily. Young suckers treated as cuttings grow on strongly. *Grafting* Recently has been grafted onto *G. robusta* using the mummy and whip grafts. Plants are growing well but it is too early to confirm long-term compatibility.

Horticultural features *G.* sp. aff. *chrysophaea* is an intriguing plant with extremely beautiful, bright flowers set against very concave spreading leaves. Erect forms appear to sucker freely in the garden; such plants are not dense and combine attractively with other plants. Flowering extends over many months and the flowers are well patronised by nectar-feeding birds. It is ideal for planting under trees and large shrubs.

Grevillea cirsiifolia C.F.Meisner (1848)
Plate 73

in J.G C.Lehmann (ed.), *Plantae Preissianae* 2: 253 (Germany)

Varied-leaf Grevillea

Named for the resemblance of the leaves of this species to those of the genus *Cirsium* (Asteraceae); Latin *folium* (a leaf). SIR-SEE-EYE-FOAL-EE-A

Type: Swan R. colony, W.A., 1844–45, J.Drummond coll. 3, no. 267 (holotype: NY).

Prostrate **shrub** forming mats 0.5–2 m across. **Branchlets** angular, silky, secund. **Leaves** 5–16 cm long, 5–30 mm wide, patent to ascending, vertically orientated, shortly petiolate, commonly pinnatifid, coarsely serrate, sometimes entire and lightly toothed, linear to obovate; upper surface silky becoming ± glabrous, midvein and lateral veins evident; margins shortly recurved; lower surface densely silky, midvein prominent. **Conflorescence** 6.5–11 cm long, erect, pedunculate, terminal or axillary, lax, conico-cylindrical, simple or occasionally branched; peduncle silky; rachis silky to villous; bracts 1.8–3.2 mm long, narrowly triangular to obovate, silky outside, often persistent at anthesis. **Flower colour**: perianth creamy white outside, bright yellow inside; style pale yellow. **Flowers** acroscopic; pedicels 5–12.5 mm long, silky; torus c. 1 mm across, ± square, straight; nectary absent; **perianth** 3–4 mm long, 1–2 mm wide, persistent, undilated, geniculate from c. half-way, oblong, silky outside, glabrous inside, the dorsal tepals first separated above the curve, then reflexed laterally and everted, inner surface forming a broad platform beneath the looped style, all tepals otherwise cohering before and after anthesis; limb revolute, globular, tomentose, firmly enclosing style end long after exposure of inner perianth, not relaxed after anthesis; **pistil** 7–8 mm long; stipe 0.5 mm long, filamentous; ovary silky, appearing sessile but borne on a slender stipe enclosed in torus; style stout, flattened, silky at base, otherwise glabrous, exposed except at base and style end before anthesis; geniculate above ovary and again below style end; style end a narrow flange; pollen presenter lateral, orbicular, convex to flat. **Fruit** 10–11.5 mm long, 6.5–7 mm wide, oblique, round or obovoid, silky, wrinkled; pericarp c. 1 mm thick. **Seed** not seen.

Distribution W.A., in scattered localities from Darkan to Mt Lindesay. *Climate* Mild, dry summer; cold, wet winter. Rainfall 950–1000 mm.

Ecology Found in winter-damp to wet swampy areas in open Jarrah or Wandoo woodland or shrubland in sandy or gravelly loam. Flowers late spring to summer. Fire response unknown. Flowers have a bee-pollination syndrome but pollinator unknown.

Variation There is some variation in leaf length. A Small-leaved and a Large-leaved form have been selected for horticulture.

Small-leaved form Has smooth, green leaves to c. 10 cm long with few to many obtuse, wedge-shaped to rounded lobes. Flowers in dense, erect conflorescences.

Large-leaved form Has dull grey-green leaves to c. 18 cm long, mostly entire or with occasional linear lobes in the upper third of the leaf. Flowers in loose trailing to ascending conflorescences.

Major distinguishing features Habit prostrate, mat-forming; leaves pinnatifid, vertically oriented on secund branchlets; conflorescence conico-cylindrical; torus minute, straight, enclosing a filamentous stipe; nectary absent; perianth zygomorphic, glabrous and bright yellow inside, the inner surface exposed at anthesis; ovary hairy.

Related or confusing species Group 33 but appears to have no close relative. In *G. leptobotrys* (Group 8) the flower opens in the same distinctive way and appears superficially similar but differs in its glabrous ovary and pink flowers.

Conservation status 3EC-i. Apparently rare and present in low numbers. Occurs in at least one nature reserve.

Cultivation *G. cirsiifolia* has been grown in cool-wet, warm-dry and to a limited extent in summer-rainfall climates. Excellent, long-lived specimens have been grown at Burrendong Arboretum, N.S.W., and in western Vic. and it has also been cultivated

73A. *G. cirsiifolia* Delicate yellow flowers (P.Olde)

73B. *G. cirsiifolia* Habit in cultivation, Stawell, Vic. (N.Marriott)

G. cirsiifolia

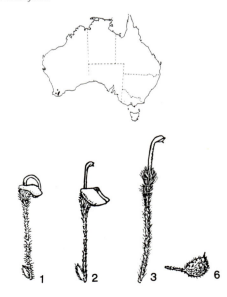

by enthusiasts in W.A. and in the U.S.A. For best results it requires moist but well-drained acidic sandy loam or gravelly loam, perhaps with some peat or humus, and shaded from afternoon sun, but it tolerates both full sun and partial shade. Although it has proved hardy to frosts to at least -6°C and withstands hot, dry spells, it often does not seem to prosper, being rather slow-growing with leaves and branches dying back. The cause is probably fungal attack but trials continue in an effort to solve the problem. Certainly, few plants in cultivation have so far achieved the size of the beautiful metre-wide mats that can be found in the wild. The Long-leaved form, only recently introduced to cultivation, is proving more vigorous. Summer watering, especially in dry conditions, induces attractive new growth. Potted plants can look most attractive especially if grown in a mix with plenty of peat or humus and slow-release fertiliser. Native plant nurseries occasionally stock this species, mainly in Vic.

Propagation *Seed* Rarely sets seed and as a result is not usually propagated by this method. *Cutting* Strikes well from firm to fairly hard cutting material, especially when taken in mid-spring or early autumn. *Grafting* Has been grafted successfully onto *G.* 'Poorinda Royal Mantle' and *G. robusta* using the top wedge technique. Although this has induced more vigorous growth, it has not reduced the problem associated with dying foliage, especially with the Small-leaved Form.

Horticultural features *G. cirsiifolia* is a delightful mat-forming plant with bright green, silver-backed leaves of interesting shape and almost vertical to the stems. The most appealing feature of this species, however, is the beautiful, soft-yellow upright racemes of vanilla-scented flowers mainly around the edges of the plant in late spring. When cultivated under good conditions, it will often flower continuously from late spring through to autumn, attracting native bees and other flying insects to the flowers. Apart from the obvious use as a ground cover, it also makes a charming rockery plant and has been successfully used as a spillover plant for breaking the outline of rockery walls.

Grevillea coccinea C.F.Meisner (1855)

Hooker's Journal of Botany & Kew Garden Miscellany 7: 76 (England)

The specific name alludes to the prominent red style and is derived from the Latin *coccineus* (scarlet). KOK-SIN-EE-A

Type: L. Tjilberup, 7 miles from Mt Manypeak, W.A., 22 Nov. 1840, L.Preiss 711 (lectotype: NY) (McGillivray 1993).

A decumbent to sprawling **shrub**, sometimes erect and pine-like with upswept branches, 1–3 m high. **Branchlets** terete, silky. **Leaves** 2.5–12.5 cm long, 1–4.5 mm wide, spreading to erect, crowded, usually curved, shortly petiolate, simple, linear or obovate, leathery, acute to obtuse, mucronate, pungent; upper surface smoothly or angularly convex or flat, usually with 3–5 longitudinal ribs, glabrous with minute pitting to silky or sparsely so, the hairs often with a sparkling sheen; margins entire, angularly to vertically refracted about an edge vein; lower surface bisulcate or sometimes exposed beside midvein and tomentose, midvein prominent. **Conflorescence** 2.5–6.5 cm long, erect, pedunculate, terminal, secund, unbranched, not exceeding leaves; peduncle silky to pubescent; rachis angular, silky to tomentose; bracts 2.4–4.8 mm long, narrowly triangular to linear-subulate, silky to tomentose, falling before anthesis. **Flowers** acroscopic; pedicels 1.3–2.1 mm long, silky to tomentose; torus 1–1.5 mm across, ± straight, oblong; nectary oblong to thickly U-shaped, entire or lipped; **perianth** 8–10 mm long, 2–3.5 mm wide, ovoid, silky to villous outside, glabrous inside, cohering except along dorsal suture; limb revolute, globular to spheroidal, enclosing style end before anthesis; **pistil** 19–23.5 mm long; stipe 0.5–1.6 mm long, villous, partly adnate to torus; ovary villous; style glabrous, papillose or wrinkled on ventral side, at first exserted at or below curve, straight to slightly reflexed at anthesis, soon becoming swept back; style end hoof-like; pollen presenter straight or slightly oblique, oblong-elliptic to ± circular, flat to convex, umbonate. **Fruit** 10.5–16 mm long, 5.5–8 mm wide, erect on curved pedicel, faintly ridged, ovoid, tomentose to pubescent with red-brown striping; pericarp 0.5–1 mm thick. **Seed** 10 mm long 3 mm wide elliptic; outer face convex, smooth with a flange-like border, shortly winged at each end; inner face channelled about margin with an elliptic concave centre; margin recurved.

Major distinguishing features Leaves simple, entire, linear to obovate, indumentum sometimes with a sparkling sheen; margins angularly revolute; conflorescence erect, terminal, secund; floral bracts attenuate; pedicels 1–2 mm long; perianth zygomorphic, tomentose outside, glabrous inside; ovary shortly stipitate to almost sessile, villous; pollen presenter straight to slightly oblique; fruit with reddish stripes.

Related or confusing species Group 35, especially *G. concinna* which differs in its deflexed shorter (1–2 cm long) conflorescence, ovate floral bracts and more oblique pollen presenter.

Variation Two subspecies are recognised, based principally on perianth size and indumentum.

Key to subspecies

Perianth 1–2 mm wide, the outer surface silky to tomentose; torus 1–1.5 mm across .... subsp. **coccinea**
Perianth 3–3.5 mm wide, the outer surface woolly; torus 1.6 mm across .... subsp. **lanata**

Grevillea coccinea C.F.Meisner subsp. coccinea
Plate 74

Shrub 1–3 m high. **Leaves** 2.5–12.5 cm long, 1–3(–4.5) mm wide. **Flower colour**: perianth greenish cream to pale brown becoming pinkish; style red. **Flowers**: bracts 2.4–4.8 mm long; **perianth** 8 mm long; **pistil** 19–23.5 mm long; stipe (0.5–)0.7–1.6 mm long.

Synonyms: *G. hewardiana* C.F.Meisner (1856); *G. concinna* var. *racemosa* G.Bentham (1870).

Distribution W.A., in south coastal and near-coastal areas from around Mt Manypeak E to the Ravensthorpe Ra. and S to the coast. *Climate* Hot, dry summer; cool, wet winter. Rainfall 350–650 mm.

Ecology Grows in heath or low to tall shrubland, in grey or gravelly sand over heavier loam or in shallow granitic loam. Flowers winter to summer. Fire response unknown. Probably pollinated by birds.

Variation There are two distinct leaf forms.

Broad-leaf form In the west of the subspecies' range, the leaves are 3–4.5 mm broad and ashy grey with the undersurface sometimes exposed; this form is a bushy shrub to 1.5 m with ascending branches.

Narrow-leaf form Plants from the Ravensthorpe Range and south to near Hopetoun have narrower, often longer leaves (1–1.3 mm wide, 6–12.5 cm long) that are mid to dark green; this is a pine-like shrub to 3 m tall.

Conservation status Not presently endangered.

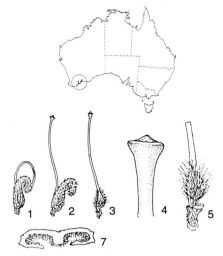

G. coccinea subsp. *coccinea*

Cultivation Although it has not been widely cultivated, *G. coccinea* subsp. *coccinea* has so far shown that it prefers a cool to warm, dry climate. It grew well at Rylstone, N.S.W., for many years and is thriving in several gardens in northern Vic. Greatest success has been achieved when planted in full sun, in acidic to neutral sand or gravelly loam. While it tolerates frosts to at least -4°C and is hardy to extended dry conditions, it demands excellent drainage and many plants have been lost even in soils of average drainage. Well-grown plants are naturally dense but some specimens do benefit from occasional tip pruning and light applications of slow-release fertiliser in spring. Summer watering is usually resented after establishment and may cause rapid collapse if given in the heat of the day. Nurseries and seed suppliers do not usually stock this species.

Propagation *Seed* Should succeed from seed using the standard peeling treatment. *Cutting* Strikes only with difficulty using firm, young growth taken in early autumn. *Grafting* Has been grafted successfully onto *G. robusta*, G. 'Poorinda Anticipation' and G. 'Poorinda Royal Mantle' using the top wedge and mummy methods.

74A. *G. coccinea* subsp. *coccinea* Close-up of flowers and foliage, Broad-leaf form (N.Marriott)

74B. *G. coccinea* subsp. *coccinea* Close-up of flowers and foliage, Narrow-leaf form. Mt. Desmond, W.A. (N.Marriott)

74C. *G. coccinea* subsp. *coccinea* Shrub in natural habitat near Pallinup River, W.A. (M.Hodge)

Horticultural features A most attractive plant. The Broad-leaf form is a dense, rounded, ashy grey shrub and the Fine-leaf form an upright, dense pine-like shrub to 3 m with dark green leaves. Each makes an interesting specimen or low screen plant and is worth considering in large gardens because of the bright red, bird-attracting flowers borne over a long period. The greyish green or dark green foliage also provides an interesting contrast.

Early history *G. coccinea* was listed for sale at Rule's Nursery, Sydney, as early as 1857, although there is some doubt about the identification. The species may have been correctly identified as there are also reports of it being despatched from Sydney Botanic Gardens to France in 1853.

Grevillea coccinea subsp. lanata
P.M.Olde & N.R.Marriott (1993)

Nuytsia 9: 277 (Australia)

The specific name is derived from the Latin *lanatus* (woolly), in reference to the perianth. LANN-ARE-TA

Type: Middle Mt Barren, Reserve 24048 [Fitzgerald R. National Park], W.A., 16 July 1970, A.S.George 10104 (holotype: PERTH).

Shrub 1.3 m tall. **Leaves** 8–12 cm long, 1.8–2.5 mm wide, narrowly obovate. **Flower colour**: perianth greenish cream to pale brown becoming pinkish; style red. **Flowers**: bracts 6 mm long; **perianth** 8–10 mm long; **pistil** 20 mm long; stipe 0.4–0.5 mm long.

Distribution W.A., restricted to a few peaks in Fitzgerald River National Park. *Climate* Hot to warm, dry summer; cool, wet winter. Rainfall 350–650 mm.

Ecology Grows in sandy loam on rocky quartzite hills. Flowers winter to summer.

Conservation status 2EC-t.

Cultivation Subsp. *lanata* has been named only recently and is unknown in cultivation. It should respond to conditions similar to those for subsp. *coccinea*.

Propagation *Seed* Should succeed from seed using the standard peeling treatment. *Cutting* and *grafting* Untested.

Horticultural features Subsp. *lanata*, has woolly conspicuous flowers, and should prove an attractive addition to the garden.

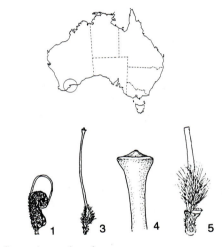

G. coccinea subsp. *lanata*

74D. *G. coccinea* subsp. *lanata* Conflorescence and foliage (N.McQuoid)

Grevillea commutata F.Mueller (1868)
Plate 75

Fragmenta 6: 207 (Australia)

The epithet, derived from the Latin *commutare* (to change, alter), may be a reference to the variable leaves which may be simple or divided. KOMM-YOU-TART-A

Type: Port Gregory, W.A., A.Oldfield, (lectotype: PERTH) (McGillivray 1993).

Synonyms: *G. pinnatisecta* F.Mueller ex G.Bentham (1870); *G. hakeoides* subsp. *commutata* (F.Mueller) D.J.McGillivray (1986).

Spreading, much-branched, open to dense **shrub** 1.5–3 m high, sometimes almost prostrate in exposed conditions. **Branchlets** angular, ridged, silky to tomentose. **Leaves** 3–12 cm long, 2–10 mm wide, spreading to ascending, ± sessile, linear to obovate, simple or lobed, obtuse, mucronate; lobes linear, < 3 mm wide; upper surface glabrous or sparsely silky, midvein and acutely angled lateral venation conspicuous; margins usually entire, recurved; lower surface silky, hairs either white or rusty, midvein prominent. **Conflorescence** erect, terminal, unibracteate (involucral bracts 2 mm long, 2 mm wide, ovate, villous), pedunculate, simple to 3-branched; unit conflorescence 1–2.5 cm long, 1.5–3 cm wide, ovoid or umbel-like, dense; peduncle and rachis silky-tomentose; bracts 0.5–0.8 mm long, ovate-cymbiform, tomentose outside, falling early. **Flower colour**: perianth and style greenish white to cream, or pale pink. **Flowers** adaxially oriented; pedicels 2.5–5 mm long, tomentose; torus 0.8–1 mm across, straight to slightly oblique; nectary inconspicuous, semi-circular to V-shaped, entire; **perianth** 3–5 mm long, 1.5 mm wide, oblong-ovoid, silky outside, bearded inside about ovary, the tepals first separating along dorsal suture, then all separated, the dorsal tepals reflexing and exposing inner surface before anthesis, afterwards free to base and soon shrivelled; **pistil** 10–15 mm long, glabrous; stipe 1.8–2 mm long; ovary ovoid to globose, stipitate; style at first exposed to base, then elongated and looped upwards, strongly incurved after anthesis; style end flattened; pollen presenter oblique, obovate, convex, umbonate. **Fruit** 10–13.5 mm long, 6–7.5 mm wide, oblique, oblong-elliptic, markedly and discontinuously wrinkled or warty, ridged, glabrous; pericarp c. 0.5 mm thick. **Seed** 9–9.5 mm long 2.5–3 mm wide, narrowly oblong; outer face convex, smooth, minutely scaly; inner face channelled.

Distribution W.A., from 70 km N of the Murchison R. S to the Greenough R. (S of Geraldton), and inland to Yuna. *Climate* Hot, dry summer; mild, wet winter. Rainfall 400–440 mm.

Ecology Grows in various habitats including heath and sand dunes, in low to tall shrubland. Soil varies from deep yellow to red neutral to alkaline sand to shallow loam over rock or red loam. Regenerates from seed after fire. Flowers winter to spring. Pollinator unknown, but attracts many insects.

Variation The species varies in leaf division, lobe width and flower size.

Typical form This form has showy ovoid conflorescences up to 20 mm long. The pistil is usually

G. commutata

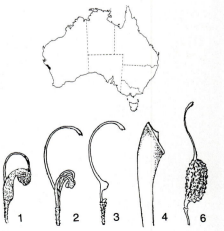

75A. *G. commutata* Close-up of conflorescences (N.Marriott)

75B. *G. commutata* Close-up of ovoid conflorescences (N.Marriott)

more than 12 mm long. Most leaves are simple and fairly broad, although divided leaves are sometimes found. It occurs from about Northampton to north of the Murchison R. and from the coast inland to Yuna.

Divided-leaf form This form occurs mainly around Geraldton and further south and until recently was known as *G. pinnatisecta*. Most of its leaves are divided into 2 or 3 linear lobes. It has similar ovoid to almost dome-shaped conflorescences which, at 8–12 mm long, are slightly shorter than the Typical form. The pistil is also shorter (10–12 mm).

Major distinguishing features Leaves simple or sparingly divided, the undersurface exposed; conflorescence ovoid to umbel-like, the rachis > 5 mm long, often up to 20 mm long; perianth zygomorphic, silky outside, bearded inside; pistil glabrous; fruit ovoid, rugose.

Related or confusing species Group 16, especially *G. argyrophylla*, *G. brachystachya*, *G. diversifolia* and *G. hakeoides*. In *G. argyrophylla* the pistil is 4–7 mm long. *G. diversifolia* has smooth fruits. *G. brachystachya* and *G. hakeoides* differ in having the leaf undersurface obscured by the margin.

Conservation status Not presently endangered.

Cultivation *G. commutata* has occasionally been cultivated in eastern and southern Australia and experienced growers report that it is hardy, tolerant of frosts to at least -6°C and extended dry conditions. Soil should be slightly acid to alkaline, sand or sandy loam, and well-drained. If situated in full sun, it will grow rapidly into a well-rounded, compact shrub requiring little maintenance. In soils of even average drainage, especially in humid, summer-rainfall climates, it may succumb to root-rot during wet weather. In this regard, the Divided-leaf form appears to be more susceptible than the Typical form. Established plants rarely require summer watering or fertilising. Both forms make excellent pot plants for a number of years if grown in a well-drained, neutral mix with some slow-release fertiliser. Nowadays, this species is rarely available in nurseries but it is being maintained in cultivation by enthusiasts.

Propagation Seed Sets many seeds that normally germinate readily (average 27 days) given the standard peeling treatment. The seedling leaves are divided, later ones mostly entire. Growth is rapid.

Cutting Firm, young growth strikes well at most seasons, especially during the warmer months but preferably in early autumn. Grafting Has been grafted successfully onto G. 'Poorinda Royal Mantle' and *G. robusta* using top wedge grafts. Long-term compatibility is untested.

Horticultural features *G. commutata* is an attractive, dense shrub with dark green leaves with a silvery underside, a feature especially noticeable in windy weather. The ovoid conflorescences are quite large and their delicate pink or white colour often attracts attention, especially when blooms are just beginning to open. Its occurrence in coastal situations from Dongara to Kalbarri suggests that it possibly tolerates airborne salt, making it useful for front-line coastal planting. It makes a useful screen or feature plant in the landscape and its flowers attract both birds and insects. The rough fruits are also a horticultural feature.

Hybrids Occasionally hybridises in the wild with *G. argyrophylla* where the two species occur together, especially in disturbed sites.

General comments D.J.McGillivray (1993) placed this taxon as a subspecies of *G. hakeoides*, a species to which it is undoubtedly most closely related. We prefer to retain it as a separate species mainly because of its generally larger flowers and fruits, and ovoid conflorescences. In addition, the leaves have the undersurface exposed, giving the species a very distinctive appearance.

Grevillea concinna R.Brown (1810)

Transactions of the Linnean Society 10: 172 (England)

Red Combs, Elegant Grevillea

The specific epithet refers to Robert Brown's impression of this species and derives from the Latin *concinnus* (neat, pretty, elegant). KONN-SINN-A

Type: Lucky Bay, [east of Esperance, W.A.], 11 Jan. 1802, R.Brown 3342 (lectotype: BM) (McGillivray 1993).

Dense, spreading **shrub** 0.6–1.6 m high. **Branchlets** terete, silky. **Leaves** 2–7 cm long, 1–4.5 mm wide, ascending, shortly petiolate, usually simple, subterete to linear or obovate, sometimes 2- or 3-lobed, acute with short mucro, sometimes pungent; juvenile leaves sometimes divided; upper surface convex, smooth, minutely pitted, glabrous or sparsely silky, midvein visible; margins usually entire, smoothly revolute, either enclosing lower surface including midvein or not; lower surface usually bisulcate, sometimes quite obscured, sometimes exposed, silky, midvein prominent. **Conflorescence** 1–2 cm

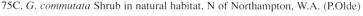
75C. *G. commutata* Shrub in natural habitat, N of Northampton, W.A. (P.Olde)

long, usually deflexed, pedunculate, terminal, simple, secund, dense; peduncle and rachis silky to tomentose; bracts 0.8–3 mm long, spreading, ovate-apiculate to triangular, silky, falling before anthesis. **Flowers** acroscopic; pedicels 1–3 mm long, silky; torus 1–1.5 mm across, ± straight; nectary arcuate, conspicuous; **perianth** 6–8 mm long, 2–3 mm wide, ovoid, silky to tomentose outside, glabrous inside, cohering except along dorsal suture; limb revolute, globular or spheroidal, enclosing style end before anthesis; **pistil** 23–26 mm long; stipe 1–2 mm long, villous; ovary villous; style glabrous, exserted first below curve on dorsal side and looped upwards, clearly erect, afterwards straight and slightly undulate above middle, dilating smoothly and evenly c. 1–2 mm before scarcely swollen style end; pollen presenter oblique, elliptic, flat, umbonate. **Fruit** 10–14.5 mm long, 5–8 mm wide, oblique on curved pedicels, ovoid, tomentose, red-blotched or striped, faintly ribbed; pericarp c. 0.5 mm thick. **Seed** 9–10 mm long, 3–4 mm wide, elliptic, marginally ridged on both sides, shortly winged at both ends.

Major distinguishing features Leaves usually simple, entire, rarely bi- or trifid, subterete to obovate; conflorescence 1–2 cm long, deflexed, secund; floral bracts ovate; perianth zygomorphic, silky outside, glabrous inside; ovary villous; stipitate; style undulate above middle, pollen presenter oblique; fruit red-striped.

Related or confusing species Group 35, especially *G. coccinea* which can be distinguished by its longer, erect conflorescence and erect pollen presenter.

Variation Two subspecies are recognised.

Key to subspecies

Either hairs on leaf undersurface straight, or midvein obscured by revolute margins; leaf margins smoothly revolute; upper surface convex .... subsp. **concinna**

Hairs on leaf undersurface curled, the midvein always exposed; upper surface rising to an angular midrib, the leaf margins angularly refracted .... subsp. **lemanniana**

Grevillea concinna R.Brown subsp. concinna Plate 76

Leaves mid-green, usually the lower surface exposed and white silky with straight hairs, upper surface convex, glabrous, glossy. **Flower colour**: perianth silvery or yellowish green; style pink to bright red, rarely yellow. **Perianth** silky.

Distribution W.A., confined to a small coastal belt between Cape Le Grand and Lucky Bay. *Climate* Warm to hot, dry summer; cool, wet winter. Rainfall 700–750 mm.

Ecology Grows in granitic sand and sandy loam, often right to the high water mark, in exposed, open shrubland. Flowers winter to spring in the wild but most of the year in cultivation. Fire response unknown. Bird-pollinated.

Variation A relatively uniform subspecies.

Conservation status 2 RC-t. A very restricted distribution but reserved in Cape Le Grand National Park.

Cultivation Subsp. *concinna* has been grown successfully in cool-wet, warm-dry, and sometimes summer-rainfall climates such as in Sydney. It is widely grown by enthusiasts in Vic., south-east S.A., N.S.W. and inland Qld. It is hardy to frost to at least -6°C and tolerates extended, dry conditions. Most growers regard it as a tough, adaptable plant. Although it occurs in coastal areas in sand, it is just as hardy in inland areas grown in heavy clay or loam as long as it has good drainage and a fairly sunny situation. Soil can be acidic to slightly alkaline and it will survive in dappled shade although under these conditions it tends to become weak, leggy and somewhat less floriferous. Pruning is rarely required, nor does it need fertiliser or summer watering. For many years it makes an attractive pot plant but will eventually become too large. It is occasionally available from native plant nurseries, mainly in Vic.

Propagation **Seed** Sets medium quantities of seed that germinate well when peeled or soaked. *Cutting* Strikes reasonably well from medium-hard wood especially when taken in early autumn or spring. *Grafting* Successfully grafted to a number of rootstocks such as *G. robusta*, G. 'Poorinda Royal Mantle', G. 'Coastal Glow' and G. 'Ned Kelly' using the top wedge technique.

Horticultural features Subsp. *concinna* is a very hardy plant that flowers profusely for many months

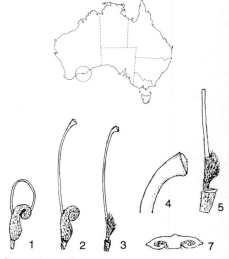

G. concinna subsp. *concinna*

76A. *G. concinna* subsp. *concinna* Pale-flower variant, Cape Le Grand NP, W.A. (N.Marriott)

76B. *G. concinna* subsp. *concinna* Close-up of flowers and foliage (N.Marriott)

76C. *G. concinna* subsp. *concinna* Shrub in natural habitat, Cape Le Grand NP, W.A. (N.Marriott)

of the year. The pink to red deflexed conflorescences are quite short but very distinctive and the flowers have unusual erect pistils. Its long-flowering habit and dense silvery green foliage provide a haven for birds, which often nest in its branches or feed on the nectar-filled flowers. Generally it will branch right to the ground, making a low screen that effectively blocks light and acts as a weed suppressant. It can also be used as an unusual feature plant because, although it has a normal rounded shape in the centre, many prostrate branches radiate around its base. This long-lived and reliable species is useful in many ways for commercial and residential landscaping and as it is also undamaged by salt-laden winds, it could be used in front-line coastal planting.

History Subsp. *concinna* was one of the first Western Australian grevilleas known to science, a specimen being collected by Robert Brown in 1802. William Baxter is credited with collecting it first for cultivation in 1823/24. Robert Sweet reports that seeds collected at King George Sound, W.A., (the nearest locality but clearly not the collection site) were sent by Baxter to Mackay's nursery in London and were successfully raised. It came to be recognised there as a hardy greenhouse plant. The Sydney Botanic Gardens despatched several plants to other colonies from 1850 but it was probably first cultivated there much earlier, possibly having been introduced by Baxter in 1828/29.

Grevillea concinna subsp. lemanniana
(C.F.Meisner) D.J.McGillivray (1986)

Plate 77

New Names in Grevillea *(Proteaceae)*: 4 (Australia)

Based on *G. lemanniana* C.F.Meisner (1856) in A.P. de Candolle, *Prodromus* 14: 366.

The subspecific epithet honours Charles Morgan Lemann (1806–1852), who had a large herbarium now located in the Herbarium of Cambridge University. LEE-MAN-EE-ARE-NA

Type: Swan River colony, W.A., late 1840s, J.Drummond coll. 5 no. 405 (lectotype: NY) (McGillivray 1993).

Leaf margins angularly or vertically refracted; midvein of upper surface ridge-like; lower surface with curled hairs. **Flower colour**: perianth reddish cream or reddish brown; style bright red. **Perianth** tomentose.

Distribution W.A.; from Cheyne Bay E to Mt Ragged and N to Needilup and N of Ravensthorpe. *Climate* Hot, dry summer; cool, wet winter. Rainfall 500–700 mm.

Ecology Occurs in open heath and open woodland in grey sand, granitic loam, sandy loam and sand over laterite. Flowers winter to spring. After fire regenerates from seed. Pollinated by nectarivorous birds.

Variation This subspecies is quite variable in habit. Different populations grow from 30 cm to 1.5 m high. Erect, shrubby forms can be found in distinct populations near low, spreading ones. Leaves vary in length and width. Plants with broader, sometimes divided leaves can be found at the eastern end of the range near Mt Ragged. North of Ravensthorpe, plants with bluish-glaucous leaves have been collected.

Conservation status Not presently endangered.

Cultivation This relatively hardy plant tolerates a

G. concinna subsp. *lemanniana*

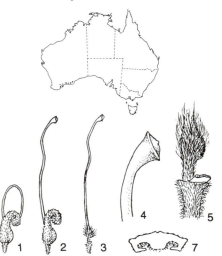

wide range of soil types and climates and has been grown successfully in cool-wet, warm-dry and, with limited success, summer-rainfall climates. Although frost tender when young, once established it will withstand frosts to at least -6°C. It is hardy to extended dry conditions. It requires a fairly well-drained site and grows and flowers best in full sun, although it will tolerate partial shade. Best results have been obtained in shallow, acidic, sandy loam over gravelly clay; growth is then rapid and flowering prolific, and it does not need watering or fertiliser. It responds to judicious pruning to improve shape and density and makes an excellent pot plant, especially the lower, decumbent forms. It is becoming more available in specialist native plant nurseries, especially in Vic. where for many years it was sold as *G. baxteri*.

Propagation Seed Sets small quantities of seed that normally germinate well peeled or soaked. *Cutting* Results have been variable; sometimes strikes easily, at others difficult and slow. Greatest success has been achieved using firm, new growth in late summer to mid-autumn. *Grafting* Grafts readily onto a wide range of rootstocks with which it remains compatible over a long period. Success has been achieved with G. 'Poorinda Royal Mantle', G. 'Bronze Rambler', G. 'Moonlight', G. 'Ned Kelly' and G. 'x rosmarinifolia nana'.

Horticultural features This widely grown plant is deservedly popular in cultivation with its glowing-red toothbrush flowers displayed in profusion for many months of the year. Even when not in flower,

77. *G. concinna* subsp. *lemanniana* Glaucous form. NE of Ravensthorpe, W.A. (P.Olde)

its grey-green to blue-green foliage makes it an attractive shrub. It is strongly bird-attractive and an extremely valuable plant for commercial and residential landscaping in a wide range of soils and situations. It makes an excellent ground cover or screen plant with the added bonus of spectacular red flowers.

Grevillea confertifolia F.Mueller (1855)
Plate 78

Transactions of the Philosophical Society of Victoria 1: 22 (Australia)

Grampians Grevillea

The specific epithet alludes to the crowded leaves on the Type specimen (rather than on the species as a whole) and is derived from the Latin *confertus* (crowded) and *folium* (a leaf). KON-FERT-I-FOLE-EE-A

Type: Mt William and towards Mount Zero, The Grampians, Vic., 1854, F.Mueller (lectotype: MEL) (McGillivray 1993).

A prostrate **shrub** spreading to 1.5 m, sometimes decumbent, to 40 cm high, sometimes erect, open, to 1(2) m. **Branchlets** angular, silky. **Leaves** 1–4.5 cm long, 0.7–3 mm wide, ascending to spreading, sometimes crowded, sessile, simple, linear, acute, pungent; upper surface convex, sparsely silky becoming ± glabrous, midvein prominent; margins entire, angularly refracted about an intramarginal vein usually to midvein below; lower surface often obscured, silky when exposed, midvein prominent. **Conflorescence** 1.5–3 cm long, 2–3 cm wide, erect or decurved, sessile, simple, dense, partially secund, dome-shaped, terminal but sometimes 2-flowered conflorescences in upper leaf axils; rachis tomentose to villous, curved; bracts 1.5–4.5 mm long, imbricate in bud, ovate-acuminate, silky to villous, falling before anthesis. **Flower colour**: perianth reddish purple; style pink to pink-mauve or mauve-red. **Flowers** adaxially acroscopic: pedicels 3–5 mm long, silky; torus 0.8–1.3 mm across, straight; nectary U-shaped, toothed; **perianth** 5–7 mm long, 1–2 mm wide, oblong, silky outside, bearded inside just above ovary, cohering except along dorsal suture, the dorsal tepals reflexing slightly before anthesis, free to beard and independently rolled back after anthesis; limb recurved, ovoid, enclosing style end before anthesis; **pistil** 10.5–12.5 mm long; stipe 0.9–1.5 mm long, glabrous; ovary globular, glabrous; style glabrous, at first exserted below curve on dorsal side, angularly incurved after anthesis, geniculate near apex, papillose or minutely hairy in apical 5 mm; style end flat, plate-like; pollen presenter oblique, elliptic, conical, umbonate. **Fruit** 11–12 mm long, 4–5 mm wide, erect, narrowly ovoid/cylindrical, glabrous, wrinkled; pericarp c. 0.3 mm thick. **Seed** 8 mm long, 3 mm wide, narrowly elliptic with an apical elaiosome; outer face convex, smooth; margin revolute; inner face channelled.

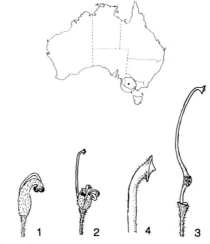

G. confertifolia

Distribution Western Vic., confined to a few localities in The Grampians. A collection from the Bolangum Ra. SW of St Arnaud is not supported by any other collection and is probably a recording error. *Climate* Hot, dry summer; cool, wet winter. Rainfall 650–1100 mm.

Ecology Occurs in either open to dense woodland, along creek banks in sand, or among sandstone rocks in exposed situations on upper slopes and mountain tops in peaty sand. Flowers spring to early summer. Regenerates from seed after fire. Pollinator unknown.

Variation There are three distinct horticultural forms.

Upright form Usually found in gullies and along scrubby creek banks, this form grows to 1 m, sometimes to 2 m. It has narrowly linear, pungent leaves that are not crowded, creating an open, upright plant. Flowers are mauve-pink.

Decumbent form This form also grows along creek banks and streams as well as in more open seepage situations, usually on hilltops. It has a spreading habit, growing to 40 cm high and 1–1.5 m wide. Foliage and flowers are similar to the Upright form.

Major Mitchell Plateau form As the name implies, this form alludes to the locality on which it grows. This is an exposed, subalpine region and the plants grow in shallow pockets of peaty sand on and between the exposed slabs of sandstone bedrock as well as beside the soaks and small streams on the mountain top. Plants are prostrate and spread to 1 m across with far broader (2–3 mm wide) oblong-elliptic leaves than any other form. These are usually grey-green and contrast with the silky white branchlets. Flowers are also distinct, being an attractive pink, although occasionally pale mauve-pink forms occur. On Mt William, an intermediate between this form and the Decumbent form occurs.

Major distinguishing features Leaves simple, entire, margins refracted about a marginal vein; conflorescence sessile, secund-dome-shaped; torus straight; perianth zygomorphic, hairy on both surfaces; pistil 10–12.5 mm long, glabrous except the apical few mm of the style.

Related or confusing species Group 21, especially *G. linearifolia*, *G. molyneuxii* and *G. sericea*. *G. linearifolia* differs in the style having hairs extending < 1 mm from the end. *G. molyneuxii* has red flowers with a pistil 19–21 mm long. *G. sericea* has penduculate conflorescences and flowers with pistils 14–19 mm long.

Conservation status 2RC-t. Locally common in a few areas in The Grampians National Park.

Cultivation *G. confertifolia* has been cultivated for many years in Vic., S.A., Tas. and the A.C.T. where it has proved very hardy. It resists extended dry conditions and is hardy to frosts to at least -10°C. On Mt William and Major Mitchell Plateau it is often covered with snow for short periods during winter. While it grows well in a wide range of conditions, greatest success is achieved when planted in a sunny to semi-shaded site in well-drained, acidic sandy loam. Flowering often ends as the soil dries out in early summer but watering can extend the season and is beneficial during the first one or two summers or during extended dry periods. This applies in particular to the Major Mitchell Plateau form. It appears to require a cold, frosty winter to induce good flowering and does not flower well in a warm, summer-rainfall climate. Regular tip-pruning, especially with the Upright form, is recommended to improve shape and density. Fertiliser is rarely necessary. The Decumbent and Major Mitchell Plateau forms make excellent tub specimens if planted in a medium to large tub. Several forms are usually available from specialist native plant nurseries, particularly in Vic.

Propagation *Seed* Sets plentiful seed that germinates well if pretreated. It self-sows in the garden without any evidence of hybridisation. *Cutting* Firm, young growth taken at most times of the year generally strikes readily. *Grafting* Has been grafted successfully onto *G. robusta* using both the approach and top wedge techniques.

Horticultural features *G. confertifolia* is a relatively insignificant, small-leaved plant when not flowering, but in full bloom is quite spectacular producing masses of bright pinkish mauve terminal racemes. The shape of the conflorescence is reminiscent of strawberries and many people know this plant as the Strawberry Grevillea. It is a long-lived plant that makes a valuable addition to public and private gardens alike. Its profuse flowering in a colour uncommon in grevilleas makes a most attractive display. The Upright form is a rather open plant but the Decumbent form is often used in

78A. *G. confertifolia* Flowers and foliage (N.Marriott)

78B. *G. confertifolia* Plant in natural habitat, The Grampians, Vic. (N.Marriott)

78C. *G. confertifolia* Major Mitchell Plateau form, in cultivation, Stawell, Vic. (N.Marriott)

massed plantings as a shrubby groundcover, being particularly suited to heavy soil and wet sites.

History This species was reportedly grown in the Melbourne Botanic Gardens in 1858.

Grevillea coriacea D.J.McGillivray (1975)
Plate 79

Telopea 1: 19 (Australia)

The specific name was derived from the Latin *corium* (leather), hence *coriaceum* (of leathery texture), in reference to the leaves. KORRY-AY-SEE-A

Type: 58 miles (93 km) from Coen towards Moreton Telegraph Station, Qld, July 1968, C.H.Gittins 1820 (holotype: NSW).

Slender **tree** 4–15 m tall with irregularly oriented foliage. **Branchlets** terete, silky to tomentose. **Leaves** 12–30 cm long, 4–10 mm wide, pendulous to spreading, leathery, simple, elongate-linear, acute to obtuse; upper surface convex, ± glabrous, or sprinkled with silky hairs, midvein obscure; margins entire, strongly and loosely recurved or revolute; lower surface silky, midvein and intramarginal veins prominent; acutely angled lateral venation faintly evident. **Conflorescence** erect, pedunculate or sessile, terminal on short branchlets, often simple, sometimes 2- or 3-branched; unit conflorescence 8–15 cm long, 5.5–8.5 cm wide, cylindrical, lax, open; peduncle silky; rachis glabrous; bracts not seen, falling very early. **Flower colour**: perianth and style white to cream, the style turning pinkish; green in bud. **Flowers** basiscopic; pedicels 6–10 mm long, glabrous; torus 2–3 mm across, oblique; nectary U-shaped, smooth, inconspicuous; **perianth** 10–11 mm long, 3–4 mm wide, oblong/ovoid, slightly dilated at base, erect, glabrous, glaucous outside, bearded inside near base, otherwise glabrous, cohering except along dorsal suture; limb revolute, globular, enclosing style end before anthesis; **pistil** 33–42 mm long, glabrous; stipe 2.5–3.5 mm long; ovary globular; style elongate, at first exserted from near curve on dorsal side, gently incurved, dilating evenly into style end; pollen presenter very oblique, oblong, broadly conical. **Fruit** 21–28 mm long, 18–24 mm wide, horizontal to pedicel, lens-shaped, glabrous, glaucous, smooth; pericarp 2–3 mm thick. **Seed** 18–20 mm long, 18–24 mm wide, orbicular to broadly elliptic, winged all round with conspicuous raphe.

Distribution Qld, known mainly from the Mareeba area from Petford to Mt Molloy and Mt Mulligan; it also extends north-west of Cooktown inland from Bathurst Bay. *Climate* Tropical monsoonal with humid, wet summer and mild to cool, dry winter. Rainfall 800–1200 mm.

Ecology Forms part of the small tree layer in layered forest or savannah woodland in stony shale on dry hills, occasionally on deep clay flats. Flowers winter to early spring. Regenerates from seed. Pollinator unknown, probably birds but possibly also bats.

Variation A fairly uniform species with little variation.

Major distinguishing features Small tree; leaves simple, entire, leathery, linear with loosely revolute margins; conflorescence cylindrical, lax, mostly simple; perianth zygomorphic, glabrous outside, pilose within; pistil 33–42 mm long; ovary glabrous, shortly stipitate; pollen presenter broadly conical; fruit compressed; seed winged all round.

Related or confusing species Group 3, especially *G. parallela*, which may be distinguished by the (usually 3) parallel veins on the lower surface of the leaves and has generally thinner leaves, denser, shorter (6–10 cm long) and branched (3–7) conflorescences and flowers with a shorter pistil (13–25 mm long). Flowering of *G. parallela* usually commences later where the two species occur together.

Conservation status Not presently endangered.

Cultivation *G. coriacea* has been grown successfully as far south as Brisbane and would probably adapt to climates as far south as Sydney where its close relative, *G. parallela*, has been successfully grown. Its hardiness and adaptability have not been tested to date but it will tolerate light frosts and dry periods, although it is susceptible to extended cold

G. coriacea

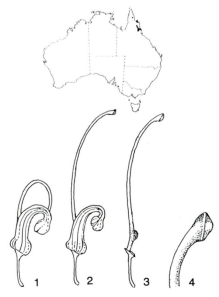

79A. *G. coriacea* Close-up of conflorescence and foliage. (F. & N. Johnston)

79B. *G. coriacea* growing naturally near the Palmer River, Qld (L.Murray)

periods. It requires a warm to hot, sunny position in acidic well-drained and friable soils where it will grow into a slender, open, small tree. It tolerates partial shade but does not grow vigorously in these situations. Flowering tends to be high on the plant but pruning may induce flowering branchlets to grow lower. Once established, it seems to be fairly long-lived. It is occasionally available from nurseries specialising in native plants, mainly in Qld.

Propagation *Seed* Generally grown from seed treated by nicking down one edge. Seed should preferably be fresh and sown in the season following collection. Soaking for 24–36 hours in warm water will aid germination. *Cutting* Not generally grown from cuttings as it is difficult to find suitable wood. *Grafting* Has been grafted successfully onto *G. robusta* using the whip graft method.

Horticultural features *G. coriacea* has long, willowy, thick leaves with rolled margins and makes an interesting foliage plant if a small tree is needed. Leaves are sometimes pendulous but often stick out at odd angles, presenting a somewhat irregular aspect to the plant. It has very showy, large, cream conflorescences in winter and early spring and would make an interesting addition to gardens in hot, summer rainfall areas. The individual flowers are very large and have conspicuous styles. It is reasonably long-lived in cultivation and has no known pests or diseases apart from leaf-chewing insects. It attracts many species of honey-eating birds which patronise the nectar-filled flowers. With a tall, narrow habit, *G. coriacea* would be ideal for a narrow bed beside a path or building. As it comes from very harsh, dry, stony hillsides, it should prove hardy in dry inland and coastal areas of northern Australia.

Grevillea corrugata P.M.Olde & N.R.Marriott (1993) Plate 80

Nuytsia 9: 247 (Australia)

The specific name is derived from the Latin *corrugatus* (crumpled), in reference to the strongly wrinkled fruit surface. KOR-RUG-ARE-TA

Type: SSE of Bindoon, W.A., 4 Oct. 1992, P.M.Olde 92/230 (holotype: NSW).

An open **shrub** 1.5–2 m tall. **Branchlets** slightly angular, openly to sparsely spreading-tomentose. **Leaves** 4–6 cm long, up to 9 cm wide, ascending, sessile, usually biternate, sometimes tripartite or pinnatipartite with the lobes trifid to bifid; leaf base 2–3 mm wide, winged, the segment between the lowest lobes and the axis of attachment reducing in length on successive leaves; lobes 2–2.5 cm long, 0.7–0.8 mm wide, linear to subulate, pungent, the basal lobes patent to spreading; upper surface glabrous or sparsely hairy, the midvein depressed in a longitudinal channel; lower surface bisulcate, mostly enclosed by margins but sometimes exposed especially at sinuses, glabrous or with a few curled hairs, midvein prominent and rounded; margins loosely but angularly refracted. **Conflorescences** erect, subsessile, simple or 2-branched, usually terminal on short axillary branchlets and subtended by a basal vegetative branchlet, sometimes axillary; unit conflorescence 1–2.5 cm long, 2 cm wide, ovoid to subglobose, relatively dense; peduncle and rachis loosely tomentose, angular; bracts 3.5–4 mm long, 3 mm wide, ovate-cymbiform, tomentose outside, glabrous inside except near apex, falling before anthesis. **Flower colour**: perianth and style white to creamy white. **Flowers** regular, glabrous; pedicels 7–9 mm long, filamentous, spreading; torus c. 0.5 mm across, ± straight; nectary sublinguiform extending c. 0.25 mm above toral rim; **perianth** 4 mm long, 0.7 mm wide, obovoid below limb, constricted below globose limb; **pistil** 3.5–4 mm long; stipe 1.5 mm long, flexuose; ovary c. 1 mm long, globose; style constricted for c. 0.2 mm above ovary, dilating to c. 0.5 mm wide, the dilation ovoid tapering to a fusiform style end; pollen presenter c. 0.6 mm long, 0.4 mm wide, truncate-conical, its base scarcely broader than style. **Fruit** 7–11 mm long, 3–6 mm wide, 3–5 mm deep, ± perpendicular to stipe, oblong-ellipsoidal, corrugate; pericarp 0.8 mm thick. **Seed** 7 mm long, 3 mm wide, obovoid; outer face smooth, convex; inner face convex with a waxy border; margin shortly recurved.

Distribution W.A., near Bindoon. *Climate* Hot, dry summer; cool, wet winter. Rainfall c. 800 mm.

Ecology Grows in gravelly loam on a road verge in partially cleared eucalypt woodland. Flowers late winter to early spring. The species is probably insect-pollinated. Regenerates from seed after fire.

Variation A uniform species.

Major distinguishing features Branchlets with a spreading indumentum; leaves divaricately divided, the leaf base winged on adult leaves; floral bracts ovate, hairy; floral rachis woolly; perianth regular, glabrous; pistil glabrous; style constricted above the ovary, dilated then tapering to the style end; style end fusiform; fruit strongly wrinkled.

Related or confusing species Group 1, especially *G. curviloba* and *G. paniculata*, from both of which it differs in its unique winged leaf base.

Conservation 2E recommended. The species was discovered in 1992 but a full search of the locality was not made as it is mostly private property. There are about 10–20 plants at the Type locality.

Cultivation *G. corrugata* has only recently been introduced to cultivation at Mount Annan Botanic Garden, N.S.W., and in western Vic., but young plants have already proved vigorous and robust. Judging from its natural habitat, this species will thrive in full sun or semi-shade in neutral to slightly alkaline, sandy to gravelly loam. Well-drained soil is suggested and growers should allow for a 2-metre

G. corrugata

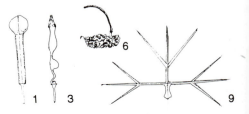

80A. *G. corrugata* Close-up of flowers (P.Armstrong)

80B. *G. corrugata* Flowering branch (P.Armstrong)

80C. *G. corrugata* Flowering habit, S of Bindoon, W.A. (B.Ball)

Grevillea costata C.A.Gardner ex A.S.George (1974)   Plate 81

Nuytsia 1: 370 (Australia)

The specific name, derived from the Latin *costatus* (ribbed), in reference to the ribs on the fruit. KOSS-TAR-TA

Type: Murchison R., W.A., 30 Aug. 1931, C.A.Gardner 2597 (holotype: PERTH).

A dense, compact **shrub** 0.5–1.5 m tall. **Branchlets** angular, ridged, silky. **Leaves** 1.5–4.5 cm long, c. 1 mm wide, ascending, sessile, simple, straight, linear, rigid, acute, pungent; upper surface flat to convex, silky, soon ± glabrous, midvein evident; margins entire, revolute to midvein; lower surface bisulcate, silky in grooves, midvein prominent. **Conflorescence** erect, pedunculate, terminal or in upper axils, simple or few-branched; unit conflorescence c. 1.5 cm long, 2–2.5 cm wide, umbel-like, few-flowered; peduncle and rachis silky; bracts 0.5 mm long, minute, linear, silky outside, falling before anthesis. **Flower colour**: white with pinkish tinges on the style. **Flowers** adaxially oriented; pedicels 4–10 mm long, slender, silky; torus c. 1 mm across, ± straight; nectary semi-circular, entire or toothed; **perianth** 3–4 mm long, 1 mm wide, slender, cylindrical, silky outside, bearded inside at level of ovary, otherwise glabrous; tepals similar in shape, at first separating along dorsal suture and exposing style to base, then all free, the dorsal tepals reflexing and everting inner surface before anthesis, afterwards free to base and rolling back to level of ovary; limb revolute, globular to spheroidal, enclosing style end before anthesis; **pistil** 7–9.5 mm long, ± glabrous; stipe 0.4–1 mm long; ovary globose; style exposed to base before anthesis, refracted above ovary, angularly incurved after anthesis, geniculate near apex, granulate or with few erect hairs just behind the flat, plate-like style end; pollen presenter ± straight to slightly oblique, orbicular, flat, umbonate. **Fruit** 9.5–11 mm long, 6.5–7 mm wide, erect on pedicel, ovoid, conspicuously ribbed, glabrous; pericarp 0.5–1 mm thick. **Seed** 9–10 mm long, oblong, smooth; margins revolute.

Distribution   W.A., restricted to the bed and river banks of the lower reaches of the Murchison R. *Climate* Hot, dry summer; mild to warm, wet winter. Rainfall 400–440 mm.

Ecology   Grows in deep mounds of alluvial sand or in alluvial sand over rock in or close to the bed of the seasonal Murchison R. It may be smashed by winter floods but reshoots strongly from its base, usually on the upstream side. Flowers winter to spring. Regenerates from lignotuber or epicormic buds after fire. Pollinator unknown, probably insect.

Variation   A uniform species.

Major distinguishing features   Branchlets angular, ribbed; leaves simple, entire, straight, the undersurface bisulcate, midvein inconspicuous on upper surface; conflorescence few-flowered, umbel-like; perianth zygomorphic, hairy on both surfaces; ovary glabrous; style end granulate; fruit longitudinally keeled.

Related or confusing species   Group 21, especially *G. inconspicua* and *G. christinae*. *G. inconspicua* is a much-branched, silvery shrub with terete branchlets and vertically descending leaves with a conspicuous midvein on the upper surface and usually kinked near the base. *G. christinae* has elliptic, flexible leaves with the undersurface visible, and glabrous. The pistil has a distinct groove on the ventral side and a prominent keel on the dorsal side.

Conservation status   2RC. A species with a very restricted distribution although common in its habitat.

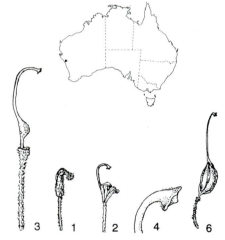

G. costata

Page 2

spread when planting. Light frosts should be tolerated but the effect of summer humidity and rain is unknown. Pruning would be necessary to retain a good shape. Further experience of this species in cultivation will permit more positive comment. The species is not commercially available.

Propagation   *Seed* Sets prolific quantities of seed in the wild which should germinate well if pre-treated in the recommended manner. *Cutting* Firm, young growth taken in early spring or autumn strikes readily. *Grafting* Untested.

Horticultural features   *G. corrugata* is an attractive, rather soft-looking shrub with greenish grey, slightly prickly leaves with spreading lobes, distributed somewhat distantly along the branchlets. In late winter and early spring, masses of scented, white, lacy flowers swell in the leaf axils on short branchlets, and sometimes along all the upper branches. Its spreading habit would make it suitable as a mid-dense screen or feature plant. The green, wrinkled fruits set abundantly in late spring are also both interesting and attractive.

81A. *G. costata* Flowers and foliage (N.Marriott)

81B. *G. costata* Shrub in natural habitat, Murchison River bridge, W.A. (C.Woolcock)

81C. *G. costata* Fruits (P.Olde)

Cultivation This species has been in cultivation for many years and is very long-lived in cool-wet or hot-dry climates. Plants over 30 years old grow at Dave Gordon's property at Glenmorgan, Qld, and at Alby Lindner's in western Vic. It is untested in summer-rainfall climates but similar success can be expected. It is relatively undemanding of drainage and, once established, never needs watering or other maintenance apart from occasional pruning to retain its naturally compact habit. A well-drained, sunny to semi-shaded site would be ideal and it grows in sand, loam, and well-drained clay with equal ease. It has proved hardy to extended drought and frosts to at least -6°C without ill-effect. It is not widely available through nurseries or seed suppliers.

Propagation *Seed* Sets copious seed that germinate well. Average time for germination 27 days (Kullman). *Cutting* Strikes readily using firm, new growth especially when taken in early autumn or early spring. *Grafting* Untested.

Horticultural features *G. costata* has narrow, grey-green leaves and masses of delightfully perfumed, white flowers during winter and spring. It is extremely long-lived in the garden and makes an attractive, compact, low screen or rockery plant in the landscape. While not spectacular, in full flower it is quite a charming plant with white flowers delicately veiling every branch. It is mainly a plant for collectors but could be useful to landscapers seeking a hardy, low shrub for massed planting.

General comments This species has been incorrectly sold for many years, particularly in Vic., as *G. inconspicua*.

Grevillea crassifolia K.Domin (1923)
Plate 82

New Additions to the Flora of W.A.; *Mémoires de la Société Royale des Sciences de Bohême* 1921–22 (2): 10–11 (Czechoslovakia)

The specific epithet is derived from the Latin *crassus* (thick) and *folium* (a leaf), in reference to the leathery leaves. KRASS-I-FOAL-EE-A

Type: probably between Cranbrook and Warrungup, W.A., 1910, A.A.Dorrien-Smith (holotype: K).

An open, twiggy **shrub** to 1 m. **Branchlets** angular to terete, tomentose to glabrous. **Leaves** 0.5–3.2 cm long, 2–6 mm wide, spreading to ascending, subsessile, simple, elliptic, acute to obtuse, mucronate; upper surface sparsely tomentose, soon granulate, glabrous, the midvein scarcely evident; lower surface velvety, midvein prominent; margins shortly recurved to revolute, entire. **Conflorescence** terminal on short branchlet or axillary, 2–4-flowered, sessile; rachis silky; bracts 1.2 mm long, narrowly triangular, villous, falling before anthesis. **Flower colour**: perianth and pistil red. **Flowers** abaxially oriented; pedicels 4–6 mm long, tomentose; torus very oblique; nectary linguiform, conspicuous, bidentate; **perianth** 5–8 mm long, 2–4 mm wide, oblong with a conspicuous basal dilation, strongly recurved at throat, prominently ribbed, sparsely tomentose outside, bearded just above base inside, sparsely tomentose elsewhere, cohering before anthesis except flared open between curve and limb, exposing style end before anthesis; limb nodding to revolute, globose, the segments keeled, tomentose; **pistil** 7–9 mm long; ovary subsessile, villous; stipe 0.5–1 mm long; style tomentose, not exserted before anthesis, scarcely or not longer than perianth after anthesis; style end clavate, conspicuous; pollen presenter concave, lateral. **Fruit** 13 mm long, 5 mm wide, ovoid, erect, pubescent; pericarp 0.5 mm thick. **Seed** 8 mm long, 2 mm wide, narrowly elliptic with apical elaiosome; outer face rugose; inner face obscured by revolute margins.

Distribution W.A.; known from a few populations in and near the Stirling Ra. and towards Cape Riche. *Climate* Cool, wet winter; warm, dry summer. Rainfall 600–800 mm.

Ecology Grows on mountains, hillsides and lower

G. crassifolia

slopes in gravelly loam, usually in low shrubland. Flowers winter to summer. Fire response unknown but probably killed as it is strongly seed regenerative. Pollinator unknown, probably small birds.

Variation A uniform species with little variation.

Major distinguishing features Leaves simple, entire, elliptic, leathery, granulate above, velvety below; conflorescence few-flowered; torus very oblique; nectary conspicuous, linguiform, bidentate; perianth zygomorphic, oblong and dilated at the base, red; ovary subsessile, densely hairy; style scarcely to not exceeding perianth; pollen presenter concave.

Related or confusing species Group 23, especially *G. depauperata* and *G. fasciculata*. *G. fasciculata* has an ellipsoidal perianth (dilated about the middle). *G. depauperata* has a perianth with a distinct basal sac but has a much longer pistil (> 10 mm long) that clearly exceeds the perianth.

Conservation status 2RC recommended.

Cultivation *G. crassifolia* has been cultivated successfully at Mount Annan Botanic Garden, N.S.W., and has proven quite hardy, especially in mild climates. It should tolerate light frosts and be reasonably drought-hardy, although some underlying soil moisture would be necessary. It can be grown as an undershrub but performs best in full sun in well-drained, neutral to acidic sand or loam. Tip pruning would avoid a tendency to straggly growth.

Propagation *Seed* Untested. Sets plentiful seed and should grow well by this method, given some pre-treatment. *Cutting* Cuttings should strike reasonably well using firm, new growth. *Grafting* Untested.

Horticultural features *G. crassifolia* is a much-branched, twiggy shrub which, under conditions of low competition, achieves a spread of some 50 cm and displays its small red flowers most attractively. Although flowering is somewhat subdued, flower colour and small leaves contrast well and flowers are quite noticeable. Birds should be strongly attracted. Ideal for large landscapes or specialist gardens in need of an interesting fill-in or low-growing feature.

General comments *G. crassifolia* was treated as a form (Element 1) of *G. fasciculata* by McGillivray (1993), but it differs sufficiently in its leaves, flowering habit and perianth shape to be recognised as a distinct species.

Grevillea crithmifolia R.Brown (1830)
Plate 83

Supplementum Primum Prodromi Florae Novae Hollandiae exhibens Proteaceas Novas: 23 (England)

The specific name recalls the similarity to the foliage of the European Sea Samphire, *Crithmum maritimum*; Latin *folium* (a leaf). KRITH-MI-FOAL-EE-A

Type: Swan River, [W.A.], 1827, C.Fraser (lectotype: BM) (McGillivray 1993).

A dense, spreading **shrub** to 2.5 m, sometimes prostrate or nearly so. **Branchlets** terete, villous. **Leaves** 2–3 cm long, 0.7–10 mm wide, ascending to spreading, crowded, ± sessile, bifid or trifid, sometimes pinnatipartite with 5 short, weakly pungent, linear lobes c.1 mm wide, occasionally simple; upper surface sparsely pubescent, soon glabrous, midvein obscure; margins revolute to midvein; lower surface bisulcate, villous to pilose, especially the prominent midvein. **Conflorescence** 1.5–3 cm long, 1.5–3 cm wide, erect, shortly pedunculate or sessile, usually terminal on short branchlet, unbranched, dense, umbel-like; peduncle and rachis villous; bracts 4.5–7 mm long, ovate, conspicuous, imbricate in bud, villous outside, falling before anthesis. **Flower colour**: perianth and style creamy white or soft pink. **Flowers** acroscopic to irregularly oriented; pedicels 6–8 mm long, filiform, sparsely silky; torus c. 0.5 mm across, straight to slightly oblique; nectary V-shaped; **perianth** 2–4 mm long, 1 mm wide, ± cylindrical, strongly curved, glabrous outside, sparsely villous inside; tepals separating first along dorsal suture, then all free, the dorsal tepals reflexing laterally to reveal inner surface before anthesis, afterwards all free to base; limb revolute, globular, enclosing style end before anthesis; **pistil** 4.8–6 mm long, glabrous; stipe 1–1.5 mm long; ovary globose, smooth but becoming tuberculate; style filiform, exposed to base before anthesis, afterwards retrorse to S-shaped; thickening before dilating suddenly at style end; pollen presenter oblique, conical with a basal rim. **Fruit** 12–15 mm long, 10 mm wide, erect, ± oblong, conspicuously muricate; pericarp 1–2 mm thick. **Seed** 7.5–10 mm long, 3–4 mm wide, elliptic; outer face smooth, convex; inner face with a prominent central ridge, channelled around edge; margin recurved.

Distribution W.A., from Mandurah to Wanneroo and an isolated occurrence at Dongara. *Climate* Hot, dry summer; cool, wet winter. Rainfall 400–850 mm.

Ecology Grows in calcareous sand and coastal limestone soil in coastal scrub, eucalypt and banksia woodland and in open coastal heath. Flowers late winter to spring. Sprouts from a lignotuber after fire. Pollinated by insects.

Variation *G. crithmifolia* varies in habit, degree and coarseness of leaf division and flower colour. There is also considerable variation in leaf colour, from grey to green. A number of selections are used by landscapers in Perth (Marion Blackwell, pers. comm.) but these are not listed here.

Typical form The Typical form has white flowers, sometimes with a touch of pink especially in young flowers about to release their styles. It grows to c. 2.5 m tall and as wide. Foliage can be dull green or grey. Fasciation of the flowers frequently occurs in this form, especially in cultivation.

Pink-flowered form This form is a horticultural selection and not a distinct population. It tends to have a less vigorous habit than the common white-flowered form, and rarely grows taller than c. 1–1.5 m. It has distinctive reddish pink branchlets and delicate, all-pink flowers.

Prostrate form Another horticultural selection that remains quite prostrate and forms a dense, vigorous mat. One prostrate plant in cultivation has been recorded over 6 m across. It makes a useful spill-over plant. Foliage is slightly coarser than the Typical form, but the flowers are the same.

Mounded form This form is a dense, low mounded shrub to c. 0.5 m high and 2 m wide. Foliage tends to be intermediate between the Typical and Prostrate forms but the flowers are typical.

Major distinguishing features Leaves 2–3 cm long, forked or pinnatipartite with 2–5 short, linear lobes; conflorescence terminal, dome-shaped; perianth

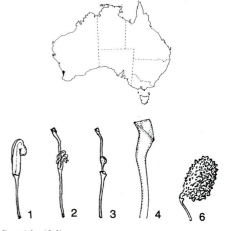

G. crithmifolia

82. *G. crassifolia* Flowers and foliage (M.Hodge)

zygomorphic; pistil glabrous, pollen presenter erect conical; fruit muricate, the tubercles 0.5–1.5 mm high.

Related or confusing species Group 8. *G. murex* (Group 16) may also be closely related but differs in its smaller leaves (4–10 mm long), angular branchlets, its late-recurved perianth and its flat pollen presenter.

Conservation status Not presently endangered.

Cultivation *G. crithmifolia* has been grown successfully in climates as different as Darwin and Adelaide, Sydney and Perth; it has also been cultivated in the U.S.A. While it is sometimes short-lived in summer-humid climates, it tolerates frosts to at least -6°C as well as drought. It needs a well-drained situation but is not fussy about soil type, growing well in sand, loam and heavy soil, either acidic or alkaline. A position in full sun will produce a dense, vigorous shrub but it will tolerate partial shade. Pruning to shape may sometimes be required. Stems may sometimes be attacked by a destructive borer that causes branches to die or may even kill the whole plant if infestation occurs near the crown of the roots. Fasciation of flowers is commonly reported in cultivation. The species is widely available in nurseries throughout Australia.

Propagation *Seed* Sets large quantities of seed that germinate well given the standard peeling treatment. In 1983, following a wildfire that burnt Mr Ken Stuckey's magnificent native garden near Millicent, S.A., hundreds of seedlings of this species germinated in the sandy soil. Average germination time 23 days (Kullman). *Cutting* Strikes readily from cuttings taken at most times of the year, with optimal times being early autumn and early to mid spring. *Grafting* Has been grafted onto *G. robusta* using both top wedge and cotyledon grafts.

Horticultural features *G. crithmifolia* is a robust shrub, ideal either as a large, dense screen or dense groundcover (Prostrate form). It has attractive foliage, most forms having grey-green, short, deeply divided leaves that give the appearance of small fingers. In spring, it is massed with white to pink flowers. The Pink-flowered form is not as robust as the normal shrubby form but is the most attractive when in full flower. The fruits also have a most unusual horned surface that always arouses interest. The flowering period does not last as long as many other grevilleas and out of season flowers are uncommon, but the intensity of flowering compensates for this. The flowers are strongly scented with a heavy nectar smell typical of some plume grevilleas and are probably pollinated by moths or other insects.

Hybrids A reputed hybrid with *G. preissii* is often sold as G. 'Magic Lantern' in Sydney. However, the presumed influence of *G. crithmifolia* in this plant is incorrect.

History *G. crithmifolia* was introduced to English

83A. *G. crithmifolia* Typical flower colour and foliage. Cultivated at Shepparton, Vic. (P.Olde)

83B. *G. crithmifolia* Unusual follicles covered with horn-like projections (N.Marriott)

83C. *G. crithmifolia* Pink-flowered form. Cultivated at Stawell, Vic. (N.Marriott)

glasshouses in 1840 and was also grown in Florence in 1855.

Grevillea crowleyae P.M.Olde & N.R.Marriott (1993)   Plate 84

Nuytsia 9: 271 (as *crowleyi*) (Australia)

The epithet honours Mrs Valma Crowley, amateur naturalist of Darkan, W.A., who, with her friend Mrs Janice Smith, also of Darkan, discovered the new species in a gravel pit while bushwalking near her home. Mrs Crowley drew our attention to this species and forwarded specimens. The ongoing concern of the two discoverers for its conservation has enabled this species to be preserved in its natural habitat. CROW-LEE-EYE

Type: near Dardadine, NE of Darkan, W.A., 26 Sept. 1991, P.M.Olde 91/234 (holotype: PERTH).

Shrub 0.5–1.5 m high, dense, spreading to 1.5 m wide, becoming spindly; bark grey, rough. **Branchlets** terete, occasionally angular, tomentose to pubescent, not striate. **Leaves** mostly 3–7 cm long, grey to grey-green, ascending, pinnatisect, subsessile but appearing petiolate (distance from leaf base to first lobe 10–20 mm); leaf lobes 3–7, 10–42 mm long, 0.8 mm wide, narrowly linear, often dipleural, acute, scarcely pungent; upper surface white tomentose to pubescent when young, becoming ± glabrous, smooth or slightly grooved beside midvein; lower surface bisulcate; margins smoothly or angularly revolute to midvein; venation obscure except midvein on undersurface; texture firmly papery. **Conflorescence** 2–5 cm long, terminal, secund, dense, simple, erect or occasionally decurved, pedunculate, scarcely exceeding foliage; peduncle 2.5–4 mm long, bracteate (bracts 3–4.5 mm long); rachis 1.2–1.5 mm thick at base, villous to woolly; bracts 2 mm long, ovate-acuminate, ascending to spreading, persistent, imbricate in bud, villous to tomentose throughout, glabrous at base inside.

Flower colour: perianth grey; style maroon-black, sometimes red. **Flowers** acroscopic; pedicels 1.5–2.5 mm long, villous; torus 1.2–1.5 mm across, straight to oblique, projecting further on ventral side; nectary tongue-shaped, slightly lipped, entire, patent. **perianth** 7–8 mm long, 2–3 mm wide at base, obliquely ovoid, tomentose outside, glabrous inside, cohering except along dorsal suture; limb villous, revolute, globose to spheroidal; style end enclosed before anthesis; **pistil** (23–)34–38 mm long; stipe 0.1–0.5 mm long, villous; ovary subsessile, villous; style exserted below curve on dorsal side and looped upwards, afterwards elongate and gently incurved, hairs extending (3–)5–10 mm along style above ovary, otherwise glabrous; style end gradually dilated c. 1 mm from end; pollen presenter convex to subconic, ± orbicular, oblique. **Fruit** 13–16 mm long, 6–9 mm wide, erect or reflexed on straight or incurved pedicel, oblong-ellipsoidal, slightly compressed, glandular-tomentose with conspicuous red or brownish blotches or striping; pericarp c. 0.3 mm thick throughout. **Seed** 12 mm long, 5 mm wide, 2 mm deep, oblong-elliptic; outer face convex, smooth with a prominent submarginal ridge all round; margin undulate, entire, pale; inner face concave with membranous, gill-like platelets radiating c. 2 mm from margin towards a central, more open area 6 mm long, 1.5 mm wide.

Distribution W.A., confined to the Type locality where only 10 plants remain, most in a disturbed gravel pit. Two were found in nearby undisturbed bushland. *Climate* Cool, wet winter; warm to hot, dry summer. Rainfall 800–1000 mm.

Ecology Grows in *Eucalyptus wandoo* forest in heavily lateritised loam. Flowers Aug.–Nov. Regenerates from seed after fire. Probably bird-pollinated.

Variation A relatively uniform species.

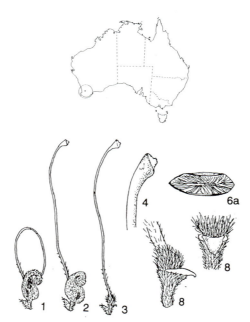

G. crowleyae

84A. *G. crowleyae* Flowering habit (P.Olde)

84B. *G. crowleyae* Plant in natural habitat, Dardadine, W.A. (P.Olde)

84C. *G. crowleyae* Flowers and foliage (P.Olde)

Major distinguishing features Leaves pinnatisect, the lobes 0.8 mm wide and closely aligned, venation of upper surface obscure; conflorescence usually erect, sometimes deflexed, secund; peduncle 2–4 mm long; perianth zygomorphic, glabrous inside; pistil 34–38 mm long; ovary villous; style with hairs extending up to 10 mm from ovary; fruit with mixed glandular and 2-armed hairs; seeds with radially oriented gill-like membranes on inner face.

Related or confusing species Group 35, especially *G. calliantha* and *G. hookeriana*. *G. hookeriana* has a pistil < 21.5 mm long. *G. calliantha* has leaf lobes > 1 mm wide, hairs on the style restricted to the basal 3 mm and more open conflorescences.

Conservation status Suggested 2EC-t. The Type locality is currently being considered as a nature reserve. *G. crowleyae* was propagated in September 1991 from cutting material sent to the Royal Botanic Gardens annexe at Mount Annan, N.S.W., as well as by interested growers locally in the Darkan area. It is recommended that no further collecting pressure be exerted on the natural population.

Cultivation There has been little experience in cultivating this newly discovered species but some inferences can be drawn from its natural habitat. It likes an open, non-competitive situation and performs best in full sun in acidic to neutral, gravelly loam. It will survive in overhead shade but in such situations becomes spindly and sparsely leaved. Soil should be reasonably moist but well-drained. Pruning is unnecessary as it forms a naturally compact habit. Drought and frost tolerance are unknown but summer humidity might cause problems. Plants in pots demand excellent drainage and low-phosphorus fertiliser.

Propagation *Seed* Sets reasonable quantities of seed in the wild which should germinate reliably with some pre-treatment. *Cutting* Strikes with difficulty from the wild. Firm, new growth from cultivated plants should do better. *Grafting* Untested.

Horticultural features In general aspect, *G. crowleyae* resembles *G. calliantha* but has much finer leaf and flower features. The grey, finely divided foliage is offset by prominent, terminal flowers that, although they are black, stand out clearly against the foliage and are most attractive. Birds will be attracted to the plant and it can be recommended as a feature plant or contrast plant in the shrubbery.

Grevillea cunninghamii R.Brown (1830)
Plate 85

Supplementum Primum Prodromi Florae Novae Hollandiae exhibens Proteaceas Novas: 23 (England)

Aboriginal: 'andan' (Kwini) (source: J.Kardady).

Named in honour of the botanist-explorer Allan Cunningham (1791–1839). KUNN-ING-HAM-EE-EYE

Type: Montague (Montagu) Sound, [W.A.], 7 Sept. 1820, A.Cunningham (lectotype: BM) (McGillivray 1993). The collection was gathered during King's Third Voyage in the *Mermaid*.

Erect, open, spindly **shrub**, sometimes a spreading, multi-stemmed shrub 1.2–3 m tall. **Branchlets** terete, glabrous, glaucous. **Leaves** 4–9 cm long, 3–5.5 cm wide, spreading, sessile, simple, ± ovate, stem-clasping with heart-shaped leaf base; juvenile leaves much longer; upper and lower surfaces similar, concolorous, glabrous, glaucous, prominently penninerved; margins pungently toothed, flat. **Conflorescence** 2.5–4.5 cm long, ± glabrous, erect, on long peduncle, axillary, unbranched, few-flowered, dome-shaped to subglobose, scarcely to not exceeding foliage; peduncle slender, 9–25 mm long, sometimes with scattered hairs; rachis glabrous; bracts 0.3–0.4 mm long, minutely ovate, sparsely tomentose with ciliate margins, falling very early. **Flower colour**: perianth red with a yellowish limb; style red. **Flowers** adaxially oriented; pedicels 2–2.5 mm long; torus 1.5–2 mm across, oblique, extended further on dorsal side; nectary conspicuous, U-shaped, smooth; **perianth** 4–5 mm long, 3–4 mm wide, ovoid, strongly curved, dilated near base, glabrous outside, bearded inside above level of ovary, cohering except along dorsal suture; limb revolute, spheroidal, impressed at margins; **pistil** 8–9.5 mm long, glabrous; stipe 1.2–2.2 mm long, adnate and lateral with torus; ovary ovoid to oblong; style, flattened, quite broad, ventrally grooved, not exserted before anthesis, afterwards straight and not far exceeding perianth; style end partially exposed before anthesis through dorsal suture, scarcely to not expanded; pollen presenter lateral, obovate, ± flat, umbonate. **Fruit** 9–10.5 mm long, 6 mm wide, horizontal to pedicel, oblong, rugose, glabrous, glaucous; pericarp c. 1 mm thick. **Seed** 5.5–7.5 mm long, 2.5–3.5 mm wide, compressed-ellipsoidal; outer face smooth, convex; inner face convex, smooth with narrow marginal wing all round with an apical elongation.

Distribution W.A., on and near the Kimberley coast and adjacent islands. *Climate* Hot, wet summer; warm, dry winter. Rainfall 800–1200 mm.

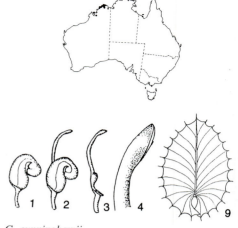

G. cunninghamii

85A. *G. cunninghamii* Close-up of foliage and flowers (P.Olde)

85B. *G. cunninghamii* Plants in natural habitat, NW of Kalumburu, W.A. (P.Olde)

Ecology Grows in gravelly sand on rocky sandstone hills and bauxite plateaus, generally in open shrubland. Flowers late autumn to early spring. Regenerates from seed after fire. Pollinator unknown, probably small honey-eating birds.

Variation A relatively uniform species.

Major distinguishing features A ± glabrous shrub; leaves simple, concolorous, ovate; margins flat, with pungent teeth, the base amplexicaul; conflorescences axillary, subglobose on slender peduncle; torus very oblique; perianth zygomorphic, bearded inside; pistil glabrous; stipe lateral on dorsally extended torus; style end not expanded; pollen presenter lateral.

Related or confusing species Group 10, which includes *G. adenotricha* and *G. longicuspis*, both of which have non-clasping leaf bases.

Conservation status Not presently endangered.

Cultivation *G. cunninghamii* has not been tested on its own roots in cultivation and no indication can be given of its hardiness or adaptability. Its habitat indicates that it would be suited to a tropical climate in a frost-free situation with a preference for sandy soil, probably among rocks. It would most likely tolerate summer humidity and rainfall and would grow best in full sun. Grafted plants have flourished in Brisbane and northern N.S.W., producing robust free-flowering shrubs of great beauty.

Propagation *Seed* Fresh seed germinates readily. *Cutting* Untested, but probably firm, young growth taken in summer would be successful. *Grafting* Has been grafted successfully onto *G. robusta*. Performs strongly on this rootstock.

Horticultural features *G. cunninghamii* is an extremely attractive plant with bright green to somewhat glaucous, beautifully sculptured leaves and delicate red flowers dotted between. Although the leaves are prickly, when planted towards the back of a garden, the visually more interesting features of the species may be appreciated. Tip pruning would improve a natural tendency to lankiness. A worthy addition to any tropical garden, this species has potential in large landscapes or as a feature in smaller gardens.

Grevillea curviloba D.J.McGillivray (1986)

New Names in Grevillea *(Proteaceae)*: 4 (Australia)

Replacing the name *G. vestita* var. *angustata* C.F.Meisner (1845).

The specific epithet alludes to the sometimes-curved leaf lobes and is derived from the Latin *curvus* (curved) and *lobus* (a lobe). KER-VEE-LOW-BA

Type: Swan R. colony, W.A., late 1830s, J.Drummond coll. 1 no. 622 (lectotype: G-DC) (McGillivray 1993).

A spreading, irregular **shrub** to 2 m with weakly ascending to erect flowering branches. **Branchlets** angular, sparsely silky, brittle. **Leaves** 1–5 cm long, < 1.5 cm wide, ascending, shortly petiolate, crowded, obovate with apical teeth or subpinnatisect, sometimes simple and entire; primary lobes 5–20 mm long, triangular or narrowly so, sometimes with secondary or tertiary divison, scarcely pungent; upper surface glabrous or sprinkled with hairs, distinctly grooved along midvein and lateral veins; margins smoothly revolute to recurved; lower surface glabrous or with scattered hairs, midvein prominent, rounded. **Conflorescence** erect, shortly pedunculate, usually axillary, sometimes terminal, simple or few-branched, often in dense panicles; unit conflorescence 1–3 cm long, 3 cm wide, ovoid to subcylindrical; peduncle tomentose; rachis glabrous to sparsely tomentose; bracts 1.3–3 mm long, imbricate, ovate, glabrous except ciliate margins, falling at or before anthesis. **Flowers** regular, glabrous: pedicels 7–10 mm long, filamentous; torus c. 0.5 mm across, straight; nectary oblong, lipped; **perianth** 3–6 mm long, 0.8–1 mm wide, oblong-obovoid below limb, ellipsoidal just before anthesis, sometimes lightly bearded inside; limb erect, globular; tepals separating below limb before anthesis, afterwards rolling back independently to base; **pistil** 3.5–6.5 mm long; stipe 1.4–2 mm long, flexuose; ovary globose; style straight, constricted above ovary for c. 1 mm, then strongly dilated, narrowing just under style end; pollen presenter conical, erect, the base broader than the style. **Fruit** 10–13 mm long, 6–9 mm wide, horizontal to stipe on curved pedicel, oblong-ellipsoidal, rugose; pericarp c. 1 mm thick. **Seed** 7–9 mm long, 3–3.5 mm wide, obovoid to ellipsoidal; outer face convex, smooth, shiny; inner face convex, faintly scarred around margin; margin scarcely recurved.

Major distinguishing features Branchlets glabrous or sparsely hairy; leaf upper surface smooth with obscure venation, the midvein channelled, the margin smoothly revolute; undersurface exposed beside midvein and at sinuses, with prominent veins, glabrous; pedicels 7–10 mm long; floral bracts > 1 mm long, ovate, glabrous but ciliate; nectary prominent; perianth actinomorphic, glabrous outside, sometimes sparsely hairy inside; style constricted above the ovary and again beneath the style end; fruit oblong-ellipsoidal, rugose; pericarp c. 1 mm thick.

Related or confusing species Group 1, especially *G. vestita*, *G. corrugata* and *G. rara*. *G. corrugata* has much broader leaves (to 9 cm wide) with a winged base, prominent veins on upper surface and angularly refracted margins. *G. rara* differs in its shorter pedicels (c. 5 mm), closely revolute leaf margins and densely hairy branchlets. *G. vestita* differs in its villous branchlets, persistent floral bracts and smooth fruit.

Two subspecies are recognised.

Key to subspecies

Leaves usually pinnatifid, sometimes subpinnatisect, with broad sinuses and the undersurface clearly exposed; simple leaves or primary leaf lobes usually obovate-cuneate with ascending lobes, sometimes linear, > 1.5 mm wide; secondary division usually lacking   subsp. **curviloba**

Leaves secund-pinnatisect, the undersurface either not visible or inconspicuously so at sinuses or beside midvein; primary leaf lobes narrowly linear to subulate, 0.8–1.2 mm wide; secondary division usually present   subsp. **incurva**

Grevillea curviloba D.J.McGillivray subsp. **curviloba**   Plate 86

Leaves 1.5–5 cm long, mostly pinnatifid, obovate-cuneate, coarsely divided either with apical teeth or more deeply cleft with strongly ascending lobes, rarely simple, entire, linear; simple leaves and lobes mostly 1.5–2 mm wide, triangular to narrowly so, occasionally deep cleft; undersurface exposed over most of its area, glabrous or sparsely tomentose. **Flower colour**: perianth and style white or creamy white. **Pistil** 3.5–4.5 mm long.

Synonym: *G. diversifolia* var. *rigida* C.F.Meisner (1856).

Distribution W.A., confined to a small area near Bullsbrook. *Climate* Warm, dry summer; cool, wet winter. Rainfall 500–830 mm.

Ecology Grows in open heath in winter-wet, deep peaty sand with other medium-sized shrubs. Flowers spring. Regenerates from seed after fire. Pollinated by insects, possibly native bees.

Variation There is some variation in the leaves and habit of this subspecies. Some plants are quite prostrate with broad dark green leaves, a habit retained in cultivation. Other plants are erect shrubs to 2 m with grey-green leaves.

Conservation status 2E. Restricted to degraded road verges.

Cultivation Subsp. *curviloba* has not been widely cultivated but initial results indicate that it is relatively hardy in most conditions. Moist, well-drained soil is optimal for its long-time survival, although once established it tolerates extended dry periods. It will survive frosts to at least -6°C. Neutral to slightly acidic sandy soil or sandy loam is preferred. Some protection from strong winds is desirable and tip-pruning will result in a more dense, vigorous plant. It responds well to light fertiliser low in phosphorus. It has been grown successfully in Sydney, Melbourne and western Vic. It is sometimes sold by specialist nurseries.

Propagation *Seed* Sets numerous seed that germinate well when peeled. Average germination time 30 days (Kullman). *Cutting* Strikes readily and quickly from cuttings using firm, young growth taken at any season. *Grafting* Untested.

86A. *G. curviloba* subsp. *curviloba* Foliage (N.Marriott)

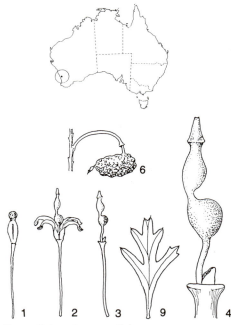

G. curviloba subsp. curviloba

86B. *G. curviloba* subsp. *curviloba* Close-up of flowers (N.Marriott)

Horticultural features The bright green new foliage of subsp. *curviloba* makes it an ideal, sought-after foliage contrast plant. Prostrate forms make beautiful ground covers, especially suitable in home situations as luxuriant spill-over plants or for weed suppression. Flowering is extremely impressive, being in a massed display, attracting many insects by the sweet scent. Upright forms are rapid, hardy growers, ideal as a dense background or screen plant.

Grevillea curviloba subsp. incurva
P.M.Olde & N.R.Marriott (1993)

Plate 87

Nuytsia 9: 243 (Australia)

The subspecific epithet alludes to the curved leaf lobes on this subspecies and is derived from the Latin *incurvus* (curved inwards) INN-KER-VA

Type: Muchea, W.A., 26 Sept. 1992, P.M.Olde 92/108 (holotype: NSW).

Leaves 1.8–5.2 cm long, secund, tripartite to subpinnatisect, the first lobe arising 6–32 mm from leaf base; primary lobes 3–5 narrowly linear, sometimes the lower lobes bi- or trisect; ultimate lobes 7–20 cm long, 0.8–1.2 mm wide, strongly incurved, narrowly linear to subulate, weakly pungent; lower surface bisulcate, the lamina obscured or almost so by margin.

Flower colour: perianth and style white or creamy white. **Pistil** 4–6.5 mm long.

Synonyms: The names *G. biternata* and *G. tridentifera* have been incorrectly used for this species.

Distribution W.A., confined to an area between Muchea and Badgingarra. *Climate* Warm, dry summer; cool, wet winter. Rainfall 500–830 mm.

Ecology Grows in open heath in winter-wet, deep peaty sand at sites with a high water table. Flowers spring. Regenerates from seed after fire. Pollinated by insects, possibly native bees or wasps.

Variation There is some variation in the habit of this subspecies which ranges from prostrate to erect.

Conservation status 2E. Restricted to degraded road verges.

Cultivation Subsp. *incurva* is an extremely adaptable plant, growing well in cold-wet and hot-dry as well as subtropical climates. Once established, it withstands dry conditions provided there is adequate subsoil moisture, tolerates frosts to at least -7°C and has been grown successfully in all States of Australia. In summer-rainfall climates, such as Sydney and Brisbane, it does not enjoy a good reputation because mature plants, after early vigour, are sometimes prone to sudden death. It can be cultivated in a wide range of sites but best results occur when planted in full sun in well-drained but slightly moist, acidic sand or sandy loam. Under these conditions, it grows rapidly and flowers profusely. In the wild, it is seasonally inundated and may stand in shallow water for several months in winter. It responds to light applications of slow-release fertiliser and to summer watering. If desired, upright branches can be removed to maintain a prostrate habit, although this tends to reduce the flowering potential. Due to its vigour, it is not suitable for long-term pot culture, although it makes a superb green cascade when planted in a large, open-ended pipe or drum. Readily available at most nurseries, often as *G. biternata*.

Propagation *Seed* Sets numerous seed that germinate well when peeled. Average germination time 30 days (Kullman). *Cutting* Strikes readily and quickly from cuttings using firm, young growth taken at any time of the year. Material should be selected from prostrate stems as there is some evidence that cuttings taken from upright branches produce more upright open bushes. *Grafting* Using *G. robusta* rootstock, it makes an interesting

G. curviloba subsp. incurva

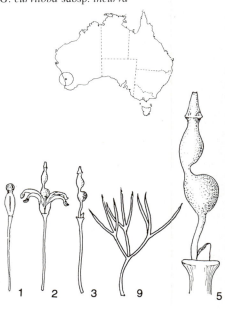

87A. *G. curviloba* subsp. *incurva* Flowers and foliage (N.Marriott)

87B *G. curviloba* subsp. *incurva* Flowering habit, Muchea, W.A. (C.Woolcock)

weeping standard. Otherwise grafting is rarely necessary.

Horticultural features In the wild, subsp. *incurva* usually grows as a vigorous, sprawling shrub to 2.5 m high and wide. It tends to be vegetatively dense at the base, sending up erect, floral branches beyond the foliage. These open branches persist and flower prolifically from year to year. In cultivation, it is often grown simply as a ground cover. Foliage is a rich, bright green giving the plant a fresh appearance and setting off the profuse, creamy white flowers in spring. It has been used extensively for large commercial and government landscapes such as around airports, universities and along freeways. It is a long-lived species except in summer-humid climates and makes a dense groundcover as well as a medium to large screen plant. Flowers are sweetly perfumed.

Hybrids Presumed parent of the hybrid G. 'White Wings'.

General comments The names *G. biternata* and *G. tridentifera* have been incorrectly applied to this subspecies.

Grevillea cyranostigma D.J.McGillivray (1975)
Plate 88

Telopea 1: 20 (Australia)

Carnarvon Grevillea, Green Grevillea

So named because the prominent, pointed pollen presenter reminded the author of the distinctive nose in profile of Cyrano de Bergerac, French writer and satirist. SI-RAR-NO-STIG-MA

Type: Mt Playfair, Qld, c. 1890–1895, Mrs Biddulph (holotype: MEL).

A much-branched, spreading **shrub** 0.5–2 m tall ?with lignotuber. **Branchlets** thin, terete, silky. **Leaves** 2.5–5 cm long, 0.5–1 cm wide, ascending, shortly petiolate, occasionally sinuate, simple, oblong to elliptic, acute to obtuse, mucronate; upper surface glabrous, glossy, midvein slender, evident, intramarginal veins and acutely angled lateral venation faintly evident; margins entire, flat to shortly recurved; lower surface silky, midvein prominent. **Conflorescence** decurved, pedunculate, simple or 2-branched, axillary and terminal; unit conflorescence 2–3.5 cm long, 2–4 cm wide, open, hemispherical to dome-shaped, few-flowered, apical flowers opening first; peduncle and rachis silky; bracts 1 mm long, 0.3 mm wide, linear, silky inside and out, falling very early. **Flower colour**: perianth translucent green; style green, sometimes turning red. **Flowers** adaxially acroscopic; pedicels 4–5.5 mm long, slender, glabrous or sparsely hairy; torus c. 1.5 mm across, oblique; nectary kidney-shaped, cupped, smooth or lightly toothed; **perianth** 7–8 mm long, 3 mm wide, ovoid, ± glabrous or sparsely hairy, white bearded inside near base, at first separated along dorsal suture, the tepals then separating in upper half before anthesis, afterwards ventral tepals curling back as an independently opposed pair much further than dorsal tepals; limb nodding, ovoid with a point, enclosing style end before anthesis; **pistil** 16.5–17.5 mm long; stipe 1.4–2 mm long; ovary globose, glabrous; style at first exserted along dorsal suture and bowed outwards, gently incurved after anthesis, glabrous or sparsely hairy, the style end minutely hairy and scarcely dilated; pollen presenter oblique or lateral, oblong, flat; stigma off-centre, extending c. 1 mm from the face. **Fruit** 14–15 mm long, 8 mm wide, erect, narrowly ovoid to ellipsoidal, smooth or slightly viscid; pericarp 0.5–1 mm thick. **Seed** 10 mm long, 2 mm wide, obovoid; outer face smooth, convex; inner face channelled around rim with a flat centre; margin recurved.

Distribution Qld, restricted to a small area around the Carnarvon Range. *Climate* Hot, wet to dry summer and warm, moist to dry winter. Rainfall 600–700mm.

Ecology Rocky slopes or creek banks in protected situations in open eucalypt forest in sandy soil over sandstone. Flowers winter to spring. After fire regenerates from seed but may sometimes be lignotuberous. Pollinated by birds.

Variation A relatively uniform species.

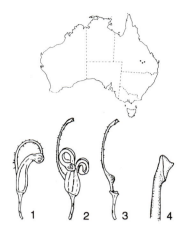

G. cyranostigma

Major distinguishing features Leaves simple, entire, upper surface glossy; perianth zygomorphic, glabrous or sparsely hairy outside, the limb nodding; pistil glabrous except style or style end minutely hairy; pollen presenter lateral with conspicuous stigma.

Related or confusing species Group 21 but very distinct. Both *G. sericea* and *G. victoriae* seem related, the former differing in its non-glossy leaves and more hairy perianth, while the latter also has a much hairier perianth.

Conservation status 3RC. Conserved in Carnarvon National Park.

Cultivation *G. cyranostigma* adapts to a wide variety of climates but is rarely cultivated. It will grow in cool-wet, warm-dry, and summer-rainfall areas with equal ease and has been grown successfully as far south as Melbourne, but these days is grown mainly by collectors in the Brisbane area. It tolerates light frosts but not long dry periods. A warm, sheltered site in partial shade with summer moisture best suits it. It does not seem particular about soil but acidic, moist sand or loam would be ideal. Pruning may occasionally be necessary, as it grows naturally into a rounded but rather open shrub. Nurseries rarely stock this species.

Propagation *Seed* Untested, but seed is likely to germinate readily especially if peeled. *Cutting* Propagates readily from cuttings of firm, young growth taken at any season. *Grafting* Has been grafted sucessfully onto *G. robusta* using both the whip and mummy techniques.

Horticultural features *G. cyranostigma* is a dense, rounded shrub for many years, becoming more open with age. The shiny-bright, pale to dark green upper surface of the leaves contrasts markedly with the silky white undersurface, but unfortunately the foliage tends to blend with the clusters of translucent green flowers. These are strongly attractive to honey-eating birds, and flowering may continue over a long period. In other respects this species is mainly for the collector, although it makes an attractive medium-sized foliage plant, useful in landscaping as a low, dense screen or massed planting suited to both coastal and inland climates.

88A. *G. cyranostigma* Close-up of flowers (M.Hodge)

88B. *G. cyranostigma* Plant in natural habitat at The Hump, Mt Moffat NP, Qld (L.Murray)

88C. *G. cyranostigma* Flowers and foliage (P.Olde)

Grevillea decipiens D.J.McGillivray (1986)
Plate 89

New Names in Grevillea *(Proteaceae)* 4 (Australia)
The specific epithet is derived from the Latin *decipere* (to deceive, pretend). This species is very closely related to *G. oligantha* and easily confused with it. DEE-SIP-EE-ENS

Type: 1 mile (c. 1.6 km) S of Mt Gibbs, W.A., 31 July 1969, A.S.George 9454 (holotype: PERTH).

Dense, compact **shrub** to 1.5 m; with ascending to erect branches. **Branchlets** terete, tomentose. **Leaves** 5–9 cm long, 1–1.6 mm wide, ascending, shortly petiolate, simple, linear, firm; apex obtuse-mucronate; upper surface slightly convex, prominently 7–9-veined, sparsely pubescent between ridges; margins entire, angularly revolute, closely abutting midvein; lower surface bisulcate, sometimes silky in grooves, midvein prominent. **Conflorescence** erect, usually axillary, sometimes terminal on short branchlets, rarely on old stems, sessile, few-flowered, umbel-like; rachis < 1 mm long, tomentose; bracts 0.5 mm long, oblong, villous outside, sometimes persistent to anthesis. **Flower colour**: perianth red; style pale orange to red with green tip. **Flowers** adaxially oriented; pedicels 4–5 mm long, silky; torus 1.3–2.1 mm across, oblique; nectary arcuate to V-shaped, prominent, the margin thick, entire; **perianth** 8–10 mm long, 2 mm wide, ovoid-cylindrical, dilated near base, sharply narrowed at neck, densely brown-silky outside, bearded inside about ovary, otherwise silky; cohering to anthesis except along dorsal suture, the tepals free to curve and rolled back independently after anthesis; limb revolute, globular, sometimes apiculate, prominently keeled, enclosing style end before anthesis; **pistil** 17–20.5 mm long, glabrous; stipe 1.3–2.2 mm long; ovary subtriangular, the base truncate; style at first exserted from below curve, afterwards gently incurved; style end flattened; pollen presenter very oblique to lateral, obovate, flat to slightly concave. **Fruit** 10–11.5 mm long, 3–4.5 mm wide, erect on stipe, narrowly subtriangular-ovoid, attenuated, smooth; pericarp c. 0.5 mm thick. **Seed** (immature) 7 mm long, 1 mm wide, narrowly ellipsoidal, with a subapical swelling; outer face convex, smooth to faintly rugose, sparsely pubescent; margin strongly recurved; inner face minutely tesselated, scaly with a narrow waxy wing along one side drawn into a short, apical, excurrent wing c. 1 mm long.

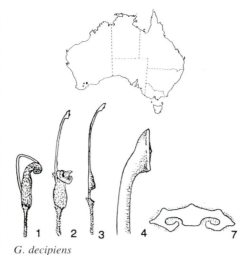

G. decipiens

Distribution W.A. Confined to inland areas from Ongerup to Frank Hann NP. *Climate* Hot, dry summer; mild-cool, wet winter. Rainfall 350–400 mm.

Ecology Grows in depressions in granite loam, sand, or silty clay mixed with limestone chips in low to tall shrubland and mallee woodland often in association with *Melaleuca* spp. Flowers winter to spring. Pollinated by birds. Regenerates from seed.

Variation A morphologically uniform species.

Major distinguishing features Leaves simple, entire, narrowly linear with several longitudinal ridges on upper surface; margins angularly revolute, closely abutting the midvein below; conflorescences few-flowered (< 6 per cluster); torus very oblique; perianth zygomorphic, densely brown-silky outside, bearded inside; pistil glabrous; ovary subtriangular; pollen presenter ± lateral.

Related or confusing species Group 15, especially *G. oligantha* and *G. sparsiflora*. *G. oligantha* has thinner, generally broader leaves with shortly recurved margins that usually do not conceal the undersurface. A narrow-leaved form that occurs around Mt Burdett is difficult to distinguish but, in fresh specimens at least, the lamina is visible between the margin and the midvein (in dried specimens, the leaves curl and the lower surface is enclosed by the smoothly rounded leaf margin). *G. sparsiflora* has an open white-silky outer perianth surface and shorter leaves (2–6 cm long) without clear ridging on the upper surface.

Conservation status 3VC. This wide-ranging species is rare in agricultural areas, where it occurs mainly on road verges. It is relatively common in the eastern part of its range, west of Lake King and extending to Frank Hann NP.

Cultivation *G. decipiens* was introduced to cultivation in 1986 but few firm guidelines can be given. In its natural habitat, it grows in poorly drained sand or alkaline, silty clay-loam in areas with regular winter frosts and low rainfall. Initial trials have shown it to be fairly slow-growing but adaptable, growing well in dry conditions in both light and heavy soils. Plants grown in tubs have proven resilient and grow attractively for many years. Garden conditions as recommended for most grevilleas should be sufficient to produce a compact, well-rounded shrub and this species should thrive in a site wet in winter and dry in summer. Unless regularly pruned when young, plants tend to be sparse of foliage on the lower branchlets with age.

Propagation *Seed* Sets plentiful seed in the wild and in the garden. Pre-germination treatment

89A. *G. decipiens* Flowers and foliage (N.Marriott)

89B. *G. decipiens* growing in *Melaleuca* scrub near Varley, W.A. (N.Marriott)

89C. *G. decipiens* Flowers and foliage (P.Olde)

improves results. *Cutting* Strikes well from cuttings using firm, new growth in spring. *Grafting* Grafts successfully onto *G. robusta* using the whip graft.

Horticultural features *G. decipiens* is a decorative shrub with erect branches and stiff, narrowly linear, blunt leaves. The foliage is dull green and relatively undistinguished, with a tendency to be denser at the branch tips than on the lower parts. Its reddish flowers are borne mainly inside the bush, in sparse few-flowered clusters in the leaf axils. From a horticultural point of view, flowering is somewhat inconspicuous except perhaps in winter at the peak flowering time. Although this may be more of a collector's plant, it is very popular with honey-eating birds, flowering heavily throughout the colder months. The neat, compact habit, especially when young, may be of value to landscapers seeking tough plants suitable for a low screen or massed planting or as a foil for more brilliantly coloured species. Furthermore, its ability to grow in winter-wet and alkaline soil should make it popular with gardeners having to cope with such conditions.

Hybrids Around Ongerup, *G. decipiens* appears to hybridise naturally with *G. oligantha* in disturbed roadside habitats.

Grevillea decora K.Domin (1921)

Plate 90

Bibliotheca Botanica 89: 589 (Germany)

The specific name is derived from the Latin *decorus* (beautiful, handsome). DEK-OR-A

Type: Mt Remarkable, near Pentland, Qld, Feb. 1910, K.Domin 2758, 2871 (holotype: PR).

Synonym: *G. goodii* subsp. *decora* (K.Domin) D.McGillivray (1986).

Open or dense **shrub** or spreading **tree** 2–5 m tall, 2–4 m wide. **Branchlets** angular, silky. **Leaves** 10–18 cm long, 1.5–4 cm wide, ascending, petiolate, simple, elliptic to oblong-lanceolate, concolorous to slightly discolorous, acute to obtuse, mucronate; upper surface silky, penninerved, midvein, intramarginal veins and reticulum evident or conspicuous; margins entire, flat or slightly recurved; young leaves conspicuously rusty; lower surface similar, silky, midvein and penninervation prominent. **Conflorescence** erect, pedunculate, terminal, simple or branched, barely exceeding foliage; unit conflorescence 5–10 cm long, secund-obovoid, dense, opening first from apex; peduncle and rachis rusty-silky; bracts 3–4 mm long, ovate with recurved apex, silky outside, falling before anthesis. **Flower colour**: perianth dull pinkish red with silver or rusty hairs; style pinkish red. **Flowers** basiscopic: pedicels 10–12 mm long, rusty silky; torus 6–8 mm long, very oblique; nectary U-shaped; **perianth** 10–12 mm long, 5–7 mm wide, ovate in side view, faintly ribbed, slightly dilated at base, depressed between dorsal and ventral tepals particularly near base, narrowing sharply at the curve, white and rusty silky outside, pilose inside, cohering except along dorsal suture, all tepals free to curve and loosely twisted after anthesis; limb erect and very conspicuous in young bud, soon revolute, spheroidal with depressed apex, the margins slightly impressed; **pistil** (36–)50–55 mm long; stipe c. 8 mm long, silky, lateral and adnate to torus; ovary villous; style elongate, at first exserted from base on dorsal side and looped upwards, afterwards gently incurved, villous, soon glabrous or remaining sparsely villous in upper half; style end flattened, glabrous; pollen presenter conspicuous, lateral, obovate, flat or slightly convex. **Fruit** 8–18 mm long, 8–11 mm wide, erect, subglobose, tomentose with reddish stripes and blotches; pericarp 1–2 mm thick. **Seed** 6–10 mm long, 4–6 mm wide, obovate with an oblique, apical wing; outer face convex, smooth; inner face flat; margin slightly recurved with a waxy border.

G. decora

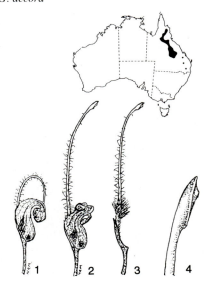

Distribution Qld; widely distributed on and near the Great Dividing Range; Aramac, Barcaldine, Clermont, Croydon, the Expedition Range and the Burra Range between Pentland and Torrens Creek, extending to the coast in isolated patches near Townsville (K.Fisher, pers. comm.). *Climate* Generally dry, mainly summer rainfall and cool to mild, dry winter. Rainfall 500–800 mm.

Ecology Grows mainly on well-drained sandstone ridges, in sand or sandy clay. Flowers autumn to spring. Killed by fire and regenerates from seed only. Pollinated by birds.

Variation There is some variation in flower size and colour, and in habit. A clear-red-flowered plant has been noted in the wild (D.Hockings, pers. comm.).

Small-flowered form Plants near Laura have pink flowers and a low habit, rarely exceeding 1.5 m (P.Radke, pers. comm.). D.J.McGillivray has also observed that plants from the Laura area have much shorter pistils (40 mm) and smaller fruits (8 mm). Infraspecific recognition of this form may be warranted after further study.

Major distinguishing features Single-stemmed tree with angular branchlets; leaves simple, the undersurface densely silky, some hairs rusty, venation with lateral veins at 45–65° to midvein; conflorescence secund, loose, scarcely exceeding leaves, the flowers reversely oriented, opening first from apex; torus very oblique; perianth zygomorphic, densely silky outside, pilose within; ovary villous, long-stipitate; style loosely villous; fruit with reddish stripes and blotches.

Related or confusing species Group 19, especially *G. glossadenia*, *G. goodii* and *G. pluricaulis*, the last two being the most closely related. *G. glossadenia*, known for many years as *G.* sp. aff. *decora*, has orange flowers in few-flowered, umbel-like conflorescences and more acutely angled lateral leaf venation (at c. 20–45° to midvein). Both *G. goodii* and *G. pluricaulis* are lignotuberous shrubs, the former prostrate and trailing, the latter erect and multi-stemmed. The ovary indumentum of *G. pluricaulis* is closely appressed, while in *G. goodii* the upper part of the style is glabrous.

Conservation status Not presently endangered.

Cultivation *G. decora* has been grown quite widely in Qld and with limited success as far south as Sydney, although grafted plants have flourished there. An especially memorable specimen was cultivated for many years by Merv Hodge near Brisbane. Growers' reports indicate that it is intolerant of heavy frosts which severely cut back its foliage, although mature specimens will eventually grow out of it. It prospers in an open, sunny position, preferably protected from strong or cold winds, in well-drained, acidic sandy or gravelly loam. In these conditions, it will grow into a dense, symmetrical shrub. Plants in dense shrubberies become straggly, often losing the lower branches and foliage. While it successfully endures dry conditions, it responds vigorously to watering, particularly in summer, but otherwise rarely requires attention once established. Seed and nursery catalogues sometimes list this species but it is not widely available commercially. Members of the Society for Growing Australian Plants are cultivating it in the Brisbane area.

Propagation *Seed* Sets copious amounts of seed that germinate readily if soaked overnight or peeled before sowing. *Cutting* Plants have been grown from cutting with some difficulty, generally using half-hard wood taken in early summer. These are allowed to callus and will initiate roots over the heat of summer. *Grafting* Has been grafted successfully onto *G. robusta* using the top wedge technique. The best results are obtained when the scion has its axillary buds in a 'just active' stage. The rootstock and scion are long-term compatible.

Horticultural features *G. decora* is a handsome, even spectacular, spreading shrub or small tree, worthy of a place in any garden. The impressive rusty, new growth contrasts strikingly with the luxuriant silvery green mature leaves, and curious, pinkish red conflorescences are displayed prominently at the ends of the branches. The racemes seem somewhat exotic because the flowers are turned in the direction opposite to most species, although this feature does occur in a number of tropical species. Individual flowers are large and continuously lure birds in search of nectar. It could be used as a feature plant or screen plant in a large garden or park and should be more widely cultivated.

General comments Although D.J.McGillivray (1986, 1993) placed this as a subspecies of *G. goodii*, we consider that it should remain as a separate species, being distinct in numerous characters including habit, indumentum of the leaf undersurface, perianth, style and ovary.

Grevillea decurrens A.J.Ewart (1917)
Plate 91

A.J.Ewart & O.B.Davies, *The Flora of the Northern Territory*: 83 (Australia)

Aboriginal: 'andjengerrer' (Mayali). Ripe seeds were eaten.

The specific epithet is derived from the Latin *decurrere* (to run down), an allusion to the base of the leaflet, the lamina of which continues past its junction with the rachis and onto the rachis itself.
DEE-KURR-ENS

Type: N of 15°S lat., N.T., Sept. 1911, W.S.Campbell (holotype: MEL).

Small **tree** 2–3 m high; bark dark grey or black. **Branchlets** terete, ridged, glabrous to silky. **Leaves** 15–40 cm long, ascending, conspicuously petiolate, detaching readily, concolorous, usually pinnatipartite, sometimes with secondary division of lower lobes; lobes 3–18 cm long, 1–6.2 cm wide, elliptic, obtuse to acute, decurrent; upper surface glabrous or sparsely silky, midvein and secondary veins conspicuous; margins flat; lower surface glabrous to silky with prominent midvein, secondary venation evident; venation coarse, longitudinally oriented. **Conflorescence** decurved, sometimes erect, axillary or terminal, pedunculate, simple or branched; unit conflorescences 2.5–3.5 cm long, lax, conico-cylindrical to partially secund, not exceeding foliage; peduncle sparsely to densely silky; rachis glabrous; bracts 1.2–2 mm long, ovate-acuminate, glabrous except ciliate margins, falling before anthesis. **Flower colour**: perianth creamy white turning pink; style creamy pink with greenish tip. **Flowers** at first acroscopic, reversing orientation before anthesis through twisting of pedicels, glabrous; pedicels 5–7 mm long; torus 7–7.5 mm long, ± lateral; nectary elongated U-shaped, smooth, entire; **perianth** 8–10 mm long, 3 mm wide, ovate, depressed between ventral and dorsal tepals, faintly ribbed, glabrous and glaucous outside, sparsely bearded at base inside, cohering except along dorsal suture before anthesis, afterwards all free to curve; limb revolute, spheroidal, the segments slightly ribbed, enclosing style end before anthesis; **pistil** 32–35 mm long, glabrous; stipe 10–12 mm long, lateral and basally adnate to torus; ovary globular; style at first exserted at base on dorsal side and looped upwards, gently incurved above; style end flattened; pollen presenter lateral, obovate, convex. **Fruit** 24–33 mm long, 21–28 mm wide, oblique, spheroidal, rarely almost round, glabrous, faintly rugose; pericarp 6–8 mm thick. **Seed** 8 mm long, 6 mm wide, obovoid to elliptic, smooth; outer face convex; inner face concave; seed body central within a broad, membranous wing 5–6 mm wide.

Distribution N.T., from Melville Is. and Darwin S to c. 16S at long. 132E and E to Mainoru. W.A., in the Kimberley, W to Derby. *Climate* Summer hot with monsoonal rains or dry; winter mild and dry. Rainfall 350–1200 mm.

90A. *G. decora* Branched conflorescence (P.Olde)

90B. *G. decora* Plant in natural habitat, Burra Range, Qld (M.Hodge)

G. decurrens

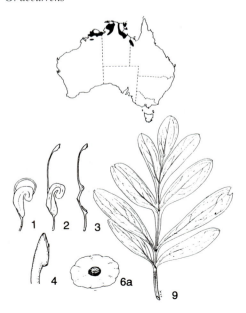

91A. *G. decurrens* Plant in natural habitat, Dorisvale Rd, N.T. (K. & L. Fisher)

Ecology Grows in low, open woodland in gravelly or sandy loam. Flowers summer to autumn, sometimes continuing to winter. After low-intensity fire regenerates from epicormic buds and from seed. Pollinated by nectar-feeding birds.

Variation There is some variation in leaf lobe width. Plants in W.A. tend to have narrower lobes, somewhat approaching *G. heliosperma* in size, with less prominently decurrent leaf lobes and, in narrow-leaved forms S of Kalumburu, very fine venation. The coarsest lobes occur on plants in the Pine River area, N.T. At least one collection from near Top Springs, N.T., has smaller, round fruits, but most plants in this area conform to the normal fruit type.

Major distinguishing features Open, single-stemmed tree to 3 m; leaf lobes decurrent at base; veins coarse, of varied size; conflorescence secund, lax; flowers pale pink, mainly in winter, reversely oriented by twisting of pedicel at anthesis; torus lateral; perianth strongly zygomorphic, glabrous outside; limb globose, smooth or slightly keeled; pistil glabrous; ovary clearly stipitate; fruit large, ellipsoidal; pericarp 6–8 mm thick; seed central with broad membranous wing.

Related or confusing species Group 5, especially *G. heliosperma* which differs in its more erect, tree-like habit to 6 m tall, its red flowers appearing in winter, and leaves with secondary and sometimes tertiary division, the lobes not decurrent, < 1 cm wide with fine veins all of ± similar size.

Conservation status Not presently endangered.

Cultivation *G. decurrens* has been widely and successfully cultivated in subtropical places such as Darwin and Brisbane and points between. Grafted plants have even survived for a limited time as far south as Sydney. It tolerates light frosts and dry conditions but, while moderately hardy in most conditions, does not survive heavy frosts or extended cold weather. An open situation without root competition is preferred but some shade may do no harm. Well-drained, acidic sand or friable loam is suitable but it may be slow to establish as it seems to spend some years thickening its trunk before launching into top growth.

Propagation *Seed* Grows readily from fresh seed, treated by either nicking or soaking for 24 hours in warm to hot water. Sow in spring. *Cutting* Untested but likely to be difficult. *Grafting* Has been grafted successfully onto *G. robusta*.

Horticultural features *G. decurrens* is a relatively unshapely small tree but has attractive pale green to bluish green broadly lobed leaves crowded at the crown. The cream to pale pink flowers tend to mingle unobtrusively within. Nonetheless, they are quite large and, without being spectacular, are usually noticeable. Flowering is usually in summer and extends over many months, in cultivation sometimes all year. The blooms are often visited by nectar-feeding birds. Many large greenish fruits are set, sometimes while the plant is still flowering, and provide an additional feature of interest. It makes a reasonable specimen plant in tropical and subtropical conditions, rarely reaching more than 3 m, and thus is ideal where a small, long-lived tree is required.

General comments *G. decurrens* was placed in synonymy under *G. heliosperma* by D.J.McGillivray (1993), but in our opinion is specifically distinct. The two species grow together over large areas of the N.T. and maintain their distinctions in morphology and flowering time. North of the King Edward River in W.A., some populations have most of the morphological features of *G. heliosperma* but share with *G. decurrens* an autumn flowering period; further study is needed to resolve the taxonomy in this area.

91B. *G. decurrens* Flowers and foliage (M.Hodge)

Grevillea deflexa F.Mueller (1883)

Plate 92

Melbourne Chemist and Druggist Suppl. 5: 72 (Australia)

Mulga Spider Flower

The specific epithet is derived from the Latin *deflexus* (bent aside, deviated), in reference to the typically deflexed conflorescences. DEE-FLEX-A

Type: near the Gascoyne R., W.A., 1882, J.Forrest & J.Pollak (lectotype: MEL) (McGillivray 1993).

Synonym: *G. ninghanensis* C.A.Gardner (1964).

A low, domed **shrub** to 30 cm or an open, erect, bushy to straggly shrub to c. 1.5 m high. **Branchlets** terete, tomentose or villous. **Leaves** 1–9 cm long, 1–10 mm wide, ascending to spreading, sessile, simple and entire, linear, elliptic, ovate or obovate, rigid, acute, pungent; upper surface silky, sometimes glabrous, midvein either lacking or an impressed groove; margins smoothly revolute; lower surface silky, midvein prominent. **Conflorescence** 2–6 cm long, 3–4 cm wide, on spreading to decurved peduncle, axillary, cauline, or rarely terminal on short branchlets, simple, open, secund; peduncle and rachis silky to villous; bracts 0.5–1 mm long, narrowly triangular, appressed villous outside, falling before anthesis. **Flower colour**: all red with a yellow limb, sometimes all yellow; after anthesis, the yellow anthers are conspicuous on red-flowered forms. **Flowers** acroscopic, the basal flowers sometimes oriented abaxially on retrorse pedicels; pedicels 3–8 mm long, silky to villous; torus 1.5–2 mm across, oblique, nectary cup-, U- or shovel-shaped, smooth or toothed; **perianth** 10–15 mm long, 2.5–3.5 mm wide, ovoid, ribbed, sparsely hairy outside, villous inside above level of ovary, cohering except along dorsal suture before anthesis, afterwards all free to curve, ventral tepals curling back much further than dorsal tepals; limb erect in late bud, revolute at anthesis, spheroidal, densely silky to villous, the style end partially exposed just before anthesis; **pistil** 17–24 mm long; stipe 1.5–3.7 mm long, glabrous to villous; ovary villous; style villous at base, otherwise glabrous with scattered hairs near apex, before anthesis exserted from dorsal suture and looped outwards, afterwards straight or slightly incurved; style end flattened; pollen presenter ± lateral, obovate, flat to slightly convex, umbonate; stigma prominent. **Fruit** 13–15 mm long, 5 mm wide, erect, oblong-ellipsoidal, villous or sparsely so; pericarp c. 0.5 mm thick. **Seed** 8 mm long, 2 mm wide, obovate to linear with prominent apical and basal wing; outer surface smooth, the apex strongly transversely ribbed; inner face minutely pitted; margins revolute.

Distribution W.A.; from Laverton to the upper Gascoyne R., S to Ninghan and W to c. 115°E longitude. ***Climate*** Hot, dry summer; mild, occasionally wet winter. Rainfall 150–250 mm.

Ecology Grows in open mulga country in depressions, often along creek lines, in red sand or calcareous clay loam, occasionally in granitic sand. Often found close to the base of other plants. Flowers winter–spring. Apparently lignotuberous. Pollinated by birds.

Variation Varies from a low, branching shrub to an open lanky shrub c.1.5 m high. Leaves also vary considerably in size and width, and from glabrous and green to silky and silver-grey. Flowers show some variation in size and colour. Some plants have conflorescences more markedly deflexed than others but this feature is variable even on the same plant. The following two forms are selected from the extremes of the species' range.

Short-leaved form (syn. *G. ninghanensis* C.Gardner) This form varies in habit from small and decumbent to erect and up to c. 1 m high. Leaves are 1–2.5 cm long, up to 1.5 mm wide and are silky grey to blue-green. The perianth is usually sparsely silky. Occurring around Sandstone and Mt Magnet and north to Cue, it is commonly found as an undershrub.

Long-leaved form An erect, lanky or bushy shrub with broader and longer (up to 9 cm long) broadly lanceolate to elliptic leaves, usually with densely villous branchlets. This form usually has larger and showier flowers with a spreading outer perianth indumentum. It occurs in the Laverton–Leonora area. A yellow-flowered form has been collected on the Canning Stock Route.

Major distinguishing features Leaves simple, entire, pungent, smooth on the upper surface; conflorescence axillary on peduncle 1–4 cm long, usually deflexed, secund, open; torus oblique; perianth zygomorphic, hairy within; ovary villous, stipitate; pollen presenter lateral.

Related or confusing species Group 25, especially *G. yorkrakinensis* which is distinguished by its shorter, few-flowered conflorescence and its more crowded and consistently narrower leaves (c. 1 mm wide).

G. deflexa

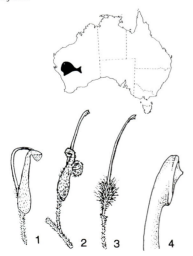

Conservation status Not presently endangered. Widespread in arid areas of the State.

Cultivation *G. deflexa* can be cultivated in a wide range of inland areas. It has been successfully grown at Burrendong Arboretum, N.S.W., western Vic., S.A. and inland Qld. It tolerates frosts to at least -6°C and demonstrates extreme drought hardiness but dislikes wet, cold or cloudy conditions, and this limits its suitability for cultivation to inland areas. It succeeds best in a hot, sunny situation with good drainage in slightly acidic to alkaline sand or gravelly loam. Plants from some populations have proven more amenable to cultivation than others. In all but sandy soil, fertiliser is unnecessary and summer watering can be harmful. Regular tip-pruning is essential to maintain density and shape. It makes an interesting pot specimen when planted in a well-drained, gravelly mix in medium-sized pots and placed in a sunny situation but will collapse if regularly over-watered. Nurseries rarely stock this species.

Propagation *Seed* Seed is seldom available and as a result it is rarely propagated by this method. The standard peeling treatment improves germination rates. Average germination time 28 days (Kullman).

Cutting Strikes readily from firm to semi-hard wood especially when taken in the warm months. Struck cuttings require a very open potting mix and should not be placed in over-large pots. *Grafting* Has been grafted successfully onto *G. robusta* and *G.* 'Poorinda Royal Mantle' using whip grafts. In some cases (e.g. the Short-leaved form), the scion emits a resinous exudate which causes the union to part; the combination fails to establish a good cambial layer and the scion falls off or dies.

Horticultural features *G. deflexa* is a species of many different forms, most of them somewhat irregular in habit. The spectacular red and yellow flowers contrast strikingly with the foliage, especially in forms with silvery leaves. The racemes commonly arise directly from the branchlet or hang from every leaf axil, occasionally on arching branches, sometimes clustered in the upper foliage, at other times almost concealed inside the plant or at the base. The beauty of pure yellow flower forms is more muted and flowers on these plants tend to be less conspicuous. Selections of the Long-leaved form have broad, silvery grey, pointed leaves, and its somewhat spindly habit with branches oddly angled and its exquisite conflorescences make it an

92A. *G. deflexa* Flowering low shrub between Leonora and Agnew, W.A. (M.Hancock)

92B. *G. deflexa* Short-leaved form, formerly known as *G. ninghanensis*, near Mt Magnet, W.A. (C.Woolcock)

92C. *G. deflexa* Flowers on Long-leaved form (N.Marriott)

92D. *G. deflexa* Yellow-flowered plant, Canning Stock Route, W.A. (N.Marriott)

unusual feature plant in the garden, but each form has its attraction. Low forms make interesting border plants or rockery specimens while taller forms could be tried as screen or specimen plants. The large flowers provide a copious nectar supply and will entice honey-eating birds.

General comments There are grounds for the continued recognition of *G. ninghanensis* based on the habit, perianth indumentum, and leaf size and shape. As we have not studied all populations we have adopted McGillivray's treatment while further research is conducted into the variation.

Grevillea delta (D.J.McGillivray) P.M.Olde & N.R.Marriott
Plate 93

The Grevillea Book 1: 181 (Australia)

Based on *Grevillea thelemanniana* subsp. *delta* D.J.McGillivray, *New Names in* Grevillea *(Proteaceae)* 15 (1986, Australia).

Delta, the fourth letter of the Greek alphabet, was selected as the specific epithet. Prior to its formal description as a subspecies, this species was known to its author as Race D of *G. thelemanniana*. DELL-TA

Type: NE of Mt Lesueur, W.A., 13 Oct. 1974, A.S.George 12910 (holotype: PERTH).

Bushy, spreading **shrub** 1–2 m high, 1–3 m wide. **Branchlets** terete, with an open, spreading indumentum of white hairs up to 3 mm long. **Leaves** 1–2.2 cm long, ascending to spreading, shortly petiolate but often appearing sessile, semi-circular in gross outline, pinnatisect; lobes 1.5–12 mm long, 0.3–0.8 mm wide, linear-oblong, strongly spreading, the lower lobes often retrorse, always with secondary division, sometimes all lobes with secondary division; juvenile leaf lobes c. 2.5 mm wide; upper surface openly villous, midvein evident; margins revolute enclosing undersurface except midvein; lower surface bisulcate, loosely villous, midvein prominent. **Conflorescence** 2–4 cm long, mostly erect, sometimes deflexed, pedunculate, simple, terminal, loosely oblong-secund; peduncle openly tomentose; rachis tomentose; bracts 2.2 mm long, narrowly triangular, silky-villous outside, falling well before anthesis. **Flower colour**: perianth and style red; style end green. **Flowers** acroscopic; pedicels 2.5–4.5 mm long, sparsely silky; torus 1 mm across, slightly oblique; nectary conspicuous, cushion-like to thickly arcuate; **perianth** 5–7 mm long 0.8–2.5 mm wide, ovoid, dilated at base, sparsely silky to almost glabrous outside, pubescent inside, cohering except along dorsal suture; limb revolute, loosely villous, subcubic to spheroidal, emarginate in side view, the segments faintly ribbed; **pistil** 27 mm long, glabrous; stipe 3 mm long, flattened, incurved; ovary triangular; style before anthesis exserted from top of curve and looped upwards, gently incurved afterwards; style end hoof-like, exposed before anthesis; pollen presenter 1.4 mm long, oblique, convex. **Fruit** 10–12 mm long, 3–4 mm wide, erect, narrowly oblong-acuminate with a very conspicuous, subbasal ridge from dorsal to ventral face, extending markedly beyond ventral face. **Seed** not seen.

Distribution W.A., confined to the Mt Lesueur area. *Climate* Hot, dry summer; moist, cool winter. Rainfall 480–520 mm.

Ecology Grows in moist situations, usually along creek lines or in depressions in sandstone-derived loam, sandy loam or clay loam in mallee heath. Flowers winter–spring.

Major distinguishing features Branchlets with a pilose indumentum, the hairs mostly exceeding 1 mm long; leaves mostly < 2 cm long, bipinnatisect with strongly spreading lobes; floral bracts c. 2 mm long; pedicels sparsely hairy; perianth zygomorphic, sparsely hairy outside, hairy inside; limb sparsely hairy; ovary subtriangular on incurved, flattened stipe; pollen presenter oblique; fruits with conspicuous basal ridging.

Related or confusing species Group 14, especially *G. humifusa*, *G. preissii* and *G. thelemanniana*. *G. humifusa* is a prostrate species with long branchlets and has glabrous pedicels and perianth with scattered hairs on the limb. *G. preissii* differs in its branchlet indumentum of short, dense or open hairs, in its strongly ascending primary leaf lobes and its conico-secund conflorescences. *G. thelemanniana*

G. delta

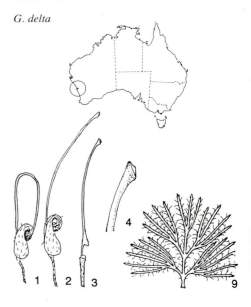

has a sparse to appressed leaf and branchlet indumentum, leaves simple or once-divided and ovate floral bracts c. 0.5 mm long.

Variation Shows little variation in leaf size and shape and in shrub habit.

Conservation status 2RC. This species has a very restricted distribution near Mt Lesueur.

Cultivation *G. delta* was unknown in cultivation until its introduction by the authors in 1986. Although it prefers a cool, wet climate, early plantings in summer-wet locations such as Sydney and Brisbane have succeeded but appear less reliable than related species. It has proved tolerant of hot-dry climates provided there is an adequate subsoil moisture, but drought tolerance is untested and possibly low. It tolerates frost to at least -3°C. Initial trials suggest a preference for well-drained, neutral to slightly acidic heavy loam, in full sun or partial shade. Occasional thorough summer watering may be beneficial. Becoming available in specialist native plant nurseries, mainly in Vic. and Perth.

Propagation *Seed* Sets plentiful seed but not yet tested. *Cutting* Strikes readily from cuttings taken of firm, young growth or semi-hard wood in spring. *Grafting* Has been grafted successfully onto *G. robusta* using whip grafts.

Horticultural features This distinct, attractive species grows rapidly into an open shrub with showy, red flowers and hairy fan-shaped leaves. Flowering is prolific and continuous over a long period. An added feature is the glossy, bronze-red fruits clustered on the rachis after flowering. It appears to be popular with birds. Grafted plants have proved extremely successful, flowering long after the normal season.

General comments The decision to recognise this taxon at specific rank was taken after examination of dried specimens and extensive field studies. The twice-divided leaves and the indumentum of leaves and flowers appear to us to separate it significantly from *G. thelemanniana*.

Grevillea depauperata R.Brown (1830)
Plate 94

Supplementum Primum Prodromi Florae Novae Hollandiae exhibens Proteaceas Novas: 21 (England)

The specific epithet refers to the habit of the plant which Robert Brown saw, presumably, as sparse and open and is derived from the Latin *depauperatus* (starved, reduced). DEE-PORE-PER-ARE-TA

Type: inland from King George Sound, [W.A.], 1828–29, William Baxter (lectotype: BM) (McGillivray 1993).

Synonym: *G. brownii* C.F.Meisner (1845).

A spreading, irregular **shrub** to 0.8 m high, sometimes prostrate. **Branchlets** angular, tomentose. **Leaves** 1–3 cm long, 1.5–10 mm wide, spreading, shortly petiolate, elliptic to ovate or oblong-elliptic, simple, acute to obtusely mucronate; upper surface flat or longitudinally concave, glabrous or sparsely silky, wrinkled to granulate, sometimes smooth and glossy, midvein indistinct or faintly grooved; margins entire, smoothly recurved or revolute; lower surface white-velvety, midvein prominent. **Conflorescence** erect, shortly pedunculate, crowded, terminal usually on short branchlets or axillary in upper axils, simple or few-branched; unit conflorescence few-flowered, umbel-like; peduncle and rachis villous; bracts 1.5–2 mm long, 1 mm wide, narrowly triangular, tomentose outside, persistent at anthesis. **Flower colour**: perianth and style red, dull orange-red or bright orange. **Flowers** acroscopic: pedicels 5–8.5 mm long, tomentose; torus 1.5–2 mm across, very oblique; nectary

93A. *G. delta* Flowers and foliage (N.Marriott)

93B. *G. delta* Flowering shrub, in cultivation, Glasshouse Mtns, Qld (P.Olde)

G. depauperata

94A. *G. depauperata* Massed flowering of the prostrate form (N.Marriott)

94B. *G. depauperata* Flowering branches (C.Woolcock)

prominent, tongue-like, lobed; **perianth** 7–9 mm long, 2–4 mm wide, irregularly oblong-ovoid, dilated at base, tomentose outside, bearded inside above level of ovary, otherwise silky, cohering except along dorsal suture from limb to curve; style end exposed before anthesis; limb nodding, ovoid with a point, cohering after anthesis; **pistil** 11–15.5 mm long, villous; stipe 1.6–2.1 mm long; ovary densely villous; style not exserted before anthesis, clearly exceeding perianth after anthesis, tomentose, thick above ovary, then narrowing, terminating along back of club-shaped style end; pollen presenter lateral, orbicular, concave. **Fruit** c. 15 mm long, 6 mm wide, oblique, ovoid with recurved acute apex, faintly ribbed, glabrous; pericarp c. 0.5 mm thick. **Seed** 9 mm long, 2.5 mm wide, obovoid and slightly recurved towards cushion-like apex which bears a terminal wing to 3 mm long; outer surface convex, smooth and shiny; inner face obscured; margins strongly revolute with a narrow wing along one side.

Distribution W.A.; occurs from the Frankland R. to W of the Stirling Ra., S to the Porongurup Ra. and extending nearly to Albany. *Climate* Warm, dry summer; cool, wet winter. Rainfall 700–800 mm.

Ecology Grows in yellow to grey sand or sandy loam over gravelly clay in eucalypt woodland with a medium to dense understorey. Flowers autumn–spring. Killed by fire and regenerates from seed. Pollinated by birds.

Variation Varies in habit from a rounded or scrambling shrub usually to c. 0.5 m to a dense, prostrate ground cover. Some forms have noticeably erect branches, while others flower relatively more prolifically. Leaf size and shape also show some variation. The following forms are recognised.

Prostrate orange form While most plants have bright red flowers, this form, common in cultivation but of unknown wild origin, has massed, bright orange flowers and glabrous, undulate leaves. Cultivated plants of this form have proven difficult to maintain in good health for long periods, going a sickly yellow after several years in most cases. The cause of this is not yet known. This problem is usually encountered in Vic. and is possibly related either to soil problems or viral infection. When growing well, it makes a spectacular display in full flower.

Prostrate red form Only recently introduced to cultivation, this has massed large clusters of most showy red flowers for many months.

Upright form An erect open shrub to c. 0.6 m with fiery orange-red flowers.

Major distinguishing features Branchlets angular; leaves simple, entire, elliptic to ovate, the undersurface velvety; conflorescence few-flowered; torus very oblique; nectary linguiform, erect; perianth zygomorphic, oblong, dilated at base; ovary densely hairy, shortly stipitate; style end exceeding perianth; pollen presenter concave.

Related or confusing species Group 23, especially *G. fasciculata* and *G. crassifolia*, both of which differ in their shorter pistil (< 10 mm long).

Conservation status Not listed as endangered but much of its former habitat has been reduced by clearing for agriculture. Occurs in the Porongurup NP.

Cultivation *G. depauperata* grows reliably in a wide range of climatic conditions and has been known in cultivation for many years. It is grown in N.S.W., Vic. and W.A. and has proven hardy in most conditions, even succeeding as a tub plant on its own roots in Qld. Frosts down to at least -6°C cause no damage to foliage and it will survive moderately dry conditions. If grown in well-drained, acidic sandy loam in full sun or partial shade, it forms a compact, well-shaped plant that rarely needs pruning. Light fertilising can assist vigour and appearance. Summer watering is beneficial in extended dry periods. It is an excellent pot plant when grown in a well-drained mix with slow-release fertiliser. Sometimes available in native plant nurseries, often as *G. brownii*.

Propagation *Seed* Sets numerous seeds that germinate readily when given the standard peeling treatment. *Cutting* Strikes readily from firm, young growth taken at most times of the year. *Grafting* Has been grafted successfully onto *G. robusta* using the top wedge technique.

Horticultural features A good, free-flowering specimen of *G. depauperata* creates considerable visual impact in the garden. Bright red or orange flowers cover the plant in spring and almost obscure the dark green foliage. In cultivation, it frequently flowers from autumn right through winter to early spring and consequently provides a welcome splash of colour in the quiet time of the garden year. This small shrub makes an excellent rockery or edging plant while the prostrate forms are often used as a feature ground cover. Unfortunately, some plants flower poorly and have quite dull red flowers. Stock selection is, therefore, quite important. The species is suitable for massed planting and is good for attracting nectar-feeding birds.

General comments The change of name for this well-known species is unfortunate but was necessitated by the rules of botanical nomenclature which require that the first name of a species take precedence. It is ironic that the name given to the species by Robert Brown (*G. depauperata*) supersedes a name (*G. brownii*) with which Meisner honoured him.

Grevillea didymobotrya C.F.Meisner (1856)

A.P. de Candolle, *Prodromus Systematis Naturalis Regni Vegetabilis*: 14: 386 (France)

The specific epithet alludes to the twin-branched conflorescences of the type and is derived from the Greek *didymos* (twin, double) and *botrys* (in botany, an inflorescence). DID-IM-O-BOT-REE-A

Type: Swan R., W.A., c. 1847, J.Drummond coll. 4, no. 280 (holotype: NY).

Synonym: ?*Anadenia filiformis* S.L.Endlicher (1838)

An erect **shrub** 1–3 m tall with ascending to erect branches. **Branchlets** terete, silky. **Leaves** 1.5–14 cm long, 1–3 mm wide, ascending, shortly petiolate, stiff, simple, unifacial, narrowly linear to ± terete, sometimes narrowly elliptic; the tip acute, often decurved, sometimes pungent; upper and lower surfaces ± the same, all silky to glabrous with silky hairs in grooves, longitudinally ridged, without midvein. **Conflorescence** erect, usually pedunculate, terminal, rarely axillary, simple or few-branched; unit conflorescence 3.5–6 cm long, narrowly cylindrical, dense; peduncle and rachis silky; bracts 0.5–1 mm long, ovate, silky outside, usually falling before anthesis. **Flowers** acroscopic; pedicels 1–2 mm long, silky; torus c. 0.5 mm across, straight; nectary inconspicuous; **perianth** 2–3 mm long, 0.5 mm wide, ± cylindrical, strongly curled in bud, silky outside, glabrous inside, at first separating along dorsal suture, all free below limb before anthesis, the dorsal tepals slightly reflexing, all free to base and detaching quickly after anthesis; limb conspicuous, revolute, globular to ovoid; **pistil** 3–6 mm long, glabrous; stipe 0.1–0.6 mm long; ovary globose; style filamentous, at first exserted below curve and looped upwards, strongly recurved above ovary to S-shaped after anthesis; style end slightly expanded; pollen presenter erect to oblique, conical with basal rim. **Fruit** 5–8.5 mm long, 3 mm wide, erect or oblique on the pedicels, triangular/obovoid often with a dorsal attenuation; pericarp c. 0.3 mm thick. **Seed** 3–5 mm long, narrowly triangular, curved over at the apex.

Major distinguishing features Leaves simple, narrowly linear, longitudinally ridged, unifacial or bifacial; conflorescence branched, the units cylindrical; torus straight; nectary inconspicuous; flowers curled in bud, the limb revolute at anthesis; perianth zygomorphic, silky outside, glabrous inside; pistil glabrous; style S-shaped, strongly recurved above the ovary; pollen presenter conical; fruits obovoid.

Related or confusing species Group 3, especially *G. makinsonii*, which differs in its glabrous outer perianth surface. *G. biformis* is also closely related (Group 2) but can be distinguished easily by the perianth limb which is erect prior to anthesis.

Variation Two subspecies are recognised.

Key to subspecies (from McGillivray 1993: 142)
Leaves unifacial, subterete to linear
    subsp. **didymobotrya**
Leaves dorsiventral, narrowly obovate to elliptic, or occasionally obovate or sublinear; venation more prominent on the lower surface   subsp **involuta**

Grevillea didymobotrya C.F.Meisner subsp. didymobotrya   Plate 95

Leaves unifacial, subterete to linear. **Flower colour**: perianth generally bright yellow, with styles turning red.

Distribution W.A., widespread in low rainfall areas from Shark Bay SE to Balladonia and S almost to Ravensthorpe. *Climate* Hot, dry summer; mild, wet winter. Rainfall 200–350 mm.

Ecology Grows in sand, sandy loam and gravel in open mallee shrubland and mixed heathland areas. Flowers (winter–)spring–summer. Regenerates from seed after fire. Insect-pollinated.

Variation There is considerable, albeit subtle variation within subsp. *didymobotrya*. Flower colour can vary from cream to bright yellow, while the leaves show variation in length, shape and colour. McGillivray (1993) noted the presence of narrow-leaved plants near Mt Holland, variation in the angle of the style end and in the location of the stigma.

Silver-leaved form North of Southern Cross, we have recently found a population with silvery-grey leaves with a dense silky indumentum. The leaves are usually terete but occasionally elliptic and are rarely longer than c. 2 cm. It forms a low, erect shrub to 1.5 m and grows in sand over laterite.

Conservation status Widespread in inland areas. Not presently endangered.

Cultivation Subsp. *didymobotrya* is a plant suitable for dry, inland climates and has been cultivated successfully in western Vic., Canberra and Burrendong Arboretum, N.S.W. It survives frosts to at least -6°C and is drought-hardy but does not adapt to humid and very wet conditions. In the right climate, it is easy to grow in full sun or partial shade in very well-drained, acidic sand or gravelly loam. Plants can suddenly collapse and die during wet weather if drainage is anything less than perfect. This is even more critical when establishing young plants which must be kept almost dry. Experienced growers report that fertiliser and summer watering are harmful and that established plants require no maintenance apart from occasional trimming to maintain shape. It makes a delightful subject for pot culture when planted in a sunny position but a very well-drained, gravelly mix with slow-release fertiliser must be used. It is not usually available in nurseries but seed is sometimes available from specialist seed suppliers.

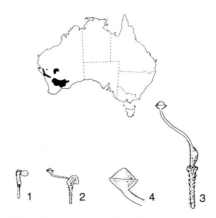

G. didymobotrya subsp. *didymobotrya*

Propagation **Seed** Sets copious seed in the wild, less in cultivation, which germinate well especially when soaked or nicked before sowing. Average germination time 39 days (Kullman). *Cutting* Strikes well from firm, new growth taken in the warm months. Great care must be taken when potting on struck cuttings to ensure that the potting mix is very open and free draining. *Grafting* Early attempts at grafting onto rootstocks hardy in coastal conditions, e.g. G. 'Poorinda Royal Mantle', have been unsuccessful. It is, however, compatible with *G. robusta*, with one grower in Melbourne having superb specimens over 15 years old. (Ian Mitchell, pers. comm.).

Horticultural features With its narrow spikes of bright yellow flowers and its narrow, ascending, dull green leaves, subsp. *didymobotrya* is a most desirable plant for dry, inland gardens. Close examination of the flowers as they mature reveals subtle changes in their structure and colour: after fertilisation, the style rolls back and down and turns reddish, the perianth detaches to reveal the small, green ovary swelling to maturity. Small triangular fruits form in abundance. All forms make attractive feature plants and, in full flower, are extremely ornamental. The flowers, which have a sweet fragrance, are borne on branching panicles above the foliage. The combination of leaves and flowers render the silver-leaved form worthy as a feature plant in dry-area native gardens.

General comments Visually, the silver-leaved form appears distinct but its position in relation to other variation observed by McGillivray requires further study before infraspecific recognition is warranted. Should the type of *Anadenia filiformis* Endlicher (1838) prove to be the same as *G. didymobotrya* then the name would have priority, necessitating a name change for this species.

95A.*G. didymobotrya* subsp. *didymobotrya* Silver-leaved form, N of Bullfinch, W.A. (P.Olde)

95A. *G. didymobotrya* subsp. *didymobotrya* Typical flowering habit, N of Murchison River, W.A. (C.Woolcock)

95B. *G. didymobotrya* subsp. *didymobotrya* Plant in natural habitat, Kalbarri NP, W.A. (P.Olde)

Grevillea didymobotrya subsp. involuta D.J.McGillivray (1986)   Plate 96

New Names in Grevillea *(Proteaceae)* 4 (Australia)

Subspecific epithet derived from the Latin *involutus* (rolled in), in reference to the leaf margins. INN-VOLL-OO-TA

Type: between the Moore and Murchison Rivers, W.A., Sept. 1901, E.Pritzel 608 (holotype: PERTH).

An erect, open, silvery **shrub** to 1.5 m tall. **Branchlets** terete, silky. **Leaves** 1–7 cm long, 3–7 mm wide, bifacial, ascending, shortly petiolate, sinuate, simple, obovate to elliptic, acute, decurved; upper surface densely silky, concave, the midvein obscured; margin entire, incurved or involute; lower surface silky, with 7–11 longitudinal ridges. **Flower colour**: perianth yellow, the styles turning red. **Conflorescence**, **fruit** and **seed** similar to subsp. *didymobotrya*.

Distribution W.A.; on the plains between Geraldton and Mullewa. *Climate* Hot, dry summer; mild, wet winter. Rainfall 300–450 mm.

G. didymobotrya subsp. *involuta*

96. *G. didymobotrya* subsp. *involuta* Close-up of flowers, foliage and fruit (P.Olde)

Ecology Grows in low heath and open, banksia shrubland in yellow sand. Flowers spring. Fire kills the plants and regeneration is from seed. Pollinated by insects.

Variation Shows minor variation in leaf shape and size and degree of inrolling of the leaf margins.

Conservation status 2E. Restricted to small areas, often on road verges.

Cultivation Subsp. *involuta* is unknown in cultivation on its own roots. From its habitat it can be assumed that it should be resistant to frosts and tolerate extended dry periods. Initial attempts should concentrate on providing a situation similar to its natural habitat; a hot, sunny position in deep, well-drained, acidic to neutral sand. The silky, silvery leaves are prone to fungal attack in humid or shady conditions. Summer watering and fertiliser are probably unnecessary and possibly harmful. It would make a delightful specimen in a medium-sized tub in a warm, sunny situation.

Propagation Seed Untried but probably as for subsp. *didymobotrya*. Cutting Initial attempts have proved extremely frustrating. Leaves tend to drop quickly and cuttings rot within days. Needs to be propagated in a 'dry' hot-house or cold-frame (without mist). Cuttings must be fresh, firm new growth, preferably taken from cultivated plants. Grafting Has been successfully grafted onto *G. robusta* using the mummy technique.

Horticultural features Subsp. *involuta* is an exceptionally beautiful plant. It has a well-formed, dense habit with elegant, ascending branches and intriguing, silvery-grey foliage. The unusual, incurved, boat-shaped leaves are crowded and interesting even when flowers are absent. The showy, bright yellow racemes mass above the foliage in dense panicles, usually throughout spring, and provide a picturesque contrast. The flowers are also sweetly perfumed. It has undoubted potential as a specimen plant for the inland garden.

Grevillea dielsiana C.A.Gardner (1942)
Plate 97

Journal of the Royal Society of Western Australia 27: 169 (Australia)

Diels' Grevillea

Named in honour of F.Ludwig E.Diels (1874–1945), a distinguished German botanist who visited W.A. in 1900–01 and described many grevilleas. DEALZ-EE-ARE-NA

Type: near the Murchison R., W.A., 20 Aug. 1931, C.A.Gardner 2590 (holotype: PERTH).

A compact prickly **shrub** to c. 2 m. **Branchlets** angular to terete, ridged, shiny, glabrous, sometimes sparsely silky, sometimes glaucous. **Leaves** 3–8 cm long, ascending, shortly petiolate, usually divaricately three-forked to biternate; lobes 0.5–2.5 cm long, c. 1 mm wide, rigid, linear-subulate to trigonous, extremely pungent; upper surface ± glabrous or sparsely silky, sometimes glaucous, longitudinally finely veined, midvein an impressed groove; margin angularly revolute to midvein; lower surface bisulcate, silky in the two grooves, midvein prominently exserted, angular. **Conflorescence** spreading or decurved, pedunculate, terminal, often on short branchlets, simple or few-branched; unit conflorescence 3–4 cm long, 6 cm wide, secund-hemispherical, lax; peduncle and rachis sparsely silky to glabrous; bracts 1–1.5 mm long, ovate, glabrous with ciliate margins, falling before anthesis. **Flower colour**: perianth and style usually red, but ranging from orange-yellow, brilliant orange and scarlet to deep red; style often pale in upper half. **Flowers** acroscopic: pedicels 5–9 mm long, glabrous; torus 2–3 mm across, oblique; nectary semi-annular, the margin thick, entire and ± flat; **perianth** 7–8 mm long, 3.5 mm wide, strongly curled in bud, oblong, undilated at base, glabrous outside, pilose inside to curve; tepals strongly ridged, first separating along dorsal suture, the dorsal tepals reflexing slightly before anthesis, then all loose to free below limb before anthesis, afterwards separated to base; limb revolute, spheroidal, the segments medially ridged, enclosing style end before anthesis; **pistil** 26.5–36 mm long, glabrous; stipe 1.8–3.3 mm long, much thinner than style, inserted perpendicular to torus; ovary globular; style exserted through dorsal suture at curve and looping upwards before anthesis, afterwards somewhat lax, ± straight or gently incurved, thickened towards apex; pollen presenter very oblique to lateral, elliptic, convex, umbonate. **Fruit** 10–14 mm long, 5–7 mm wide, oblique to the stipe, oblong-ellipsoidal, rugulose, glabrous; pericarp c. 1 mm thick. **Seed** 7–10 mm long, 3 mm wide, oblong-ellipsoidal, slightly thickened at apex; outer face convex, rugulose; inner face convex, channelled around margin; margin recurved.

Distribution W.A.; in the northern sandplain from Geraldton almost to Shark Bay and inland to Mullewa. *Climate* Hot dry summer; mild, wet winter. Rainfall 350–425 mm.

Ecology Occurs on open sandplain among low, thick scrub in yellow sand or gravelly sand over loam. Flowers winter–spring. Killed by fire and regenerates from seed. Pollinated by birds.

Variation *G. dielsiana* is a fairly uniform species. Plants in some populations have glaucous leaves. There is some variation in flower colour, the typical

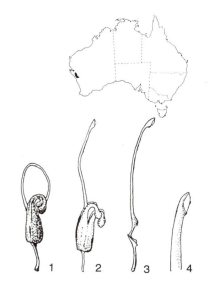

G. dielsiana

97A. *G. dielsiana* Commonly cultivated orange-flowered form (P.Olde)

97B. *G. dielsiana* Massed flowering (M.Hodge)

97C. *G. dielsiana* Shrub in natural habitat, N of Murchison River bridge, W.A. (C.Woolcock)

colour being red. Horticultural selections listed below have been introduced.

Orange-flowered form A form of open habit with orange-yellow flowers and pale green, non-glaucous foliage is widely grown. Flowers may be pure yellow in shaded conditions. It is generally much easier to cultivate and hardier than most other forms of the species. Although its origin is uncertain, it is believed to have come from a swarm of different colour forms growing in one patch near Ajana (W.R.Elliot, pers. comm.).

Glaucous form This selection has rich red flowers and glaucous, blue-green leaves. It is generally a low shrub to c.1 m with a rounded, compact habit. In cultivation, it is sometimes shy to flower and short-lived although other selections of plants with red flowers are spectacular, free-flowering and long-lived. Care should be taken in selecting superior forms for cultivation.

Scarlet-orange-flowered form Common in the wild, this form has non-glaucous foliage and a dense, shrubby habit. It has proved to be relatively hardy in cultivation.

Major distinguishing features Leaves divaricately once or twice divided with linear-subulate, pungent lobes; conflorescences secund-hemispherical, lax; torus oblique; perianth zygomorphic, oblong, glabrous outside, hairy inside; pistil glabrous; stipe perpendicular to the torus; pollen presenter lateral; fruit oblong-ellipsoidal.

Related or confusing species Group 8. Closest to *G. teretifolia* which differs in its shorter pistils (< 20 mm long).

Conservation status Although much of its former habitat has been cleared for agriculture, the species is secure in several National Parks.

Cultivation All commonly grown selections of *G. dielsiana* except the Glaucous form are adaptable and relatively easy to grow. It tolerates extended dry conditions, and at the Australian National Botanic Gardens, Canberra, survives frosts to at least -11°C. It has been grown successfully in most States of Australia as well as in California, U.S.A., including coastal, summer-rainfall and dry, inland climates. A full-sun position in deep, well-drained, acidic sand or friable, gravelly loam is preferred. In both sun and shade, plants tend to become leggy and need tip pruning from an early age; they may also need regular, hard pruning to retain shape. Established plants generally do not need summer watering. Fertiliser will improve foliage density but this may be at the expense of shortening life and reduced flowering. The Glaucous form, found commonly in the wild, remains something of an enigma and a challenge. It demands almost perfect drainage and will not tolerate summer watering or humidity, and may die after a summer storm, even when all other conditions are correct. It can also be shy to flower in cultivation and seems to need a cold winter to induce good bud set. Several forms are available in specialist nurseries.

Propagation *Seed* Sets small quantities of seed that germinate well when given the standard peeling treatment. Average germination time 33 days (Kullman). *Cutting* Can be difficult. Good results have been obtained from firm to slightly hard new growth taken from cultivated plants in early autumn, winter or spring. High concentrations of root hormones (4000–8000 ppm IBA powder) applied to the base improve strike rates. *Grafting* Has been grafted successfully onto *G. robusta*, *G. rosmarinifolia* and *G.* 'Poorinda Royal Mantle' using many differing techniques. The grafts appear to be compatible.

Horticultural features *G. dielsiana* is regarded by many as one of the loveliest of all grevilleas, and would be far more widely appreciated were it not for its extremely prickly foliage. This unattractive feature limits its potential in the garden, although it would make an effective barrier plant. Small birds sometimes use the foliage for protection and for nesting. Weeding around its base can also be a hazard because fallen leaves become quite rigid and needle-like and always seem to find an exposed finger. Notwithstanding the drawbacks of its foliage, the real charm of this species lies in its glorious flowers which cover the plant from late winter. The individual flower clusters of brightly coloured, long-styled flowers are genuinely striking in their beauty. Nectar-feeding birds are regular visitors. It can be recommended as a specimen plant for the home garden, with limited potential for public landscaping.

Grevillea diffusa F.Sieber ex K.Sprengel (Jan.–July 1827)

In C.Linnaeus, *Systema Vegetabilium* 16th edn 4 (2), *Curae posteriores* 46 (Germany)

Red Flower Grevillea

The specific epithet is derived from the Latin *diffusus* (spread out), in reference to the spreading habit. DIFF-USE-A

Type: precise locality unknown, N.S.W., 1823, F.W.Sieber 36 (lectotype: K) (McGillivray 1993).

A low, compact, much-branched **shrub** to 50 cm, sometimes open, partly weeping, to 2 m. **Branchlets** terete to markedly angular, silky to pubescent. **Leaves** 1–13 cm long, 1–6 mm wide, often secund, shortly petiolate, simple, narrowly elliptic, oblong-lanceolate to linear, acute; upper surface silky to glabrous and smooth, sometimes shiny, midvein and intramarginal veins evident with occasional granules; margins entire, flat to slightly recurved or revolute; lower surface silky, sometimes with a few ascending hairs, midvein prominent. **Conflorescence** erect, often on decurved or pendulous peduncle, rarely sessile, terminal, simple or few-branched; unit conflorescence 1–2 cm long, 1–2 cm wide, dense to open, globular to spheroidal; peduncle (0.4–0.6 thick) glabrous to silky; rachis silky to villous; bracts 0.5–3.5 mm long, narrowly elliptic, silky to villous, falling late but usually before anthesis. **Flowers** acroscopic: pedicels 2–4 mm long, silky to villous; torus 0.8–1.6 mm across, square, slightly oblique; nectary squarish to U-shaped, smooth; **perianth** 3–6 mm long, 1–1.5 mm wide, oblong, sometimes dilated at base, silky or sparsely so outside, bearded inside c. level of ovary, cohering except along dorsal suture before anthesis, afterwards the tepals free to beard and independently rolled down; limb revolute, spheroidal, villous, enclosing style end before anthesis; **pistil** (6–)10–11.5(–13.5) mm long; stipe 1.5–2.5 mm long, flattened; ovary oblong-ovoid; style glabrous except sparse hairs near pollen presenter, at first exserted in lower half and looped upwards, angularly incurved, geniculate c. 1–2 mm from flattened, plate-like style end; pollen presenter oblique, oblong, convex, umbonate. **Fruit** 11–16 mm long, 3.5–5 mm wide, erect, oblong-ellipsoidal, smooth; pericarp c. 0.3 mm thick. Seed not seen.

Major distinguishing features Leaves simple, entire, the undersurface silky, the upper surface minutely pitted, the intramarginal veins minutely granular, the lateral venation usually obscure on upper surface; conflorescence subglobose in dried specimens, sometimes loosely so in live specimens, dull maroon-red; torus ± straight, square; perianth zygomorphic, oblong, silky outside, bearded within; pistil glabrous except minute hairs on the style end; fruit smooth.

Related or confusing species Group 21, especially *G. capitellata*, *G. evansiana*, *G. oldei* and *G. sericea*. *G. capitellata* differs in its longer, wider, villous leaves and thicker peduncle. *G. evansiana* has a rusty perianth limb, shiny upper leaf surface with smooth intramarginal veins and a more prominent nectary. *G. oldei* has weak, flexible branchlets and leaves conspicuously 3-veined on the upper surface with an open, spreading indumentum of loose, long hairs on the under surface. *G. sericea* has a secund conflorescence, usually of pink flowers, with a pistil ≥ 14 mm long.

Variation Two geographically disjunct subspecies are here recognised.

Key to subspecies

Outer perianth surface densely hairy; longest peduncles ≤ 1.5 cm long, densely hairy, slender to stout (0.7-1.2 mm thick); leaf under surface silky **subsp. diffusa**

Outer perianth surface sparsely silky; longest peduncles > 2 cm long, glabrous or with scattered hairs, filamentous (0.4-0.6 mm thick); leaf under surface sparsely silky **subsp. filipendula**

Grevillea diffusa F.Sieber ex K.Sprengel subsp. **diffusa**   Plate 98

Synonym: *G. sericea* var. *diffusa* (F.Sieber ex K.Sprengel) G.Bentham (1870)

Branchlets terete, silky. **Leaves** 1–13 cm long (mostly < 7 cm long), silky on the under surface; margins shortly recurved to loosely revolute. **Conflorescence** very dense, within or close to foliage. **Flower colour**: perianth dull, black-red or red; style red.

Distribution N.S.W., from the Georges R. S and SW to the Cordeaux Dam area. A disjunct population has been found in the Colo area. *Climate* Hot, wet summer; cool to cold, wet or dry winter. Rainfall 1000–1200 mm.

Ecology Occurs in low shrubland or eucalypt forest in sandy loam, sometimes with lateritic intrusions. Flowers winter to early summer. Regenerates from seed after fire. Pollinator unknown, probably insect. Introduced honey-bees often visit the flowers and may be effective pollinators.

Variation Several distinct forms overlap in the wild but are horticulturally distinct. There is variation in habit, leaf size, indumentum and colour.

Short-leaved form (Type form) The Type form is found widely in the Menai, Heathcote, Appin and Cordeaux Dam areas and is notable for its short (usually < 4 cm long), narrowly elliptic (< 5 mm wide) leaves, bearing an appressed, white indumentum on the undersurface. The wiry branchlets are frequently secund and down-arching. This form has a grey-green appearance and while some plants can be found up to 1 m tall, it is usually a domed, much-branched shrub c. 50 cm high. It is found mostly on dry, gravelly ridges and sandy hillsides.

Long-leaved Form This form occurs widely in Royal NP, south of Sydney, but extends south from the Georges R. to around Helensburgh. It is an erect, open or mid-dense shrub to 2 m with relatively thin, green leaves 5–10 cm long. Flowering is somewhat sparse at times and conflorescences are sometimes borne on longer peduncles than the Type form. It is usually found along creek lines or in sheltered, moist situations under eucalypts but appears to be only a clinal variant.

Conservation status Restricted but not endangered.

Cultivation Subsp. *diffusa* is a very adaptable, hardy plant that has been cultivated successfully in the eastern States of Australia since at least the early 1980s. It tolerates light frosts and extended, dry periods and is unaffected by summer humidity. It thrives in an open, well-drained, gravelly loam or sandy, acidic loam in full sun but does not adapt well to puggy clay. Plants grown in partial shade may need pruning to prevent them becoming leggy, but in normal conditions it is a compact, dense plant. Summer watering is appreciated and light fertiliser improves appearance and vigour. Native plant nurseries in the Sydney area often carry this species.

Propagation *Seed* Sets numerous seed that germinate readily. *Cutting* Readily propagated from cuttings taken at any time of the year using firm, new growth. *Grafting* Has been grafted successfully onto *G. robusta* using the top wedge technique.

Horticultural features Subsp. *diffusa* is an attractive foliage plant with grey-green leaves with a whitish indumentum. Under most conditions it develops into a dense, compact plant and should be more widely used in landscaping, especially in the Sydney area where it is relatively free from pests and requires no maintenance once established. The maroon-red flowers are often hidden within the foliage and even when exposed are not showy. Although they produce copious nectar, they are not noted for their ability to attract birds.

Hybrids Hybrid swarms involving *G. sericea* sometimes occur in the wild, usually in disturbed areas. Flowers of these hybrids are often very brightly coloured and usually larger and more open

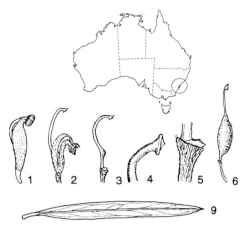

G. diffusa subsp. *diffusa*

98A. *G. diffusa* subsp. *diffusa* Type form beside road, Wedderburn, N.S.W. (P.Olde)

98B. *G. diffusa* subsp. *diffusa* Flowers and foliage of Type form, Heathcote, N.S.W. (N.Marriott)

98C. *G. diffusa* subsp. *diffusa* Long-leaved form, near Helensburgh, N.S.W. (P.Olde)

than those of the parents. *G. diffusa* is one of the known parents of the garden hybrids G. 'Shirley Howie' and G. 'Poorinda Hula'.

Grevillea diffusa subsp. filipendula
D.J.McGillivray (1986)   **Plate 99**

New Names in Grevillea *(Proteaceae)*: 4 (Australia)

Mt White Grevillea

The subspecific epithet is derived from the Latin *filum* (a thread) and *pendulus* (hanging down), in reference to the long peduncle from which the conflorescence hangs.

Type: between Calga and Mt White, N.S.W., 14 Sept. 1968, D.J.McGillivray 3097 (holotype: NSW).

Branchlets angular, glabrous to sparsely silky. **Leaves** 5-15 cm long (mostly > 7 cm long), long linear, sparsely silky to glabrous on undersurface, margins angularly refracted. **Conflorescence** moderately dense to loose, usually on long peduncle well clear of foliage. **Flower colour**: perianth red or pink, white beard evident at anthesis; style dark red.

Distribution N.S.W., restricted to the central coast between the Hawkesbury R. and Gosford. ***Climate*** Hot, wet summer; cool, wet or dry winter. Rainfall 1000-1200mm.

Ecology Grows in moist, low eucalypt woodland with dense understorey or in open heath, often on exposed sandstone outcrops, in grey or gravelly sand. Flowers all year but mainly in winter and spring. Regenerates from seed after fire. Pollinator unknown.

G. diffusa subsp. *filipendula*

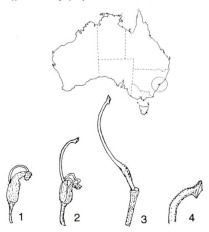

Variation A uniform subspecies. Variations in leaf length and width are usually habitat-related, being longer in shady sites. In these situations plants are also usually open and taller.

Conservation status Suggested 2R?. Population size, distribution and conservation status require assessment.

Cultivation Subsp. *filipendula* has been cultivated widely at least since the early 1980s on the east coast from Brisbane to Vic. Experienced growers report that, while adaptable in most situations and resistant to frost to at least -6°C, it is not drought-tolerant. Short periods of dryness are withstood but it appreciates some artificial watering, especially in summer. It prospers in a wide range of garden soils but grows best if planted in well-drained acidic, sand in full sun or partial shade. In this situation, it will develop into an attractive, compact shrub which requires little pruning or fertilising. While it accommodates bad drainage better than some, it cannot be recommended for this situation. It makes an excellent pot plant in a well-drained mix with slow-release fertiliser. In general nurseries, it is often propagated and sold as the Mt White form of *G. capitellata*.

Propagation *Seed* Untested but likely to germinate readily. *Cutting* Strikes readily from cuttings of firm, young growth at any time of the year. *Grafting* Untested but rarely necessary.

Horticultural features Subsp. *filipendula* is valued as one of the most graceful grevilleas in cultivation. Its compact, dense habit and long, dark green leaves borne on reddish branchlets are widely appreciated. However, its greatest asset is its exquisite flowering habit. The pendulous clusters of red flowers are disported on fine, branching peduncles which veil the plant in thread-like delicacy. This long-lived, hardy plant should be more widely planted in commercial landscapes as a feature plant or massed display and it has already gained widespread acceptance among home gardeners.

Hybrids Occasionally hybridises with forms of *G. linearifolia* on the boundaries of its habitat. Flowers of the hybrid are bright pink.

99A. *G. diffusa* subsp. *filipendula* Flowers on long peduncles (N.Marriott)

99B. *G. diffusa* subsp. *filipendula* Plant in natural habitat near Calga, N.S.W. (P.Olde)

Grevillea dimidiata F.Mueller (1863)
Plate 100

Fragmenta Phytographiae Australiae 3: 146 (Australia)

Caustic Bush

Aboriginal name: Warlubum (Gurindji) Kalkaringi. The species has ceremonial significance, exudate from the viscid fruit being used to produce decorative scars on the skin.

The specific epithet is derived from the Latin *dimidiatus* (halved), an allusion to the leaves appearing as only half a leaf. DIM-ID-EE-ARE-TA

Type: Roper R., N.T., 1862, F.Waterhouse, J.McDouall Stuart's expedition (holotype: MEL).

A small, scrappy **tree** 2–6 m high, glabrous on all parts except a few hairs in leaf axils. **Branchlets** stout, terete, glaucous. **Leaves** 9–33 cm long, 1–7.5 cm broad, detaching readily, concolorous, unifacial through movement of midvein towards upper margin within a few leaves from base, erect to ascending, petiolate, leathery, glaucous, simple, sickle-shaped to semi-obovate; margins flat, entire, strongly undulate, the surfaces longitudinally and irregularly wrinkled but without conspicuous venation; first juvenile leaves dorsiventral. **Conflorescence** glabrous, erect, terminal, paniculate, sometimes with secondary branching; unit conflorescences 4–11 cm long, 3 cm wide, cylindrical, loose; peduncle angular, with triangular base; bracts 3 mm long, overlapping in bud, ovate-acuminate, falling soon after expansion of rachis. **Flower colour**: perianth and style creamy white, sometimes with a pinkish tinge. **Flowers** acroscopic; pedicels 2–5.5 mm long; torus c. 2 mm across, concave, very oblique; nectary U-shaped; **perianth** 10–12 mm long, 1–2 mm wide, oblong-cylindrical with ribbed tepals, glaucous, smooth or papillose on inside in lower half; tepals first separating to base along dorsal suture and from each other below limb where reflexing slightly to expose inner surface before anthesis, ultimately all free to base and loose; limb revolute, ovoid to round with inconspicuous ribbing, firmly enclosing style end before anthesis; **pistil** 16.5–21 mm long; stipe 4.5–7.5 mm long; ovary globose; style at first looped outwards from dorsal suture in upper half, markedly refracted about ovary, almost perpendicular to line of pedicel, straight or slightly kinked at middle after anthesis, smoothly dilated into the narrowly flanged style end; pollen presenter an erect cone with a very oblique base. **Fruit** 16–23 mm long, 14–20 mm wide, oblique on pedicel, often persistent, lens-shaped to spheroidal with short apiculum, glaucous, smooth or slightly wrinkled, caustic-viscid when young; exocarp flaking off with age; pericarp c. 1 mm thick. **Seed** 13 mm long, 10–11 mm wide, flat, irregularly ovate, ridged submarginally, narrowly winged all round.

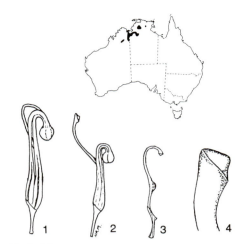
G. dimidiata

Distribution Qld, N.T., W.A. and adjacent offshore islands. Widely distributed in the monsoonal areas south to about 16°S latitude and west to 126°E longitude. *Climate* Summer hot and wet; winter mild and dry. Rainfall 600–1200 mm.

Ecology Grows in low, open savannah or grassy eucalypt woodland, in wide shallow depressions or creek lines in coarse, red-brown calcareous soil, sometimes in light, sandy soil or in stony loam on hillsides. Flowers winter–spring. Regenerates from epicormic buds after low-intensity fire. Insect-pollinated. The leaves have stomates on both surfaces (A.Salkin, pers. comm.).

Variation A fairly consistent species with little variation other than leaf size. Leaves of plants from the Port Douglas area, N.T., are up to 30 cm long and 7.5 cm wide.

Major distinguishing features Leaves unifacial, concolorous, sickle-shaped, glabrous, glaucous, the margin undulate; conflorescence glabrous, paniculate, the racemes cylindrical; flowers glabrous; torus very oblique; perianth zygomorphic, narrow, sometimes sparsely papillose within; pistil 16.5–21 mm long, geniculate about ovary; pollen presenter conical; fruit viscid when young; seeds winged all round.

Related or confusing species Group 3, especially *G. mimosoides* which has narrower, elliptic leaves and smaller flowers (pistil 5–10.5 mm long).

Conservation status Not presently endangered.

Cultivation *G. dimidiata* is a recent introduction to cultivation, mainly through a few enthusiasts in Qld in whose subtropical gardens it has performed well to date. To judge from its natural habitat, it is probably frost-tender, although low overnight winter temperatures would be experienced in some areas and provenance might be important in this respect. It should tolerate a wide range of soil types, from acidic to possibly slightly alkaline. An open, sunny position in well-drained soil, without overcrowding or shade, would seem to be favoured. It is probably drought-hardy, but better-looking plants should result from summer watering. Many northern species spend a number of years thickening the lower trunk before spurting into active, tall growth. In western Vic., grafted plants in tubs have been grown to flowering, being placed in a sunny, sheltered, frost-free site in winter.

Propagation *Seed* Untested but should germinate freely. Treatment by soaking in warm to very hot water for c. 24 hours prior to sowing should improve results. *Cutting* Untested but probably difficult. *Grafting* Has been grafted successfully onto *G. robusta* using whip and mummy grafts.

Horticultural features In the wild, *G. dimidiata* can often look an untidy, scrappy tree, but this should not deter the potential grower because good conditions invariably produce better-looking plants in cultivation. The bluish green foliage is decorative and the leaves, while often tatty and chewed in the wild, are extraordinary. They resemble elephant ears with a wavy margin and have been measured up to c. 30 cm long. The flowers are equally impressive:

100A. *G. dimidiata* Flowering habit, near Victoria River crossing, N.T. (M.Hodge)

100B. *G. dimidiata* Close-up of conflorescences (T.Blake)

100C. *G. dimidiata* Tree in natural habitat, E of Victoria River crossing, N.T. (K. & L. Fisher)

prominent branching panicles of creamy white, sweetly scented flowers. Certainly worth testing, this unusual small tree should suit tropical gardens, but children (and adults) should not handle the sticky fruits as they may cause severe blisters.

Grevillea diminuta L.A.S.Johnson (1962)
Plate 101

Contributions from the National Herbarium of New South Wales 3: 95 (Australia)

Brindabella Grevillea

The specific epithet is derived from the Latin *diminutus* (made small). This species was once considered a small form of *G. victoriae* which it closely resembles.   DIM-IN-YOU-TA

Type: near Mt Franklin, Brindabella Ra., A.C.T., 11 Oct. 1956, R.D.Hoogland 6279 (holotype: NSW).

Spreading, dense **shrub** 0.3–1 m tall, up to 3 m wide. **Branchlets** angular, silky. **Leaves** 0.5–2 cm long, 3–9 mm wide, ascending to spreading, petiolate, leathery, simple, elliptic,

G. diminuta

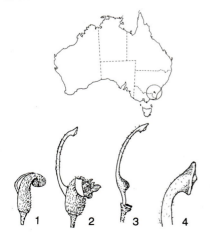

101A. *G. diminuta* Flowers and foliage (N.Marriott)

101B. *G. diminuta* Buds (P.Olde)
101C. *G. diminuta* Plants in natural habitat, Mt Franklin, A.C.T. (P.Olde)

obtuse, mucronate; upper surface silky, ageing glabrous and glossy, midvein scarcely evident; margins entire, shortly refracted; lower surface silky white, rusty when young, midrib usually faintly prominent and hairy. **Conflorescence** 2.5–5 cm long, 2.5–3 cm wide, decurved, pedunculate, terminal and axillary, simple or few-branched; unit conflorescence 2.5–3 cm long, open, cylindrical; peduncle and rachis silky; bracts 0.7–1.8 mm long, triangular, silky, falling before anthesis. **Flower colour**: perianth rusty brown outside, bright red inside; style purplish red. **Flowers** acroscopic, the basal flowers abaxially oriented; pedicels c. 2 mm long, silky; torus c. 1.5 mm wide, squarish, oblique to straight; nectary kidney-shaped, smooth; **perianth** 6–7 mm long, 2.5 mm wide, oblong-ellipsoidal, sharply narrowed at the curve, silky brown outside, bearded inside above ovary, otherwise glabrous, cohering except along dorsal suture before anthesis, afterwards free to beard and independently rolled back; limb nodding, spheroidal, enclosing style end before anthesis; **pistil** 10–11 mm long; stipe 1–1.5 mm long, refracted from line of pedicel; ovary ellipsoidal; style scarcely bowed out beyond dorsal suture before anthesis, reflexed above ovary and smoothly incurved after anthesis, glabrous with occasional scattered hairs, ventrally grooved; style end conspicuous, flattened, with silky hairs on dorsal side; pollen presenter very oblique to lateral, obovate, convex with a surrounding flange. **Fruit** 10–15 mm long, 6–9 mm wide, slightly oblique, ellipsoidal, glabrous; pericarp c. 1 mm thick. **Seed** 8–9 mm long, 3 mm wide; outer face strongly convex, minutely hairy; inner face minutely pitted; margin revolute.

Distribution N.S.W. and A.C.T., mainly in the Brindabella Ra. *Climate* Subalpine: mild, dry summer and cold, wet winter, often with ice and snow. Rainfall 600–800 mm.

Ecology Occurs in subalpine woodland in gullies and on scree slopes, in skeletal, gravelly granitic loam. Flowers spring–summer. Regenerates from seed after fire. Pollinated by birds.

Variation A fairly uniform species.

Major distinguishing features Leaves simple, entire, elliptic, glabrous and glossy above; densely hairy below including midvein; perianth zygomorphic, densely hairy outside, bearded inside; pistil 10–11 mm long; ovary glabrous; style sparsely hairy especially towards apex.

Related or confusing species Group 21, especially *G. victoriae* which usually has much larger leaves and a pistil 16–26 mm long.

Conservation status Not presently endangered.

Cultivation Grown widely in south-eastern Australia from Adelaide to Sydney, including Tas. *G. diminuta* has proved both hardy and adaptable, tolerating cold winters including ice and snow, and hot, dry summers. It is unaffected by frosts to at least -11°C, and indeed, this period of extreme cold may be necessary for good bud initiation. Certainly, in summer-humid, coastal climates, it is extremely shy to flower and while it grows into a satisfactory foliage shrub, the lack of flowers makes the effort seem hardly worthwhile at times. Even in cooler areas, an unusually dry winter can cause considerable bud drop. Open or partially shaded conditions and well-drained, acidic to neutral soil result in compact, attractive plants requiring little maintenance. It is not fussy about soil texture, growing equally well in sand and clay. Summer watering induces vigorous growth. Native plant nurseries in Canberra and Vic. regularly carry this species. It should do well in N.Z.

Propagation *Seed* Sets numerous seeds that should germinate readily given the standard peeling treatment. *Cutting* Grown mainly by cutting, it strikes well using firm, new growth taken from spring to autumn. *Grafting* Untested.

Horticultural features *G. diminuta* is a shapely, compact plant with a strong spreading habit. Its small, elliptic leaves have a grey undersurface and a glossy, green upper surface, imparting a grey-green aspect to the plant. The pendulous racemes of rusty-brown flowers produce an impressive display. The flowers, highlighted by the prominent bright red inner part of the flower, attract numerous honey-eating birds which parade regularly through the branches in search of nectar. In landscaping, it is used as a ground cover and as a massed border planting. It is long-lived in cultivation and can be used confidently as a structure plant.

Grevillea dimorpha F.Mueller (1855)
Plate 102

Transactions of the Philosophical Society of Victoria 1: 21 (Australia)

Flame Grevillea, Olive Grevillea

The specific epithet is derived from the Greek *dimorpha* (with two forms), a reference to the variable leaves. DIE-MORE-FA

Type: in the Serra and Victoria Ranges, The Grampians, Vic., Nov. 1853, F.Mueller (lectotype: MEL) (McGillivray 1993).

Synonyms: *G. dimorpha* var. *latifolia* F.Mueller (1855); *G. dimorpha* var. *angustifolia* F.Mueller (1855); *G. dimorpha* var. *linearis* C.F.Meisner (1856); *G. dimorpha* var. *lanceolata* C.F.Meisner (1856) *G. oleoides* var. *dimorpha* (F.Mueller) G.Bentham (1870); *G. speciosa* subsp. *dimorpha* (F.Mueller) D.J.McGillivray (1986).

Erect **shrub** 1–3 m high, occasionally prostrate to decumbent, 0.4–1(–4) m wide. **Branchlets** conspicuously angular, ribbed, silky. **Leaves** 5–15 cm long, 0.15–4 cm wide, erect to spreading, sessile or shortly petiolate, sometimes leathery, simple, entire, narrowly linear to broadly elliptic or obovate, acute to obtusely mucronate; upper surface glabrous, silky at base, midvein evident with conspicuous, acutely angled venation on upper surface of broad leaves; margins shortly recurved; lower surface silky, midvein prominent, glabrous. **Conflorescence** erect to decurved, axillary or cauline, sessile or shortly pedunculate, usually simple; unit conflorescence c. 1 cm long, 2–4 cm wide, dome-shaped to secund; peduncle and rachis tomentose; bracts 1.5–2.5 mm long, linear, tomentose, falling before anthesis. **Flower colour**: perianth and style red. **Flowers** abaxially oriented on retrorse pedicels; pedicel 4–8 mm long, tomentose; nectary inconspicuous, V-shaped; torus oblique, 1.5 mm across; perianth 10–15 mm long, 2–3 mm wide, oblong-ovoid, red-tomentose outside, densely bearded inside about middle, papillose elsewhere, cohering except along dorsal suture before anthesis, afterwards free to beard and independently rolled down; limb revolute, spheroidal, enclosing style end before anthesis; **pistil** 21–26 mm long; stipe 2–4 mm long, arising on outer toral rim, glabrous, somewhat flattened; ovary ovoid, glabrous, much broader than stipe and style; style exserted from near base and looped out before anthesis, afterwards gently incurved, sprinkled with erect hairs especially near tip; pollen presenter almost lateral, round to obovate, convex. **Fruit** 12–16 mm long, 5–6.5 mm wide, oblique to erect, ellipsoidal, granular; pericarp wall c. 0.5 mm thick. **Seed** 8.6–10.2 mm long, 2.3–2.8 mm wide, narrowly elliptic to oblong, minutely hairy, with an apical pulvinus; outer face convex; inner face channelled, sometimes obscured by revolute margin.

Distribution Vic., confined to the southern and central ranges of The Grampians. *Climate* Hot, dry summer; cool, wet winter. Rainfall 650–1000 mm.

Ecology Usually occurs on the upper slopes of foothills and ridges in open eucalypt woodland, occasionally in heath-woodland and eucalypt forest among large boulders; usually in well-drained sand, stony loam or sand over loam. Flowers spring–summer, sometimes extending to late autumn. Regenerates from seed after fire. Pollinated by birds.

Variation Varies in habit as well as leaf size, width and shape. Some horticultural selections deserve mention.

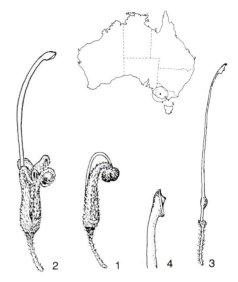

G. dimorpha

102A *G. dimorpha* Narrow-leaved form (N.Marriott)

102B. *G. dimorpha* Shrub in natural habitat, The Grampians NP, Vic. (N.Marriott)

102C. *G. dimorpha* Flowering branch of the Tall form (N.Marriott)

Narrow-leaved form A shrubby to decumbent variety with linear, acute leaves 1–3 mm wide on which the fiery, red flowers are openly displayed. It occurs in the southern areas of The Grampians from Yarram Gap to Mt Abrupt.

Broad-leaved form There are numerous populations with medium to broad leaves, mostly growing as medium-sized upright shrubs in the southern and central areas of The Grampians. Many intermediates occur between this and the Narrow-leaved form, and they grow together at Cassidy Gap.

Robust form This is a very large form to 2.5(–4) m high and 2.5 m wide with broad, grey-green leaves, unlike all other forms that have dark green leaves. It occurs in a small area on the slopes of Fyans Creek in the central region of The Grampians, well north of all other populations.

Tall form Confined to Teddy Bear Gap area of The Grampians, this form is a few-branched shrub to c. 2.5 m and has distinctive, very large, broad leaves but is sparsely foliaged in the lower half. It is very similar in habit and leaf size to some broad-leaved populations of *G. oleoides*.

Decumbent form This form can be prostrate or decumbent and spreads to 1 m wide. It has small to medium elliptic leaves and is confined to the Jimmy Creek area.

Major distinguishing features Branchlets sharply angular; leaves 5–15 cm long, simple and entire, midvein glabrous on underside; conflorescence axillary or stem-borne, 8–10-flowered, not exceeding foliage; perianth zygomorphic, densely bearded within; ovary ovoid, slightly wider than style and stipe; style sparsely hairy; pollen presenter almost lateral.

Related or confusing species Group 21, especially *G. oleoides*, *G. speciosa* and *G. victoriae*. *G. oleoides* generally has a longer pistil (28–36 mm) with a longer stipe and more flowers per raceme (12–16). Its ovary is the same width as the style. *G. speciosa* has leaves shorter than 3.5 cm and terminal, pedunculate conflorescences on prominent branches. In *G. victoriae* the indumentum covers the midvein on the leaf undersurface.

Conservation status 2RC. Confined to a limited area in the southern and central Grampians where it is nowhere extensive, although often locally common.

Cultivation *G. dimorpha* was first cultivated in Vic. in 1858, no doubt as the result of Mueller's efforts, and has proven very hardy and reasonably adaptable, succeeding in cold-wet, warm-dry and summer-rainfall climates so long as humidity is not excessive. It has been grown successfully in many areas of Vic., N.S.W., Qld, Tas. and S.A. as well as in the U.S.A. Best results are achieved when planted in a partially shaded site in well-drained, acidic sand or gravelly loam. It responds to occasional, deep, summer watering but this is not essential except in extended, dry periods. Frosts to at least -6°C have been recorded without ill effect. Fertilising is rarely necessary but plant appearance is improved by the occasional light pruning. Once established, plants will grow happily beneath large trees. Both the Narrow-leaved and Decumbent forms make very attractive pot plants when grown in a medium to large tub. Several forms are available from specialist native plant nurseries especially in Vic.

Propagation *Seed* Sets numerous seeds that germinate well if peeled or soaked. *Cutting* Firm, young growth usually strikes well at most times of the year. *Grafting* Has been successfully grafted onto *G. robusta*.

Horticultural features *G. dimorpha* is a very pleasing plant with attractive, dark green leaves that are markedly silver on the undersurface. Some of the compact forms offer outstanding value; when young they have distinctive, reddish hairs on the branchlets and foliage which makes new growth most attractive. Fiery red flowers arise in profusion in the leaf axils and directly on the branches and are particularly prominent on the Narrow-leaved form. All forms are very popular with honey-eating birds. This long-lived species is a valuable addition to the garden either as a feature plant or a medium screen plant. The Tall form is excellent for a narrow bed along a drive or path.

General comments There is undoubtedly a close relationship between *G. dimorpha* and *G. oleoides* and more distantly with *G. speciosa*. McGillivray (1986) combined these taxa as subspecies of *G. speciosa*, but we have reverted to specific rankings because of the extreme disjunction of the population of *G. dimorpha* (over 1,000 km from *G. oleoides* and *G. speciosa*), its shorter pistil, more conspicuous ovary, sessile or very shortly pedunculate axillary conflorescences which are also at times cauline, its more oblique pollen presenter and the reddish hairs on new growth.

Grevillea disjuncta F.Mueller (1868)
Plate 103

Fragmenta Phytographiae Australiae 6: 206 (Australia)

The specific epithet is derived from the Latin *disjunctus* (disjunct, separated). The reference is obscure and refers perhaps to disjunct localities which may have been known to Mueller or perhaps to the conflorescences scattered on the Type specimen, perhaps to the shape of the perianth, the limb of which appears disjointed. DISS-JUNK-TA

Type: Salt [Pallinup] R., W.A., 18 Aug. 1860, G.Maxwell (lectotype: MEL) (McGillivray 1993).

A low, spreading, dome-shaped **shrub**, 30 to 60 cm tall. **Branchlets** terete, tomentose or silky, sometimes roughened by persistent leaf bases and peduncles. **Leaves** 0.7–2(–6) cm long, 1–3 mm wide, ascending, sometimes crowded, sessile, stiff, simple, subterete to narrowly linear, weakly pungent; upper surface smooth, longitudinally ridged, glabrous or sparsely silky, the base densely silky; margins entire, angularly revolute, enclosing midvein; lower surface obscured, silky in groove. **Conflorescence** 0.5–2 cm long, erect, sessile, axillary, crowded usually in pairs in axils, sometimes arising on old stems, usually simple, sometimes few-branched; unit conflorescences 0.5 cm long, 2 cm wide, 2–4-flowered, often arising successively from same peduncle; peduncle and rachis villous, persistent; bracts 0.5–1 mm long, triangular, appressed villous outside, persistent to anthesis. **Flower colour**: perianth pale orange to bright red with a yellow or red limb; style pale to reddish pink with green tip. **Flowers** abaxially oriented; pedicels 5–10 mm long, sparsely silky; torus 1.5 mm across, very oblique to lateral; nectary V-shaped; **perianth** 7–8 mm long, 2–3 mm wide, oblong-ovoid, slightly dilated at base, abruptly contracted at throat, silky to tomentose outside, bearded inside near base, otherwise sparsely silky, cohering except along dorsal suture; limb revolute, spheroidal, villous, the style end partially exposed before anthesis; **pistil** 19.5–28 mm long; stipe 0.5 mm long, inserted on upper toral margin; ovary ± sessile, villous; style exserted below curve on dorsal side and looped upwards before anthesis, afterwards gently incurved, villous at base, ± glabrous near apex, straight, terminating along back of flattened style end; pollen presenter lateral to very oblique, obovate, convex. **Fruit** 10–12 mm long, 4 mm wide, erect, ± ovoid, sparsely

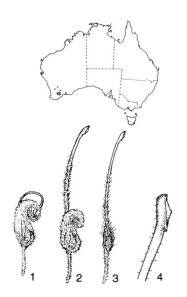

G. disjuncta

villous; pericarp c. 0.5 mm thick. **Seed** 9–9.5 mm long, 2–2.5 mm wide, minutely pubescent all over, elliptic with an apical wing 1–1.5 mm long; outer face convex with an apical transverse band; inner face convex; margins strongly revolute with a narrow waxy border.

Distribution W.A., mainly restricted to the Dumbleyung–Nyabing–Pingrup area with isolated populations near York, Kellerberrin and Stokes Inlet. *Climate* Hot, dry summer; mild to cool, wet winter. Rainfall 350–600 mm.

Ecology Grows in low scrub, heathland and open eucalypt woodland, in granitic loam, gravelly or sandy loam. Flowers winter–spring. Fire response unknown but probably killed and regenerates from seed. Pollinated by birds.

Variation Two forms are recognised.
Typical form This form occurs in the Stirling Ra.–Lake Grace area and generally has shorter leaves (mostly < 2.5 cm long).
Long-leaved form (formerly *G. disjuncta* subsp. *dolichopoda* Smooth-leaf form). This form is known from a few collections from the Hopetoun area. It generally has longer leaves (up to 6 cm long).

Major distinguishing features Leaves crowded, simple, smooth, ribbed, the undersurface completely enclosed by angular margins; conflorescences axillary or on old wood, crowded along the branchlet; torus very oblique, c. 1.5 mm across; perianth zygomorphic, hairy on both surfaces; ovary densely villous, subsessile; pollen presenter lateral.

Related or confusing species Group 25, especially *G. dolichopoda* and *G. haplantha*. *G. dolichopoda* differs in its granular leaves, usually on secund branchlets. *G. haplantha* is a robust shrub 1–2 m tall with erect branches, finely granulose, villous leaves and a densely villous perianth c. 12 mm long and 3 mm wide.

Conservation status Not presently endangered.

Cultivation *G. disjuncta* is a relatively tough species that withstands cold-wet winters including frosts to at least -6°C and endures extended dry periods. It grows well in Vic., in less humid areas of N.S.W., and in south-eastern S.A. In a sunny situation in well-drained, neutral to acidic sand or loam, it forms a nicely rounded shrub that rarely requires maintenance apart from occasional pruning or watering. In a sunny position, it makes a comely potted specimen but, like many grevilleas, resents too much fertiliser or a poorly drained soil. It is occasionally available in nurseries, mainly in Vic., where it has sometimes been sold as *G. haplantha*.

Propagation *Seed* Sets small quantities of seed that germinate well if given the standard peeling treatment. *Cutting* Firm, new growth strikes readily at most times of the year, optimal results being achieved in early autumn. *Grafting* Grafts successfully onto *G. robusta* using the wedge and mummy methods.

Horticultural features *G. disjuncta* is an attractive, small shrub with short, narrow, dull to grey-green leaves and orange to red flowers crowded along the stems among the foliage. In cultivation, flowering commences in autumn and continues till late spring. The copious flowers provide a good nectar supply to honey-eating birds in the colder months. Its low, dome-shaped habit make it useful for massed planting or as a feature plant in a rockery. While it is long-lived in low-humidity climates, it has not been tested in summer-rainfall areas where it may be short-lived.

Grevillea dissecta (D.J.McGillivray) P.M.Olde & N.R.Marriott (1993)   Plate 104

Nuytsia 9: 282 (Australia)

Based on *Grevillea pilosa* subsp. *dissecta* D.J.McGillivray, *New Names in* Grevillea *(Proteaceae)* 12 (1986, Australia).

The specific epithet refers to the deeply divided leaves and is derived from the Latin *dissectus* (deeply divided). DISS-ECK-TA

Type: Lake Barker area, W.A., 13 Feb. 1973, W.H.Butler (holotype: PERTH).

A low, rounded, prickly **shrub** to c. 1 m. **Branchlets** terete, silky. **Leaves** 1–2 cm long, spreading, ± sessile, the rachis decurved, divaricately pinnatisect or bi or trisect, sometimes simple, linear, sometimes the secondary lobes bi- or trisect; lobes 7–15 mm long, 1–1.5 mm wide, narrowly linear to subulate, pungent; upper surface glabrous, glaucous, the basal stalk sometimes silky, midvein prominent to evident; margins angularly revolute; lower surface bisulcate, silky in grooves, midvein prominent. **Conflorescence** erect, pedunculate, terminal, simple or few-branched; unit conflorescence 0.5–2 cm long, 4 cm wide, secund, umbel-like; peduncle silky, subtended by leaf-like bracts; rachis sparsely silky to glabrous; floral bracts 3–4 mm long, ovate, sparsely silky to glabrous outside with ciliate margins, falling before anthesis. **Flower colour**: perianth rose-pink with a creamy white limb; style pinkish red. **Flowers** acroscopic; pedicels 10–15 mm long, sparsely villous, filamentous; torus 2–4 mm across, 2–2.2 mm broad, oblique; nectary patelliform, adnate to torus; **perianth** 10 mm long, 4 mm wide, erect, oblong, dilated at base, with sparse mixed glandular and 2-armed hairs outside, bearded inside, cohering except along dorsal suture and between dorsal and ventral suture at curve; limb revolute, spheroidal, impressed at margins, openly villous, enclosing style end before anthesis, cohering afterwards; **pistil** 18–20 mm long; stipe inserted ± at 90° to torus, rising 0.5–1 mm from rim, adnate to inner torus for 2–3 mm; ovary villous; style exserted first at curve and looped slightly upwards, straight to slightly incurved after anthesis, villous; style end club-shaped; pollen presenter very oblique to lateral, convex. **Fruit** 10–12 mm long, 7.5 mm wide, erect, globose to ellipsoidal, sparsely hairy; pericarp c. 0.5 mm thick. **Seed** 7 mm long 3 mm wide, ellipsoidal, narrowly winged all round; outer face convex, smooth; inner face with a raised, oblong-elliptic centre surrounded by a narrow, marginal, waxy channel.

Distribution W.A., in an area between Moorine Rock and Lake Barker and south to Mt Holland area. As much of this area remains uncleared and contains few roads, the species may be more common than collections indicate. *Climate* Hot, dry summer; cool, wet winter. Rainfall 240–270 mm.

103A. *G. disjuncta* Flowers and foliage (N.Marriott)

103B. *G. disjuncta* Flowering shrub (N.Marriott)

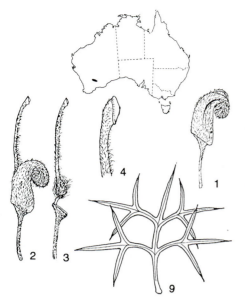

104A. *G. dissecta* Conflorescences and foliage (P.Olde)

104B. *G. dissecta* Conflorescence and foliage (M.Hodge)

104C. *G. dissecta* Plant in natural habitat, N of Mt Holland, W.A. (P.Olde)

G. dissecta

Ecology Grows in low mallee scrub and heath in sandy or gravelly loam. Flowers spring–summer. Regenerates from seed after fire. Pollinated by ground-hopping honey-eaters.

Variation A species with little variation. Some plants have more conspicuously blue-green foliage than others.

Major distinguishing features Branchlets silky; leaves divaricately pinnatisect, sometimes with secondary division of lower lobes; lobes linear-subulate, pungent, margins revolute; upper surface glabrous; undersurface 2-grooved; rachis decurved; conflorescence pedunculate, terminal; torus very oblique; perianth oblong-ovoid, sparsely hairy outside, densely bearded within; stipe perpendicular to torus; ovary villous.

Related or confusing species Group 26, especially *G. pilosa* subsp. *redacta* which differs in its flat, toothed leaves with a glandular indumentum on the upper surface.

Conservation status 3RC.

Cultivation This species, introduced to cultivation in 1986, has proven moderately hardy and adaptable, withstanding extended dry conditions and frosts to at least -4°C. It likes a sunny position in well-drained, acidic sand or gravelly loam but demands perfect drainage, rapidly succumbing to root-rot in poorly drained or heavy soil. Summer watering often proves fatal. Fertiliser and pruning are unnecessary. It makes a most attractive pot specimen for larger tubs, responding to light applications of slow-release fertiliser low in phosphorus. Grafted plants have done well in less favourable conditions. Rarely available from nurseries at present.

Propagation *Seed* Seed treated by peeling or soaking germinates well. *Cutting* Firm, young growth usually strikes readily at most times of the year. *Grafting* Has been grafted successfully onto *G. robusta* and G. 'Poorinda Royal Mantle' using top wedge and whip grafts.

Horticultural features *G. dissecta* is a most attractive, strongly arching, small to medium shrub with eye-catching, blue-grey divided leaves, which contrast with its bright, rose-pink flowers. This compact, low, rounded shrub is eminently suitable as a feature or accent plant in small gardens or large commercial landscapes, especially in dry-climate inland gardens. The prickly leaves are sometimes a disadvantage if too close to pathways but provide excellent protection for birds seeking both nectar and shelter.

Grevillea diversifolia C.F.Meisner (1845)

in J.G.C.Lehmann (ed.), *Plantae Preissianae* 1:547 (Germany)

Valley Grevillea

Some forms of this species have both divided and simple leaves; the specific epithet derives from the Latin *diversus* (different or diverse) and *folium* (a leaf). DIE-VER-SI-FOAL-EE-A

Type: Vasse R., W.A., 13 Dec. 1839, L.Preiss 697 (lectotype: NY) (McGillivray 1993).

Erect, dense **shrub** 1–6 m tall, rarely prostrate. **Branchlets** angular, reddish, silky, the hairs sometimes sparse or absent. **Leaves** 3–8.5 cm long, 2–11 mm wide, ascending, ° sessile, pliable, acuminate, simple, linear to lanceolate to obovate, sometimes apically toothed or more deeply divided into 2 or 3 oblong lobes 2–4 mm wide; upper surface glabrous with conspicuous midvein and lateral venation; margins often entire, shortly recurved usually about an intramarginal vein; lower surface silky, glabrous or sparsely silky, midvein prominent. **Conflorescence** erect, shortly pedunculate, simple or few-branched, terminal and axillary; unit conflorescence 0.5–1 cm long, 1–1.5 cm wide, dense, umbel-like; peduncle and rachis silky; bracts c. 0.5 mm long, triangular, villous outside, falling before anthesis. **Flowers** adaxially oriented; pedicels 3–6 mm long, tomentose; torus 0.5–1 mm across, ° straight; nectary semi-circular or cushion-like, smooth; **perianth** 2–3 mm long, 0.8 mm wide, oblong, tomentose outside, bearded inside about ovary, otherwise papillose; tepals separated along dorsal suture before anthesis, then all separate below limb to c. halfway, the dorsal tepals slightly reflexed, all free to beard and rolling back independently after anthesis; limb revolute, ovoid, enclosing style end before anthesis; **pistil** 6.5–9.5 mm long, glabrous; stipe 0.6–1.3 mm long, somewhat flattened; ovary oblong-ellipsoidal; style at first exserted on dorsal side and looped upwards, straight becoming sharply hooked c. 1 mm before the small, plate-like style end after anthesis; pollen presenter oblique, round, flat. **Fruit** 10–13 mm long, 5 mm wide, oblique, ellipsoidal to cylindrical, glabrous and smooth; pericarp c. 0.3 mm thick. **Seed** 9 mm long, 2 mm wide, oblong-ellipsoidal with an apical wing, minutely pubescent; outer face convex; inner face flat to slightly convex; margin revolute.

Major distinguishing features Leaves simple or divided, exposed on the undersurface; conflorescence regular, umbel-like; floral rachis silky; torus ± straight; perianth zygomorphic, undilated, densely hairy outside, hairy inside; pistil glabrous; fruit smooth.

Related or confusing species Group 16, especially *G. manglesioides* and *G. papillosa*. Both *G. manglesioides* and *G. papillosa* have short, secund conflorescences, a glabrous floral rachis and a sparse open indumentum on the outside of the perianth.

Variation Two subspecies are recognised.

Key to subspecies

Leaf undersurface glabrous or with scattered hairs; simple leaves thin, soft, 3–11 mm wide   subsp. **diversifolia**

Leaf undersurface silky; simple leaves relatively stiff, 1.5–6 mm wide   subsp. **subtersericata**

Grevillea diversifolia C.F.Meisner subsp. diversifolia   Plate 105

Leaf undersurface glabrous or with scattered hairs; simple leaves thin, soft, 3–11 mm wide. **Flower colour**: perianth white, pale cream; style red.

Distribution W.A., in the Darling Ra. from near Mundaring Weir S to Donnybrook, with an isolated record from the Vasse R. *Climate* Hot to warm, dry summer; cool, wet winter. Rainfall 800–1200 mm.

Ecology Usually grows in forest or tall woodland in sites that are moist for most of the year, (beside creeks, near swamps), in damp, sandy-peat soil or heavy clay loam. Flowers winter to summer. Regenerates from seed after fire. Pollinator unknown.

Variation This subspecies varies from a prostrate ground cover forming dense, layered mats c. 2 m across to the more usual large, dense shrub to 6 m. Leaves may be simple or divided although both are often found on the same bush.

Conservation status Not presently endangered.

Cultivation This adaptable shrub thrives in cold-wet, warm-dry and even summer-rainfall climates and has been grown widely in Vic., as well as occasionally in N.S.W. and inland Qld. In cultivation, it will withstand frosts to at least -6°C. It flourishes in moist but well-drained, acidic sand or loamy soil and tolerates wet, almost boggy soil extremely well. It is equally suited to both full sun and full shade and adjusts to quite dry soil in summer. As it grows naturally into a leafy, dense shrub, pruning is rarely required except to control size. It can, however, be pruned regularly and therefore makes a useful hedge plant. Summer watering extends flowering and keeps it looking lush but fertiliser is unnecessary due to its natural vigour in various conditions. Occasionally available from specialist native plant nurseries and seed suppliers.

Propagation Seed Sets small quantities of seed that germinate readily especially when treated by soaking or nicking. *Cutting* Strikes readily from firm, new growth at most times of the year. *Grafting* Untested. Possibly a valuable rootstock.

Horticultural features Subsp. *diversifolia* is valued mainly as a foliage plant for its very dense, bright green to bluish leaves. While its flowering is quite heavy in early spring, the conflorescences are rather dull, small and fairly inconspicuous among the foliage. The delicate flowers are slightly perfumed but do not attract birds. It makes an ideal screen or fill-in plant and is useful in mass public plantings

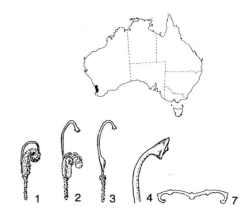

G. diversifolia subsp. *diversifolia*

105B. *G. diversifolia* subsp. *diversifolia* Flowering branch (P.Olde)

105A. *G. diversifolia* subsp. *diversifolia* Close-up of flowers (P.Olde)

105C. *G. diversifolia* subsp. *diversifolia* Plant in natural habitat, Mundaring Weir, W.A. (C.Woolcock)

because its foliage is both attractive and lush. It is long-lived in cultivation and resistant to most root pathogens which often destroy the root systems of more attractive grevilleas, and it appears to have no major pests and diseases.

Grevillea diversifolia subsp. subtersericata D.J.McGillivray (1986)
Plate 106

New Names in Grevillea *(Proteaceae)* 5 (Australia)

The subspecific epithet alludes to the silky hairs on the undersurface of the leaves and is derived from the Latin *subter* (beneath) and *sericatus* (covered in hairs). SUB-TER-SERRIK-ARE-TA

Type: Camfield, Broke Inlet, W.A., 10 May 1974, A.S.George 11788 (holotype: PERTH).

Synonym: *G. manglesioides* var. *angustissima* G.Bentham (1870).

Leaf undersurface silky; simple leaves relatively stiff, 1.5–6 mm wide. **Flower colour**: perianth white or cream; style red.

Distribution W.A., from near Albany to Broke Inlet.
Climate Warm, dry summer; cool, wet winter. Rainfall 800–1200 mm.

Ecology Occurs on moist, sandy creek banks, usually in eucalypt shrubland or swamp vegetation. Flowers winter to spring. Regenerates from seed after fire. Pollinator unknown.

Variation Varies from a low, compact shrub 2 m

G. diversifolia subsp. *subtersericata*

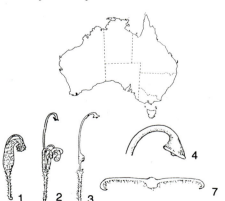

106A. *G. diversifolia* subsp. *subtersericata* Flowers and foliage; note undersurface of leaves (P.Olde)

tall to a large, dense shrub to 5 m. Leaves may be simple or divided.

Conservation status 3V. Much of its former habitat in the Albany region has been cleared for agriculture.

Cultivation Although not widely cultivated, subsp. *subtersericata* has been grown in western Vic. where it has proved both easy to grow and adaptable. It tolerates frosts to at least -6°C and dry conditions. As it has a natural habitat similar to subsp. *diversifolia*, it would require similar conditions in cultivation, although it may be less tolerant of shady conditions. It is not available in nurseries.

Propagation *Seed* Should germinate well from seed nicked along one side or soaked for 24 hours prior to sowing. *Cutting* Grows readily from cuttings taken of firm, new growth at any time of the year. *Grafting* Untested.

Horticultural features Similar in most ways to subsp. *diversifolia*, this subspecies has even less prominent flowers, often being borne closer to the stems and thus more hidden in the leaves. While not really striking enough to be considered as a feature plant it could be useful in mass screening or landscaping public areas where flowers are relatively unimportant or perhaps where drainage is questionable. It should be long-lived in cultivation.

Grevillea dolichopoda (D.J.McGillivray) P.M.Olde & N.R.Marriott (1993)
Plate 107

Nuytsia 9: 291 (Australia)

Based on *G. disjuncta* subsp. *dolichopoda* D.J. McGillivray, *New Names in* Grevillea *(Proteaceae)* 5 (1986, Australia)

The specific epithet refers to the long torus and is derived from the Greek *dolichos* (long) and *podion* (a base or foot). DOLLY-KO-PODE-A

Type: c. 21 km by road N of Ongerup, W.A., 26 June 1976, D.J.McGillivray 3521 & A.S.George (holotype: NSW).

Divaricately-branched **shrub** to c. 60 cm high. **Branchlets** terete, silky to tomentose to villous, usually secund. **Leaves** (1–) 2–5 (–7) cm long, 1–2.5 mm wide, erect to spreading, sessile, simple, linear to ± terete, stiff, weakly pungent to obtuse, mucronate; upper surface glabrous, scabrous to granulate, sometimes the granules subdued (Newdegate area), sometimes with faint longitudinal ridges, otherwise midvein obscure; margin entire, smoothly or angularly revolute; lower surface usually unisulcate, obscured by revolute margins, rarely

106B. *G. diversifolia* subsp. *subtersericata* Plant in natural habitat, Napier Creek, W.A. (P.Olde)

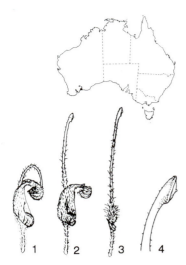

G. dolichopoda

exposed, silky with prominent midvein. **Conflorescence** axillary or cauline, 2–4-flowered, often from old peduncle (peduncle absent to c. 1 mm long); rachis c. 1 mm long, villous; bracts 0.5–0.9 mm long, narrowly triangular, tomentose outside, persistent at anthesis. **Flower colour**: perianth red and orange; style red with green tip. **Flowers** abaxially oriented; pedicels 5–9 mm long, sparsely silky; torus 2–3 mm across, oblique; nectary conspicuous, V-shaped, toothed at ends, the walls rising 0.5–0.8 mm above the torus; **perianth** 5–7 mm long, 2.2–3 mm wide, dull red, oblong with basal dilation, much narrowed and geniculate at throat, sparsely silky to almost glabrous outside, densely bearded inside, cohering except along dorsal suture; limb rusty-orange, tomentose, revolute, spheroidal, distant from perianth, partially exposing style end before anthesis; **pistil** 23–24 mm long; stipe 0.5–1 mm long, inserted on dorsal rim of torus; ovary villous; style exserted below curve and looped upwards before anthesis, afterwards gently incurved, villous at base, sparsely hairy to glabrous near apex; pollen presenter lateral, round to oblong, convex. **Fruit** 12 mm long, 5–6 mm wide, erect, ovoid, sparsely villous; pericarp c. 0.3 mm thick. **Seed** 7–8 mm long, 2–3 mm wide, oblong-elliptic with a short apical wing, sparsely pubescent; margin strongly revolute.

Distribution W.A., in two disjunct areas, one between Nyabing and the Gairdner R. and one in the Lake Varley–Ravensthorpe area. *Climate* Hot, dry summer; cool, wet winter. Rainfall 350–500 mm.

Ecology Found in open, low mallee shrubland and low, sandy heathland often over laterite, on quartzite ridges, occasionally in clay near granite outcrops and in granitic loam. Flowers autumn to spring. Regenerates from seed after fire. Pollinated by birds.

Variation There is some variation. Plants from the Newdegate area generally have shorter leaves, c. 1.5–2 cm long.

Major distinguishing features Leaves simple, scabrous or granular, sometimes with faint longitudinal ridging, linear, the margins revolute and enclosing the lower surface including the midvein; conflorescence axillary, few-flowered; torus 2.5–3 mm across, very oblique; perianth zygomorphic, sparsely silky outside, bearded inside; ovary subsessile, villous; style villous at the base becoming glabrous.

Related or confusing species Group 25, especially *G. disjuncta* and *G. haplantha*. *G. disjuncta* differs in its ridged, crowded leaves and its shorter torus (1.5 mm across). *G. haplantha* is a robust, dense shrub 1–2 m tall with ascending branches, larger, more densely hairy flowers (perianth c. 12 mm long), with a conspicuously stalked ovary (stipe 0.6–1.3 mm long).

Conservation status Widespread but often confined

107A. *G. dolichopoda* Flowering branch (N.Marriott)

107B. *G. dolichopoda* Plant in natural habitat, SW of Newdegate, W.A. (N.Marriott)

to road verges in the north of its range because of clearing for agriculture.

Cultivation *G. dolichopoda* has been successfully cultivated in Vic. and at Burrendong Arboretum, N.S.W., for many years, although it has become less readily available at nurseries in recent years. Although full sun is preferred, it has performed reliably in most situations except dense shade, in various soils including well-drained, acidic to neutral sand, loam and clay. Frosts to at least -5°C do not damage the foliage and the plant responds to light pruning. It is sometimes sold as *G. haplantha*.

Propagation *Seed* Sets plentiful seed that germinate reasonably well, especially when nicked or soaked. *Cutting* Grows well from cuttings of firm, young growth at most seasons. *Grafting* Grafts successfully onto *G. robusta* using both the whip and mummy techniques.

Horticultural features *G. dolichopoda* has a surprising charm and is valued mostly for its unusual foliage. New growth is an attractive coppery green which soon turns pale green and contrasts vividly with the older, darker foliage. The leaves are so neatly arrayed in an upright manner along all the small branches that one is struck by their arrangement. Large bright to dull red flowers are borne in masses along the branches in autumn and winter when little else is in flower and are valuable in attracting birds. With a compact, low habit, it is suitable for consideration as a rockery or border plant and, being relatively long-lived in cultivation, worthy of much wider use in both public and private landscapes.

General comments *Grevillea dolichopoda* was included in *G. disjuncta* as a subspecies by McGillivray (1986), but from our observations of this taxon in the field and in herbarium studies, it is at least as distinct as *G. haplantha*, another closely related species, and should therefore be recognised at species rank.

Grevillea donaldiana K.F.Kenneally (1988)
Plate 108

Western Australian Naturalist 17: 115 (Australia)

The specific epithet honours Donald J. McGillivray, botanist, whose revision of the genus *Grevillea* was published in 1993. DON-AL-DEE-ARE-NA

Type: bank of the Sale River, Kimberley, W.A., 16 May 1986, K.F.Kenneally 9676 (holotype: PERTH).

Spreading **tree** to 10 m with greyish-brown, fissured bark. **Branchlets** terete to angular, reddish-brown, appressed-pubescent. **Leaves** 5–12 cm long, 4–10 mm wide, ascending, shortly petiolate, concolorous, simple, elliptic, sometimes slightly falcate, acute with a broad mucro; both surfaces similar, silky, rusty when juvenile, the midvein prominent to c. halfway with fine, longitudinally oriented lateral veins; margins entire, flat. **Conflorescence** erect, pedunculate, terminal, paniculate; unit conflorescence c. 7 cm long, 2 cm wide, cylindrical, dense; peduncle and rachis silky; bracts 1.5 mm long, ovate acuminate, silky outside, persistent to anthesis. **Flower colour**: perianth and style white. **Flowers** acroscopic; pedicels 0.5–1 mm long, glabrous; torus c. 0.5 mm across, straight, bearing 6 ovate-attenuate appendages of unequal length regularly spaced around the rim; nectary prominent, 3/4 annular; tepals 3; **perianth** 3–4 mm long, 0.8 mm wide, cylindrical, glabrous, cohering except along dorsal suture; limb revolute, ovoid, enclosing style end before anthesis; **pistil** 9–12 mm long, glabrous; stipe 0.5–0.7 mm long; ovary globose; style exserted from base of perianth and looped outwards, geniculate about half way after anthesis, at length straight; style end scarcely dilated; pollen presenter erect, conical with its base slightly oblique and c. the same width as style end. **Fruit** 22 mm long, 19 mm wide, oblique, ellipsoidal, apiculate, glabrous, rugulose, the exocarp soon exfoliating; pericarp woody, 3–6 mm thick. **Seed** (not seen, described from the inner pericarp) elliptic, flat, winged all round, the wing and testa with minute short grooves.

Distribution W.A., near the Sale R., N of Derby in the Kimberley. *Climate* Hot, wet summer; warm, dry winter. Rainfall 600–800 mm.

Ecology Common on scree below cliff faces and in steep, rocky crevices beside the Sale R. Flowers autumn to winter. Fire response unknown. Probably insect-pollinated.

Variation A uniform species.

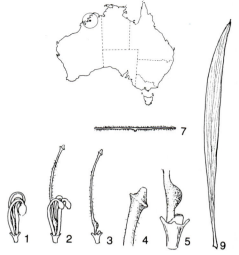

G. donaldiana

Major distinguishing features Leaves simple, entire, concolorous, lanceolate, silky; margins flat; conflorescence cylindrical, branched; torus straight with c. 6 petal-like appendages evenly spaced around the torus; perianth zygomorphic, glabrous, of 3 tepals; pistil glabrous; stipe very short; pollen presenter erect, conical; fruits thick-walled; seed winged all round with minute, brush-like markings on the surface.

Related or confusing species Group 4, a small group of easily distinguished species (see key).

Conservation status 2R. A species known only from one locality in the Kimberley where it is reasonably plentiful.

Cultivation *G. donaldiana* is unknown in cultivation and comments on its needs are therefore speculative. A position in well-drained, gravelly

108A. *G. donaldiana* Flowering branch (K.Coate)

108B. *G. donaldiana* Close-up of conflorescences (M.Pieroni)

loam in full sun is recommended, with a likely preference for acidic to neutral soil. Its tolerance of temperate climates is unknown but it is unlikely to endure long cold spells or frost. It should do well in tropical or subtropical gardens protected from strong winds.

Propagation Not yet attempted.

Horticultural features G. donaldiana is an impressive small tree with silvery-green leaves and dense clusters of white, strongly sweet-scented flowers held on large prominent racemes above the foliage. The leaves are relatively large and attractive and together with the rusty new growth make this species an excellent foliage plant, potentially useful in many areas of tropical landscaping. It would make an ideal specimen plant or screen plant and could be considered as a small street tree.

Grevillea drummondii C.F.Meisner (1845) Plate 109

in J.G.C.Lehmann (ed.), *Plantae Preissianae* 1: 536 (Germany)

Drummond's Grevillea

Named after the Scotsman, James Drummond (1784–1863), whose early botanical collections were so important for the description of new species in W.A. DRUM-ON-DEE-EYE

Type: near the Swan River, W.A., 1844–45, J.Drummond coll. 3 no. 335 (lectotype: NY) (McGillivray 1993).

Spreading, dense **shrub** 0.3–1 m tall. **Branchlets** terete, villous. **Leaves** 1–4 cm long, 0.2–0.6 cm wide, ascending, crowded, simple, narrowly elliptic to obovate, pliable, acute; upper surface openly villous, midvein evident; margins entire, recurved, ciliate; lower surface villous, midvein prominent. **Conflorescence** c. 0.5 cm long, 1 cm wide, erect or decurved, shortly pedunculate, simple, terminal on short branchlets, not exceeding foliage, ± globose to umbel-like, 6–8-flowered; peduncle and rachis villous; bracts 0.5–1.5 mm long, ± triangular, villous, persistent at anthesis. **Flower colour**: perianth cream turning pink to red; style cream turning pinkish red, the end green. **Flowers** acroscopic: pedicels 1–2 mm long, glabrous; torus < 1 mm across, oblique; nectary U-shaped, smooth; **perianth** 5–6 mm long, 2 mm wide, ellipsoidal, ribbed, glabrous outside, bearded inside near curve, cohering except along dorsal suture from limb to curve, the style end exposed before anthesis; limb nodding, spheroidal or depressed cubic, prominently ridged; **pistil** 4–6 mm long; stipe 1–1.5 mm long, villous; ovary villous; style not exserted beyond perianth, straight, stout, sparsely villous; style end broadly expanded; pollen presenter lateral, round, concave. **Fruit** 12–14 mm long, 5–6 mm wide, erect, ovoid, sparsely villous, ridged; style persistent; pericarp c. 0.5 mm thick. **Seed** not seen. 6 mm long, elliptic, margins tightly inrolled (Keighery 1991).

Distribution W.A., in a limited area in the Darling Ra. north of Perth, from Bindoon to Bolgart. *Climate* Hot, dry summer; cool, wet winter. Rainfall 600–800 mm.

Ecology Grows in gravelly brown loam on lateritic rises in open to thick dryandra scrubland and eucalypt woodland. Flowers winter–spring. Regenerates from seed after fire. Pollinator unknown but in cultivation flowers are visited by birds.

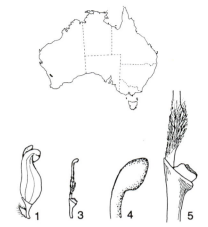

G. drummondii

109A. *G. drummondii* Plant in natural habitat, S of New Norcia, W.A. (N.Marriott)

109B. *G. drummondii* Close-up of flowers (M.Pieroni)

109C. *G. drummondii* Young and old conflorescences (P.Olde)

Variation A fairly uniform species.

Major distinguishing features Branchlets villous; leaves simple, entire, narrowly elliptic to obovate, villous, ciliate; conflorescence 6–8-flowered; torus oblique; perianth zygomorphic, glabrous outside, hairy within; ovary villous, stipitate; pollen presenter lateral, concave.

Related or confusing species Group 23, especially *G. centristigma* and *G. pimeleoides*, both of which have a hairy outer perianth surface and yellow flowers.

Conservation status 2VC. Mostly confined to private land and disturbed road verges.

Cultivation In the wild, *G. drummondii* is a tough species but has proved somewhat touchy in cultivation. It has been established in a few gardens in the drier areas of Vic. and N.S.W., where it has survived frost to at least -4°C and severe drought. In coastal areas, it has collapsed in high humidity and excessive moisture, usually in summer. Best results so far have been achieved when it is grown in well-drained, acidic to neutral gravelly loam in partial shade. Acceptable growth habit is achieved without fertiliser except in sandy, poor soil when growth may be spindly. Regular tip-pruning will improve shape and density. It makes an attractive potted specimen in a well-drained potting mix with slow-release fertiliser. Members of the Grevillea Study Group are cultivating this species but it is not yet widely available from nurseries or seed suppliers.

Propagation Seed As yet untested but should germinate readily when peeled. *Cutting* Strikes readily from cuttings of firm, young growth at most times of the year. Care should be taken not to overwater cuttings as they can easily rot off. *Grafting* Initial attempts to graft this species have been unsuccessful. Recent success has been achieved using G. 'Poorinda Royal Mantle' and the hybrid known as G. 'rosmarinifolia nana', both using top wedge grafts.

Horticultural features *G. drummondii* is a delightful, small to medium shrub with softly hairy, dark green foliage that is pleasurable to the touch. Its conspicuous cream to white flowers are crowded in the leaf axils and age to pink or red, creating a multi-colour effect as the flowering season progresses. It would make an attractive rockery plant in a semi-shaded situation or an interesting massed border plant. Honey-eating birds are attracted to the flowers.

110A. *G. dryandri* subsp. *dryandri* (K. & L. Fisher)

Grevillea dryandri R.Brown (1810)

Transactions of the Linnean Society of London, Botany 10: 175 (England)

Dryander's Grevillea

Aboriginal: 'burrun burrun' (Gupapuynga); 'andjamgu' (Gagadju).

Named after Jonas Dryander (1748–1810), bibliographer and botanist, who succeeded Daniel Solander as librarian to Joseph Banks in 1782. DRY-AND-RYE

Type: islands a, b, c, g, g2, j3, h, Gulf of Carpentaria, [N.T.], Nov.–Dec. 1802, R.Brown 3345 (lectotype: BM). Robert Brown visited a number of islands in the Gulf during the circum-navigation of Australia with Flinders' expedition in 1802–03. It is unknown on which his specimens were collected.

A spreading to sprawling **shrub**, 0.3–1 m high with conspicuous, emergent, branching conflorescences. **Branchlets** angular to terete, silky to glabrous. **Leaves** 6–20 cm long, erect, petiolate, secund, pinnatipartite; leaf lobes linear, 1–5 mm wide; upper surface glabrous or sparsely silky, sometimes villous, midvein faintly evident, sometimes with oblique longitudinal wrinkles; margin recurved to revolute; lower surface exposed or enclosed by margin to midvein, silky, midvein and sometimes intramarginal veins prominent. **Conflorescence** glabrous, erect, emergent from foliage on long peduncle, terminal, usually branched, occasionally simple; unit conflorescences 10–60 cm long, secund, open; peduncle and rachis glabrous and glaucous; bracts 1.7–2.9 mm long, ovate, falling before anthesis. **Flowers** acroscopic in young bud, reversing orientation by twisting of the pedicels before anthesis, rolling down and under the rachis after anthesis; pedicels 5–10 mm long; torus ± 2 mm across, oblique; nectary U-shaped, entire or toothed; **perianth** 10–12 mm long, 3 mm wide, ovoid, ribbed, narrowing markedly at the curve, glabrous and glaucous outside, bearded inside near curve, otherwise glabrous, cohering except along dorsal suture before anthesis, afterwards free to curve and loose; limb revolute, subcubic, very conspicuously keeled, enclosing style end before anthesis; **pistil** 40–52 mm long, glabrous; stipe 8.5–10.5 mm long; ovary globose, scarcely prominent, rarely with a few hairs; style exserted from just below curve on dorsal side and looped upwards, gently incurved, narrowing and terminating in a suddenly dilated, flattened style end; pollen presenter oblique, oblong-elliptic, convex. **Fruit** 7.5–15 mm long, 7 mm wide, erect or oblique, ± ovoid to ellipsoidal, glabrous, viscid; pericarp c. 0.5 mm thick. **Seed** 8 mm long, 3 mm wide, obovoid; outer face convex, rugulose; inner face concave; margin recurved, shortly winged.

Major distinguishing features Leaves pinnati-partite with secund, narrowly linear to linear lobes; conflorescence branched, emergent above foliage, glabrous; torus oblique; perianth zygomorphic, with strongly keeled limb, the flowers reversely orienting during development of raceme; pistil glabrous; ovary long-stipitate.

Related or confusing species Group 5. A distinctive species readily distinguished from related species.

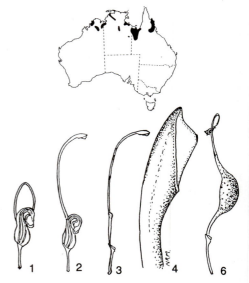

G. dryandri subsp. *dryandri*

Key to subspecies

Fruit glabrous; habit lignotuberous; leaf lobes 5–35, up to 6 mm wide   subsp. **dryandri**

Fruit hairy; habit single-stemmed; leaf lobes 39–71, c. 1 mm wide   subsp. **dasycarpa**

Grevillea dryandri R.Brown subsp. dryandri   Plate 110

Plant lignotuberous. **Leaf** lobes 5–35, up to 6 mm wide. **Flower colour**: perianth creamy white to pale pink with a cream limb; style cream or red with a green tip; floral rachis deep purple, (green in cream-flowering forms). **Fruit** glabrous.

Synonyms: *G. rigens* A.Cunningham ex R.Brown (1830); *G. callipteris* C.F.Meisner (1856).

Distribution Transcontinental across northern Australia. W.A., in the Kimberley and adjacent islands. N.T., widespread in the north, extending inland almost to Tennant Creek. Qld, widely distributed between Mt Isa and the Herberton area (Cloncurry, Julia Creek, Camooweal.) *Climate* Summer hot and wet; winter mild and dry. Rainfall 350–1400 mm, mainly in summer.

Ecology Grows in open savannah woodland and on sandstone ridges or slopes, often regenerating in disturbed areas, usually among spinifex or grasses in skeletal sandy soil, heavy clay-loam, gravelly loam or deep, red sand. Flowers autumn to winter. Regenerates from seed and lignotuber after fire. Pollinated by birds.

Variation Subsp. *dryandri* has two colour forms, creamy white or cream and red. There is some variation in leaf indumentum and flower size.

Wyndham form Three collections of a form with villous, grey leaves and flowers with larger perianths have been made in the Wyndham area of W.A. This form appears to have much horticultural merit and would be worth introducing to tropical gardens.

Conservation status Not presently endangered.

Cultivation Subsp. *dryandri* will grow reasonably well in subtropical climates as far south as Brisbane where it flowers well. Strangely, plants from the Herberton area have been particularly difficult to establish in cultivation because they tend to succumb to fungal diseases of the foliage. It prefers a warm, open, sunny position and summer watering, but does not tolerate frost well, nor does it survive continuous cold weather. Although a heavy soil

110B. *G. dryandri* subsp. *dryandri* White-flowered form, Hell's Gate, NW Qld (A.Foster)

110C. *G. dryandri* subsp. *dryandri* Flowers and foliage (A.Fairley)

would be suitable in dry climates, it seems to prefer a sandy gravelly soil in cultivation. Nurseries in Darwin and specialist Qld nurseries usually stock this subspecies and it is often listed in seed catalogues.

Propagation *Seed* Grows readily from seed sown when fresh and nicked down one side. Soaking in warm water for 24 hours will also aid germination. In cool climates, it will damp off easily if conditions are not kept clean. *Cutting* Cuttings of firm, semi-mature wood will strike if placed in hot, humid conditions. Cuttings taken in late spring to summer do best. The percentage strike is often low in cool climates where seed is the preferred method. *Grafting* Top-wedge grafts, approach grafts and cotyledon grafts have all proved successful using *G. robusta* and *G.* 'Poorinda Royal Mantle' as rootstocks. Grafting offers great hope for growers in more temperate climates, where *G. dryandri* is at best regarded as a touchy species when grown on its own roots.

Horticultural features Subsp. *dryandri* is a spectacular plant in the garden where it flowers over a long period during the winter months. Being lignotuberous, it is also very long-lived once established. The prominently displayed, brightly coloured flowers and the ferny foliage combine to make this one of the most desirable species for cultivation and it would make an ideal feature plant in any subtropical garden. It has one of the longest conflorescences of any species of *Grevillea* and its flowers are strongly attractive to nectar-feeding birds.

History Seed of this species was sent to Kew in 1802 by Peter Good but no record has been found of cultivation there. It was first cultivated in Qld in 1875.

Grevillea dryandri subsp. dasycarpa
D.J.McGillivray (1986) **Plate 111**

New Names in Grevillea *(Proteaceae)*: 5 (Australia)

Aboriginal: 'andadjek' (Mayali); flower heads were sucked for nectar.

The subspecific epithet is derived from the Greek *dasys* (shaggy, hairy) and *carpos* (a fruit), in reference to the indumentum of the follicle. DAISY-CAR-PA

Type: c. 10 km ESE of the mouth of Deaf Adder Gorge on S side, N.T., 18 July 1978, D.J.McGillivray 3937 & C.R.Dunlop (holotype: NSW).

Erect, single-stemmed **shrub** 0.5–3 m. high. **Leaf** lobes 39–71, c. 1 mm wide **Flower colour**: perianth pink, orange-red or deep red; style red. **Fruit** densely covered with erect, glandular and multicellular hairs.

Distribution N.T., widely distributed in the northern section of the Territory. *Climate* Hot, wet summer; warm, dry winter. Rainfall 800–1600 mm.

Ecology Grows on sandstone cliffs and in gorges, in grey sand or rocky soil in thickets. Flowers autumn–winter, sometimes extending to spring.

Variation Varies in habit, sometimes attaining the status of a small tree.

Conservation status Not endangered. Conserved in Kakadu NP.

Cultivation Subsp. *dasycarpa* is readily grown as far south as northern N.S.W. where it flowers well and regenerates from seed in the garden. It is widely cultivated in Darwin gardens. A grafted plant was a spectacular feature of the Shaw garden in Brisbane for many years. Both pink- and red-flowered forms have flowered at Bulli, south of Sydney, grown against a north-facing wall (H. & D.Saunders, pers. comm.). A grafted plant has been maintained successfully in a pot at Ocean Grove on the Victorian coast although it was protected from winter frosts on a sheltered verandah (D.McKenzie, pers. comm.). It is not resistant to frost or continuous cold weather but is otherwise a hardy species in temperate and subtropical climates. It does best in well-drained, acidic sandy soil in full sun. Summer watering is beneficial but not essential except in very dry conditions. Although it is a compact, rounded plant, old flowering canes may need pruning to maintain shape. If placed in a warm to hot position, protected from cold winds and frosts, it makes an excellent tub plant even in southern

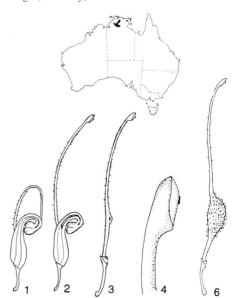

G. dryandri subsp. *dasycarpa*

climates, although strictly it should be held in a warm, dry glasshouse over winter. Commercial nurseries in Darwin and north Qld frequently stock this subspecies and it has become more common around Brisbane in recent years.

Propagation *Seed* Both cultivated and wild plants set copious seed that germinate readily. Nicking or soaking for 24 hours prior to sowing improves germination rates. *Cutting* Strikes with reluctance from cuttings of firm, new growth taken in summer. *Grafting* Grafts with difficulty onto *G. robusta*. It is important to use scion material with just-active axillary growth, otherwise grafts will not shoot. Best season to graft is summer. Cotyledon grafts have proved the easiest method, although whip grafts are

110B. *G. dryandri* subsp. *dryandri* Fruits (M.Hodge)

111B. *G. dryandri* subsp. *dasycarpa* Flowering branch (M.Hodge)

111C. *G. dryandri* subsp. *dasycarpa* Hairy fruits (M.Keech)

also quite successful and ensure selection of form and colour.

Horticultural features A magnificent, robust plant with fine, ferny foliage and prominent, emergent branching conflorescences. Its flowers range from creamy white to peach, pink and red, and when different colour forms are grown together, the impact is stunning. A further feature is the follicles that set in great numbers. The conspicuous, erect hairs that cover these are eye-catching especially when back-lit by the sun. In cultivation this appears to be the hardier of the two subspecies and would make an outstanding plant for the tropical garden, being both dense and floriferous. In cultivation, it flowers from late autumn to early spring and readily attracts birds. The foliage is deceptively prickly and hairs on the leaves can cause skin irritations. Accordingly, it is wise to plant it well away from areas to be used for access.

Grevillea dryandroides C.A.Gardner (1934)

Journal of the Royal Society of Western Australia 19: 81 (Australia)

Phalanx Grevillea

The resemblance of the foliage to that of some species of *Dryandra* inspired the specific epithet. DRY-AND-ROY-DEES

Type: near Ballidu, W.A., 22 Sept. 1931, C.A.Gardner 2711 (lectotype: PERTH) (McGillivray 1993).

Clumping, suckering **shrub** 10–50 cm high with leafless trailing peduncles to 1 m long. **Branchlets** terete to angular, ridged, villous. **Leaves** 7–14 cm long, slightly incurved, ascending to spreading, sessile, secund, pinnatipartite; lobes 5–14 mm long, 1–2 mm wide, closely aligned, linear to obovate, uncinate; upper surface glabrous to openly woolly, longitudinally ridged, midvein prominent; margins angularly refracted; lower surface bisulcate, villous in the grooves or when exposed, midvein prominent. **Conflorescence** erect, pedunculate, usually branched; unit conflorescence secund, dense; peduncle sparsely silky to villous; rachis silky; bracts 1–2 mm long, ovate to lanceolate, silky or sparsely so outside, ciliate, falling before anthesis. **Flowers** acroscopic; pedicels 1–2 mm long, silky; torus 1–1.5 mm across, oblique to almost straight; nectary conspicuous, linguiform with recurved apex; **perianth** 6–8 mm long, 2–3 mm wide, ovoid, silky outside, glabrous inside, cohering except along dorsal suture; limb revolute, ovoid-ellipsoidal, enclosing style end before anthesis; **pistil** 17–23 mm long; stipe 0.5–1.2 mm long, sparsely silky; ovary appressed-villous; style villous at base becoming glabrous, exserted first just below curve and looped upwards, after anthesis straight or kinked about middle; style end dilated; pollen presenter erect, elongate-cylindrical, bulbous at base, usually tapering to a point. **Fruit** 14–16.5 mm long, 8.5 mm wide, oblique, oblong, pubescent sometimes with reddish stripes; pericarp c. 0.5 mm thick. **Seed** 7 mm long, 2.5 mm wide, oblong-ellipsoidal; outer face convex, rugulose; inner face flat, channelled around margin; margin recurved with a papery or waxy border.

Major distinguishing features Low, clumping, suckering habit; leaves pinnatipartite, the lobes secund, usually linear; conflorescence on long trailing peduncle, branched, secund; nectary tongue-shaped; perianth zygomorphic, hairy outside, glabrous within; ovary shortly stipitate, villous; style glabrous in upper third; pollen presenter elongate-cylindrical with bulbous base; fruit with irregular patches of reddish hairs.

Related or confusing species Group 35, especially *G. thyrsoides* which differs in its entirely hairy style and ± flat pollen presenter.

Variation Two subspecies are recognised.

Key to subspecies

Most leaf lobes < 10 mm long, glabrescent; pistil 17–18 mm long; ovarian stipe < 1 mm long subsp. **dryandroides**

Most leaf lobes > 12 mm long, persistently hairy; pistil 17–23 mm long; ovarian stipe 1–1.5 mm long subsp. **hirsuta**

Grevillea dryandroides C.A.Gardner subsp. dryandroides   Plate 112

Lightly suckering **shrub** to 50 cm tall, usually forming colonies of up to 4 'plants', or solitary. **Leaves** dull yellow-green; rachis glabrous; lobes 5–10(–15) mm long, glabrescent. Unit **conflorescence** 3–4 cm long; pedicels 1–1.5 mm long; torus oblique, at c. 30°. **Flower colour**: perianth pink to orange-pink with a grey-green limb; style red or pink with green tip. **Perianth** 6–7 mm long; **pistil** 17–18 mm long; ovarian stipe 0.5–0.7 mm long

Distribution W.A., confined to the Ballidu area. *Climate* Hot, dry summer; cool, moist winter. Rainfall 300–400 mm.

Ecology Occurs in open heathland and banksia woodland, usually in yellow sandy loam over laterite or in shallow sandy loam over clay. Flowers winter to summer. Regenerates from seed or sucker after fire or disturbance. Bird-pollinated.

Variation The typical subspecies shows little variation.

Conservation status 2E. Extremely rare and endangered. Almost all populations are confined to degraded road verges where they are damaged by road works and swamped by exotic grasses. Urgently in need of positive conservation measures.

Cultivation Subsp. *dryandroides* can be grown in hot-dry and cool-wet climates and has been cultivated successfully in Vic. Grafted plants have thrived in Sydney and Brisbane. This species is proving quite adaptable and hardy in cultivation, tolerating frost to at least -4°C and extended dry conditions, but performance in humid summer-rainfall conditions remains questionable. A full sun position in a built-up bed of acid to neutral, free-draining, gravelly or sandy loam enhances the likelihood of success; it grows naturally in deep, yellow, sandy soil. Once established, it should not

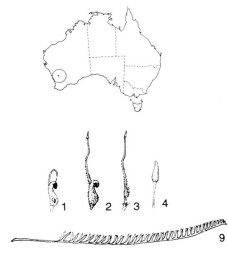

G. dryandroides subsp. dryandroides

require summer watering, but light applications of slow-release, low-phosphorus fertiliser are beneficial and increase growth rate. It makes an excellent tub specimen when planted in an open, gravelly mix. It is presently only occasionally available from nurseries. A few members of the Grevillea Study Group are preserving it in cultivation.

Propagation *Seed* Sets few seed and as a result there has been little experience with this method of cultivation. Kullman reports germination in about 33 days. *Cutting* Firm, young growth gives a fairly good strike rate in spring or early autumn. *Grafting* Has been grafted onto a number of rootstocks, the most common being G. robusta, G. 'Poorinda Anticipation' and G. 'Poorinda Royal Mantle' using the top wedge and mummy methods. Although grafting increases its hardiness and adaptability, it removes the ability to sucker. Most grafted plants have not flourished as yet.

Horticultural features G. dryandroides is an enchanting, small plant, even when not in flower, with its bold, fish-bone leaves and low, clumping habit. The lovely pink to red toothbrush flowers are borne on leafless stems and trail appealingly along the ground around the edge of the shrub. In massed flower, the wreath-like effect makes an unusual contrast with the dull green leaves.

General comments As this species suckers, it shoots vigorously after fire or light damage to the surface soil. On verges graded for roadworks, it has regenerated vigorously. An urgent program should be undertaken to propagate and recolonise areas in W.A. where this species could survive.

Grevillea dryandroides subsp. hirsuta
P.M.Olde & N.R.Marriott (1993)   Plate 113

Nuytsia 9:270 (Australia)

The subspecific epithet is derived from the Latin *hirsutus* (hairy), in reference to the leaf indumentum.
HER-SUE-TA

Type: 2.6 km N of Cadoux, W.A., 25 Sept. 1980, J.Briggs 645 (holotype: PERTH).

A vigorously suckering **shrub** to 30 cm high, usually forming colonies of up to 50 or more 'plants'. **Leaves** grey; rachis appressed-villous; lobes (8–)12–35 mm long, persistently hirsute with crisped hairs. Unit **conflorescence** 5.5–10 cm long, conico-secund; pedicels 1–2 mm long. **Flower colour**: perianth pink to orange-pink (rarely yellow) with a grey-green limb; style red or pink (rarely yellow) with green tip. **Perianth** 7–8 mm long; **pistil** 17–23 mm long; ovarian stipe 1–1.5 mm long.

Distribution W.A., relatively widespread in the wheatbelt between Cadoux and Corrigin. *Climate* Hot, dry summer; cool, moist winter. Rainfall 300–400 mm.

Ecology Occurs in open heathland and banksia woodland, usually in yellow sandy loam over laterite or in shallow sandy loam over clay. Flowers spring–summer. Regenerates from sucker or seed after fire. Pollinated probably by birds but may also be by native marsupials.

Variation There are differences in leaf lobe length and width, in conflorescence length and in flower colour. Plants from Jubuk appear the most robust, with generally longer leaves and conflorescences and pale orange flowers. An extremely robust form from Cadoux with ashy grey leaves to 20 cm long but normal flowers has recently been introduced in

112A. *G. dryandroides* subsp. *dryandroides* Close-up of flowers and foliage (P.Olde)

112B. *G. dryandroides* subsp. *dryandroides* Habit, Ballidu W.A. (N.Marriott)

G. dryandroides subsp. hirsuta

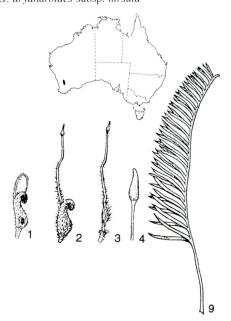

Perth, W.A., and Vic. where it is usually sold as the 'Giant-leaf form'. Recently, a population was discovered near Kellerberrin with flowers ranging from yellow through orange to pink or red. In the Lake Mears area, the conflorescences are usually larger than elsewhere.

Conservation status 2EC. Many populations are on degraded road verges. Substantial populations have recently been discovered in Durakoppin Reserve.

Cultivation Subsp. *hirsuta* can be grown in hot-dry and cool-wet climates and has been successfully cultivated in Vic. as well as in Perth and near Manmanning, W.A. Grafted plants have succeeded in Sydney and Brisbane. This species is proving quite adaptable and hardy in cultivation, tolerating frost to at least -4°C and extended dry conditions. Performance in humid summer-rainfall conditions remains questionable. It prefers full sun in a built-up bed of acidic to neutral, free-draining, gravelly or sandy loam; in its natural habitat, the yellow, sandy soil is quite deep. Once established, it should not require summer watering but light applications of slow-release, low-phosphorus fertiliser are beneficial and increase growth rate. It makes an excellent tub specimen when planted in an open, gravelly mix but is presently only occasionally available from nurseries. In Vic., grafted plants (on *G. robusta*) of several forms are becoming widely available; most popular is the 'Giant leaf form'.

Propagation *Seed* Sets little seed in cultivation. Germination time 33 days (Kullman–1 seed). *Cutting* Firm, young growth gives a fairly good strike rate at most times of the year. *Grafting* Has been grafted successfully onto a number of rootstocks, the most common being *G. robusta*, G. 'Poorinda Anticipation' and G. 'Poorinda Royal Mantle' using the top wedge method. Compatibility in *G. robusta* is uncertain. *Tissue culture* Successful propagation by tissue culture has been achieved by a member of the Grevillea Study Group and the technique offers the chance of quickly increasing stocks of this species, for although it strikes from cuttings reasonably well, it is a small, little-branched plant giving very little cutting material annually.

Horticultural features Subsp. *hirsuta* is an enchanting, small plant, even when not in flower, with its bold, fish-bone leaves and low, clumping habit. The lovely pink to red toothbrush flowers are borne on leafless stems and trail appealingly along the ground around the perimeter of the shrub where they sit awaiting the attention of ground-hopping honeyeaters. In massed flower, the wreath-like effect makes a wonderful contrast with the silvery or dull green leaves. Due to its small size and uncertain longevity, it has limited landscaping appeal, although a mass planting in a built-up rockery would be most spectacular should it ever become widely available. It blooms almost continuously in cultivation.

Grevillea dryophylla N.Wakefield (1956)
Plate 114

The Victorian Naturalist 73: 74 (Australia)

Goldfields Grevillea

The specific epithet alludes to the oak-like foliage and is derived from the Greek *dryo* (of an oak) and *phyllon* (a leaf). DRY-OH-FILL-A

Type: Kangaroo Flat (near Bendigo), Vic., Nov. 1934, A.J.Tadgell (holotype: MEL).

A spreading to erect **shrub** 0.3–1.5 m high. **Branchlets** angular to terete, silky. **Leaves** 3–6.5 cm long, 2.5–5 cm wide, ascending, petiolate, ± ovate, sometimes simple and toothed, more usually pinnatifid, with broad triangular or smoothly rounded, pungent lobes 5–10 mm wide, prominently penninerved both sides; upper surface usually sparsely pubescent, rarely glabrous; margins shortly recurved, sometimes about an intramarginal vein; lower surface sparsely to moderately pilose. **Conflorescence** 1–4 cm long, 1.5 cm wide, erect to decurved, pedunculate, terminal, simple, ± secund; peduncle stout, tomentose; rachis tomentose; bracts 1–2 mm long, ovate, silky outside, persistent at anthesis. **Flower colour**: perianth green, brown, light maroon, fawn; style dull yellow, buff, pink, red. **Flowers** acroscopic; pedicels 1.5–3 mm long, tomentose; torus 1–1.5 mm across, straight to slightly oblique; nectary U-shaped; **perianth** c. 6 mm long, 1.5–2.5 mm wide, ovoid, silky outside, glabrous inside, cohering except along dorsal suture; limb revolute, globular, enclosing style end before anthesis; **pistil** 13.5–15.5 mm long; stipe 1.1–1.6 mm long, silky; ovary silky; style glabrous except a few basal hairs, at first exserted below curve on dorsal side and looped upwards, straight to incurved after anthesis but strongly curved c. 4 mm from the hoof-like style end; pollen presenter oblique, oblong-elliptic, convex. **Fruit** 9–12.5 mm long, 5–6 mm wide, oblique on incurved stipes, oblong, tomentose, longitudinally red-striped; pericarp c. 0.5 mm thick. **Seed** 6.5–7.5 mm long, 2.5 mm wide, oblong with acute apices; outer face convex smooth; inner face flat; margin recurved with waxy border.

Distribution Vic.; confined to central and western regions from St Arnaud east to the Bendigo–Castlemaine area. ***Climate*** Hot, dry summer; cool to mild, wet winter. Rainfall 400–650 mm.

113A. *G. dryandroides* subsp. *hirsuta* Flowers and foliage (N.Marriott)

113B. *G. dryandroides* subsp. *hirsuta* Orange flowers (N.Marriott)

G. dryophylla

114A. *G. dryophylla* Flowers and foliage (N.Marriott)

114B. *G. dryophylla* Bendigo form (N.Marriott)

Ecology Grows in stony and sandy clay in dry eucalypt forest, especially ironbark forest on low hills and ridges. Flowers winter–spring. Regenerates from seed after fire. Pollinated by birds.

Variation A most variable species with each population showing differences in leaf size, shape, degree of division and colour. Flower colour also varies considerably although they are usually fawn or dull red. The following forms are sufficiently distinct to warrant horticultural recognition.

St Arnaud form A form with reasonably small, holly-like leaves that are an attractive blue-grey in better forms. Flowers normally have red styles. It occurs in several areas in the dry, stony hills around St Arnaud.

Bendigo form This form has most attractive large, entire or coarsely divided, bronze-green to dull green leaves. Unfortunately, the flowers are generally dull, usually having fawn styles. It is widespread in dry, stony soil around Bendigo.

Kingower form In this form, the leaves are usually coarsely to deeply divided. Foliage is grey-green and the flowers usually have dull red styles. It occurs in the hills around Kingower.

Major distinguishing features Leaves usually ± ovate, pinnatifid, with broad, triangular lobes 5–10 mm wide, upper surface with short, curly hairs, lamina of undersurface visible beneath indumentum; conflorescence secund, terminal, on a stout peduncle; peduncle and floral rachis tomentose; torus straight to slightly oblique; perianth zygomorphic, glabrous within; ovary silky, stipitate; fruit incurved, with reddish blotches or stripes.

Related or confusing species Group 35, especially *G. floripendula*, *G. ilicifolia* and *G. microstegia*. *G. floripendula* has conflorescences borne on thin, wiry, glabrous peduncles. *G. ilicifolia* differs in its longer pistil (19–25 mm long). *G. microstegia* has leaf lobes < 5 mm wide and villous branchlets.

Conservation status Fairly widespread and common throughout its range and not presently endangered, although gold mining operations are affecting some populations.

Cultivation *G. dryophylla* is a hardy, reasonably adaptable species in cultivation and succeeds in cool-wet, warm- to hot-dry and summer-rainfall climates where summer humidity is not excessive. It withstands drought and is hardy to frosts to at least -7°C. Best results are achieved when planted in full sun or semi-shade in well-drained, acidic gravelly to sandy loam or clay. Watering and fertilising are not normally required, nor usually is pruning unless planted in a shady site where it may become a little open. It is an attractive foliage plant for pot culture when planted in a medium to large tub. Native plant nurseries occasionally stock this species.

Propagation *Seed* Sets numerous seeds that germinate well when nicked or peeled. *Cutting* Firm, young growth normally strikes readily at most times of the year. *Grafting* Has been grafted successfully onto *G. robusta* using the whip and mummy techniques.

Horticultural features *G. dryophylla* is a most handsome foliage plant with grey to olive-green or bronze-green, coarsely lobed holly-like leaves. It has short, toothbrush conflorescences that are sometimes profuse, but the flowers, generally dull red or fawn, are often unnoticed. Honey-eating birds are attracted to their good nectar supply. In the wild, it survives for many years in dry, rocky sites under large trees. In the garden, it is extremely valuable for planting in similar areas, flourishing where many other plants cannot survive under conditions of strong root competition. Plants in such situations need regular watering until established. It can be planted as a low screen and is useful in a massed planting or as a specimen plant.

Hybrids There are several records of this species hybridising with *G. alpina* in the wild. One of these hybrids is a lovely, free-flowering plant that has been recently introduced into cultivation but, as yet, remains unnamed.

Grevillea elbertii H.Sleumer (1955)

Blumea 8: 2 (Holland)

Vernacular: 'lampia' (Luwu language), Malili tribe.

Named after J.Elbert who was associated with the geographical expedition in 1909–10 during which the species was discovered. ELL-BERT-EE-EYE

Type: Mt Sangia, Kabaena Island, Celebes, 22 Oct. 1909, per J.Elbert 3475 (holotype: L). The collection was apparently gathered by Grundler.

Tree 10–17 m tall. **Branchlets** terete, striate, silky brown. **Leaves** 5–12(–18) cm long, 2.5–7 cm wide, petiolate, oblong to broadly elliptic, simple, leathery, the apex rounded, retuse or emarginate; upper surface silky becoming glabrous and shiny, penninerved; margins entire, shortly recurved, undulate; lower surface pubescent, golden-brown especially when young, glabrescent, midvein and penninervation prominent. **Conflorescence** paniculate, pedunculate, terminal, sometimes simple and axillary in upper axil; unit conflorescence 6–10 cm long, cylindrical, dense; peduncle and rachis silky; bracts 1–1.5 mm long, ovate, silky to villous, falling before anthesis. **Flower colour**: perianth pale green to white; style ?white. **Flowers** basiscopic; pedicels 2–3 mm long, silky, stout; torus c. 1.5 mm across, ± straight; nectary inconspicuous, toothed, U-shaped; **perianth** 10–12 mm long, 2 mm wide, narrowly cylindrical, silky outside with scattered rusty hairs, papillose inside, pubescent in lower half, otherwise glabrous; limb revolute, globular; tepals ribbed; **pistil** 18–19 mm long, glabrous; stipe 2–2.5 mm long; ovary ovoid; style erect to curved; style end flattened; pollen presenter lateral, convex, obovate. **Fruit** 2.5–3 cm long, 2 cm wide, erect to oblique on pedicel, compressed-ellipsoidal; pericarp c. 2.5 mm thick. **Seed** not seen.

Distribution Sulawesi, Indonesia; occurs on Kabaena and Malili Islands and Paka Lampla. *Climate* Tropical; summer hot and wet; winter mild and dry. Rainfall 2000–3000 mm.

Ecology Occurs from sea level to c. 1000 metres, in dry monsoonal rainforest on crystalline slate in hilly country. Flowers spring–winter. Probably killed by fire. Pollinator unknown.

Variation Apparently a uniform species.

Major distinguishing features Arborescent habit; leaves simple, entire; lamina broad, shiny on upper surface; conflorescence paniculate; torus straight;

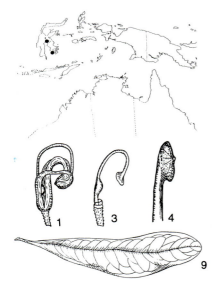

G. elbertii

perianth zygomorphic, undilated, hairy inside and out; pistil glabrous; pericarp thick-walled.

Related or confusing species Group 8. Closely related to *G. helmsiae* which differs in its conflorescence not exceeding the leaves and its fruit horizontal to the stipe with a follicle wall < 2 mm thick.

Conservation status Uncertain.

Cultivation This species is unknown in cultivation in Australia but should enjoy conditions common to most monsoonal species: deep, well-drained soil rich in humus, regular watering, and a position protected from strong winds. It is unlikely to tolerate frost or drought. Since it is found near the sea in some areas it is likely to be hardy and responsive to good garden conditions and may be able to withstand salt-laden winds.

Propagation *Seed* Untested. Soaking for 24 hours or nicking the testa should assist germination.

G. elbertii illustration (C.Woolcock)

Cutting Cuttings of firm, young growth should strike successfully, particularly during the warmer months. *Grafting* Untested.

Horticultural features *G. elbertii* is a small tree with large, glossy, dark green leaves that have a golden brown hairy undersurface. The leaves contrast attractively with the hairy, bronze new growth. Large panicles of greenish or greenish white flowers are carried above the foliage and are conspicuous in full flower. This poorly known species would make an interesting addition to the garden in tropical and subtropical areas and should be re-collected with the aim of introducing it to cultivation.

Grevillea elongata P.M.Olde & N.R.Marriott (1994)   Plate 115

The Grevillea Book 1: 175 (Australia)

The specific epithet is derived from the Latin elongatus (elongated), in reference to the long floral rachis. EE-LONG-GAR-TA

Type: Tutunup Rd, Ruabon, W.A., 10 Oct. 1991, P.M.Olde 91/271 (holotype: NSW).

Shrub 1.5–2 m high, 2–2.5 m wide. **Branchlets** red, erect, glabrous or sparsely silky, terete with longitudinal ribbing. **Leaves** 2.5–5 cm long, glabrous, divaricately tripartite to pinnatipartite, sometimes with some or all lobes again bi- or tri-partite, sessile or shortly petiolate, the first lobe 12–30 mm from leaf base; leaf lobes often of unequal length, 5–30 mm long, c. 0.8 mm wide, subulate, trigonous, pungent; upper surface smooth, venation obscure; lower surface, bisulcate with midvein prominent; texture firmly papery to leathery. **Conflorescence** terminal or axillary, sessile or shortly pedunculate, simple or few-branched; unit conflorescence 2–5.5 cm long, open, cylindrical; development acropetal; peduncle tomentose; rachis c. 1 mm wide, sometimes sparsely pubescent at the base, otherwise glabrous; bracts 2.8–3.4 mm long, 3–4 mm wide, imbricate, ovate, glabrous except ciliate margin, persistent almost to anthesis; pedicels 2.2–3.2 mm long, glabrous; torus 0.3 mm wide, oblique at c. 10–15º. **Flower colour**: perianth and style white throughout, the bracts cream. **Flowers** glabrous; **perianth** 3.5 mm long, 0.5 mm wide, actinomorphic, oblong-obovate constricted below limb, erect; all tepals separating and rolling back at anthesis; limb 1 mm long, 1.2 mm wide; **pistil** 4.5 mm long; stipe 1.2 mm long, flexuose, filamentous; ovary 1 mm long, globose; style constricted just above ovary, dilating abruptly to 0.5 mm thick, gradually tapering to 0.3 mm wide at base of pollen presenter;

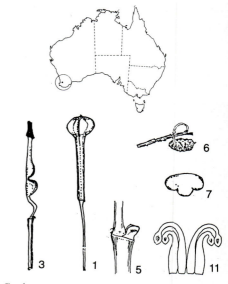

G. elongata

pollen presenter 0.7 mm high, 0.4 mm wide at base, erect with base slightly oblique, faintly rimmed, truncate-conical to subcylindrical. **Fruit** 8 mm long, 3.5 mm wide, 4 mm deep, oblique, rugulose; pericarp 0.2 mm thick at centre face of the suture. **Seed** not seen.

Distribution W.A., restricted to the Ruabon–Busselton area. *Climate* Winters cold and wet; summers cool to warm, dry. Rainfall 800–1000 mm.

Ecology Occurs in poorly drained grey sand in scrubby heath, often beside creeks. Flowers Oct.–Nov.(–Dec.). Regenerates from seed after fire. Many small beetles have been seen on the flowers. Pollinated by insect.

Variation A morphologically uniform species.

Major distinguishing features Branchlets silky to glabrous; leaves divaricately bipinnatisect, glabrous and smooth above; lobes subulate, pungent, often of unequal length; conflorescence elongate; floral bracts broadly ovate, papery; perianth actinomorphic, glabrous; pistil glabrous; style constricted above the ovary then again dilated; pollen presenter

115A. *G. elongata* Flowering habit (P.Olde)

115B. *G. elongata* Flowering branch (P.Olde)

115C. *G. elongata* Plant in natural habitat, Ruabon, W.A. (P.Olde)

truncate conic to subcylindrical; fruits rugulose, thin-walled.

Related or confusing species Group 1, especially *G. paniculata* which differs in its leaves channelled on the upper surface, smaller floral bracts (c. 1 mm long) and globose conflorescence with shorter floral rachis (c. 5 mm long), longer pedicels and deeply wrinkled fruits.

Conservation status Suggested 2E. This species is rapidly being depleted by development and is in urgent need of survey.

Cultivation *G. elongata* was introduced to cultivation in 1986, and although plants grew vigorously for several seasons, they were eventually lost. Judging from its natural habitat, it will tolerate a relatively high rainfall climate and regular frosts, possibly to as low as -5°C. It is unlikely to be drought-tolerant and consistent subsoil moisture will be essential. Summer humidity is likely to be resented and flowering may be affected in subtropical climates. Poor drainage may be tolerated but is not recommended. The species is too robust and prickly for pot culture but will respond to normal well-drained potting mixes with low-phosphorus, slow-release fertiliser. Some pruning may be necessary in cultivation to maintain shape.

Propagation *Seed* Untested. *Cutting* Grows readily from cuttings of firm, new growth. *Grafting* Untested.

Horticultural features *G. elongata* has some merit in horticulture, being extremely floriferous. The relatively (for this group) long conflorescences are very attractive, especially when young as the expanding buds shed their bracts. The reddish branchlets provide an interesting contrast. It has uses as a feature or contrast plant, especially with its white flowers which provide an eyecatching focus when in full bloom. Because the foliage is slightly prickly, it should not be grown near paths or walkways although it could be useful in discouraging foot traffic in some circumstances. The species is not available from nurseries.

General comments This species was included by McGillivray (1993) in *G. paniculata*, as the Busselton form.

Grevillea endlicheriana C.F.Meisner (1845)   Plate 116

in J.G.C.Lehmann (ed.), *Plantae Preissianae* 1: 546 (Germany)

Spindly Grevillea

Named after Stephan Endlicher (1804–1849), Austrian Professor of Botany at Vienna, who described many Australian plant species. END-LICK-ER-EE-ARE-NA

Type: Darling Range, W.A., 25 July 1839, L.Preiss 698 (lectotype: LD) (McGillivray 1993).

Synonym: *G. filifolia* C.F.Meisner (1845)

A spindly to dense, erect **shrub** 0.5–2.5 m tall with ± leafless floral branches emergent above foliage. **Branchlets** elongate, angular to terete, ridged, silky. **Leaves** 3–13 cm long, 0.5–3(–5) mm wide, ascending, shortly petiolate, simple, linear to subterete, pliable, acute, usually uncinate; upper surface silky grey, midvein scarcely evident; margins flat or incurved, entire; lower surface silky, midvein obscure, or faintly evident in broad leaves. **Conflorescence** on sparsely leaved branches above foliage, erect, terminal or axillary, branched or simple, pedunculate; unit conflorescence c. 1 cm long, 1 cm wide, globose to subcylindrical; peduncle silky; rachis villous to almost glabrous; bracts imbricate in bud, 1.5–3.6 mm long, ovate, villous to glabrous with ciliate margins outside, falling before anthesis. **Flower colour**: perianth creamy pink to pale pink; pedicels pinkish red turning pale pink. **Flowers** adaxially acroscopic; pedicels 2.2–4.5 mm long, glabrous; torus 0.8–1 mm across, square, oblique; nectary scarcely evident above torus, U-shaped; **perianth** 3–4 mm long, 1–1.5 mm wide, oblong-cylindrical, glabrous outside, villous inside, cohering except along dorsal suture, separating below limb to almost half way before anthesis, afterwards independently rolled down to almost half way; limb revolute, spheroidal to subcubical, ribbed; **pistil** 7.5–12 mm long, glabrous; stipe 1–1.5 mm long; ovary globose, conspicuous; style exserted just below curve on dorsal side and looped upwards before anthesis, afterwards strongly incurved, at length straight, filiform; style end enclosed before anthesis, flattened, attached at its base to style; pollen presenter very oblique to lateral, obovate, ± flat. **Fruit** c. 8 mm long, 6 mm wide, horizontal to pedicel, round to ellipsoidal, glabrous, faintly wrinkled; pericarp c. 1 mm thick. **Seed** 6.5–8 mm long, 2.2–3 mm wide, obovoid; outer face convex, rugulose, horizontally crimped near apex; inner face ± obscured by margins, conspicuously rugose, with a papery ruff; margin revolute bordered by a waxy rim.

Distribution W.A., on the Darling Scarp east of Perth and near Mogumber, Wannamal and Wongan Hills. *Climate* Hot, dry summer and cool to mild, wet winter. Rainfall 500–900 mm.

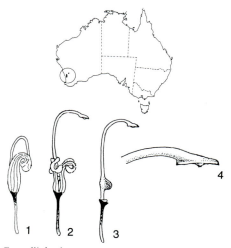

G. endlicheriana

Ecology In the Darling Ra. occurs on granitic or lateritic slopes in well-drained loam in open to dense mixed scrub. At Wongan Hills it grows in grey, sandy soil. At Mogumber, it grows in yellow to grey sand or gravelly loam, usually in dense thickets in eucalypt open woodland. Flowers winter to early summer. Following low-intensity fire, it sometimes sprouts vigorously from epicormic buds on the lower branches and trunk. Pollinated by insects.

Variation A reasonably uniform species but with at least one major variant.

Fine leaf form This form is found in the northern part of the species' range with collections from Wattengutten Hill near Wongan Hills and Wannamal. It has subterete basal leaves or the margins very strongly involute (cf. linear to narrowly elliptic leaves in the typical variety). Flowering is also more sparse and scattered and the flowers lack the suffused pink tones of the typical variety which extends south from Mogumber to Perth. Some collections from the northern part of the range have a shorter pistil (7.5–9 mm long).

Major distinguishing features Spindly to bushy shrub with emergent floral branches; leaves simple, smooth and silky on both surfaces, usually uncinate,

116A. *G. endlicheriana* Flowers and fruit (P.Olde)

116B. *G. endlicheriana* Flowering habit, E of Mogumber, W.A. (N.Marriott)

116C. *G. endlicheriana* Fine-leaf form, Wattengutten Hill, W.A. (P.Olde)

116D. *G. endlicheriana* Massed flowering habit (P.Vaughan)

midvein obscure; perianth zygomorphic, undilated, glabrous outside, bearded within; pistil glabrous; pollen presenter flat to convex, lateral; fruit subglobose.

Related or confusing species Group 17, especially *G. acacioides* which differs in its longitudinally ridged leaves.

Conservation status Not presently endangered.

Cultivation *G. endlicheriana* has proved a hardy species, growing reliably in a wide range of climates in Australia, including cool-wet, warm-dry and summer-rainfall climates; it is also successfully cultivated in the U.S.A. This species is hardy to extended dry conditions, provided there is sufficient subsoil moisture, and withstands frosts to at least -6°C. Given a sunny, well-drained position in acidic to neutral gravelly loam or sand, it grows rapidly into a compact, healthy shrub, requiring little maintenance apart from occasional summer watering. Straggly plants respond to light applications of fertiliser and light trimming and its appearance is often improved by the annual removal of the large, almost leafless, flowering branches. This should be done immediately after fruiting, otherwise the following year's flowering may be reduced. It is widely available at native plant and some general nurseries.

Propagation *Seed* Sets numerous seed that germinate well given the standard peeling treatment. Young plants have uncharacteristically broad leaves for the first year or two. Germination times average

30 days (Kullman). *Cutting* Firm, young growth normally strikes readily at any time of the year. *Grafting* Not tested.

Horticultural features *G. endlicheriana* is an erect shrub of variable habit with ascending branchlets adorned with silvery leaves and masses of cream to pale pink flowers prominently displayed on long, emergent almost leafless branches. The flowering habit is delicate and imparts a graceful character to the plant. Plants in the wild sometimes look inferior to those in cultivation, being often smaller and of spindly habit. Around Mogumber, however, dense thickets of bushy plants are a most impressive sight. This long-lived species has great potential as a bold feature plant in landscapes where it is most effective as a mass planting or, if used to line paths, to create an avenue effect. Flowers are sweetly, sometimes strongly, scented and attract native bees and other flying insects. Small, reddish black fruits dot the branches after flowering and are most attractive in their own right. Due to its size it is not suited to normal pot culture but would make a very bold accent plant in large tubs or open-ended pots.

General comments Most herbarium collections of *G. endlicheriana* do not have the lower leaves, which may be either subterete as in our Fine leaf form or narrowly elliptic to linear as in the typical form. The flowering branches often have terete leaves whether or not the basal leaves are narrowly elliptic. Whether *G. filifolia* C.F.Meisner is the same taxon as our Fine leaf form is uncertain as it may be simply a collection with terete floral leaves. Further research is needed to resolve the taxonomy of *G. endlicheriana*. We cannot confirm that the northern or smaller-flowered form cited by McGillivray exists other than in a few herbarium collections.

Grevillea erectiloba F.Mueller (1876)
Plate 117

Fragmenta Phytographiae Australiae 10: 44 (Australia)

The epithet refers to the habit of the leaves and is derived from the Latin *erectus* (erect) and *lobus* (a lobe). ERR-EKT-EE-LOBE-A

Type: between Ularing and Mount Jackson, W.A., 17–20 Oct. 1875, J.Young (holotype: MEL). Young was the collector on the Elder Exploring Expedition across Australia led by Ernest Giles.

Dense, rounded, blue-green **shrub** 1–2 m high, 2–4 m wide. **Branchlets** terete, silky to glabrous. **Leaves** 5–12 cm long, ascending to erect, petiolate, stiff, pinnatipartite; lobes 3.5–8.2 cm long, 0.6–0.9 mm wide, erect to strongly ascending, closely aligned, ± terete, diplopleural, with faint, longitudinal ribs, pungent; upper surface glabrous to sparsely silky, rarely with erect glandular hairs, venation obscure; lower surface obscured, the midvein prominent, glabrous or with scattered hairs. **Conflorescence** erect to decurved, shortly pedunculate, terminal or axillary, simple or few-branched, not exceeding foliage; unit conflorescence usually 8–10-flowered, umbel-like to subglobose, loose, outermost flowers opening first; peduncle and rachis silky; bracts 1–1.5 mm long, ovate, silky outside, falling long before anthesis. **Flower colour**: perianth green turning through orange to red; style purplish red with a pale pink style end. **Flowers** basiscopic; pedicels 6–11 mm long, silky; torus c. 3.5 mm across, cup-shaped, broadly elliptic to obovate, oblique; nectary V-shaped, cup-like, lining torus; **perianth** 10–13 mm long, 4–5 mm wide, oblong-ovoid, dilated at base, glabrous or sometimes sparsely glandular-pubescent outside, loosely villous and prominently to faintly ribbed inside; tepals cohering except along dorsal suture and between dorsal and ventral tepals at the curve; limb revolute, spheroidal, the apex and margins of the segments impressed; **pistil** 25.5–35 mm long; stipe 3.5–4.7 mm long, aligned with pedicel, openly villous, more densely on dorsal side, adnate to torus at its base; ovary very large, globular, densely villous; style at first exserted at curve and looped upwards, gently incurved after anthesis, loosely villous; style end partially exposed before anthesis,

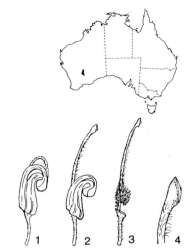

G. erectiloba

conspicuously dilated, glabrous at apex; pollen presenter very oblique, orbicular, flat to slightly convex. **Fruit** 8–10 mm long, c. 7 mm wide, erect to slightly oblique on pedicel, ellipsoidal-round with apical attenuation, villous especially at ends; pericarp c. 1 mm thick. **Seed** 5–5.5 mm long, 2.5–3 mm wide, ellipsoidal; outer face convex, smooth or faintly wrinkled, mottled; inner face similar, flat, with a raised, narrow marginal wing 0.8 mm wide, slightly extended at each end into a short, oblique extension c. 1 mm long.

Distribution W.A., confined to semi-arid country north of Southern Cross. *Climate* Hot, dry summer; cool to warm, moist winter. Rainfall 225–275 mm.

Ecology Grows in dry, stony or gravelly loam in tall shrubland on lateritic rises. Flowers late spring–summer. Regenerates from seed after fire. Pollinated by nectar-feeding birds.

117A. *G. erectiloba* Flowers and foliage (M.Hodge)

117B. *G. erectiloba* Flowers and foliage (N.Marriott)

117C. *G. erectiloba* Plant growing in gravel pit, Johnson Rock, W.A. (N.Marriott)

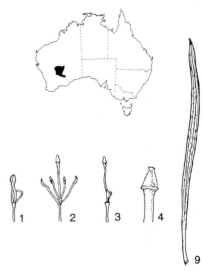

G. eremophila

Variation A uniform species.

Major distinguishing features Leaves bluish green, pinnatipartite with 5–12 closely aligned, stiff, erect subterete lobes; conflorescence umbel-like to subglobose, borne within foliage; apical flowers opening first; perianth zygomorphic, oblong to ovoid, glabrous or minutely glandular outside, hairy inside, separating between dorsal and ventral sutures at curve; ovary villous, stipitate; pollen presenter very oblique.

Related or confusing species Group 26, especially *G. georgeana* which can be distinguished by its divaricately twice-divided leaves and conspicuous cylindrical conflorescence.

Conservation status 3RC. Restricted to stony rises but present in reasonable numbers.

Cultivation *G. erectiloba* was introduced to cultivation by members of the Grevillea Study Group in 1988. Unfortunately, as with many grevilleas restricted to laterite, it has proved rather difficult to establish on its own roots. Young plants have done best when grown in well-drained acidic gravelly sand or loam, in full sun. Even then, they have mostly succumbed after thunder storms or summer watering. A hot dry summer appears essential. Fortunately the species grafts well onto *G. robusta* and several other species, the resultant plants growing very strongly and flowering well when placed in a warm, sunny site. Grafted plants have been established from inland Vic. to south-eastern Qld. They form extremely dense shrubs and have been exposed to frost to at least -3°C without ill effect. Such plants are becoming available at specialist native nurseries.

Propagation *Seed* Plants set good quantities of seed. Nicking or peeling the testa should aid germination. *Cutting* Cuttings of firm, new growth taken in the warmer months have proved difficult. Strict cleanliness should be observed as cuttings suffer fungal attacks in humid, hothouse conditions. Rooted cuttings must be potted up in a very free-draining soil. *Grafting* Has been grafted successfully onto *G.* 'Poorinda Royal Mantle', *G.* 'Poorinda Anticipation', *G. olivacea* and *G. robusta*. The combination with *G. robusta* is proving the most successful after 5 years of trials.

Horticultural features *G. erectiloba* is an intriguing plant with great horticultural potential. It has erect, finely divided, blue-green foliage with contrasting bright green new growth and grows naturally into an attractive, rounded shrub. In late spring, large numbers of unusually coloured flowers appear that, although somewhat hidden by the foliage, are nonetheless quite prominent due to their vibrant colour. The range of flower colour from almost iridescent green in the bud stage through orange to deep red at flowering creates a magnificent multicoloured conflorescence. It makes an ideal screen or large feature plant, strongly attractive to birds, and, with such potential to please, should become more popular now that hardy, grafted plants are becoming available.

Grevillea eremophila (L.Diels) P.M.Olde & N.R.Marriott (1994)   Plate 118

The Grevillea Book 1: 176 (Australia)

Based on *G. integrifolia* var. *eremophila* L.Diels, *Botanische Jahrbücher für Systematik* 35: 156 (1904) (Germany).

The subspecific epithet is derived from the Greek *eremia* (a desert) and *philo-* (fond of), a reference to the inland areas where it occurs. ERR-EM-OFF-ILL-A

Type: Marmion, 25 km S of Menzies, W.A., 28 Oct 1901, L.Diels 5159, (lectotype: B) (McGillivray 1993).

Synonym: *G. integrifolia* var. *grandiflora* S.Moore (1920).

Erect **shrub** 1.5–3 m high. **Branchlets** silky or sparsely so, sometimes glabrous. **Leaves** (juvenile) broadly elliptic (up to 22 mm wide), the upper surface glabrous to sparsely silky or villous, the midvein obscure, the lower surface similar with prominent midvein and reticulum, margins flat to angularly infracted; adult leaves unifacial, 7–16 cm long, 1–5 mm wide, ascending to erect, sessile, simple, linear to narrowly obovate, straight, leathery, acute to obtuse with short, blunt point; margins entire, flat to undulate; venation longitudinally parallel-ribbed, the ribs silky and obscure or glabrous and prominent with silky grooves, ribs on one surface usually 3–5 and unequal in width, on the other up to 10 and equal in width. **Conflorescence** erect, pedunculate, terminal or axillary usually in upper axil, simple or branched; unit conflorescence 7–15 cm long, 1–1.5 cm wide, cylindrical; flowers opening ± synchronously, sometimes from apex first; peduncle glabrous; rachis yellow, glabrous; bracts 1.5 mm long, ovate, acuminate, silky-tomentose, usually with white hairs only, usually falling early, sometimes persistent to anthesis. **Flower colour**: perianth and style creamy white. **Flowers** regular, glabrous, adaxially oriented; pedicels 2.2–4.8 mm long; torus c. 0.4 mm across, straight; nectary absent, sometimes present as a short, horn-like structure; **perianth** (below limb) 3–3.5 mm long, 0.5 mm wide, oblong; tepals channelled on inside, with obscure to prominently exserted V-shaped labia below anthers, before anthesis separating to base and bowed out below cohering limb, afterwards all free; limb 2.2 mm long, ellipsoidal, the segments slightly flanged at margins, an obscure midrib at base of each; distance from base of anthers to narrowed section of tepal 0.5–1 mm; **pistil** 7.5–10 mm long, glabrous; stipe 0.5–1.2 mm long; ovary 1 mm long, lateral, obovoid with 2 apical, ventral lobes; style 6–8 mm long, 0.25 mm thick, minutely granular, deeply grooved, before anthesis exserted near base and folded outwards, afterwards noticeably kinked about middle, becoming straight to undulate; style end 0.5–0.7 mm across, straight; pollen presenter 1.5–1.8 mm long, cylindrical-urceolate; stigma cup-shaped. **Fruit** 12–16 mm long, 1.5–2.5 mm wide, oblique, narrowly obovoid-cylindrical with slightly recurved apex, smooth, often persistent; pericarp 0.2–0.5 mm thick. **Seed** 4–8.5 mm long, 1.2–2 mm wide, obovoid or narrowly so, rugose to smooth; apex and base with a membranous band 0.5–1 mm wide; outer face convex; inner face flat.

Distribution W.A., in an area bounded by Comet Vale, Beacon, Narembeen and Lake Cronin. *Climate* Hot, dry summer; cold, wet winter. Rainfall 200–300 mm.

Ecology Grows in deep, yellow sand in sand heath where it is a dominant species. Flowers late spring–summer. Pollinated by insect. Regenerates from seed.

Major distinguishing features Leaves and branchlets silky or sparsely so; leaves unifacial, linear to narrowly obovate, prominently ribbed; conflorescence cylindrical; bracts ovate, c. 1.5 mm long; rachis yellow, glabrous; flowers regular, glabrous; ovary bilobed; fruit narrowly obovoid-cylindrical.

Related or confusing species Group 2, especially *G. biformis* and *G. ceratocarpa*, the former differing in its smaller fruits and usually hairy rachises, the latter in its spreading foliar and branchlet indumentum and longer floral bracts. There is a very close relationship with both taxa.

Variation There is some variation in leaf width among populations. One collection from near Narembeen has persistent floral bracts.

Conservation status Not presently endangered.

Cultivation *G. eremophila* has not been grown in cultivation to our knowledge. A well-drained, deep, acidic to neutral sand or sandy loam in a protected position in full sun should produce good results.

118A. *G. eremophila* Flowers and foliage (P.Olde)

118B. *G. eremophila* Habit and habitat, near Mt Churchman, W.A. (P.Olde)

Frosts to at least -6°C are experienced in the wild and it should prove drought hardy once established. No watering or fertiliser recommended after establishment. Not available from nurseries.

Propagation *Seed* Sets large quantities of fruit. Fresh seed pretreated by nicking should germinate well. *Cutting* Firm, young growth from cultivated plants should strike reasonably easily from early spring. *Grafting* Untested.

Horticultural features *G. eremophila* has perhaps the longest inflorescences in the *G. integrifolia* group. These are displayed prominently above the green foliage and are very bright. There is excellent potential for this species as a feature or screen plant in both home or public landscapes. Further testing is necessary before recommendations can be made.

General comments McGillivray (1993) treated *G. eremophila* as a form of *G. integrifolia* subsp. *ceratocarpa*. We view this classification as being too broad and consider that some formalised ranking is warranted. It shares most features with *G. ceratocarpa* but there is a very close morphological approach to some populations of *G. biformis*, especially those with glabrous, yellow rachises. The possibility that *G. eremophila* is conspecific with *G. ceratocarpa* is high but morphological differences and overlapping distribution warrant closer investigation. We have refrained from making the combination (at subspecific rank) until more field work and statistical analysis is completed and the phylogenetic relationship of all taxa in the *G. integrifolia* group is clarified.

Grevillea erinacea C.F.Meisner (1855)
Plate 119

Hooker's Journal of Botany & Kew Gardens Miscellany 7: 74 (England)

Hedgehog Grevillea, Standback

The specific epithet alludes to the prickly habit of the plant and is derived from the Latin *erinaceus* (a hedgehog). ERR-INN-ACE-EE-A

Type: between the Murchison R. and Dandaragan, W.A., 1851–52, J.Drummond coll. 6 no. 186 (holotype: NY).

A prickly, spreading **shrub** 0.5–1.5 m tall with arching branches. **Branchlets** terete, pubescent-tomentose. **Leaves** 1–3.5 cm long, ascending, sessile or shortly petiolate, trifid or sometimes pinnatipartite and divaricately divided, sometimes the lobes secondarily divided; lobes 0.5–2 cm long, 0.5–1 mm wide, subterete to subulate, pungent; upper surface glabrous or sparsely silky, venation obscure; lower surface bisulcate, silky in grooves, midvein prominent. **Conflorescence** erect, shortly pedunculate, simple or branched, terminal or axillary; unit conflorescence 1–1.5 cm long, 1 cm wide, open, subglobose; peduncle silky; rachis tomentose; bracts 0.4–0.8 mm long, ovate, tomentose outside with ciliate margins, falling before anthesis. **Flower colour**: perianth greyish green in bud, turning cream to creamy white; style white. **Flowers** regular; pedicels (4.5–)7–10 mm long, sparsely hairy; torus c. 0.5 mm across, straight; nectary semicircular, smooth; **perianth** 3–4 mm long, 0.8 mm wide, oblong-obovoid to ellipsoidal below curve, the tepals separating and bowing out below limb before anthesis, silky outside, glabrous inside; tepals splitting to base and rolling back independently after anthesis; limb erect, globular, densely silky; **pistil** 2.5–4.5 mm long, glabrous; stipe 0.7–2.2 mm long, filamentous, flexuose; ovary globose, conspicuous; style constricted just above ovary, dilating to c. 0.5 mm wide, narrowing just before style end; pollen presenter erect, conical. **Fruit** 8–10 mm long, 6 mm wide, oblique to perpendicular on curved pedicel, glabrous, oblong-ellipsoidal, faintly wrinkled; pericarp c. 0.5 mm thick. **Seed** not seen.

Distribution W.A., from Ellendale (SE of Geraldton) south to Arrowsmith and Cervantes. *Climate* Hot, dry summer; cool to warm, wet winter. Rainfall 350–450 mm.

Ecology Occurs in grey to yellow sand and lateritic gravel in open or dense, low heathland or tall shrubland. Often found as a coloniser of disturbed areas such as road verges. Flowers winter–spring. Regenerates from seed after fire. Pollinated by insects.

G. erinacea

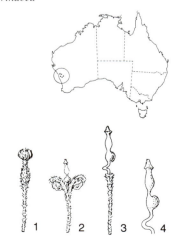

Variation A morphologically uniform species.

Major distinguishing features Branchlets pubescent; leaves trifid to divaricately twice-divided with linear-subulate, pungent lobes; conflorescence globose; perianth regular, silky outside; pistil glabrous; style constricted above ovary, then dilated; style end conical.

Related or confusing species Group 1. All other species in this group have glabrous flowers.

Conservation status Not presently endangered.

Cultivation *G. erinacea* is an adaptable, hardy species, tolerant of cool-wet and warm-dry climates, and has been successfully cultivated on its own roots at Burrendong Arboretum, N.S.W., and in western Vic. It seems to dislike summer humidity since potted plants held for over 3 years at Bulli, N.S.W., at length succumbed. Grafted plants, however, have flourished even in subtropical climates. It can endure frosts to at least -4°C without ill effect and is drought-hardy. A full-sun situation in extremely well-drained, slightly acidic to alkaline sand or gravelly loam provides the optimal conditions for cultivating this species. Once established it does not require additional maintenance apart from an occasional trimming to shape. It is seldom available even at specialist nurseries.

Propagation *Seed* Sets copious seed that germinate readily. Nicking the testa improves the germination rate. *Cutting* Varying results are obtained from cuttings, some forms striking more readily than others. Firm, young growth taken in the warmer months gives best results. *Grafting* Has been grafted successfully onto *G. robusta* using the mummy, whip and top wedge methods.

Horticultural features *G. erinacea* is probably best regarded as a collector's plant although it does have some attractive features. It has a rather wild-looking natural habit with irregularly spreading arching branches. For a few weeks in early spring most of these grey-green branches are covered with creamy white, scented flowers that almost conceal the foliage. It has potential as a barrier plant or bizarre feature plant but, because of its unpleasantly prickly foliage, should not be planted near paths.

Grevillea eriobotrya F.Mueller (1876)
Plate 120

Fragmenta Phytographiae Australiae 10: 44 (Australia)

Woolly Cluster Grevillea

The specific epithet is derived from the Greek *erion* (wool) and *botrys* (an inflorescence), a reference to the woolly flowers. ERR-EE-OH-BOT-REE-A

Type: near Mount Churchman, W.A., spring 1875, J.Young (holotype: MEL). Collected on Ernest Giles' third expedition across Australia.

Synonym: *G. victoriae* A.Morrison (1912)

A dense **shrub** c. 3 m high, 3–4 m wide. **Branchlets** erect, terete, silky. **Leaves** 8–18 cm long, 2 mm wide, ascending, shortly petiolate, leathery, usually simple and entire, linear, to bi- or tripartite, with pungent linear lobes; upper surface longitudinally ridged without distinct midvein, glabrous or with appressed hairs in grooves; margins angularly revolute to midvein below; lower surface bisulcate, densely woolly in grooves, midvein prominent. **Conflorescence** erect, shortly pedunculate, terminal, rarely axillary, simple or branched; unit conflorescence 6–8 cm long, 2 cm wide, cylindrical, dense, opening irregularly often along one side first; peduncle silky; rachis villous, stouter than peduncle; bracts 5–6 mm long, conspicuous and overlapping in bud, ovate, villous outside, persistent to anthesis. **Flower colour**: perianth and pistil white to creamy white. **Flowers** directed at right angles to rachis; pedicels 2.5–4.5 mm long, villous; torus c. 1 mm across, oblique; nectary U-shaped, smooth; **perianth** 3–4 mm long, 2 mm wide, undilated, oblong but shape concealed by woolly

119A. *G. erinacea* Flowering branch (P.Olde)

119B. *G. erinacea* Plant in natural habitat, NW of Mingenew, W.A. (P.Olde)

outer indumentum, strongly curved from base; tepals separating below limb before anthesis, the dorsal tepals reflexing laterally to reveal glabrous inner surface; limb revolute, globular, densely woolly, enclosing style end before anthesis, not relaxed at anthesis; **pistil** 10.5–12.5 mm long; stipe 1.1–1.3 mm long, scarcely visible; ovary villous; style villous except a glabrous ventral patch, exserted at curve on dorsal side and looped upwards before anthesis, afterwards strongly incurved in apical half; style end evenly dilated, flanged; pollen presenter oblique elliptic, conical. **Fruit** 20–23 mm long, 18–22 mm wide, horizontal to pedicel, round to ellipsoidal, minutely apiculate, densely pubescent; pericarp c.1.5 mm thick. **Seed** 11–15 mm long, 9–11 mm wide, obovoid-ellipsoidal to hemispherical, with smooth, mottled surface; outer face convex; inner face flat; membranously winged all round.

Distribution W.A., confined to an area on the eastern fringe of the northern wheatbelt around Koorda and Mukinbudin and north to Beacon. *Climate* Hot, dry summer; cool to warm, moist winter. Rainfall 225–275 mm.

Ecology Grows on the crests of low hills among mixed, tall shrubs, in yellow sand over sandy loam

G. eriobotrya

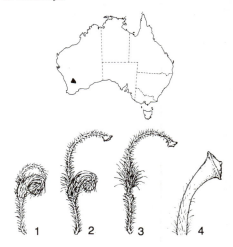

and gravelly clay. Flowers spring and summer. Regenerates from seed after fire. Probably pollinated by birds; singing honey-eaters have been seen feeding at the flowers. Many insects including moths and scarab beetles may also be involved as they are the major pollinators of the closely related *G. pterosperma* and have been seen on *G. eriobotrya* in the wild.

Variation A uniform species.

Major distinguishing features Leaves simple or divided, narrowly linear or with linear lobes, ribbed on upper surface, double-grooved below; conflorescence cylindrical, dense, woolly; torus straight; perianth zygomorphic, white, woolly outside, glabrous inside; pistil hairy all over; ovary villous, shortly stipitate; fruit globose; seed winged all round.

Related or confusing species Group 30. *G. pterosperma* and *G. albiflora* are closely related, the former distinguished by its glabrous pistil 12.5–20 mm long and its perianth never villous outside, the latter also by its glabrous pistil and the lack of ridges on the upper surface of the leaves.

Conservation status 3E. Confined to a few isolated populations. Much of its habitat has been cleared for agriculture.

Cultivation *G. eriobotrya* is scarcely known in cultivation, although a plant was grown to flowering near Mandurah south of Perth (N.Moyle, pers. comm.). It has proved extremely difficult to propagate, admittedly from limited material. Seed collected in late spring and sown fresh should be tested. From its habitat, it can be presumed that deep, well-drained neutral sand or sandy loam in full sun would be minimum requirements for successful cultivation. It is likely to be drought-resistant and frost-tolerant and may suit only dry, inland climates unless grafted onto a suitable rootstock.

Propagation Seed Apparently germinates well. Average germination time 25 days (Kullman). *Cutting* Very difficult to strike using normal techniques, as cuttings tend to rot quickly. Better results may be obtained if cuttings are placed in a 'dry' hothouse or coldframe. *Grafting* It has been grafted successfully onto G. 'Poorinda Royal Mantle', a combination that survived in Sydney for many years but did not grow. Attempts to graft onto *G. robusta* have failed..

Horticultural features *G. eriobotrya* is a spectacular species. In the wild, it grows to about 3 m tall and forms a compact, rounded shrub dominant among the surrounding shrubbery. New growth is fawnish brown and is not prominent among the dark green mature foliage, except on young plants. It has a most prolific flowering habit. Flowers are borne in dense, woolly racemes all over the bush, are delightfully perfumed and attract many insects. Urgent attempts should be made to introduce this rare species to cultivation as it has great potential for the dry-area landscape.

Grevillea eriostachya J.Lindley (1840)
Plate 121

Appendix to the first twenty-three volumes of Edwards's Botanical Register: . . . with A sketch of the Vegetation of the Swan River Colony xxxvi n. 181 (England)

Yellow Flame Grevillea, Desert Grevillea

Aboriginal: 'wama' (Yewara); 'kaliny kalinypa' (Pintupi); 'rawur rawurba', 'galigiri' (Gugadja).

The specific epithet is derived from the Greek *erion* (wool) and *stachys* (an ear of corn, in botany an inflorescence), in reference to the hairy indumentum

120A. *G. eriobotrya* Close-up of conflorescences (P.Olde)

120B. *G. eriobotrya* Flowering specimen in natural habitat, near Bencubbin, W.A. (P.Olde)

120C. *G. eriobotrya* Flowering branch (K.Alcock)

on the flowers. ERR-EE-OWE-STACK-EE-A

Type: Swan R. colony, W.A., c. 1839, J.Drummond (lectotype: CGE) (McGillivray 1993).

A **shrub** 1.5–2 m tall, bushy at base with emergent, sparsely leaved floral branches to 3 m. **Branchlets** angular to terete, ridged, tomentose. **Leaves** 5–30 cm long, ascending, shortly petiolate, leathery, mostly simple on floral branches, otherwise once or twice pinnatipartite with 2–7 narrowly linear, non-pungent lobes; upper surface usually pubescent often becoming glabrous, midvein and marginal veins evident; margins angularly revolute; lower surface bisulcate, pubescent, midvein prominent. **Conflorescence** erect, pedunculate, terminal or rarely axillary, simple or branched; unit conflorescences 7.5–20 cm long, conico-secund, dense, borne clear of foliage; peduncle villous; rachis woolly, 3.5–6 mm thick at base; bracts 3–4 mm long, elliptic with narrowed base, villous outside, usually persistent to anthesis. **Flower colour**: perianth and style green turning yellow. **Flowers** acroscopic; pedicels 1.5–4 mm long, villous; torus c. 1.5 mm across, ± straight; nectary U-shaped with a wavy margin; **perianth** 7–8 mm long, 2–3 mm wide, oblong-cylindrical, faintly ribbed, detaching soon after anthesis, loosely villous outside, glabrous inside, cohering except along dorsal suture; limb revolute, ovoid, villous, firmly enclosing style end before anthesis; **pistil** 16–22 mm long; ovary sessile, villous; style first exserted below curve on dorsal side and looped upwards, straight or gently undulate, sometimes noticeably incurved just below pollen presenter after anthesis, sparsely pubescent in upper half, dilating suddenly at style end; pollen presenter oblique, orbicular, convex to conical. **Fruit** 15–22 mm long, 9–13 mm wide, 4–7 mm thick, oblique, oblong-ellipsoidal, much-compressed, glandular-tomentose; pericarp 0.5–0.8 mm thick. **Seed** 8 mm long, 3 mm wide, oblong-elliptic with a conspicuous, flat margin, channelled around margin below, prominently winged all round.

Distribution W.A., N.T. and S.A., extending from near Uluru (Ayers Rock), N.T., into the northwest corner of S.A. In W.A. widespread from Derby S almost to Perth. *Climate* Hot, dry summer; cool to warm, wet or irregularly moist winter. Rainfall 150–500 mm.

Ecology Grows in deep red, yellow or grey sand, in low heath, tall mixed scrub and spinifex shrubland. Flowers all year, peaking in spring, but in desert regions in response to rainfall. Regenerates from lignotuber and/or seed after fire. Pollinated by birds, the following having been seen at the flowers: White-fronted Honeyeater, Brown Honeyeater; Singing Honeyeater; White-eared Honeyeater; Red Wattle Bird.

Variation A somewhat variable species. The conflorescences are usually on long, leafless peduncles above the foliage. In the north of its range and in central Australia, however, the peduncles are short, hence the flowers are closer to the foliage. Foliage varies from grey to green. Plants in southern areas tend to be of more open, spreading habit.

Major distinguishing features Habit shrubby usually with emergent floral branches; leaves or lobes long, linear, pubescent on upper surface;

G. eriostachya

121A. *G. eriostachya* Conflorescences (P.Olde)

121B. *G. eriostachya* Shrub in natural habitat, near Denham, W.A. (P.Olde)

121C. *G. eriostachya* Flowering branches (M.Hodge)

conflorescence with very stout rachis, conico-secund, on peduncle > 30 cm beyond foliage; floral bracts 3–4 mm long, persistent to anthesis; flowers yellow; torus straight; perianth zygomorphic, yellow, hairy outside, glabrous within; ovary sessile, densely hairy; style ventrally pubescent in upper half; fruit 4–7 mm thick, the pericarp 0.5–1 mm thick at centre face.

Related or confusing species Group 35, especially *G. excelsior*, *G. pteridifolia* and *G. juncifolia*. *G. excelsior* differs in its conflorescences carried on short, leafy branchlets, leaves turning yellow when dry, caducous floral bracts 6–9 mm long, orange flowers, style glabrous in the upper half, and fruit less compressed (9–12 mm broad) with a pericarp 2–3 mm thick. *G. juncifolia* has conico-cylindrical conflorescences and pedicels > 6 mm long. *G. pteridifolia* is usually a tall tropical tree, the leaves more-divided, thinner, with the undersurface exposed and silky, and has a style end with a distinctive hump.

Conservation status Not presently endangered.

Cultivation *G. eriostachya* has been grown for a number of years, mostly in dry, inland locations. It has been cultivated successfully in western Vic., inland Qld, at Millicent, S.A., and at Kings Park, W.A. It has not proved to be adaptable in cultivation and has usually been difficult to establish, although once achieved, it is both drought- and frost-hardy (to at least -6°C). A very warm to hot situation in full sun in extremely well-drained, acidic to alkaline sand or sandy loam soil is required. It will succumb rapidly to very wet conditions, especially during the warmer months when the effects of root pathogens become obvious. Tip-pruning is advised to improve shape and density, while removal of old flowering canes helps to maintain an attractive appearance; but their removal will result in poorer flowering as the flowers arise from the same flowering branches for several years. The flowers are sometimes eaten by looper caterpillars.

Propagation *Seed* Sets large quantities of seed that germinate usually in about 30 days when nicked or soaked for 24 hours prior to sowing. Seed of cultivated plants is often sterile. Seedlings are particularly prone to fungal infection and should be placed in a sunny, dry position in a free-draining mix. *Cutting* Extremely difficult to strike from cutting. *Grafting* Compatible grafts have been made with *G. banksii* using the cotyledon technique and with *G. robusta* using top wedge and approach grafts. Grafted plants are growing well at Mount Annan Botanic Garden, N.S.W. There have been compatibility problems when using *G. robusta* but these were overcome by using G. 'Poorinda Royal Mantle' or G. 'Poorinda Anticipation' as an interstock. In areas with humid summers, growers have succeeded using *G. pteridifolia* and G. 'Moonlight' as rootstocks.

Horticultural features *G. eriostachya* is valued greatly for its majestic yellow flowers that wave in branched panicles on long canes above the foliage. Flowering continues prolifically for many months with the flowers maturing from green to yellow or orange and attracting many birds that alight on the canes to search for copious nectar in the depths of each flower. At peak nectar flows the flowers drip quantities onto the ground. Regrettably, southern forms often have a scrappy, unattractive habit, although the individual leaves have an attractive ferny appearance. Plants from northern areas are very compact but possibly less adaptable to general cultivation as they occur in calcareous red sand.

Early history Although it did not persist, this species was grown in 1845 at Cambridge, England. It is thought to have been introduced by James Mangles in 1839.

General comments The nectar-bearing flowers were used by Aborigines in central Australia as a drink, either by sucking the flowers or dunking the flowers in water and sipping the resultant sweet drink. This was sometimes allowed to ferment and could intoxicate the drinker. In cooking, food was placed on branches which were laid on the fire.

Grevillea eryngioides G.Bentham (1870)
Plate 122

Flora Australiensis 5: 476 (England)

Curly Grevillea

George Bentham, in describing this species, thought the foliage bore a resemblance to that of the genus *Eryngium* (Apiaceae). ERR-ING-GEE-OY-DEES

Type: south-western W.A., undated, J.Drummond 16 (holotype: K).

A suckering **shrub** 0.5–2 m tall with emergent, floral branches. **Branchlets** terete, glabrous or sometimes with a few scattered hairs, glaucous. **Leaves** 10–20 cm long, 3–6.5 cm wide, erect, sessile, ovate to oblong in outline, pinnatifid; lobes 2–11, 1–2 cm long, c. 2 cm wide, oblong to obovate with decurrent base, obtuse with a short, curved excurrent spine; upper and lower surfaces similar, usually glabrous and glaucous, the midvein, lateral and reticulate venation more prominent on upper surface; margins flat, undulate. **Conflorescence** erect on emergent floral branches, axillary or terminal, branched; unit conflorescences 4–10 cm long, 2 cm wide, dense, obovoid, opening at apex first; peduncle and rachis glandular-pubescent; bracts 5–12 mm long, 3–9 mm wide, imbricate in bud, ovate, membranous, glandular-pubescent outside, falling before anthesis. **Flower colour**: perianth purple; style yellow with purplish black tip. **Flowers** acroscopic; pedicels 2.5–3 mm long, glandular-pubescent; torus c. 1 mm across, straight; nectary annular, cup-shaped, scarcely evident above toral rim; **perianth** 3–4 mm long, 1.5 mm wide, oblong, curved strongly, minutely and sparsely glandular-pubescent or glabrous outside, glabrous within, cohering except along dorsal suture, the dorsal tepals reflexed and everted but not separated from ventral tepals, tepals falling soon after anthesis; limb conspicuous, revolute, spheroidal, enclosing style end before anthesis; **pistil** 9–11 mm long, glabrous; stipe 2.1–2.8 mm long, thin, inserted centrally on torus; ovary globose; style exposed except at limb before anthesis, afterwards slightly incurved, dilated c. 1 mm from flattened, geniculate style end; pollen presenter lateral, obovate, flat. **Fruit** 14–21 mm long, 10–15 mm wide, horizontal to pedicel, glabrous, viscid, flat, ± round, shortly apiculate; pericarp c. 1 mm thick. **Seed** 9 mm long, 4 mm wide, elliptic, winged all round; outer face smooth with a submarginal ridge; margin flat; inner face concave channelled submarginally.

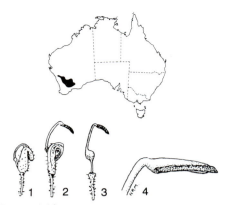

G. eryngioides

Distribution W.A., widely distributed in inland areas of the southwest from Morawa to W of Coolgardie and S to Lake King. ***Climate*** Hot, dry summer; cool to warm, wet winter. Rainfall 250–450 mm.

Ecology Grows in open heath or on rises in mixed scrub, in sand, gravelly or sandy loam, or laterite. Flowers spring to summer. Regenerates primarily

122A. *G. eryngioides* Close-up of conflorescence (M.Hodge)

122B. *G. eryngioides* showing typical suckering habit, near Ballidu, W.A. (N.Marriott)

from suckers, sometimes from seed. Pollinator unknown, possibly native bees.
Variation A uniform species with only minor variation; in most forms the foliage is blue-green but some have silver-grey leaves.
Major distinguishing features Suckering habit; leaves similar on both surfaces, blue-green, glabrous, glaucous, pinnatifid with broad, obtuse, decurrent lobes; conflorescence obovoid, on emergent viscid peduncle; floral bracts large, membranous; nectary annular; perianth zygomorphic, glabrous on both surfaces or glandular-pubescent outside; pistil glabrous; pollen presenter oblong, lateral; fruit compressed, viscid.
Related or confusing species Group 38. *G. eryngioides* appears to have no close relatives.
Conservation status Not presently endangered.
Cultivation G. eryngioides is a reasonably adaptable species, succeeding in hot, dry climates and in warm sheltered sites in cool-wet climates and has been grown in several gardens in inland Vic., at Cadwell's Arboretum, Rylstone, N.S.W., and in gardens around Brisbane. It is hardy to extended drought and frost to at least -6°C. Most success has been achieved in warm, sunny and well-drained conditions in acidic to neutral gravelly or sandy loam. In these conditions it flourishes, and once established requires no maintenance. Pruning the old flower spikes is necessary to enhance its appearance after flowering, but this considerably reduces the next year's flowering as conflorescences occur on the same leafless stems year after year. The best plant seen in cultivation grows at Rylstone where, in deep sandy loam, it had spread by suckers over an area of 10 square metres. Native plant nurseries rarely stock this species.
Propagation *Seed* Produces abundant seeds that have proved difficult to germinate even when peeled. Average germination time forty days (Kullman). *Cutting* Firm, young growth strikes well especially when taken in autumn. As the species suckers, propagation by root cutting is quite successful. Small suckers when treated as cuttings grow on readily. *Grafting* Has been grafted successfully onto *G. robusta* and G. 'Poorinda Royal Mantle' using top wedge techniques. Long-term compatibility has not been assessed.
Horticultural features This singularly curious *Grevillea* produces small plantlets from suckers at short intervals along root lines spreading in all directions around the parent plant. In the wild, especially where roadside grading stimulates sucker growth, plantlets spring up over a wide area. In cultivation, the suckering is not invasive and does not become a nuisance. The stiffly lobed, glaucous, blue-grey leaves are the plant's most attractive feature, making a spectacular foliage contrast in the garden or rockery. The conflorescences on leafless branches above the foliage are interesting but not particularly attractive; the flowers have often been described as purple but the overall appearance is a mauve-brown colour. An interesting feature of the floral canes and follicles is their profuse exudation of a viscid resin during seed formation. This is eagerly foraged by bees; seed capsules can be collected after this without causing sticky fingers. Gardeners should consider this species as a specimen plant for a well-drained garden or rockery.

Grevillea erythroclada W.V.Fitzgerald (1906)
Plate 123

Western Mail (Perth) 21 (1066): 10, 28 (Australia)
Needle-leaf Grevillea
The specific epithet is from the Greek *erythros* (red) and *clados* (a branch), in reference to the reddish branchlets. ERR-ITH-ROW-CLADE-A

Type: Isdell R. near Mt Barnett homestead, W.A., Sept. 1905, W.V.Fitzgerald 1487 (lectotype: PERTH) (McGillivray 1993).

A tall ± glabrous **shrub** or open **tree** 2.5–8 m tall with thick, corky bark. **Branchlets** stout, reddish, terete, glabrous or sparsely silky. **Leaves** 25–52 cm long, 1–2 mm broad, unifacial, ascending, leathery, sessile but appearing long-petiolate, glabrous or sparsely silky, deeply and longitudinally wrinkled, usually pinnatipartite; leaf lobes < 2 mm broad, linear or ± terete, acute, the lower lobes sometimes with secondary division; upper leaf edge channelled or fused into a pale stripe. **Conflorescence** paniculate, terminal, exceeding foliage, glabrous except sometimes the bracts; unit conflorescences 6–18 cm long, 2 cm wide, conico-cylindrical, dense; peduncle and rachis stout; bracts 1–2 mm long, ovate, sometimes ciliate, falling before anthesis. **Flower colour**: perianth and style creamy white. **Flowers** acroscopic; pedicels 1.2–2.5 mm long; torus 1–2 mm across, slightly oblique; nectary U-shaped, inconspicuous; **perianth** 3–5 mm long, 1.5 mm wide, cylindrical to oblong below curve, faintly ridged, papillose inside, at first separated along dorsal suture, sometimes tepals loose below limb before anthesis, afterwards free to base and loose; limb revolute, ovoid, enclosing style end before anthesis; **pistil** 6.5–9 mm long, glabrous; stipe 1.6–3 mm long; ovary globose; style at first exserted below curve on dorsal side and looped upwards, incurved to straight after anthesis, filamentous; style end small, scarcely dilated; pollen presenter oblique to almost straight, round, conical. **Fruit** 19–29 mm long, 13–18.5 mm wide, horizontal to pedicel, compressed-ellipsoidal, viscid with a caustic exudate when young; pericarp 0.5–1 mm thick. **Seed** 15–16 mm long, 10 mm wide, elliptic, winged all round, conspicuously so at the ends.

123A. *G. erythroclada* Flowering branch (K. & L. Fisher)

123B. *G. erythroclada* Conflorescence (D.McGillivray)

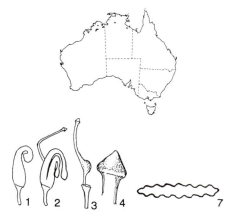

G. erythroclada

Distribution N.T., W.A. Occurs near Katherine, N.T., and in scattered populations in the east Kimberley, W.A. ***Climate*** Monsoonal: Summer hot and wet; winter warm and dry. Rainfall 500–900 mm.

Ecology Grows in red volcanic soil or poorly drained sand, in open forest or on river and creek banks. Flowers late autumn to early spring. Regenerates from seed and epicormic buds after fire. Insect-pollinated.

Variation A fairly uniform species. In some plants, the leaf lobes are somewhat flattened.

Major distinguishing features Branchlets stout, reddish; leaves unifacial, ± glabrous, pinnately divided with secondary division confined to lower lobes, the lobes < 2 mm wide, terete or flat, longitudinally wrinkled; conflorescence glabrous, paniculate, the units cylindrical; perianth limb 1.6–2 mm wide, revolute; pollen presenter conical, oblique; fruits viscid, compressed-elliptic.

Related or confusing species Group 37, especially *G. pyramidalis* which is distinguished especially by its usually narrower perianth limb and wider leaf lobes (1.2–1.7 mm).

Conservation status Not currently endangered.

Cultivation *G. erythroclada* has not been cultivated widely and little information can be given from experience on its adaptability and hardiness. The habitat indicates that it will do well in dry tropical conditions, perhaps responding well to watering. It may not be frost-tolerant and may lose vigour in cooler climates. To some degree, it tolerates poor drainage and should respond to good garden conditions especially if grown in full sun without competition from other plants.

Propagation *Seed* Fresh seed sown in early or late spring is worth trying. *Cutting* Untested but likely to be extremely difficult. *Grafting* Cotyledon grafts using freshly germinated seedlings could be tried on *G. robusta*.

Horticultural features *G. erythroclada* has most attractive, weeping, fine, foliage and forms a nice tree producing conspicuous panicles of cream flowers over long periods. Some attention should be given to the introduction of this species to suitable tropical gardens or streetscapes. Widespread cultivation is unlikely due to the caustic, sticky fruits which can cause serious burns if not handled with care.

Grevillea evanescens P.M.Olde & N.R.Marriott (1994)   **Plate 124**

The Grevillea Book 1: 181 (Australia)

The specific epithet derives from the Latin *evanescens* (quickly disappearing), in reference to the leaves quickly shedding hairs. EV-ANN-ESS ENZ

Type: Military Rd, N off Lancelin road towards Gingin, W.A., 29 Sept. 1991, P.M.Olde 91/240 (holotype: NSW).

Erect to spreading, single-stemmed **shrub** 2–3.5 m high. **Branchlets** angular or terete, glabrous or with scattered, appressed hairs. **Leaves** 2–4.5 cm long, 2.5–9 mm wide, ascending to erect, shortly petiolate, truncate to retuse with short mucro, simple, entire, oblong to obovate or elliptic; upper surface smooth, glabrous or with scattered appressed hairs, midvein and lateral veins evident and often slightly raised; margins shortly recurved; lower surface glabrous or with scattered appressed hairs, midvein and lateral veins prominent. **Conflorescence** 0.5–1.5 cm long, decurved, shortly pedunculate, terminal on very short side branches, simple, secund, mostly 8–12-flowered, scarcely to not exceeding foliage; peduncle and rachis together 18–30 mm long, with scattered hairs or glabrous; bracts 0.5 mm long, ovate, glabrous to sparsely silky outside, usually falling before anthesis, sometimes persistent. **Flower colour**: perianth pale to bright red with creamy limb; style red with prominent green apex. **Flowers** acroscopic; pedicels 2.5–4 mm long, glabrous or a few scattered hairs; torus c. 1 mm across, slightly oblique; nectary prominent, cushion-like to oblong; **perianth** 6–7 mm long, 2–3 mm wide, ovoid, dilated slightly at base, glabrous and sometimes glaucous outside, tomentose-pubescent inside, densely so about ovary, the hairs otherwise concentrated along tepal margins; tepals cohering except along dorsal suture; limb revolute, spheroidal with slight apical depression, strongly ribbed, the ribs 0.2 mm high, a few scattered hairs in young bud, otherwise glabrous, white-spotted (the spots minute, ?white trichomes); **pistil** 26–27 mm long, glabrous; stipe 4–4.5 mm long, incurved, flattened; ovary triangular, stipitate; style before

124A. *G. evanescens* Flowering branch (P.Olde)

124B. *G. evanescens* Habit of plant, W of Gingin, W.A. (P.Olde)

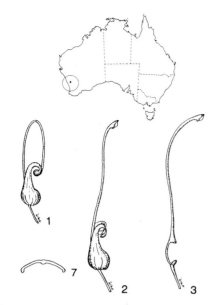

G. evanescens

anthesis exserted from curve and looped strongly upwards, afterwards gently curved; style end abruptly expanded, exposed before anthesis; pollen presenter 1.5 mm long, oblique, oblong-elliptic, convex, flanged. **Fruit** 10–11 mm long, 4–6 mm wide, erect, oblong-acuminate, faintly ridged at base, the ridges scarcely extending beyond ventral plane, rugulose; pericarp c. 0.5 mm thick. **Seed** 7 mm long, 2 mm wide, ellipsoidal with apical cushion-like swelling; outer face slightly convex, minutely pubescent; inner face flat, mostly obscured by the revolute margins; margin with a waxy border along one side drawn into an excurrent, apical elaiosome c. 1 mm long.

Distribution W.A., W of Gingin. *Climate* Summer hot, dry; winter cool to mild, wet. Rainfall 800 mm.

Ecology Grows in brown, calcareous sand on ridges in banksia woodland. Flowers winter–spring.

Major distinguishing features Single-stemmed shrub, strongly erect; branchlets glabrous or almost so; leaves mostly 3–4 cm long, 4–9 mm wide, simple, entire, linear-oblong to obovate or elliptic, retuse to emarginate or truncate; midvein and lateral veins clearly evident, sometimes raised on upper surface, glabrous or almost so below, margins shortly recurved; perianth zygomorphic, glabrous outside, hairy inside; limb with white spots, prominently ribbed; pistil glabrous; stipe flattened, incurved; ovary triangular; pollen presenter oblique; fruit with faint basal ridge.

Related or confusing species Group 14, especially *G. obtusifolia* which differs in its shorter, narrower leaves with the undersurface silky, in its spreading to prostrate habit and in its slightly shorter pistil.

Variation A morphologically uniform species. Plants in cultivation have a spreading habit. Plants at the type locality were strongly erect, virgate shrubs.

Conservation status 2E suggested. Its distribution and status should be assessed.

Cultivation *G. evanescens* has been cultivated for many years in eastern Australia which is perhaps surprising for a rare species. It was probably introduced by H.Demarz who was the collector for Kings Park and Botanic Garden, Perth, and who is the only known other collector. It is an extremely hardy, adaptable species that grows well in hot-dry, cold-wet and summer-rainfall climates. Extended dry conditions are tolerated, provided there is some subsoil moisture. It resists frosts to at least -6°C and grows readily in a wide range of conditions and soil types. Both full sun or partial shade are tolerated. A well-drained, sandy or loamy slightly alkaline soil is preferred but it will grow in acidic soil. Its strongly erect habit should not need pruning. It is occasionally sold in nurseries as *G. obtusifolia*.

Propagation *Seed* Untested. Should germinate readily if the testa is nicked or peeled before sowing. *Cutting* Firm, young growth and semi-hardwood strike readily at most times of the year. *Grafting* Has been grafted successfully onto *G. robusta* with which it is thought to be long-term compatible.

Horticultural features *G. evanescens*, in its typical form, is a most impressive plant. Its strongly erect habit, bright green almost glabrous leaves and attractive flowers have much to recommend them. It has potential for narrow gardens that require a strongy erect plant to break up walls or other unattractive structures. The flowers are strongly bird-attractive. Hardiness and longevity in cultivation suggest possible uses in road or public building landscapes.

General comments *G. evanescens* is morphologically most closely related to *G. obtusifolia*, but its occurrence in a habitat so different from that species along with its significant foliar morphology have influenced us to recognise it at specific rank. It grows with *G. preissii*.

Grevillea evansiana H.S.McKee (1953)
Plate 125

Proceedings of the Linnean Society of New South Wales 78: 49 (Australia)

Evans' Grevillea

Named after Obed Evans, curator for many years of the John Ray Herbarium, University of Sydney. EV-AN-ZEE-ARE-NA

Type: Currant Mountain Gap, 5 miles [c. 8 km] E of Olinda (E of Rylstone), N.S.W., 2 Sept. 1951, H.S.McKee & L.A.S.Johnson (holotype: NSW).

Synonym: *G. diffusa* subsp. *evansiana* (H.S.McKee) D.J.McGillivray (1986).

Divaricately branched, mounded **shrub** to 50 cm with arching branches. **Branchlets** terete, tomentose. **Leaves** 2.5–6 cm long, 0.5–1 cm wide, ascending, shortly petiolate, mucronate, simple, elliptic to linear, acuminate or obtuse; new growth rusty; upper surface ± flat, silky to glabrous and shiny, sparsely and minutely pitted, midvein and intramarginal veins evident and smooth (without granules), lateral venation obscure; margins entire, sharply and evenly refracted about intramarginal vein; lower surface silky white, midvein prominent. **Conflorescence** 1–1.5 cm long, 1.5 cm wide, erect, terminal, usually on pendulous branchlet, sessile to shortly pedunculate, simple, subglobose, very condensed; peduncle 1–1.2 mm thick, villous; rachis tomentose; bracts 3–3.2 mm long, linear, villous outside, falling before anthesis. **Flower colour**: perianth reddish black or white with white indumentum; style dark blackish red or green. **Flowers** acroscopic; pedicels 1.5 mm long, stout, tomentose; torus 0.8 mm wide, ± straight; nectary prominent, U-shaped; **perianth** 4–5 mm long, 1 mm wide, oblong, tomentose outside, bearded inside above level of ovary; limb revolute, spheroidal, densely villous, usually rusty; **pistil** 9.5–10 mm long; stipe 2 mm long, flattened; ovary oblong-ovoid, scarcely evident; style end minutely hairy. **Fruit** 10–12 mm long, 4–6 mm wide, erect, oblong-cylindrical, glabrous; pericarp 0.3 mm thick. **Seed** 8 mm long, 2.5 mm wide, compressed-obovoid with a conspicuous apical wing and a minute basal wing; outer face flat, longitudinally wrinkled with an apical collar; inner face obscured by revolute margin.

Distribution N.S.W., restricted to an area east of Olinda in the Rylstone–Kandos district. *Climate* Hot, dry summer; cold, wet winter. Rainfall 600–800 mm.

Ecology Grows in open woodland or open heath in shallow, sandstone-derived soil. Flowers spring–summer. Regenerates from seed after fire. Pollinator unknown.

Variation A relatively uniform subspecies with some variation in leaf length, indumentum and flower colour.

White-flowered form H.S.McKee collected two specimens from plants with 'green' flowers, both of which were growing near each other. These flowers had a green style, stigma and base of the perianth. We found this form again in November 1989 and the overall impression of flower colour is white.

Major distinguishing features Branchlets terete, tomentose; leaves simple, entire, elliptic, silky white below, some usually wider than 6 mm, smooth and glossy above, lateral veins obscure, marginal veins smooth; conflorescence shortly pedunculate; pedicels short (< 2 mm long), stout; nectary conspicuous (0.6–1 mm high); perianth zygomorphic, densely bearded above ovary within, the limb usually rusty.

Related or confusing species Group 21, especially *G. diffusa* which differs in its usually narrower leaves (< 5 mm wide), longer pedicels 2–4 mm long and less conspicuous nectary. The marginal veins of *G. diffusa* are also granular and the perianth limb has silky white hairs.

Conservation status 2VC. Restricted to a small area.

Cultivation Widely cultivated along the east coast of Australia and in the A.C.T. and also grown in the U.S.A., *G. evansiana* has proved hardy and adaptable in cultivation. It tolerates frosts to at least -10°C and extended dry periods. In very humid conditions, black fungal spots sometimes develop on leaves. Tolerant of a wide range of soil types, it will grow in either sandy soil or heavy loam provided it is fairly well-drained and either acidic or neutral. It prefers a sunny or open, partially shaded situation, generally without crowding competition from other plants. Pruning is not necessary and summer watering is rarely required except in harsh, dry conditions. East coast nurseries specialising in native plants usually stock this plant under the name *G. evansiana*.

Propagation *Seed* Sets good quantities of seed that should germinate when given the standard peeling treatment. *Cutting* Strikes readily from cuttings taken in spring, using firm, new growth. *Grafting* Not tested.

Horticultural features This pleasing shrub has attractive dark green and sometimes shiny leaves with contrasting silvery undersurface, sometimes grey-green in overall appearance. It has a compact, low habit and usually consists of many twiggy branches with interesting arching branchlets at the

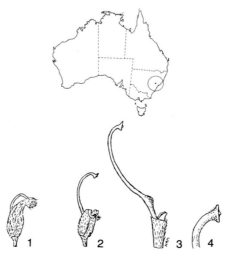

G. evansiana

125A. *G. evansiana* Close-up of flowers (N.Marriott)

125B. *G. evansiana* White-flowered plant, Dunns Swamp, N.S.W. (P.Olde)

125C. *G. evansiana* growing on rock outcrops beside Kandos Weir, N.S.W. (P.Olde)

extremities. The terminal flowers are sometimes showy even though they are usually very dark red-black. Flowers on the white-flowered form are more conspicuous but still tend to blend with the foliage. It flowers most prolifically in areas that have frosty winters. This useful, hardy plant has good potential as a groundcover, low screen or border plant, or as a massed foliage plant both on the coast and inland.

Grevillea excelsior L.Diels (1904)

Plate 126

Botanische Jahrbücher für Systematik 35: 151 (Germany)

Orange Flame Grevillea

The specific epithet is derived from the comparative of the Latin *excelsus* (lofty, high), a reference to the tree-like habit. EK-SELL-SEE-OR

Type: near Tammin, W.A., 24 Oct. 1901, L.Diels 5852 (holotype: B).

Synonym: *G. eriostachya* subsp. *excelsior* (L.Diels) D.J.McGillivray (1986).

Erect, brittle, often leaning small **tree** to 3–5(–8) m. **Branchlets** angular, ridged, tomentose. **Leaves** 5–30 cm long, ascending, shortly petiolate, leathery, occasionally simple and linear, more usually pinnatipartite with 2–7 linear lobes; upper surface ± glabrous or sometimes minutely pubescent, yellowish when dry, midvein inconspicuous; margins angularly revolute; lower surface bisulcate, minutely pubescent in grooves, midvein prominent. **Conflorescence** erect, pedunculate, terminal, simple, sometimes branched, partially borne within foliage; unit conflorescence 8–20 cm long, conico-secund, dense; peduncle usually 1–4 cm long sometimes longer on emergent forms, tomentose; rachis 3–6 mm thick at base, densely villous; bracts 6–9 mm long, ovate with basal stalk, villous outside, falling before anthesis. **Flower colour**: perianth white outside, bright orange or orange red inside; style orange. **Flowers** acroscopic; pedicels 1.5–4 mm long, villous; torus c. 1.5 mm across, straight; nectary U-shaped, smooth; **perianth** 8–10 mm long, 2–3 mm wide, oblong-cylindrical, tomentose outside, glabrous inside, cohering except along dorsal suture; limb revolute, villous, ovoid, enclosing style end before anthesis; **pistil** 20–28.5 mm long; ovary villous, sessile; style exserted below curve on dorsal side and looped upwards before anthesis, afterwards straight to undulate to slightly incurved, ± glabrous or with granules about middle, sparsely villous at base; style end evenly but suddenly dilated; pollen presenter oblique, oblong-elliptic, convex. **Fruit** 15–22 mm long, 13–17 mm wide, 9–12 mm thick, oblong-elliptic, beaked, somewhat compressed, tomentose; pericarp 2–3 mm thick. **Seed** 8–10 mm long, 3–4 mm wide, oblong-elliptic with membranous wing, smooth; inner face channelled near flat margin.

Distribution W.A., in the southern and central wheatbelt regions from Quairading to Wongan Hills, E to Coolgardie and S to around Peak Charles. *Climate* Hot dry summer; cool, moist winter. Rainfall 200–350 mm.

Ecology Occurs in medium to tall scrub and heath on flat or gently undulating plains, in grey to yellow sand and gravelly sand over loam, often dominant. Flowers early spring to late summer. Killed by fire and regenerates from seed. Pollinated by nectarivorous birds of which the following have been observed feeding at the flowers; Red Wattle Bird, Singing Honeyeater, White-fronted Honeyeater. In the wild, plants often have a distinct lean but the reason for this is unknown.

Variation A fairly uniform species. At the southeastern edge of its range, plants sometimes have conflorescences borne clear of the foliage.

Major distinguishing features Tree-like habit; leaves or lobes narrowly linear, the upper surface glabrous or minutely pubescent, drying yellow; conflorescence conico-secund, not greatly exceeding leaves; flowers orange; floral bracts spathulate, 6–9 mm long, falling before anthesis; nectary smooth; perianth zygomorphic, hairy outside,

G. excelsior

126A. *G. excelsior* Close-up of conflorescence (F. & N. Johnston)

126B. *G. excelsior* Plants regenerated in road verge, E of Lake King, W.A. (M.Hodge)

glabrous within; ovary sessile, villous; style glabrous or granular; fruit 9–12 mm thick, pericarp 2–3 mm wide.

Related or confusing species Group 35, especially *G. eriostachya, q.v.* McGillivary (1993) has observed that the floral bracts are wider and have a resinous duct near the base of the head.

Conservation status Not presently endangered. Reasonably common over a wide area. A primary coloniser of disturbed sites.

Cultivation *G. excelsior* has not been widely cultivated because of difficulties in propagation. It was grown for many years at Sid Cadwell's arboretum at Rylstone, N.S.W., at Dave Gordon's arboretum in inland Qld, and is grown by a few enthusiasts in inland south-western Australia. It tolerates heavy frosts and extended dry conditions without ill effect. Soil should be deep, well-drained sand or gravelly loam in a warm, sunny situation. Plants need protection from strong winds because their branches are brittle and easily broken and the root structure is not strong. Pruning is seldom necessary as it grows naturally into a small tree. Once established, it does not require watering or fertilising. Seed is usually available from suppliers of wildflower seed but viability may be uncertain.

Propagation *Seed* Sets copious quantities of seed in the wild, and regenerates prolifically after fire. Seed should be treated by nicking or soaking. Average germination time 56 days (Kullman). Seedlings should be placed in a sunny position in a well-drained mix and may need treatment against fungal attack. *Cutting* Extremely difficult. *Grafting* Many unsuccessful attempts have been made to graft this species. Success may be achieved only with cotyledon grafts.

Horticultural features *G. excelsior* is an outstanding plant for horticulture, bearing brilliant, orange conflorescences in great numbers, contrasting against the dark green foliage. Nectar-eating birds are attracted to its flowers. At the height of its flowering, nectar may be observed dripping to the ground. It makes an attractive specimen plant even when not in flower with its attractive, ferny foliage and symmetrical habit. Its brittleness precludes it from being suitable for general landscaping, but it makes a beautiful, specimen plant for collectors in warm-climate gardens.

General comments D.J.McGillivray (1993) treated this as a subspecies of *G. eriostachya* to which it is undoubtedly closely related. However, the two taxa maintain themselves in homogeneous populations often growing together and with few hybrids. Reproductive isolation, in addition to the small but consistent characters which separate them, indicates to us that species rank is appropriate.

Grevillea exposita P.M.Olde & N.R.Marriott (1994)   Plate 127

The Grevillea Book 1: 181 (Australia)

The specific epithet derives from the Latin *expositus* (exposed), a reference to the leaf undersurface and the more exposed conflorescences. EX-POZ-IT-A

Type: Arrowsmith R., Brand Hwy, N of Eneabba, W.A., 29 Sept. 1992, P.M.Olde 92/161 (holotype NSW).

A single-stemmed **shrub**, 1–2 m high, spreading up to 3 m wide. **Branchlets** angular or terete, densely tomentose-pubescent with glabrous striation. **Leaves** 1.5–3(–3.5) cm long, 1.2–4 mm wide, ascending to spreading, shortly petiolate, obtuse, mucronate, mostly simple, entire, oblong-linear to narrowly obovate or elliptic, usually a few bifid to trifid; upper surface smooth to obscurely punctate, glabrous or with scattered appressed hairs, the midvein sometimes evident in an obscure longitudinal channel; margins revolute to strongly so; lower surface densely silky, sometimes obscured by revolute margins, midvein prominent, glabrous. **Conflorescence** decurved or deflexed at base, pedunculate, axillary or terminal, sometimes on short branchlet, simple or 1-branched; unit conflorescence 0.5–2 cm long, secund to loosely subcylindrical, mostly 8–12-flowered, usually clearly exceeding foliage; peduncle and rachis together 15–50 mm long, sparsely to densely silky; peduncle often refracted at base; bracts 0.5 mm long, ovate, silky outside, usually falling just before anthesis. **Flower colour**: perianth bright red with creamy white limb; style red with green tip. **Flowers** acroscopic; pedicels 3.5–4 mm long, glabrous or a few scattered hairs; torus 0.5–1 mm across, slightly oblique; nectary prominent, cushion-like to oblong; **perianth** 8 mm long, 3–3.5 mm wide, narrowly ovoid, dilated slightly at base, glabrous and sometimes glaucous outside, tomentose-pubescent inside, densely so about ovary, the hairs otherwise concentrated along tepal margins; tepals cohering except along dorsal suture; limb revolute, spheroidal-subcubic, emarginate in side view, apically depressed, the segments faintly to prominently ribbed, sparsely silky; **pistil** 20–24 mm long, glabrous; stipe 2.5–3 mm long, incurved, flattened; ovary triangular, stipitate; style before anthesis exserted from curve and looped strongly upwards, afterwards gently curved; style end abruptly expanded; pollen presenter 1.8 mm long, oblique, oblong-elliptic, convex, flanged. **Fruit** 12–13 mm long, 4 mm wide, erect, oblong-acuminate, faintly ridged at base, the ridges scarcely extending beyond ventral plane, rugulose; pericarp c. 0.5 mm thick. **Seed** 7 mm long, 1.5 mm wide, oblong, with apical cushion-like swelling; outer face strongly convex, minutely pubescent; inner face flat, mostly obscured by revolute margins; margin with a waxy border along one side.

Distribution W.A., confined to the Eneabba–

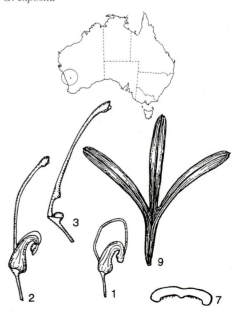

G. exposita

127A. *G. exposita* Near Eneabba, W.A. (N.Marriott)

127B. *G. exposita* Flowering habit (P.Olde)

Arrowsmith River area. *Climate* Summer hot, dry; winter cool to mild, wet. Rainfall 500 mm.

Ecology Grows along creek lines in moist to well-drained, sandy loam or in lateritic sand over clay-loam. Flowers winter–spring. Nectar-feeding birds have been observed probing the flowers. Reproduction after fire is from seed.

Variation Varies in the number of divided leaves and the degree of revolution of the leaf margin.

Major distinguishing features Single-stemmed shrub; branchlets densely tomentose-pubescent; leaves mostly 2–3 cm long, simple, entire with some leaves bifid to trifid, linear-oblong to obovate or elliptic, midvein usually clearly evident on upper surface, the undersurface silky; margins flat to shortly recurved; pedicels glabrous or almost so; conflorescences usually carried well beyond leaves; perianth zygomorphic, glabrous outside except limb, hairy inside; limb spheroidal to subcubic with prominent ribs, usually with scattered hairs; pistil glabrous; stipe flattened, incurved; ovary triangular; pollen presenter oblique; fruit with faint basal ridge.

Related or confusing species Group 14, especially *G. obtusifolia*, *G. pinaster* and *G. thelemanniana*. *G. thelemanniana* has mostly shorter and narrower leaves, a lignotuberous habit and sparsely silky to glabrous branchlets. *G. obtusifolia* differs in its entirely simple leaves with flat or shortly recurved margins, its sparsely silky to almost glabrous branchlets, in its conflorescences more enclosed in the foliage and in its mostly paler flowers; it also has a slightly smaller pollen presenter. At least one form of *G. pinaster* is similar morphologically but differs in the leaf undersurface enclosed by the margin or in the leaves mostly longer than 4 cm.

Conservation status Not presently endangered although much of its former habitat has been cleared for agriculture.

Cultivation *G. exposita* is an extremely hardy, adaptable species that grows well in most areas except arid or tropical climates. It is hardy to extended dry conditions provided there is some sub-soil moisture and resists frosts to at least -6°C. It has not been widely cultivated except in Vic. It has flourished in full sun in well-drained heavy gravelly loam. Regular tip-pruning is recommended to induce compactness. Native plant nurseries occasionally stock this species.

Propagation *Seed* Sets numerous seeds that germinate readily if treated by nicking or peeling the testa. *Cutting* Firm, young growth strikes readily at most times of the year. *Grafting* Untested. Related species graft readily onto *G. robusta*.

Horticultural features *G. exposita* is a vigorous, dense, leafy shrub with attractive bright red and white flowers that attract many nectar-seeking birds. Flowering is quite prolific and prominent, usually within the bush but away from the leaves. It appears to be long-lived in cultivation and is ideal as a screen or background plant. The leaves are somewhat darker green than the closely related *G. obtusifolia* and contrast with the white branchlets.

General comments *G. exposita* was previously included in *G. thelemanniana* subsp. *obtusifolia* by McGillivray (1986, 1993), as a population exhibiting morphological differences insufficient to warrant formalised nomenclature. Apart from the morphological characters separating it from *G. obtusifolia*, recognition of *G. exposita* as distinct is maintained in part because of its habitat which differs markedly from that of *G. obtusifolia* and closely approaches that of *G. pinaster* in the same region. *G. pinaster* in this region has pinnate leaves.

Grevillea extorris S.Moore (1899)
Plate 128

Journal of the Linnean Society, Botany 34: 221 (England)

The specific epithet is derived from the Latin *extorris* (exiled, banished), an oblique reference to the remoteness of the area in which the Type was collected. EKS-TORR-ISS

Type: between Wilson Pool and Lake Darlot, W.A., May 1895, Spencer Moore (lectotype: BM).

Dense **shrub** 1–2.5 m tall with erect branches. **Branchlets** terete, silky. **Leaves** 4–12 cm long, 1–3 mm wide, ascending, shortly petiolate, stiff, simple, linear or narrowly oblong, leathery, scarcely pungent; upper surface silky becoming glabrous, 7–11-grooved, midvein obscure; margin entire, angularly revolute, sometimes recurved; lower surface bisulcate, silky, hairs sometimes only visible in grooves, midvein prominent. **Conflorescence** erect, shortly pedunculate, axillary or arising from the stem on old wood, simple or few-branched; unit conflorescence c. 2.5 cm long, 5 cm wide, 3–12-flowered, umbel-like or partially secund, loose; peduncle and rachis silky; bracts c. 0.5 mm long, elliptic, villous outside, falling before anthesis. **Flower colour** Perianth white to pale outside, dark red or yellow inside. Style red to yellow. **Flowers** basiscopic; pedicels 2–6 mm long, silky; torus c. 1 mm across, oblique, rectangular; nectary inconspicuous, U-shaped; **perianth** 8–10 mm long, 2.5 mm wide, oblong, silky outside, white-bearded inside at level of ovary, cohering except along dorsal suture; limb revolute, spheroidal, the style end partially exposed before anthesis; pistil 28–38 mm long; stipe 2.2–3.1 mm long, silky, glabrous on ventral side; ovary appressed-villous; style exserted below limb and looped upwards before anthesis, afterwards elongate and gently incurved, ± glabrous, sparsely hairy at base and apex, straight, dilating suddenly at style end; pollen presenter oblique, convex, obovate. **Fruit** 11–15 mm long, 5 mm wide, oblique on the pedicels, ovoid to ellipsoidal, silky; pericarp c. 0.5 mm thick. **Seed** not seen.

Distribution W.A., in the semi-arid south-west beyond the fringe of the wheatbelt, extending from Mullewa east to Lake Darlot and from Cue south to Mt Jackson. **Climate** Hot, dry summer and mild winter with irregular rain. Rainfall 175–250 mm.

Ecology Mostly found in dry, stony clay and laterite,

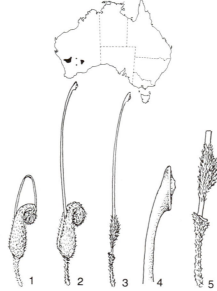

G. extorris

usually on rocky rises and outcrops where it grows in open to thick, mixed low scrub. Presumably pollinated by nectarivorous birds. Regenerates from seed and possbly from lignotuber. Flowers autumn to early spring.

Variation This species occurs over a wide range in disjunct populations. Two distinct forms are recognised.

Narrow-leaved (Typical) form Occurs between Cue and Mullewa and southeast to Mt. Jackson with an isolated population in the Lake Darlot area. This form has narrowly linear leaves 1–2 mm wide with the margin revolute to the midvein and enclosing the lower surface.

Broad-leaved form Occurs mainly from Leonora to Comet Vale and west to Sandstone. Typically, it has oblong leaves up to 3 mm wide, the margins not enclosing the lower surface. It has a grey-pubescent aspect and grows to 1.5 m.

Major distinguishing features Leaves simple, entire, linear to oblong, with 7–11 grooves on upper surface; conflorescence axillary or cauline, 3–12-flowered; torus oblique; perianth zygomorphic, hairy on both surfaces, cohering after anthesis; pistil hairy; ovary stipitate, villous; style conspicuously exceeding perianth; pericarp c. 0.5 mm thick.

Related or confusing species Group 25, especially *G. phillipsiana* which differs in its terminal conflorescences, and *G. granulosa* which has a granular upper leaf surface.

Conservation status Not presently endangered.

Cultivation *G. extorris* has not been widely cultivated, but reports from Vic. and Burrendong Arboretum, where it thrives, indicate that it is relatively hardy and grows well in cultivation in well-drained, gravelly loam. In the wild it grows vigorously in both full sun and partial shade. Pruning may be needed to maintain a compact shape, as plants shed leaves from the lower branches. Reasonably heavy frosts and drought are withstood without ill effect. It has so far done well in cool-wet and warm-dry climates and has lived in pots in summer-humid climates for many years without signs of collapse. Plants of *G. granulosa* have often been sold in nurseries as *G. extorris*.

Propagation *Seed* Untested. Plants in the wild and in cultivation appear to set few seed. *Cutting* Strikes fairly well from firm, young growth taken in the warmer months. *Grafting* Successfully grafted onto a number of rootstocks including G. 'Poorinda Royal Mantle', *G. robusta*, *G. rosmarinifolia* and the hybrid G. 'rosmarinifolia nana' using the top wedge and whip methods.

128A. *G. extorris* Plant in natural habitat, near Mt Jackson, W.A. (J.Cullen)   128B. *G. extorris* Flowers and foliage (C.Woolcock)

Horticultural features *G. extorris* is a naturally dense, shapely plant to c. 1 m with grey to grey-green foliage. Large flowers with conspicuous styles decorate the leaf axils and old wood in great profusion during winter. Flowers are brightly coloured in red or yellow, often being more conspicuous on the lower branches where foliage is less dense. Its early flowering period makes it a valuable plant both for winter colour and for birds seeking winter food. This naturally long-lived, tough species would be suitable as a screen or mass-planting in home gardens or commercial landscapes, especially in dry climates.

General comments The taxonomy of *G. extorris* requires further field and herbarium study.

Grevillea exul J.Lindley (1851)

Journal of the Horticultural Society London 6: 273 (England)

The specific epithet is derived from the Latin *exul* (an exile), a reference to the fact that this New Caledonian species was found c. 1500 km from other species of the genus known to that time. EKS-ULL

Type: east coast, New Caledonia, 1850, C.Moore (holotype: CGE).

Small, open **tree** to 10 m in sheltered sites; a large, spreading **shrub** to 4 m in exposed places. **Branchlets** terete, silky to pubescent, rusty-striate. **Leaves** 4.5–13.5 cm long, 0.5–5.7 cm wide, ascending, petiolate, simple, elliptic, obovate, rarely linear, obtuse-mucronate, leathery; upper surface sparsely silky, soon ± glabrous, with conspicuous longitudinal venation; margins entire, shortly recurved; lower surface rusty-silky, sometimes glabrous, the veins glabrous, scarcely prominent. **Conflorescence** erect, pedunculate, terminal, rarely axillary, simple or branched; unit conflorescence 5–20 cm long, dense, secund; peduncle and rachis silky, sometimes glabrous; bracts c. 1 mm long, ovate-acuminate, silky, falling before anthesis. **Flowers** acroscopic; pedicels stout, 5–9 mm long, silky, sometimes glabrous; torus c. 2 mm across, slightly oblique to straight; nectary inconspicuous to evident, U-shaped, entire; **perianth** 10–15 mm long, 2–3 mm wide, oblong-ovoid, slightly dilated at base, silky outside, sometimes glabrous, papillose inside near base or sometimes glabrous, separating first along dorsal suture before anthesis, all tepals free to base and dorsal tepals slightly everted, afterwards free, loose; limb nodding, globular, firmly enclosing style end before anthesis; **pistil** 26.5–42.5 mm long, glabrous; stipe 2.2–5.8 mm long; ovary ellipsoidal; style exserted below curve and looped outwards before anthesis, afterwards straight to slightly inflexed about middle; style end scarcely dilated; pollen presenter slightly oblique, elliptic, conical. **Fruit** 15–20 mm long, 12–15 mm wide, oblique on pedicel, lens-shaped, smooth; pericarp c. 2 mm thick. **Seed** 10 mm long, 7 mm wide, ovate, flattened, membranously winged all round.

Major distinguishing features Leaves simple, entire; conflorescence secund, 5–25 cm long; perianth zygomorphic; pistil glabrous; pollen presenter conical.

Related or confusing species Group 20. *G. exul* is most closely related to *G. robusta* which can easily be separated by its divided leaves and golden flowers.

Variation Two subspecies are recognised.

Key to subspecies

Conflorescence simple or few-branched; hairs of floral rachis without coloured contents; outer perianth surface glabrous or hairy   subsp. **exul**

Conflorescence much-branched; hairs of floral rachis red or brown; outer perianth surface hairy   subsp. **rubiginosa**

Grevillea exul J.Lindley subsp. exul
Plate 129

Synonyms: *G. macrostachya* A.Brongniart & J.Gris (1865); *G. sinuata* A.Brongniart & J.Gris (1865); *G. heterochroma* A.Brongniart & J.Gris (1865); *G. macmillanii* Guillaumin (1958)

Conflorescence simple or few-branched; hairs of floral rachis without coloured contents. **Flower colour**: perianth and style white; style end greenish. **Perianth** glabrous or hairy outside.

Distribution New Caledonia. Widespread. Var. *nudiflora* has been recorded from only a few collections in the southern and central parts of the island. *Climate* Summer hot, humid, wet; winter cool to mild, dry. Rainfall 900–3000 mm.

Ecology Occurs in low-lying valleys, on ridges and slopes from sea level to c. 1000 m, growing in serpentine schist and skeletal soil, as well as in alluvial sand beside streams and gravelly loam, usually in open forest or scrub. Flowers May to January, during the later months at higher altitudes. Fire response unknown. Pollinator unknown.

Variation There is some variation of perianth indumentum. R.Virot (1968) has divided subsp. *exul* into two varieties: var. *exul* with an appressed indumentum on the outer surface of the perianth, and var. *nudiflora* R.Virot (1968) (syn. *G. sinuata*) with the perianth glabrous outside. Further studies are needed to determine if these varieties occur in discrete populations or as variants in mixed populations.

Conservation status Not presently endangered.

Cultivation *G. exul* was grown for many years at the Royal Botanic Gardens, Sydney, before disappearing from cultivation just after the turn of the century. In 1988, officers from the Gardens collected new material, re-established it there and made material available to the Grevillea Study Group. It is now grown in gardens from Vic. to Qld. Specimens in the author's garden in western Vic. flowered magnificently after 2 years. The species has proven hardy in a wide range of conditions, including cold, wet winters as well as tropical

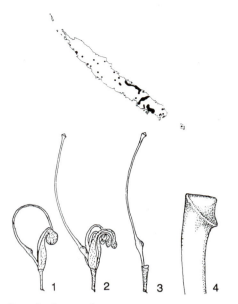

G. exul subsp. *exul*

129B. *G. exul* subsp. *exul* Flowering plant in natural habitat, Mt Koniambo, New Caledonia (P.Abell)

129A. *G. exul* subsp. *exul* Close-up of flowers and foliage (N.Marriott)

climes. It tolerates frost to -2°C and withstands dry spells with little ill effect. Plants grown on their own roots appear as hardy and vigorous as grafted ones.

Propagation *Seed* Untested but either treatment by nicking or soaking should be successful. *Cutting* Firm, young growth taken in the warmer months strikes well. *Grafting* Has been grafted successfully onto *G. robusta* using various methods.

Horticultural features Subsp. *exul* is an open, shrub or small tree with most attractive conspicuously veined, dark green leaves. Long, one-sided, creamy white conflorescences dominate the shrub in season and combine to create a most imposing plant, worthy of much greater usage in horticulture. It is long-lived and has potential in the shrubbery or commercial landscape. Birds should find the nectar-laden flowers attractive.

History *G. exul* was grown at the Sydney Botanic Gardens from 1850 till at least 1903 and apparently performed well there then, as it does now. The first material was collected by the then director, Charles Moore, during a voyage in HMS *Havannah* around the islands of the Western Pacific.

Grevillea exul subsp. rubiginosa
(A.T.Brongniart & J.A.A.Gris) R.Virot (1968)
Plate 130

Flore de la Nouvelle Calédonie et dépendences 2: 153 (France)

The subspecific epithet is derived from the Latin *rubiginosus* (rusty red), an allusion to the usual hair colour of the floral rachis. ROO-BIG-INN-OSE-A

Types: New Caledonia, Nov.–Dec. 1861, E.Deplanche 213 (syntype: P); Mont Dore, New Caledonia, c. 1855–65, E.Vieillard 1114 (syntype: P).

Synonyms: *G. rubiginosa* A.T.Brongniart & J.A.A.Gris (1865); *G. rubiginosa* var. *angustifolia* A.Guillaumin (1935); *G. exul* forma *bicolor* R.Virot.

Conflorescence much-branched; hairs of floral rachis red or brown. **Flower colour**: perianth and style creamish white. **Perianth** hairy outside.

Distribution New Caledonia, confined to the southern part of the island. *Climate* Hot, wet summer; cool to mild, dry winter. Rainfall 900–3000 mm.

Ecology Occurs from sea level to c. 1000 m., in low-lying valleys, on rocky ridges and slopes in serpentine schist and skeletal soil, as well as in alluvial sand beside streams and gravelly loam, usually in open forest or scrub. Flowers May–Jan., the later months usually at higher altitudes. Fire response unknown. Presumably pollinated by birds.

Variation There is some variation in hair colour on the floral axis and elsewhere. R.Virot divided the subspecies into two forms on this basis, but a number of specimens cannot be placed in either form and the separation appears to be unwarranted.

Conservation status Not presently endangered.

Cultivation See notes under subsp. *exul*. This subspecies has performed moderately well at the Royal Botanic Gardens, Sydney.

Propagation *Seed* Untested but treatment by either nicking or soaking should be successful. *Cutting* Firm, young growth taken in the warmer months strikes well. *Grafting* Has been grafted successfully onto *G. robusta*.

Horticultural features Subsp. *rubiginosa* is an open shrub with quite broad, dark green leaves bearing a conspicuous rusty indumentum on the undersurface. Long, one-sided, creamy white conflorescences dominate the shrub in summer and attract many birds. Like subsp. *exul*, it appears to be long-lived and could probably be grown with ease in subtropical to temperate gardens.

Grevillea fasciculata R.Brown (1830)
Plate 131

Supplementum Primum Prodromi Florae Novae Hollandiae exhibens Proteaceas Novas 20 (England)

The specific epithet alludes to the flowers clustered in the leaf axils and is derived from the Latin *fasciculus* (a small bundle). FASS-IK-YOU-LAR-TA

Type: inland from King George Sound, [W.A.], 1828–29, W.Baxter (lectotype: BM) (McGillivray 1993).

Synonyms: *G. aspera* var. *linearis* C.F.Meisner (1845); *G. meisneriana* F.Mueller ex C.F.Meisner (1854); *G. fasciculata* var. *linearis* (C.F.Meisner) K.Domin (1923); *G. fasciculata* var. *dubia* K.Domin (1923).

A low, open to spreading **shrub** 0.3–1 m tall. **Branchlets** angular, tomentose to villous. **Leaves** usually 1–5 cm long, 1–10 mm wide, ascending to spreading, sessile to shortly petiolate, simple, entire, obtuse, mucronate to acute, linear to obovate; upper surface convex, tomentose, soon scabrous, rarely smooth, midvein evident or obscure; margins smoothly recurved or rolled to midvein below; lower surface often concealed, either silky or tomentose with curled, matted hairs, midvein prominent. **Conflorescence** usually erect, sessile to shortly pedunculate, axillary or terminal on short branchlet, simple to few-branched; unit conflorescence c. 1 cm long, 2 cm wide, usually 6–10-flowered, umbel-like; floral rachis villous; bracts 0.5–2 mm long, triangular, villous outside, most falling before anthesis. **Flower colour**: perianth orange or red; style end yellow or orange. **Flowers** abaxially oriented; pedicels 4.5–9 mm long, villous; torus 0.8–1.3 mm across, oblique; nectary prominent, V-shaped to arcuate; **perianth** 4–7 mm long, 3 mm wide, ellipsoidal, prominently ribbed, contracted just below curve, densely villous to sparsely silky outside, bearded inside from c. halfway, cohering except along dorsal suture between the curve and the limb where style end exposed before anthesis; limb nodding, villous, ovoid, the segments free after anthesis; **pistil** 6.5–8.5 mm long; stipe 0.7–1.1 mm long; ovary villous; style villous, not exserted before anthesis, either scarcely or not exceeding perianth after anthesis, straight; style end conspicuous, spoon-like; pollen presenter lateral, round, concave. **Fruit** 10–14.5 mm long, 5–6 mm wide, erect, narrowly egg-shaped, tomentose; pericarp c. 0.5 mm thick. **Seed** 7–8 mm long 1.5–2.5 mm wide, oblong with an apical wing c. 2 mm long; outer face smooth with a slightly raised apical collar; inner face concealed by revolute margin.

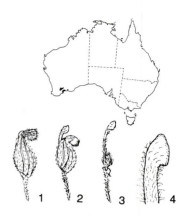

G. fasciculata

Distribution W.A., in the south-west from Cranbrook to Albany and east to Bremer Bay, centring on the Stirling Range. *Climate* Warm to hot, dry summer; cold, moist winter. Rainfall 550–850 mm.

Ecology Grows on sandplain, in mallee scrub and eucalypt woodland, in grey sand, lateritic sand or gravelly loam, sometimes on rocky hillsides. Flowers autumn to spring. Regenerates from seed after fire. Pollinated by birds.

Variation Three forms ('elements' of McGillivray, 1993) are currently recognised.

Coastal (typical) form This form ranges from low or almost prostrate to an erect, bushy shrub to 1 m, with narrowly elliptic to almost linear, dull green leaves with the margins strongly rolled. The undersurface is covered with straight, silvery hairs; the upper surface is scabrous to granular. This extremely floriferous form is found generally from Albany to Bremer Bay and inland to the Stirling Range with one collection from near Mt Barker ('elements' of McGillivray).

Robust form This exceptionally beautiful, distinctive form has longer leaves than other forms (to 9 cm). The linear leaves are silvery-grey with tightly rolled margins which, in dried specimens, obscure the lower surface and its indumentum of straight hairs. Dense, bright orange-red, gold-tipped flowers cluster over every branchlet. It occurs at the eastern end of the Stirling Range (element iii of McGillivray).

Northern form In both habit and leaf shape, this form closely resembles the Coastal form, being an open shrub to 50 cm with linear to narrowly elliptic leaves, but the undersurface of the leaves has matted, curled hairs. The flowers are similar to the Coastal form except that some plants have styles that marginally exceed the perianth. It occurs mainly on the north-western boundary of the Stirling Range near Cranbrook and to the south of the Stirlings (element iv of McGillivray).

Major distinguishing features Leaves simple, entire, 1–5(–9) cm long, linear to elliptic, hairs on undersurface straight or curled; torus oblique; perianth zygomorphic, ellipsoidal, hairy on both surfaces, cohering at anthesis; ovary densely hairy, shortly stipitate, style scarcely to not exceeding the perianth; pollen presenter lateral, concave.

Related or confusing species Group 23, especially *G. fuscolutea*, *G. depauperata* and *G. crassifolia*. The last two differ in their perianth shape which is oblong with a sharp basal dilation. *G. fuscolutea* differs in its longer torus (c. 2 mm across).

Conservation status Not presently endangered.

Cultivation *G. fasciculata* flourishes in cool-wet and warm-dry climates but most forms cultivated

130. *G. exul* subsp. *rubiginosa* in natural habitat, Yate Rd, New Caledonia (P.Abell)

131A. *G. fasciculata* Flowering branches of plant in cultivation, Stawell, Vic. (N.Marriott)

131B. *G. fasciculata* Robust form, cultivated at Vectis, Vic. (N.Marriott)

131C. *G. fasciculata* Robust form, E of Stirling Range, W.A. (P.Olde)

to date have been short-lived. Healthy plants have been recorded in many gardens in Vic., at Burrendong Arboretum, N.S.W., Australian National Botanic Gardens, Canberra, as well as at the Western Australian Herbarium and it is cultivated in California, U.S.A. All forms are hardy to frosts to at least –6°C and may also tolerate occasional light snowfalls. Some forms are more tolerant of extended dry conditions than others, depending largely on provenance. Neither summer-humid climates nor very wet conditions are well tolerated. It performs well in well-drained acidic sand or gravelly loam in the sun or dappled shade. Most forms require either summer watering or soil that retains subsoil moisture throughout the summer. Tip pruning improves density and shape and will increase the number of flowering side branches. Light applications of suitable fertiliser may produce vigorous growth although some growers report that this can shorten life span. It makes a delightful pot plant in a free-draining, peaty gravel mix with slow-release fertiliser but should be protected from strong winds. Many native plant nurseries stock this species.

Propagation Seed Sets good quantities of seed that germinate well when given the standard peeling treatment. Average germination time 44 days (Kullman). Cutting Firm, young growth strikes readily at most times of the year. Grafting Has been grafted successfully onto G. robusta and G. 'Poorinda Royal Mantle' using whip-and-tongue and top wedge grafts.

Horticultural features Probably the most attractive form of *G. fasciculata* is the Robust form, for both its decorative silvery-grey foliage and its brilliant red, yellow-tipped flowers that set the shrub alight from late winter to early summer. Other forms also make a pleasing show in the garden with flowers displayed daintily along the branchlets. Ranging from a dense ground cover to an upright or mounded shrub, it is ideal for a wide range of garden situations, performing well for several years as a rockery plant or small feature plant. Small honeyeaters visit the flowers.

Early history The species was first recorded in cultivation in England in 1873 (Nicholson). Within Australia, it was first cultivated at Melbourne Botanic Gardens in 1870.

General comments The several varieties placed in synonymy mostly correspond to the Coastal form. Further research may lead to formal names for some forms. Element i of McGillivray is here treated as a distinct species, *G. crassifolia*.

Grevillea fastigiata P.M.Olde & N.R.Marriott (1994) Plate 132

The Grevillea Book 1: 186 (Australia)

The specific epithet is derived from the Latin *fastigiatus* (clustered), an allusion to the branchlets. FASS-TIDG-EE-ARE-TA

Type: 20.8 km E of Ravensthorpe, W.A., 12 Oct. 1991, P. Olde 91/305 (holotype: NSW).

Synonym: *G. tetragonoloba* race 'e' of McGillivray (1993).

Erect sometimes pine-like, dense **shrub** 2–2.5 m high with fastigiate branchlets. **Branchlets** angular, silky-pubescent with prominently raised glabrous ridges. **Leaves** 1.8–4.3 cm long, erect to strongly ascending, shortly petiolate, sometimes simple and narrowly linear, mostly bi- or tri-partite, rarely secund-pinnatipartite with up to 7 lobes clustered at apex; lobes (0.1–)0.5–1(–2) cm long, 0.8–1.2 mm wide, tetragonous, linear, rigid, pungent; upper surface glabrous or with scattered straight to slightly wavy hairs, the midvein and marginal veins prominently raised; margins angularly refracted; lower surface obscured, bisulcate. **Conflorescence** 4.5–8 cm long, sessile or almost so; peduncle (0–2 mm long) and rachis rusty-villous; bracts conspicuous, overlapping in bud, 2–2.5 mm long, 2 mm wide, ovate to transversely oblong-elliptic with short glabrous point, tomentose-villous outside, falling before anthesis. **Flower colour**: perianth rusty, pale brown or rusty-brown; style orange-red. **Flowers** acroscopic; pedicels 1.5 mm long, rusty-villous; torus c. 0.8 mm across, straight; nectary tongue-like to patelliform, patent with recurved margin, slightly exceeding toral rim; **perianth** 5–6.5 mm long, 2–2.5 mm wide, ovoid to S-shaped, cohering except along dorsal suture, tomentose-villous outside, glabrous and strongly discolorous inside with a persistent rose-colour evident on dried flowers especially near curve; limb revolute, ovoid to globose, rusty-villous; **pistil** 19–22 mm long; ovary sessile, villous; style first exserted from just below curve and looped strongly upwards, afterwards straight to S-shaped, with sparse short erect or curled hairs on ventral side from base to apex; style end slightly expanded, flanged, enclosed by limb before anthesis; pollen presenter straight, orbicular, flat with prominent stigma. **Fruit** 10 mm long, 5.5–6 mm wide, erect on strongly incurved pedicel, ovoid-ellipsoidal but attenuate, glandular-pubescent with most 2-armed hairs falling before dehiscence; pericarp c. 0.5 mm thick.

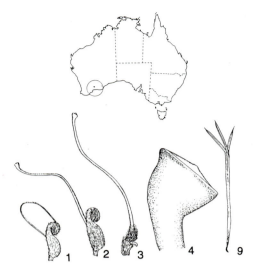

G. fastigiata

Seed 7 mm long, 3 mm wide, narrowly obovoid; outer face shiny with broad pale margin; inner face with a concave, oblong-elliptic central section encircled by a broad channel filled with a raised waxy exudate drawn at both ends into a short, excurrent wing c. 1 mm long.

Major distinguishing features Branchlets with prominent, glabrous ridges; leaves simple or with 3–5 lobes mostly < 1 cm long clustered at apex; lobes glabrous, tetragonous, narrowly linear, the undersurface enclosed by margins, rigid, pungent; floral rachis and pedicels rusty-villous; torus < 1 mm wide; perianth zygomorphic, villous outside, glabrous within; ovary villous, sessile; style sparsely hairy; fruit with predominantly glandular hairs (reddish stripes and patches not seen).

Distribution W.A., confined to a small area around the Jerdacuttup R. ***Climate*** Summer hot, dry; winter cool, wet. Rainfall 350–625 mm.

Ecology Grows in mallee heath or tall shrubland in dark brown decomposed granite loam with blue stone. Flowers spring to autumn with some flowers all year. Pollination is by nectarivorous bird. Regeneration is from seed.

Related or confusing species Group 35, especially *G. rigida* which differs in its larger flowers, non-rusty rachis, larger leaf lobes and sparsely glandular fruit. *G. fastigiata* has a unique, long-persistent rose colouration of the inner tepal surface, evident even on dried specimens.

Variation A relatively uniform species. Some specimens have mostly simple leaves.

Conservation status A code of 2E is recommended. Known only from narrow, degraded road verges.

Cultivation *G. fastigiata* has been in limited cultivation for about 10 years mainly in Vic. where it has flourished and flowered well. It tolerates cool-wet and warm-dry climates and endures frosts to at least -3°C as well as extended dry conditions. Plants do well in full sun in well drained slightly acidic to neutral loam or gravelly clay. This hardy, adaptable plant requires little maintenance once planted out and established. Occasional tip pruning and branch shortening will create a pleasing shape. Nurseries do not usually stock this species.

Propagation *Seed* Sets limited amounts of viable seed that should be pretreated by nicking. *Cutting* Firm, young growth strikes readily in spring. *Grafting* Has been grafted successfully onto a number of rootstocks including *G. robusta* and *G.* 'Poorinda Anticipation', using top wedge and mummy grafts.

Horticultural features *G. fastigiata* is a most rewarding plant valued for its prolific racemes of brownish orange flowers with bright orange-red style and rather small, fine leaves closely appressed to the branchlets which serve to further emphasise the flowering. Nectarivorous birds are strongly drawn to the flowers which continue to open for many months. It is proving extremely reliable in cultivation and deserves wider recognition among both private and commercial growers both for its intrinsic beauty and its relative rarity in the wild. A good feature or screen plant.

132. *G. fastigiata* Cultivated at Stawell, Vic. ex E of Ravensthorpe, W.A. (N.Marriott)

Grevillea fililoba (F.Mueller ex D.J. McGillivray) P.M.Olde & N.R.Marriott (1994)

Plate 133

The Grevillea Book 1: 182 (Australia)

Based on *G. thelemanniana* subsp. *fililoba* F.Mueller ex D.J.McGillivray *New Names in Grevillea (Proteaceae)* 15 (1986, Australia).

Ellendale Pool Grevillea

The specific name is derived from the Latin *filum* (a thread) and *lobus* (a lobe), in reference to the fine leaf lobes of this species. FILL-EE-LOW-BA

Type: Ellendale Pool, SE of Geraldton, W.A., 29 July 1961, R. D. Royce 6459 (holotype: PERTH).

Spreading, dense, soft **shrub** c. 1.5 m high, up to 3 m wide. **Branchlets** angular, sparsely silky, soon ± glabrous. **Leaves** 2–4.5 cm long, ascending to spreading, sessile, pinnatipartite, the basal lobes sometimes with secondary division; lobes 0.2–2 cm long, c. 0.5 mm wide, very narrowly linear, sometimes curved, becoming more spreading with age; upper surface sparsely silky, soon glabrous, midvein obscure; lower surface bisulcate, the lamina obscured, silky in grooves, midvein prominent. **Conflorescence** erect, sometimes pendulous, terminal, pedunculate, simple or few-branched; unit conflorescence c. 3 cm long, 5 cm wide, cylindrical to subglobose when young, ultimately subsecund, lax, usually exceeding foliage; apical flowers opening first; peduncle sparsely silky; rachis sparsely silky, often curved; bracts 0.8 mm long, obovate, glabrous, silky outside, falling before anthesis. **Flower colour**: perianth pink to bright red with white limb; style pink with green tip. **Flowers** adaxially acroscopic, exceeding end of rachis; pedicels 4–6 mm long, glabrous; torus 1.3–1.5 mm across, oblique; nectary adnate to stipe, cushion-like to subreniform; perianth 9–10 mm long, 3–4 mm wide, ovoid, much dilated at base, glabrous and ± glaucous outside, faintly bearded inside, cohering except along dorsal suture, the tepals loose after anthesis and soon falling; limb revolute, obtuse, spheroidal, glabrous to sparsely silky; **pistil** 24–28 mm long, glabrous; stipe 4.5–4.8 mm long, flattened with a central channel, slightly incurved; ovary subtriangular with its base truncate; style at first exserted at curve and looped upwards, gently incurved after anthesis, gradually dilated into style end; pollen presenter very oblique, elliptic, convex. **Fruit** c. 13 mm long, subtriangular, with a broad truncate basal section c. 9 mm long and an attenuated apex c. 4 mm long, 4 mm wide, erect to slightly oblique; pericarp c. 0.3 mm thick. **Seed** 7–8 mm long, 2 mm wide, oblong with a short, oblique beak at the apex; outer face convex, smooth, glossy; inner face obscured by the strongly revolute margins; margin on one side membranous.

Distribution W.A., confined to a few localities around the Greenough and Irwin Rivers E of Geraldton. ***Climate*** Summer hot, dry; winter mild, wet. Rainfall 400–475 mm.

Ecology Occurs on a range of sites from open, mixed scrub along riverbanks to low, dense scrub on crests of hills in breakaway country. Soils range

G. fililoba

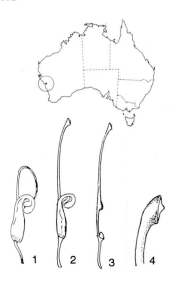

133A. *G. fililoba* Flowers and foliage (N.Marriott)

133B. *G. fililoba* Habit of plant, cultivated at Stawell, Vic. (N.Marriott)

from sandy loam to gravelly clay and rocky loam. Flowers winter–spring. Regenerates from seed after fire. Bird-pollinated.

Variation Only minor variations have been noted. A low to prostrate form has been collected by one of us (Marriott) but this habit is not maintained in cultivation. There is some variation in the thickness and degree of spreading of leaf lobes.

Major distinguishing features Leaves pinnatipartite with spreading, filamentous lobes c. 0.5 mm wide; conflorescence cylindrical to subglobose, lax, terminal, usually clearly exceeding foliage; torus oblique; nectary subreniform; perianth zygomorphic, glabrous outside except sometimes the limb, sparsely bearded within; pistil glabrous; stipe flattened, channelled, incurved; ovary triangular; fruit with a basal ridge, very thin-walled.

Related or confusing species Group 14, especially *G. pinaster*, *G. preissii* and *G. thelemanniana*. *G. thelemanniana* and *G. pinaster* differ in their simple or once-divided leaves. *G. preissii* has denser, more enclosed and more prominently secund conflorescences, coarser, strongly ascending leaf lobes (mostly > 0.8 mm wide) and a smaller pollen presenter.

Conservation status Recommended 2E. Confined to narrow, degraded road verges and river banks on private property and is not presently conserved.

Cultivation *G. fililoba* is a moderately adaptable plant that grows well in cool-wet and hot-dry climates, although in some situations it has a tendency to 'drop dead' for unaccountable reasons (a syndrome well known to growers of Australian plants). Near Perth, it thrives in local gardens and public plantings where it grows well in calcareous sand. In one Sydney garden, it has been growing for nearly 7 years in a deep sand bed. It has truly flourished in heavy, gravelly loams in western Vic. and at Burrendong Arboretum, N.S.W., but needs a sunny, well-drained situation in cold-wet areas. It has also succeeded in Brisbane and further north. It grows rapidly when planted in a warm, sunny site in well-drained, acidic to alkaline sand or sandy loam. Judicious pruning maintains shape although it is naturally a compact, dense shrub. Occasional deep summer watering extends the growth period of the plant and its flowering time, but it is proving quite hardy in extended dry conditions and withstands frosts to at least -4°C. It makes a most attractive pot plant even when not in flower and is a superior horticultural specimen. A well-drained, nutritious mix is ideal. Available at some specialist nurseries especially in Vic. and Perth (where it is sold as *G.* 'Ellendale Pool' of *G.* 'Ellendale').

Propagation *Seed* Untested although it sets numerous seeds that should germinate well. *Cutting* Firm, young growth and semi-hardwood both strike well at most times of the year. *Grafting* Has been grafted successfully onto *G. robusta* using the top and side wedge techniques.

Horticultural features *G. fililoba* is one of the loveliest grevilleas in cultivation. It produces masses of large pendulous flowers that stand out conspicuously from the shrub for many months. These are set off by its equally attractive, fine, rich green foliage which is delicate and reminiscent of lacy fern. It makes a spectacular feature plant or screen plant and deserves to be widely grown because of both its natural horticultural features and its rarity in the wild. Thought to be extinct until recently, it makes an ideal plant for horticulturists to practise conservation by cultivation.

History This species was cultivated in Santa Barbara, California, U.S.A., in 1908 where it had been introduced some 15 years earlier. Cultivation was continuous until at least 1931 and it may be grown there still. The source of its introduction is unknown. In Australia, it was unknown in cultivation until c. 1985, when it was introduced by Zanthorrea Nursery, near Perth, W.A.

General comments Following close examination of this taxon in the field, in the herbarium and in cultivation, we have decided to treat it at specific rank. It differs from all other elements of *G. thelemanniana sensu* McGillivray (1993) in its more prominently terminal, exposed, subglobose, many-flowered and somewhat lax, basipetal conflorescences, its larger, more robust perianth and style end and in its narrower, more spreading leaf lobes. Not to be confused with *G. filifolia*, here treated as a synonym of *G. endlicheriana*.

Grevillea fistulosa A.S.George (1974)
Plate 134

Nuytsia 1: 371 (Australia)

Barrens Grevillea, Pipe Leaf Grevillea

The specific epithet is derived from the Latin *fistula* (a pipe), a reference to the tube-like appearance of the leaves caused by the recurved margins. FISS-TEW-LOW-SA

Type: Middle Mt Barren, W.A., 25 Sept. 1925, C.A.Gardner 1861 & W.E.Blackall (holotype: PERTH).

Erect, mid-dense **shrub** to 2 m. **Branchlets** angular, tomentose. **Leaves** 4–9 cm long, 0.5–1.5 cm wide, ascending to erect, shortly petiolate, linear to obovate, leathery, simple, entire, obtuse, mucronate; upper surface convex, tomentose, soon glabrous but finely scabrous, midvein and intramarginal vein evident; margins smoothly revolute to recurved; lower surface densely white-felted; midrib prominent, appearing as a green stripe. **Conflorescence** 1.5 cm long, 2–3 cm wide, erect, terminal or axillary, shortly pedunculate, simple, rarely 2-branched; unit conflorescence secund to umbel-like, lax, few-flowered; peduncle and rachis villous; bracts 3–4 mm long, ovate to obovate, villous outside, falling before anthesis. **Flower colour**: perianth orange-red; style end yellow. **Flowers** acroscopic; pedicels 5–10 mm long, villous, slender; torus c. 2 mm across, oblique, cup-like; nectary inconspicuous, mostly enclosed within torus; **perianth** 6–8 mm long, 3–4 mm wide, ovoid–attenuate, contracting at neck, villous outside, glabrous inside except sometimes a line of hairs on dorsal tepal and sometimes a faint beard, cohering except at limb where flared and exposing style end before anthesis; limb nodding, ovoid, densely villous; **pistil** 6–8 mm long, markedly oblique to line of pedicel; stipe 1.5–2.2 mm long, adnate to torus; ovary villous; style not exserted before anthesis, straight, appressed-villous; style end broadly flattened, not exceeding perianth; pollen presenter lateral, round, flat to concave, spoon-like. **Fruit** 12–18 mm long, 5–7 mm wide, erect, narrowly ovoid, tomentose, ridged; pericarp c. 0.5 mm thick. **Seed** 7–9 mm long, 2.5 mm wide, oblong-obovoid with an apical elaiasome 2 mm long; outer face convex, smooth; inner face flat; margin strongly revolute.

Distribution W.A., confined to Middle Mt Barren and the Whoogarup Ra. in the Fitzgerald R. NP. *Climate* Warm to hot, dry summer; cool, wet winter. Rainfall 700–850 mm.

Ecology Grows in open heath to thick shrubland in well drained sand on steep quartzite slopes. Flowers autumn to early summer. Regenerates from seed after fire. Pollinated by birds.

Variation A uniform species.

G. fistulosa

134A. *G. fistulosa* Two-year old shrub in cultivation, Stawell, Vic. (N.Marriott)

134B. *G. fistulosa* Flowers and foliage (N.Marriott)

Major distinguishing features Branchlets tomentose; leaves 4–9 cm long, simple, entire, linear to obovate with strongly rolled margins, felted beneath; torus c. 2 mm across, very oblique; nectary obscure; perianth zygomorphic, hairy outside, glabrous or sparsely hairy within; ovary villous, shortly stipitate; style not exceeding perianth; pollen presenter concave.

Related or confusing species Group 23, especially *G. fasciculata* and *G. fuscolutea*, both of which have the torus < 1.5 mm across.

Conservation status 2RC-t. Restricted to small populations in a small area of Fitzgerald River NP.

Cultivation Introduced in the 1980s, *G. fistulosa* has not adapted well to cultivation although it has been grown successfully in some private gardens, mainly in Vic. and N.S.W. where it was established for several years at Burrendong Arboretum and in at least one Sydney garden. It seems to prefer cool-wet to cool-dry climates for most reliable results. In summer-humid climates, fungal attack on the leaves sometimes disfigures its appearance and it may succumb to root-rot disease. In a sunny to dappled shade site in extremely well-drained acid to neutral sand or gravelly loam it will rapidly produce an attractive, densely foliaged plant requiring little maintenance. It tolerates extended dry conditions and frosts to at least –3°C. Regular tip pruning may be necessary to maintain shape and density. It makes an excellent pot plant for a sunny position but the mix must be free-draining. Sometimes stocked by specialist nurseries.

Propagation Seed Sets few seed in cultivation but these germinate fairly well when given the standard peeling treatment. *Cutting* Strikes readily at most times of the year from cuttings of firm, young growth with optimal results being achieved in autumn. Once struck, cuttings grow rapidly. *Grafting* Has been grafted successfully onto G. 'Poorinda Royal Mantle' and the combination appears to be long-term compatible.

Horticultural features *G. fistulosa* is a vigorous free-flowering plant in cultivation and forms a dense, compact shrub. Its erect green leaves, distinguished by strongly rolled margins and a dull white undersurface, have a soft, papery feel and make a harmonious backdrop to the fawnish brown new growth. The orange-red flowers with gold tips are produced in open, umbel-like heads all year with a late winter–early spring flush and stand out prominently. The early flowering is useful in landscaping, providing interest and colour at an otherwise quiet season. Honey-eating birds regularly attend the flowers. On a hardy rootstock and in suitable climates, it would make an excellent feature or screen plant.

Comment Collections from Mt Lindesay previously included in this species have been segregated recently as *G. fuscolutea* q.v.

Grevillea flexuosa (J.Lindley) C.F.Meisner (1845)   Plate 135

in J.G.C.Lehmann (ed.), *Plantae Preissianae* 1: 553 (Germany)

Based on *Anadenia flexuosa* J.Lindley, *Appendix to the first twenty-three volumes of Edwards's Botanical Register*: xxxi n. 142. (1840, England).

Zig Zag Grevillea

The specific epithet is derived from the Latin *flexuosus* (zig zag), in reference to the leaf rachis.
FLEX-YOU-OWE-SA

Type: south-western W.A., late 1830s, J.Drummond (holotype: CGE).

135A. *G. flexuosa* Close-up of conflorescence (N.Marriott)

135B. *G. flexuosa* Habit of plant, Berry Reserve, Stoneville, W.A. (P.Olde)

Irregular, few-branched, ± glabrous **shrub** to 1.5 m. **Branchlets** sharply angular, glabrous. **Leaves** of flowering branches 5–10 cm long, 5–7 cm wide, elsewhere 15–26 cm long, 10–16 cm wide at base, spreading to patent, sessile but appearing petiolate, subpinnatisect; rachis markedly flexuose; primary lobes 7–18, patent to retrorse, usually distant, pinnatifid with up to 5 secondary lobes, sometimes with trifid lobing; ultimate lobes 0.5–3 cm long, 5–10 mm wide, broadly to narrowly triangular, very pungent; upper surface glabrous, the midvein and lateral veins evident; lower surface glabrous, glaucous, the midvein and laterals prominently raised; margins shortly recurved. **Conflorescence** erect, conspicuously pedunculate, terminal or in upper axil, clearly exceeding foliage, usually branched; unit conflorescence 3.5–6.5 cm long, 1 cm wide, cylindrical, very condensed with basipetal development; peduncle sharply angular, glabrous; rachis glabrous; bracts 0.7 mm long, ovate, glabrous with ciliate margins, falling very early. **Flower colour**: yellow to creamy white. **Flowers** acroscopic, glabrous: pedicels 2–2.2 mm long; torus c. 0.5 mm wide, straight, expanded at apex into 4 obtuse, subopposite lobes c. 0.2 mm long; nectary absent; **perianth** 3.5 mm long, 0.5 mm wide, oblong-cylindrical, strongly recurved in young bud, papillose inside, opening first along dorsal suture, all tepals then free below limb, the dorsal tepals reflexing to expose inner surface before anthesis, afterwards loose and irregular; limb revolute, spheroidal, the segments keeled; style end enclosed before anthesis; **pistil** 5.5–8.5 mm long, glabrous; stipe 1.2 mm long; ovary globose; style exposed but scarcely exserted before anthesis, afterwards retrorse to S-shaped; style end scarcely expanded; pollen presenter ± straight, conical with round to oblong-elliptic base.

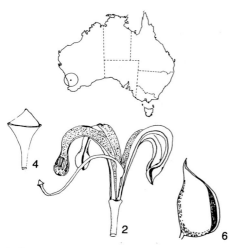

G. flexuosa

Fruit 20 mm long, 10 mm wide, erect, ovoid to ovoid–ellipsoidal, attenuate, rugulose; pericarp 1.5–2 mm thick at suture. **Seed** (cult.) 9 mm long, 3 mm wide, oblong but acute; outer face convex, smooth, slightly crimped within margin; inner face with a raised, central ridge, encircled by a broad channel; margin recurved with a short, waxy wing prominent in basal half.

Distribution W.A., confined to a few locations between Stoneville and Toodyay. *Climate* Cool, wet winter; hot, dry summer. Rainfall 800 mm.

Ecology Grows in granitic sand among granite rocks in exposed platform vegetation and low heath. Flowers winter to early spring. Fire response unknown. Pollinated by insects.

Variation A relatively uniform species.

Major distinguishing features Branchlets sharply angular, glabrous; leaves glabrous, bipinnatifid mostly with > 7 primary, retrorse lobes, the rachis flexuose; conflorescence terminal, exceeding leaves, cylindrical; floral bracts falling very early; perianth zygomorphic, glabrous outside, papillose within; pistil glabrous; style retrorse to S-shaped after anthesis; pollen presenter erect, conical; fruit c. 20 mm long; pericarp c. 2 mm thick.

Related or confusing species Group 8, especially *G. synapheae* which differs in its smaller fruits, its leaves with fewer, ascending lobes and its ± straight leaf rachis.

Conservation status 2EC-t. The recent discovery of a large population has greatly expanded the known distribution.

Cultivation *G. flexuosa* grows very rapidly in cultivation, whether raised from seed or cutting or grafted, but it has proved intolerant of less-than-perfect drainage when on its own roots. As a result, it should be grown in a free-draining, acidic sand or loam, in either sun or dappled shade. Summer watering should be avoided. Regular tip pruning is needed to improve the species' naturally open, spindly habit. Most experience has been gained from potted specimens and grafted plants. In subtropical to cool-temperate conditions, specimens grafted onto *G. robusta* have flourished. A well-drained potting mix with light applications of slow-release fertiliser produce robust, free-flowering potted plants.

Propagation *Seed* Germinates well, particularly when soaked for 24 hours before sowing. Large quantities of seed can be obtained from cultivated plants. *Cuttings* Strikes well from firm, new growth, particularly in spring or early autumn. Leaves should be considerably reduced to prevent excessive moisture loss and reduce bulk. *Grafting* Takes readily onto *G. robusta* using whip, top wedge and mummy grafts.

Horticultural features This most interesting foliage plant with conspicuous bright creamy white flowers is worth growing in any native garden. Contrast applications or feature situations are suggested but it can be grown in crowded conditions, which will tend to hide its generally open, somewhat untidy growth habit. The extraordinary basal leaves will certainly attract comment from interested gardeners. Due to the attractive foliage, ease of grafting and the rapid growth rate, grafted plants of this species are becoming available from specialist nurseries.

Early history Although first collected in 1839, it was not seen again in the wild until 1985, probably not far from Drummond's original locality. It was grown as a glasshouse plant in England and Europe for about a decade from 1840.

General comments *G. flexuosa* is closely related to *G. synapheae* but differs in its habit, its larger fruits, its extraordinary leaves and, with some qualifications and overlap, its longer conflorescence and longer pistil. Its sympatric occurrence with *G. synapheae* subsp. *synapheae* indicates that recognition of this species at specific rank is appropriate.

Grevillea floribunda R. Brown (1830)

Supplementum Primum Prodromi Florae Novae Hollandiae exhibens Proteaceas Novas 19 (England)

Rusty Spider Flower; Seven Dwarfs Grevillea

The specific epithet was derived from the Latin *floribundus* (abounding in flowers), in reference to the flowering habit. FLAW-RIB-UN-DA

Type: hills near Bathurst, N.S.W., 1817, Alan Cunningham, collected on Oxley's First Expedition (lectotype: BM) (McGillivray 1993).

A single-stemmed or lignotuberous **shrub** 0.2–2 m high. **Branchlets** angular to terete, villous. **Leaves** 1.5–8 cm long, 3–15 (–35) mm wide, ascending to spreading, shortly petiolate, simple, linear, oblong, narrowly elliptic to ovate-elliptic, obtuse, mucronate; upper surface pubescent when young, usually soon glabrous and granular, midvein and penninervation conspicuous; margins entire, recurved to revolute; lower surface villous with matted hairs or silky, midvein prominent. **Conflorescence** decurved, pedunculate, terminal, rarely axillary, usually unbranched; unit conflorescence 2–5 cm long, open, partially secund, hemispherical or shortly cylindrical; peduncle and rachis rusty villous; bracts 0.8–2 mm long, ovate to triangular, rusty villous outside, ± glabrous inside, some usually persistent to anthesis. **Flowers** acroscopic; pedicels 3–6 mm long, rusty villous; torus 2.2–3.5 mm across, straight to slightly oblique; nectary inconspicuous, saucer-like to semicircular, the margin wavy; **perianth** 5–8 mm long, 3–8 mm wide, ellipsoidal to subglobose, prominently ridged, sparsely to densely villous outside, sometimes with scattered ascending hairs, inside bearded in upper half above a glabrous, concave base; dorsal and ventral tepals sometimes conjoined for a few mm above base, otherwise cohering except along dorsal suture before anthesis, flared open between curve and limb exposing style end before anthesis; limb densely rusty villous, revolute, globular, the segments free after anthesis, limb of dorsal tepals often reflexing and aligning with ventral limbs in a straight row; **pistil** 9–16(–19.5) mm long; ovary sessile, white-villous; style villous, scarcely to not exserted before anthesis, afterwards straight to slightly incurved and clearly exceeding perianth; style end flattened; pollen presenter ± lateral, flat, ± obovate. **Fruit** 10.5–17 mm long, 6–7.5 mm wide, erect, ovoid, villous, ridged; pericarp c. 0.5 mm thick. **Seed** 12–12.5 mm long, 3.5 mm wide, oblong, with an apical elaiasome, minutely pubescent; outer face convex; inner face channelled; margin recurved.

Major distinguishing features Leaves simple, entire with relatively prominent secondary venation on upper surface; conflorescence decurved, the rachis 1–4.5 cm long; torus straight; nectary inconspicuous; perianth zygomorphic, ellipsoidal, hairy outside, hairy within but contracted into a beard in upper half; ovary densely villous, sessile.

Related or confusing species Group 25, especially *G. chrysophaea* and *G. polybractea*. *G. chrysophaea* is distinguished by the beard only in the lower half of the inner perianth surface, the floral rachis < 1 cm long and the usually longer pistil. *G. polybractea* has broad floral bracts (most > 2 mm wide) and a conspicuous, erect, oblong nectary.

Variation Two subspecies are recognised.

Key to subspecies

Perianth persistent to fruiting, usually rusty-hairy; torus 2.2–3.5 mm across, straight; pedicels, peduncle and rachis usually ≥ 1 mm thick, villous   subsp. **floribunda**

Perianth soon falling; rusty hairs on perianth either lacking or scattered among mainly white hairs; torus 1.2–1.8 mm across, slightly oblique; pedicels, peduncle and rachis c. 0.7 mm thick, tomentose   subsp. **tenella**

Grevillea floribunda R. Brown subsp. floribunda   Plate 136

Synonyms: *G. autumnalis* Lhotsky ex G.Bentham (1870); *G. sphacelata* A.Cunningham ex G.Bentham (1870); *G. ferruginea* R.Graham (1840).

Branchlets angular to terete, villous to tomentose. **Leaves** 1.5–8 cm long, 3–15(–35) mm wide; undersurface matted-villous to silky. **Conflorescence** 2–5 cm long; bracts 0.8–2 mm long, usually persistent at anthesis; pedicels 3–6 mm long. **Flower colour**: perianth greenish beneath a conspicuous rusty indumentum, noticeable especially at the bud stage; style brownish red. **Perianth** 5–8 mm long, 3–8 mm wide, pubescent-villous outside; **pistil** 9–19.5 mm long.

Distribution Qld, in inland areas from Carnarvon to Yuleba and Moonie, and N.S.W., widespread on the W slopes of the Great Dividing Ra. from the Qld border S almost to Albury. *Climate* Hot, dry summer; cool to warm, wet winter. Rainfall 300–800 mm.

Ecology Grows in open or dense scrub, sometimes in tall forest, on hillsides, often on disturbed road verges, in skeletal rocky soil or sandy soil, often over granite or clay. Flowers late autumn to spring. After fire regenerates from seed or sometimes from lignotuber. Pollinated by birds.

Variation Very variable in habit (lignotuberous or single-stemmed), indumentum, leaf shape, size of flowers and floral parts. In some forms the rusty-brown indumentum is more conspicuous and deeper in colour, especially in bud. Dwarf forms to 30 cm have been collected in the Goonoo Goonoo Forest near Dubbo, N.S.W., (P.Althofer pers. comm.) and throughout the Pilliga Scrub and in Qld around Mt

G. floribunda subsp. *floribunda*

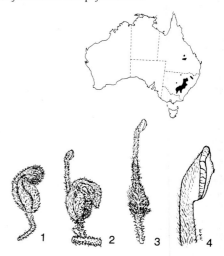

136A. *G. floribunda* subsp. *floribunda* Flowers and foliage (M.Keech)

136B. *G. floribunda* subsp. *floribunda* Conflorescence (P.Olde)

Moffatt. Elsewhere in the wild, low plants occur sporadically. A robust form of unknown origin with broadly elliptic leaves is grown in Brisbane and Sydney.

Small-flowered form This form, mainly confined to the Tingha area, N.S.W., is usually a lignotuberous shrub to c. 30 cm growing on granite. The pistil is 9–11 mm long and the peduncle and rachis are quite slender, showing a tendency towards subsp. *tenella*, but the flowers otherwise are typical, including the rusty indumentum and the perianth persistent to fruiting. There has been one collection of this form in the Pilliga, north-east of Kenebri.

Conservation status Not presently endangered.

Cultivation Subsp. *floribunda* is an extremely hardy species tolerant of a wide range of climates. Once established in the garden, it is both drought- and frost-hardy and thrives in a deep, well-drained soil. Although it prefers a dry climate, plants from some populations have been grown successfully on the coast in Sydney and Brisbane. Some forms seem more adaptable than others, and more trials are needed with selected provenances. An acid to neutral, heavy loam in dry, well-drained conditions in dappled shade under eucalypt canopy or in full sun is ideal. Plants from most populations seem to resent summer watering but there are some notable exceptions. Using a free-draining potting mix, it makes a superb potted plant for many years. Native plant nurseries occasionally stock this species and it is sometimes listed in seed catalogues.

Propagation *Seed* Can be grown successfully from seed using the standard peeling pretreatment. Best sown in spring. *Cutting* The preferred method of propagation to retain select forms. It is relatively easy to strike from cuttings of firm, new growth in spring. *Grafting* Has been grafted successfully onto *G. robusta* using the top wedge technique.

Horticultural features Subsp. *floribunda* is a most attractive shrub with unusual brown flowers, especially noticeable at the bud stage. While some forms have a pale brown indumentum on the flowers, one form, grown for many years at Cadwell's arboretum near Rylstone, had a deep chocolate indumentum. Some forms are extremely floriferous and have interesting, conspicuous flowers. It is suitable for inland climates and makes a striking landscape feature plant. Some forms will tolerate a wide variety of conditions including coastal humidity and light sandy soil provided there is a heavy texture subsoil.

Early history Seeds of this species were sent to the Royal Botanic Garden, Edinburgh, by Richard Cunningham in 1835. It first flowered in the glasshouse there in 1837 and continued to do so till at least 1839, flowering in winter. The invalid name *G. ferruginea* Graham (1840) was once misapplied to *G. floribunda*.

Hybrid Known parent of the cultivar G. 'Poorinda Refrain'.

Grevillea floribunda subsp. tenella
P.M.Olde & N.R.Marriott (1994) **Plate 137**

The Grevillea Book 1: 184 (Australia)

Crows Nest Grevillea

Subspecific epithet derived from the Latin *tenellus* (dainty or delicate), an allusion to the generally smaller, more slender floral parts when compared to subsp. *floribunda*. TEN-ELL-A

Type: Crows Nest Falls NP, Qld, 3 Sept. 1993, P.M.Olde 93/50 (holotype: NSW).

Branchlets slender, terete, tomentose to villous. **Leaves** 4–5 cm long, 2–4 mm wide; lower surface silky. **Conflorescence** 1–2.5 cm long; floral bracts c. 0.8 mm long, falling before anthesis; pedicels 4–5 mm long. **Flower colour**: perianth yellowish green turning brownish red; style brownish red. **Perianth** 3–6 mm long, 3–4 mm wide, sparsely pubescent to villous outside; **pistil** 9–12 mm long.

Distribution Qld, confined to the Darling Downs and lower Burnett district. *Climate* Hot, wet summer; cold, wet or dry winter. Rainfall 750 mm.

Ecology Grows in sandy soil in low eucalypt forest. Flowers late winter to spring. Fire response unknown. Pollinator unknown, but probably honey-eating birds.

Variation There is considerable variation.

Crows Nest Form (Typical form) This is a robust, single-stemmed shrub to c. 2 m with a spreading habit and relatively bright green flowers ageing reddish brown. It occurs in several scattered populations between Crows Nest and Inglewood.

Durong form This delightful little shrub appears to be lignotuberous and grows to c. 50 cm high, 50 cm wide. It is a compact, dome-shaped shrub with bright golden flowers, reminiscent of *G. chrysophaea*. It is known to us from only a single population but may be more widespread in its area of occurrence near Durong, Qld.

Northern form This is a rather spindly, few-branched erect plant to c. 1 m high with dull green, villous flowers. Leaf margins are strongly rolled and the pistil is very short. It has been collected near Manar Homestead, Boondooma and near Eidsvold.

Conservation status Suggested 3RC.

G. floribunda subsp. *tenella*

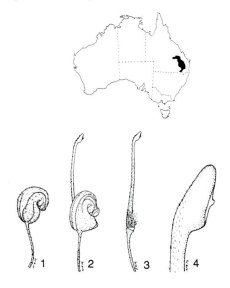

137A. *G. floribunda* subsp. *tenella* Northern form, near Mundubbera, Qld (M.Hodge)

137B. *G. floribunda* subsp. *tenella* The golden flowers of the Durong form, Qld (M.Hodge)

137C. *G. floribunda* subsp. *tenella* Flowers and foliage (P.Olde)

reference to the flowering habit. FLOOR-EE-PEN-DEW-LA

Type: Ben Major Forest Reserve, Vic., 15 Oct. 1976, R.V.Smith 76/23 (holotype: MEL).

A prostrate to sprawling **shrub** 0.3 to 1 m tall. **Branchlets** terete, tomentose. **Leaves** 2–6.5 cm long, 1.5–4 cm wide, ascending, usually petiolate, ovate to oblong–ovate in outline, bipinnatifid, sometimes deeply cleft; lobes 1–2 cm long, 7–15 mm wide, ± triangular, often with shallow, secondary lobing, or toothed, obtuse to acute, pungent; upper surface sparsely pubescent becoming ± glabrous, midvein and lateral veins visible; margins shortly recurved; lower surface with an even, sparse indumentum of curled hairs, midvein, lateral veins and reticulum prominent to evident. **Conflorescence** usually pendulous, terminal, axillary or cauline, pedunculate, simple or few-branched; unit conflorescence 3–5.5 cm long, secund, dense; peduncle 8–25 mm long, wiry, bracteate, glabrous but sparsely pubescent near rachis; rachis pubescent; bracts 2–5 mm long, ovate, silky outside, persistent at anthesis. **Flower colour**: perianth grey-green to purple-brown; style yellow, green-yellow, pale pink or light red with green tip. **Flowers** acroscopic; pedicels 2–3 mm long, silky; torus c. 1.5 mm across, oblique; nectary U-shaped, crenate; **perianth** 6–7 mm long, 2–3 mm wide, ovoid, strongly narrowed at throat, silky outside, glabrous inside, cohering except along dorsal suture; limb revolute, globular, enclosing style end before anthesis; **pistil** 12.5–16 mm long; stipe 2.2–2.9 m long, appressed-villous; ovary densely villous; style at first exserted near base on dorsal side and looped

Cultivation Subsp. *tenella* was introduced in the 1980s but little can be stated with authority about the way it should be grown or its hardiness. It has, however, been grown by enthusiasts, particularly in south-eastern Qld where, once established, it has proved quite hardy to extended dry conditions. It tolerates frosts to c. −2°C and summer rain and humidity. Best results have been achieved in sun or dappled shade in well-drained sandy or gravelly loam. Some protection from wind may be necessary as the branchlets are relatively slender. Tip pruning would undoubtedly improve shape and overall appearance. Not yet available from commercial nurseries.

Propagation *Seed* Untested, but fresh seed treated by nicking should germinate well. *Cutting* Firm, young growth strikes readily at most times of the year. *Grafting* Has been grafted successfully onto *G. robusta* using the whip and mummy methods.

Horticultural features On first seeing this plant, one is struck by the unusual yellowish green colour of the flowers and the change to red after fertilisation, sometimes seen in the one raceme. While the flowers appear very bright and attractive, the foliage is often quite sparse and unremarkable, being confined to the extremities of the branchlets. Garden conditions certainly improve the appearance of this plant which would be a useful screen or foil to more brilliantly coloured plants, or as a feature plant.

Grevillea floripendula R.V.Smith (1981)
Plate 138

Muelleria 4: 423 (Australia)

Ben Major Grevillea

The specific epithet is derived from the Latin *flos* (a flower) and *pendulus* (hanging down), in

G. floripendula

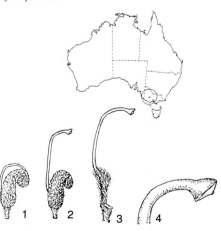

upwards, straight but geniculate near apex after anthesis, glabrous, villous at base, wrinkled on ventral side, dilating suddenly into hoof-like style end; pollen presenter oblique, oblong-elliptic, convex, umbonate. **Fruit** 8.5–11 mm long, 6 mm wide, erect to oblique, usually on incurved stipe, oblong-ellipsoidal, silky with reddish striping or patches; pericarp (Smith 1981) c. 0.5 mm thick. **Seed** 5–7 mm long, 2–3 mm wide, elliptic, compressed, with short wings at each end; outer face convex, rugulose; inner face flat.

Distribution Vic., confined to several small populations near Beaufort. ***Climate*** Hot, dry summer; cool, wet winter. Rainfall 600–675 mm.

Ecology Occurs in shallow gravelly or rocky clay in dry, sclerophyll eucalypt forest on undulating rocky hills. Flowers spring–summer. Fire response unknown. Pollinated by birds.

Variation A variable species in leaf shape and degree of division as well as flower colour. Although several other populations occur, the following forms warrant horticultural recognition.

Ben Major form This form has slightly smaller, less deeply lobed holly-like leaves and dull red to purple-red flowers. It is a spreading, semi-prostrate plant up to 3 m across and is confined to a very limited area in the Ben Major State Forest.

Musical Gully form More variable than the Ben Major form, this population has larger, usually deeply divided leaves, often with secondary division of the lobes, and dull purple-brown, yellow, yellow-green, pink or red flowers. It is generally a decumbent shrub to 1.5 m across, but some plants reach 1 m tall. Several small populations occur in State Forest near the Musical Gully Reservoir, NW of Beaufort.

Major distinguishing features Leaves bipinnatifid, the undersurface with an even, sparse indumentum of curled hairs; conflorescence secund, pendulous; peduncle bracteate, glabrous or almost so, wiry; rachis with a dense spreading indumentum; floral bracts 2–5 mm long, ovate; perianth zygomorphic, glabrous inside; ovary densely hairy, shortly stipitate; fruits reddish-striped.

Related or confusing species Group 35, especially *G. dryophylla* which has a hairy and stouter peduncle (> 1 mm thick); *G. microstegia* which differs in its smaller floral bracts (< 1.1 mm long) as well as its hairy peduncle; *G. montis-cole* subsp. *brevistyla* which has a thicker peduncle and erect fruits; and *G. steiglitziana* which has a longer style (15–20 mm) and a silky leaf undersurface.

Conservation status 2RC-i. Parts of the State Forest where the species is found have been clearfelled for replanting with pine. Urgent steps are required to conserve the remaining habitat of this species.

Cultivation *G. floripendula* has been cultivated for many years and has proven moderately hardy and adaptable, succeeding in cold-wet, cool-wet and warm-dry climates and in summer-rainfall climates as far north as Sydney. It has been grown widely in Vic. and has been established at Burrendong Arboretum, N.S.W., where it tolerates dry summers and frosts to at least –4°C. Greatest success is achieved in a sunny to semi-shaded site in well-drained, acid, gravelly or sandy loam or clay which never becomes too wet but retains sufficient moisture to support the plant in extended dry periods. Light prunings can assist to maintain shape and density, while phosphorus-free fertiliser encourages attractive new growth. It makes an attractive foliage plant for a medium to large tub, although leaves are sometimes disfigured by leaf miners. Specialist nurseries in Vic. carry it, occasionally with the label Grevillea 'Ben Major'.

Propagation *Seed* Sets numerous seeds that germinate well if pretreated. *Cutting* Firm, young growth normally strikes readily at most times of the year. *Grafting* Has been grafted successfully onto G. 'Poorinda Royal Mantle' using a top wedge graft.

Horticultural features *G. floripendula* is an intriguing plant with delightful, holly-like leaves and attractive pale green new growth, but it is the pendulous flowers that create the greatest interest. These are arranged in a short 'toothbrush' and hang on long, delicate stems which resemble thin pieces of curved wire attached to the branchlet. They are not showy in the usual sense but rather appeal as a novel discovery for those closely observing the foliage, among which they tend to be somewhat hidden. In cultivation it is a reasonably long-lived plant and makes a most pleasing groundcover, being particularly suitable for sites under trees or in stony soils. Landscapers could consider it as a mass-planting for low screen or simply planted as a single specimen.

138A. *G. floripendula* A variant with yellow styles (N.Marriott)

138B. *G. floripendula* A variant with red styles (M.Keech)

Grevillea formosa D.J.McGillivray (1986)
Plate 139

New Names in Grevillea *(Proteaceae)* 6 (Australia)

Mt Brockman Grevillea

Aboriginal name: 'andjamgo' (Mayali).

The specific name is derived from the Latin *formosus* (beautiful, handsome), probably in reference to the foliage and flowers. FOR-MOW-SA

Type: c. 6.5 km NW of El Sharana, Pine Ck Rd, N.T., 23 Jan. 1986, R.Schodde & P.Martensz AE495 (holotype: CANB).

Spreading, often sprawling, ferny-leaf **shrub** 0.3–1 m high, 1–2 m wide. **Branchlets** angular, ridged, silky. **Leaves** 6–18 cm long, spreading, ± sessile, secund, pinnatisect; lobes 3–9 cm long, 0.5–1 mm wide, 8–13, narrowly linear, acute, some lobes twice divided; upper surface silky, channelled along the conspicuous midvein; margins revolute to midvein; lower surface ± enclosed, silky in grooves, midvein prominent. **Conflorescence** conspicuous, erect, pedunculate, terminal, simple or few-branched; unit conflorescence 12–30 cm long, secund, dense; peduncle and rachis silky; bracts 1–1.5 mm long, rhombic, silky outside, falling before anthesis. **Flower colour**: green in bud, the limb turning yellow and the perianth yellow to orange; style green turning yellow then orange. **Flowers** acroscopic; pedicels 7–13 mm long, silky; torus c. 2.5 mm across, ± straight to slightly reverse oblique, cup-shaped; nectary conspicuous, semi-circular, irregularly toothed, erose; **perianth** 10–11 mm long, 2.5–3 mm wide, erect, cylindrical, slightly dilated at base, silky outside, glabrous inside, cohering except the dorsal tepals flared open below limb before anthesis; limb revolute, ovoid, densely silky, enclosing style end before anthesis; **pistil** 54–60 mm long; stipe 1.2–1.6 mm long, enclosed within torus; ovary appressed-villous, ± sessile; style glabrous, sparsely silky near ovary, at first exserted from below curve on dorsal side and looped upwards before anthesis, refracted from line of pedicels, elongate and gently incurved after anthesis, strongly incurved just below apex, conspicuously swollen on dorsal side just before expanded style end; pollen presenter oblique, elliptic, conical with a basal collar. **Fruit** 11–16 mm long, 6–7 mm wide, erect, ± round or ovoid, flattened, tomentose; pericarp c. 0.5 mm thick. **Seed** 12–13 mm long, 3 mm wide, oblong-ellipsoidal with a pale, raised marginal border, much flattened, smooth and shiny, membranously winged all round.

Distribution N.T., widespread especially in the Kakadu area around Deaf Adder Gorge but extending about 170 km south of Maningrida.
Climate Tropical. Summer hot and wet, humid; winter mild and dry. Rainfall 800–1200 mm.

Ecology Occurs in shallow sandy soil on rock outcrops in open treeless terrain, usually at the top of sandstone escarpments. Flowers summer to winter. Fire response unknown, possibly regenerates from lignotuber or epicormic buds. Pollinated by

139A. *G. formosa* Flowers and foliage (M.Hodge)

139B. *G. formosa* Flowering plant in cultivation, Brisbane, Qld (M.Hodge)

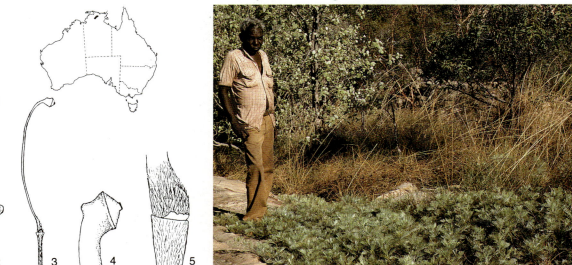
139C. *G. formosa* Growing in natural habitat at Koolpin, N.T. (B.Gunn)

G. formosa
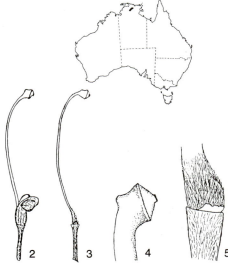

nectarivorous birds but mammals may also be involved.

Variation A species with little apparent variation.

Major distinguishing features Leaves pinnatisect with linear lobes < 1 mm wide, sometimes with secondary division; conflorescence elongate, secund, terminal, simple; torus ± straight; nectary erose; perianth zygomorphic, hairy outside, glabrous within; ovary densely hairy, subsessile; style pale yellow, elongate, humped on dorsal side near style end; pollen presenter oblique, conical with basal collar; fruit much compressed; seed winged all round.

Related or confusing species Group 35, especially *G. pteridifolia* which is usually a tree and differs in its wider leaf lobes (usually 1–5 mm wide), shorter pistil (23–36 mm) and deep orange flowers.

Conservation status Not presently endangered.

Cultivation Although found in the tropical north of Australia, *G. formosa* has adapted well to cultivation since its introduction in the early 1980s, growing on its own roots as far south as Brisbane and, when grafted, at least as far south as Wollongong, N.S.W. Grafted plants have been grown to flowering in protected sites in northern Vic. It has been cultivated widely in Darwin where it grows in the Botanic Garden as well as in public and private gardens. Magnificent plants grow in private gardens near Brisbane where light frosts cut back the foliage but it grows back in warmer weather; heavy frost would undoubtedly kill it. It tolerates fairly dry conditions, even drought. Given an open, sunny position in acidic, sandy, well-drained soil, *G. formosa* grows readily into a spectacular shrub, requiring little maintenance except occasional trimming to retain shape and size. It responds to summer watering and light fertiliser which keep it lush and attractive. Extended cold, wet conditions usually kill grafted plants. It can be grown successfully as a container plant provided the mix is well-drained and fertilised. Nurseries in Brisbane and Darwin occasionally stock this species.

Propagation *Seed* Sets prolific viable seed that germinates readily if sown fresh, preferably in summer, particularly in more southern climates. Better germination results from seed being peeled, nicked or soaked in hot water for c. 24 hours before sowing. *Cutting* Successfully grown from cuttings of firm, young growth in early summer and propagated in a hot-house. Roots do not form readily. *Grafting* Early experiments indicate that *G. formosa* is compatible with *G. robusta* and G. 'Poorinda Royal Mantle' as well as G. 'Ivanhoe' and G. 'rosmarinifolia nana'. Grafted plants have grown into beautiful, long-lived specimens.

Horticultural features *G. formosa* is one of the most spectacular of the tropical grevilleas. It forms a prostrate to decumbent shrub up to 3 m across with silvery, finely lobed foliage which alone is sufficiently attractive to justify its inclusion in the landscape; but it is the huge showy yellow conflorescences, borne on branches around the edge, creating a wreath-like effect, that are guaranteed to excite any plant lover. The flowers develop quickly, opening emerald green and turning a strong yellow or light orange. They are strongly perfumed, especially around mid-day when they drip nectar in such a flow that their popularity with birds is assured. Native bees are also frequent visitors. The seed, which sets copiously, often germinates in the garden. Use as groundcover or rockery plant is self-evident and it makes a fine feature plant. In Darwin, it has been used to great effect in traffic-island gardens.

Hybrids It has great potential as one parent in developing large-flowered plants for horticulture.

Grevillea fulgens C.A.Gardner (1964)
Plate 140

Journal of the Royal Society of Western Australia 47: 55 (Australia)

The specific epithet alludes to the waxy, glossy flowers and is derived from the Latin *fulgens* (dazzling, bright). FULL-JENZ

Type: Mt Desmond, near Ravensthorpe, W.A., 30 Aug. 1962, C.A.Gardner 14070 (holotype: PERTH).

Spreading, straggly **shrub** to 3 m with long, lax branches. **Branchlets** secund, terete, silky, some more conspicuously flower-bearing than others. **Leaves** 4–11 cm long, 1–4 mm wide, ascending, shortly petiolate, stiff, leathery, acute to obtuse, mucronate, simple, linear, entire or few-toothed; juvenile leaves pinnatifid or sparingly toothed, the lobes twisted, triangular to linear, c. 1 mm wide; upper surface pilose, soon glabrous, channelled along midvein; margins smoothly revolute to midvein; lower surface enclosed, silky in grooves, midvein prominent. **Conflorescence** erect, shortly pedunculate to sessile, axillary or terminal on short branchlet, usually 1-flowered; peduncle and rachis villous; bracts 3–9 mm long, imbricate, obovate, brown-villous outside, ciliate, falling before anthesis.

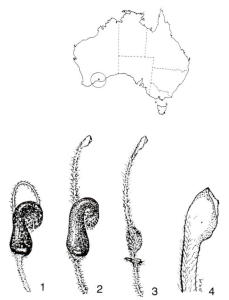

G. fulgens

140A. *G. fulgens* Close-up of flowers (N.Marriott)

140B. *G. fulgens* Plant in natural habitat, Mt Desmond, W.A. (N.Marriott)

Flower colour: perianth reddish pink or rose pink with conspicuous white hairs on the limb; style red. **Flowers**: pedicels 3–8 mm long, brown-villous; torus 2.5–3 mm across, oblique; nectary conspicuous, V-shaped, smooth; **perianth** 13–15 mm long, 5 mm wide, ± oblong, slightly dilated at base, depressed on ventral side near curve, glabrous and waxy outside, bearded inside in lower half, cohering at anthesis except along dorsal suture and between ventral and dorsal sutures at the curve, the tepals detaching soon after anthesis; limb revolute, spheroidal, all covered with long white hairs, enclosing style end before anthesis; **pistil** 23–26 mm long; stipe 0.7–1.5 mm long, villous, inserted perpendicular to torus on dorsal end; ovary densely villous; style at first exserted at curve and looped upwards before anthesis, afterwards slightly incurved, refracted slightly above ovary, villous becoming sparse near apex; style end conspicuous, club-shaped; pollen presenter almost lateral to very oblique, elliptic, convex, umbonate. **Fruit** 12–15 mm long, 6–7 mm wide, erect or slightly oblique, ovoid–attenuate, villous when young but the hairs soon rubbing off; pericarp 0.5–1 mm thick. **Seed** 6.5–7 mm long, 3.2 mm wide, ellipsoidal; outer face convex, smooth and ± shiny; inner face with a raised central elliptic section, with a broad, waxy to membranous margin drawn to a short, oblique wing at the apex and a submarginal recurved lip.

Distribution W.A., restricted to a few hills in the Ravensthorpe Range. There is also one doubtful record from the Parker Range. *Climate* Hot, dry summer; cool to warm, wet winter. Rainfall 350–400 mm.

Ecology Grows in open to dense, mixed mallee scrub in lateritic soil. Flowers spring–summer. Regenerates from seed after fire. Pollinated by birds.

Variation There is some variation in habit and in flower density and colour.

Major distinguishing features Leaves linear, simple, sparingly toothed, rarely entire; flowers solitary or paired; torus oblique; perianth zygomorphic, oblong, waxy and shiny outside, hairy inside, the limb villous; ovary densely hairy, shortly stipitate, the stipe inserted perpendicular to the torus; pollen presenter very oblique.

Related or confusing species Group 26, especially *G. involucrata* which differs in its pinnate leaves and persistent bracts in a whorl.

Conservation status 2E. Both known sites are at serious risk due to potential and actual exploitation for minerals.

Cultivation *G. fulgens* has been cultivated since about 1966 during which time it has proved to be both hardy and adaptable over a wide range of conditions. It has been grown successfully at Burrendong Arboretum, N.S.W., and throughout Vic. where it tolerates frosts down to at least –4°C. Most success is achieved in a warm to hot, sunny site in well-drained, acidic gravelly or sandy loam (pH < 7). Heavier soils should be opened up with good quantities of coarse gravel. Light applications of slow-release fertiliser are appreciated in spring especially in sandy soil but it does not like summer watering. Regular and judicious tip pruning makes the plant less straggly and prevents wind-damage to the somewhat brittle branches. It grows best where its branches can interlock with other shrubs for support as it also has a weak and shallow root system. However, allow a minimum of 1 metre between plants. It is not suitable for more than a few years in a pot due to its large, open habit, although plants have been held in tubs for a number of years at the Grevillea Study Group collection and at Austraflora Nursery, Melbourne. Some specialist native plant nurseries occasionally stock this species.

Propagation *Seed* Sets small quantities of seed that germinate well given the standard peeling treatment. *Cutting* Strikes readily from cuttings of firm, young growth in spring or early autumn. *Grafting* It has been grafted successfully onto a number of different rootstocks but it is too early to tell if the combinations are long-term compatible. Initial success has been achieved with G. 'Poorinda Royal Mantle', G. 'Poorinda Anticipation', *G. robusta*, G. 'Ned Kelly', G. 'Moonlight', and *G. rosmarinifolia*.

Horticultural features *G. fulgens* is an open, 'untidy' plant that develops long, arching, yellowish branches and eventually grows into quite a sprawling, spreading shrub unless regularly pruned. Lightly toothed or simple, dark green leaves are produced sparsely. Many bright pink waxy flowers are dotted singly or in pairs along the upper branchlets, but with age the number of flower-bearing branches is significantly reduced with most branches near the base remaining flowerless. Its open habit reduces its value as a landscape plant unless a large open screen is desired but, grown among other plants, it adds interest and diversity to the planting. In cultivation it is quite long-lived and its large, bright flowers are popular with honey-eating birds. For conservation reasons it should be widely cultivated.

Grevillea fuscolutea G.J.Keighery (1992)
Plate 141

Nuytsia 8: 228–229 (Australia)

Mt Lindesay Grevillea

The specific epithet is derived from the Latin *fuscus* (dark brown) and *luteus* (yellow), in reference to the reddish brown new growth and yellow flowers. FUSS-KOH-LOO-TEE-A

Type: Mt Lindesay, south-western W.A., 10 April 1989, G.J.Keighery 11271 (holotype: PERTH).

Erect to spreading **shrub** to 2 m. **Branchlets** terete becoming angular, white-villous. **Leaves** 1.8–7.8 cm long, 0.5–1.2 cm wide, ascending to erect, shortly petiolate, simple, linear to obovate, obtuse, mucronate; new growth lax, purple, densely villous; upper surface villous, soon glabrous, finely granulate, midvein and intramarginal vein prominent; margins entire, recurved to revolute; lower surface villous, midvein prominent. **Conflorescence** 1–1.5 cm long, 1–2 cm wide, erect to slightly decurved, axillary or terminal on short branchlet, shortly pedunculate to subsessile, simple to 2-branched; unit conflorescence secund-umbel-like, open, 4–10-flowered; peduncle and rachis villous; bracts 3–4 mm long, linear–ovate, villous outside, persistent past anthesis. **Flower colour**: dull yellow. **Flowers** acroscopic; pedicels 3–6 mm long, villous; torus 1.5–1.8 mm across, slightly oblique; nectary semi-circular, entire; **perianth** 5–6 mm long, ellipsoidal, ribbed, villous outside, glabrous inside except a conspicuous beard at the neck, cohering except flared open on dorsal side above curve with ear-like projections behind limb; limb nodding, subglobose, villous, style end exposed before anthesis; **pistil** 6–8 mm long; ovary densely villous, shortly stipitate to sessile; stipe to 0.5–0.8 mm long; style not exserted before anthesis, straight, scarcely exceeding perianth after anthesis, villous; style end broadly expanded; pollen presenter lateral, obovate, concave, spoon-like. **Fruit** 17 mm long, 6 mm wide, narrowly ovoid, ribbed, sparsely villous; pericarp c. 0.5 mm thick. **Seed** 9–9.5 mm long, 2.5 mm wide, oblong–obovoid with a waxy apical elaiasome 2.5 mm long; outer face convex with an apical collar, smooth; inner face flat; margin strongly recurved.

Distribution W.A., restricted to the upper slopes of Mt Lindesay. *Climate* Cool to warm, dry summer; cool, wet winter. Rainfall 800–1200 mm.

Ecology Grows in low, often wet heath in shallow granitic loam near exposed granite outcrops. Flowers autumn to early summer. After fire regenerates from seed. Presumably pollinated by birds.

Variation A uniform species.

Major distinguishing features Branchlets villous; leaves simple, entire, linear to obovate, mostly exceeding 4 cm long; torus slightly oblique; perianth zygomorphic, ellipsoidal with auriculate lobes on the dorsal side at the limb, hairy on both surfaces, cohering at anthesis; ovary densely hairy, subsessile; style not exceeding perianth at anthesis; pollen presenter concave; fruits ribbed.

Related or confusing species Group 23, especially *G. fasciculata* which has a shorter torus (< 1.5 mm across), leaves in most forms < 5 cm long and fruits without ribbing, and *G. fistulosa* which has tomentose branchlets, felted hairs on the leaf undersurface, pedicels (5–)7–10 mm long, the beard on the inside of the perianth absent or only faintly developed, and pale fawn new growth.

Conservation status Suggested 2E. Restricted to small populations in partly cleared area.

Cultivation Introduced in the early 1980s, *G. fuscolutea* has not been widely cultivated. It has been established successfully in both cool-wet and hot-dry climates but is sensitive to extended dry conditions. It requires a well-drained, acidic sand, gravelly loam or clay loam in full sun or semi-shade, and once established needs little maintenance. It has tolerated frosts to at least –6°C but dislikes a humid summer when fungal attack on leaves or root-rot disease can lead to sudden death. Specialist nurseries occasionally stock this species which makes a nice potted specimen if planted in a large tub in a well-watered, free-draining mix.

Propagation *Seed* Sets few seed but these should germinate if given the standard peeling treatment. *Cutting* Grows readily from cuttings using firm, new growth at most times of the year. Take care to avoid overwatering, as the hairy stems trap moisture which can lead to collapse of cuttings. *Grafting* Has been grafted successfully onto *G. robusta* and G. 'Poorinda Royal Mantle'. Grafted plants on the latter grew for many years in Sydney.

Horticultural features The main horticultural features of this species are the shaggy, white indumentum on both branchlets and young leaves and the striking purple new growth which may be produced over long periods. The dull yellow flowers are not boldly displayed, tending to be hidden among the long leaves. They nonetheless provide nectar in abundance for small birds which dart through its branches. It has been successfully used as a screen and feature plant in home landscapes.

Early history The earliest collection of this species was in 1879 by William Webb, a ticket-of-leave man. He was a noted naturalist and collected for Mueller, mainly in the Albany area where he lived and worked as a sandalwood cutter and shepherd.

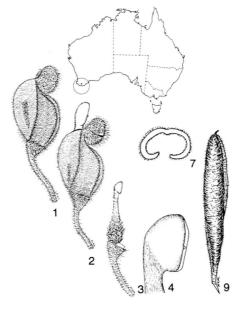

G. fuscolutea

141A. *G. fuscolutea* Shrubs in natural habitat, Mt Lindesay, W.A. (P.Olde)

141B. *G. fuscolutea* Close-up of flowers and foliage (N.Marriott)

Grevillea georgeana D.J.McGillivray (1986)
Plate 142

New Names in Grevillea *(Proteaceae)*: 6 (Australia)

Named in honour of the Western Australian botanist A.S.George, who jointly collected the Type with the author. JAW-JEE-ARE-NA

Type: on N. side of Die Hardy Ra., N of Southern Cross, W.A., 4 July 1976, D.J.McGillivray 3673 & A.S.George (holotype: NSW).

Open, prickly, spreading to rounded **shrub** 1–2 m tall, 1–3 m wide. **Branchlets** terete, silky. **Leaves** 3–7 cm long, spreading, shortly petiolate, deeply and divaricately twice-divided; lobes 4–19 mm long, 1–1.5 mm wide, linear, rigid, pungent; upper surface glabrous and glaucous, ribbed; margins angularly revolute to midvein below; lower surface bisulcate, silky in grooves, midvein prominent. **Conflorescence** erect sometimes on recurved peduncle, simple or few-branched, terminal; unit conflorescence 5–10 cm long, 4 cm wide, cylindrical, apical flowers opening first; peduncle glabrous or sprinkled with silky hairs; rachis glabrous; bracts 1.5–2.5 mm long, ovate, glabrous or sparsely silky, falling before anthesis. **Flower colour**: perianth red, rarely cream-yellow with pink tinges, with a white limb; style red, rarely yellow-cream. **Flowers** basiscopic; pedicels 9.5–11 mm long, slender, glabrous; torus 2–3 mm across, very oblique, cup-shaped; nectary inconspicuous, U-shaped, smooth; **perianth** 8–12 mm long, 3–4 mm wide, oblong-cylindrical, narrowed at neck, strongly curved, glabrous outside, bearded inside in lower half, cohering except along dorsal suture and between dorsal and ventral tepals at the curve; limb revolute, spheroidal, the four segments strongly impressed at the margins; **pistil** 20–25 mm long; stipe 2–3.5 mm long, villous along dorsal side, glabrous on ventral side; ovary villous; style exserted at the curve and looped upwards before anthesis, afterwards gently incurved to straight, sparsely villous near ovary, otherwise glabrous; style end very conspicuous; pollen presenter ± lateral, obovate, flat, slightly umbonate. **Fruit** 7–10 mm long, 6–7.5 mm wide, erect, almost round, white villous, soon glabrous in patches; pericarp c. 1 mm thick. **Seed** 7–8 mm long, 4 mm wide, oblong-elliptic; outer face convex, faintly rugulose; inner face flat, membranously bordered.

Distribution W.A.; ranges N of Southern Cross between Koolyanobbing and Diemals. *Climate* Hot, dry summer; cool to warm, occasionally moist winter. Rainfall 200–225 mm

Ecology Grows in brown loam, in open shrubland on plains or in strongly mineralised rock on slopes of ironstone hills. Flowers spring to summer. Regenerates from seed after fire. Pollinated by birds.

Variation A small-flowered form is recorded from the Mt Manning Ra. (McGillivray) but otherwise this is a relatively uniform species. Several yellow-flowered plants were discovered by the authors in the Die Hardy Ra.

Major distinguishing features Leaves deeply and divaricately twice-divided, upper surface glabrous; lower surface bisulcate; conflorescence cylindrical, opening from the apex first; torus very oblique; perianth zygomorphic, glabrous outside, sparsely hairy within; stipe inserted laterally in the torus; ovary densely villous, style glabrous.

Related or confusing species Group 26, especially

G. georgeana

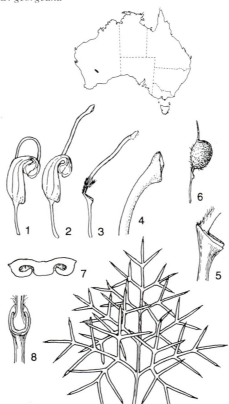

G. wilsonii which differs in its all-red flowers, hairy style and longer pistil (3–3.5 cm long).

Conservation status 3R.

Cultivation Introduced to cultivation by the authors in 1986, *G. georgeana* requires perfect drainage in an open-textured sandy or gravelly loam for success. In the wild, the soil in which it grows consists mainly of dark, strongly mineralised rock on hillsides. In summer, the heat both absorbed and radiated by these rocks must be enormous and one wonders how anything could survive there. To encourage a free-flowering habit, a situation in a warm to hot, dry site in full sun is suggested and, while it is hardy to extended dry conditions, early trials have shown it to be intolerant of even mild frosts. Summer watering and fertiliser are not recommended and regular tip-pruning may be necessary to maintain shape. Grafted plants have done well in Brisbane and Sydney in summer-humid conditions, but on its own roots it is probably suited only to low-rainfall climates. It grows reasonably well in pots but requires a well-drained mix and slow-release fertiliser. Although it is currently unavailable through native plant nurseries, a number of enthusiasts are cultivating it in Brisbane and Vic.

Propagation Seed Untested. Standard pre-treatments should be successful. Cutting Strikes fairly well from cuttings of firm, new growth in spring or early autumn. Struck cuttings need an extremely well-drained soil mix. Grafting Has been grafted successfully onto *G. robusta*, G. 'Poorinda Royal Mantle' and G. 'Poorinda Anticipation'. Early results indicate that the *G. robusta* combinations are compatible long-term and several plants with this rootstock are thriving in different situations.

Horticultural features Floristically one of the most beautiful of all grevilleas, *G. georgeana* has brilliant flowers carried on long cylindrical racemes like fiery candles all over the plant. New growth is pale rusty green but quickly develops the blue-green glaucous appearance of adult foliage which is unpleasantly prickly. In the wild, it tends to be an open plant with wide-spreading branches appearing sparsely foliaged; but the flowering is truly striking. The red and white flowers bloom for an extended period and seem unaffected by hot, dry conditions. The

142A. *G. georgeana* Flowers just before anthesis (P.Olde)

142B. *G. georgeana* Close-up of conflorescence (P.Olde)

142C. *G. georgeana* Unusual yellow-flowered variant in cultivation ex Die Hardy Range, W.A. (M.Hodge)

cream-yellow-flowered form is also quite attractive. Large numbers of honey-eating birds have been observed feeding on the flowers and it should attract similar interest in the garden, while the foliage is ideal as a refuge or nesting site for small birds. It would make an ideal feature plant or massed planting and should be tested over a wide range of situations to assess its suitability. At present, although it is eagerly sought by collectors of the genus, it cannot be recommended for commercial use with any confidence unless grafted.

Grevillea gillivrayi W.J.Hooker (1854)

Plate 143

Hooker's Journal of Botany and Kew Garden Miscellany 6: 358 (England)

Hêtre gris, Hêtre rouge

Named after John MacGillivray (died Sydney 1867), who collected the Type specimen while naturalist aboard the H.M.S. *Herald* on its voyage to New Caledonia in 1853. GILL-IV-RAY-EYE

Type: near summit, Isle of Pines [New Caledonia], 8 Oct. 1853, J.MacGillivray & W.Milne (holotype: K).

Synonyms: *G. deplanchei* Brongniart & Gris (1865); *G. vieillardii* Brongniart & Gris (1865); *G. vieillardii* var. *emarginata* Brongniart & Gris (1865); *G. acervata* S.Moore (1921); *G. comptonii* S.Moore (1921); *G. tontoutensis* Guillaumin (1959).

Bushy or open **shrub** 3–5 m high, sometimes a small tree to 10 m with long, whippy branches. **Branchlets** angular, rusty when young, silky to pubescent. **Leaves** 3.5–15 cm long, 0.2–3.5(–7) cm wide, ascending, petiolate, leathery, usually simple, ± elliptic, sometimes narrowly so, or broadly obovate, obtusely mucronate to acute; upper surface silky when young, soon ± glabrous, the midvein and penninervation evident; margins entire, flat or shortly recurved; lower surface silky or glabrous, the venation prominent. **Conflorescence** erect, pedunculate, usually terminal, simple or branched; unit conflorescence 5–15 cm long, cylindrical, dense, the flowers opening mostly from the apex, sometimes irregularly; peduncle and rachis white- or brown-silky; bracts c. 0.5 mm long, ovate, silky outside, falling before anthesis. **Flower colour:** perianth pink, salmon-pink, yellow, purplish red, or red, sometimes white with a pinkish tinge; style white to red. **Flowers** acroscopic; pedicels 2.5–5 mm long, silky to tomentose; torus 1–1.5 mm across, slightly oblique to straight; nectary U-shaped; **perianth** 8–10 mm long, 1.5–2 mm wide, oblong-cylindrical, silky outside, sometimes glabrous, pubescent to papillose inside, cohering except along dorsal suture; limb revolute, round to spheroidal, enclosing style end before anthesis; **pistil** 16–35 mm long, glabrous; stipe 1.8–5.2 mm long; ovary ovoid; style at first exserted from near base on dorsal side and looped upwards before anthesis, afterwards straight to gently incurved; style end scarcely expanded; pollen presenter oblique, broadly conical with pointed stigma. **Fruit** 15–20 mm long, 7–12 mm wide, oblique on incurved pedicels, compressed-ellipsoidal, glabrous, rugulose; pericarp 0.5–1 mm thick. **Seed** 8 mm long, 3 mm wide, oblong-elliptic, smooth, membranously winged all round.

Distribution New Caledonia, mainly in the southern part. Var. *gillivrayi* is the more widespread and occurs from the coast to an altitude of 950 m. Var. *glabriflora* is known only from Mt Dore, near Noumea. *Climate* Tropical: summer hot, humid and wet; winter mild, dry. Rainfall 900–3000 mm.

Ecology Found mainly in skeletal shaly loam on slopes, at the foot of mountains and beside streams, in open forest, scrub. Flowers all year but principally

142D. *G. georgeana* Plant in natural habitat, Die Hardy Range, W.A. (N.Marriott)

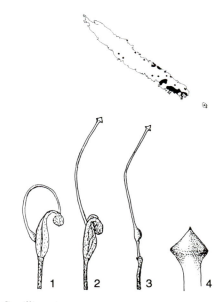

G. gillivrayi

June–Oct. Fire response unknown. Probably pollinated by birds.

Variation There is considerable variation in this species. In the *Flore de la Nouvelle-Calédonie et Dépendances*, Virot recognised two varieties with several intermediates.

var. *gillivrayi* This variety has the outer surface of the perianth pubescent and has two forms based on leaf shape:

forma *gillivrayi* Has broad oval, elliptic or obovate leaves, the apex rounded to obtuse or notched. This form is relatively common in the southern portion of the island.

forma *angustifolia* R.Virot Has either linear to narrowly elliptic or oblong leaves, the apex acute to obtuse. This form is less widely distributed, being confined mainly to the lower slopes along watercourses south from Canala to Bouloupari.

var. *glabriflora* R.Virot This rarely seen variety has bright red flowers and the outer surface of the perianth is glabrous.

Major distinguishing features Leaves simple, entire, the lamina broad, conspicuously veined; conflorescence cylindrical; torus ± straight; perianth zygomorphic, scarcely dilated, hairy or papillose within; pistil glabrous; pollen presenter conical; fruit markedly compressed; seed winged all round.

Related or confusing species Group 3, especially *G. meisneri* which differs in its longer pistil.

Conservation status Except var. *glabriflora* which appears to be rare, the species is relatively common.

Cultivation *G. gillivrayi* was introduced by P.Abell in 1987 and has been grown to flowering at Bulli, N.S.W., where it has grown vigorously to nearly 3 m without obvious problems from pest or disease. Although hardy at this latitude and exposed to a mainly temperate climate, the situation in which it grows naturally is near-coastal and suffers no frost, and therefore care must be taken when assessing its likely performance in more inland, winter-cold

143B. *G. gillivrayi* Plant growing near the sea, New Caledonia (P.Abell)

143A. *G. gillivrayi* Pink-flowered variant with fruits, New Caledonia (P.Abell)

143C. *G. gillivrayi* Close-up of the pale pink-flowered variant, New Caledonia (P.Abell)

143D. *G. gillivrayi* Plant in cultivation, Bulli, N.S.W. (N.Marriott)

situations. It is likely to be fairly hardy in tropical climates and, judging from its natural habitat, is best suited to well-drained, gravelly or sandy loam in full sun or partial shade. Its straggly habit could possibly be improved by tip-pruning or judicious pruning of untidy, long branches.

Propagation Seed Should germinate well from seed sown in the warmer months. Winged seed is known to respond to soaking in hot water and allowed to stand for c. 24 hours. *Cutting* Untested but firm, young growth should strike during the warmer months of the year. *Grafting* Has been grafted successfully onto *G. robusta* using the whip, top wedge and mummy techniques.

Horticultural features *G. gillivrayi* is a decorative shrub though in the wild it often carries many straggly branches that appear untidy. However, its brightly-coloured, cylindrical conflorescences are extremely showy and would produce a most spectacular display if mass planted in all its many colour forms. It would make an attractive addition to the tropical garden where it could be used for screening or even as a feature plant provided its untidy habit was controlled. The nectar-laden flowers should attract honey-eating birds and it may also prove an excellent parent for hybridisation.

Grevillea glabrescens P.M.Olde & N.R.Marriott (1993) Plate 144

Telopea 5: 406 (Australia)

The specific epithet is derived from the Latin *glabrescens* (becoming glabrous), in reference to the leaves and branchlets which soon shed the hairs from young growth. GLAB-RESS-ENZ

Type: c. 4 km S of El Sharana, Kakadu NP, N.T., 13°33'S, 132°30'E, 20 April 1990, A.V.Slee 2681 & L.A.Craven (holotype: CANB).

A spreading to erect, sometimes open, single-stemmed **shrub**, 0.9–2.5 m tall. **Branchlets** angular to terete, glabrescent. **Leaves** 6.5–13 cm long, 1–4 cm wide including petiole c. 10 mm long, ascending, glabrous except occasional appressed hairs on midvein, concolorous, densely pitted, oblong in gross outline; margins flat with 2–10 irregular to subopposite, triangular, sharp lobes (points 1–2 mm long, pungent or scarcely so) and sinuses deeply and broadly arcuate to rectilinear; apex acute, pungent; base cuneate; venation mixed craspedodromous with conspicuous reticulum. **Conflorescence** decurved, terminal to axillary, simple or few-branched; unit conflorescence 2–4 cm long, conico-cylindrical, loose, delicate; primary peduncle 1–2 cm long, glabrous, sometimes two peduncles arising in same axil; rachis glabrous; bracts 0.5 mm long, ovate, glabrous, margins sometimes shortly ciliate. **Flower colour**: perianth green in bud, turning white at anthesis to cream after pollination; style green turning cream with green style end. **Flowers** acroscopic; pedicels 2.5–4 mm long, glabrous; torus 1–1.5 mm across, oblique; nectary U-shaped, entire or toothed; **perianth** 6 mm long, 1.5–2 mm wide, narrowly ovoid-attenuate, glabrous outside, conspicuously papillose and mealy inside and either entirely glabrous or with occasional scattered trichomes or quite densely bearded with most trichomes < 1 mm long; dorsal tepals separating from ventral tepals, reflexing strongly at anthesis; limb spheroidal, revolute in upper half of perianth; style end partially exposed before anthesis; **pistil** 15–20 mm long; stipe 2–2.5 mm long, glabrous; style exposed almost to base, looped upwards before anthesis, afterwards strongly incurved, glabrous except tubercles or scattered trichomes extending 2–3 mm from style end; pollen presenter convex; stigma off-centre. **Fruit** 14–16 mm long, 10–12 mm wide, ellipsoidal, adaxially transverse to stipe with sutures directed outwards; pericarp 1 mm thick at centre face. **Seed** 8 mm long, 3 mm wide, ellipsoidal, flat to slightly convex on inner face with encircling, membranous wing 2–2.5 mm wide.

Distribution N.T. from Graveside Gorge S to c. 10 km S of El Sharana in Kakadu NP. *Climate* Tropical with a wet, hot summer and dry, warm to hot winter. Rainfall 1200 mm, falling mostly in summer.

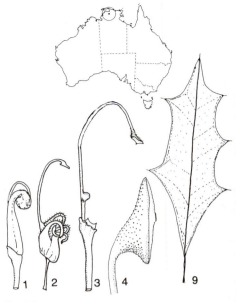

G. glabrescens

Ecology Found exclusively on the sandstone escarpment either just below the top, on the slopes and in valleys beside creeks, or in low heath at the top. It grows in sandy, rocky and shallow soils or in cracks and crevices. Flowering dependent on length of wet season. Specimens in flower have been collected in all months except Dec. In 1992, most flowering had finished by July. Fire response is unknown but regeneration is probably from seed, and some sprouting from epicormic buds following

144A. *G. glabrescens* Close-up of conflorescence (K. & L. Fisher)

144B. *G. glabrescens* Flowering branch (K. & L. Fisher)

low-intensity fires. Pollinator unknown, probably either birds or bees or both. Flowers sweetly scented.

Variation A relatively uniform species. The beard on the inner perianth surface varies somewhat.

Major distinguishing features Leaves simple, toothed, concolorous, glabrous with flat margins; conflorescence decurved, conico-cylindrical; torus oblique; perianth zygomorphic, glabrous outside, hairy or papillose within; ovary glabrous, stipitate; style far exceeding perianth, minutely hairy or granular at apex; fruit oblong-ellipsoidal; seed winged all round.

Related or confusing species Group 11, especially *G. angulata* and *G. brevis*. *G. angulata* differs in its minutely hairy leaves and longer stipe of the ovary (3–5 mm long). *G. brevis* has a shorter pistil (8–12.5 mm long) and some or all leaves entire.

Conservation status A code of 3RC is recommended.

Cultivation *G. glabrescens* is unknown in cultivation, but judging from its natural habitat should grow well in both tropical and subtropical climates. It prefers full sun in sandy soil, acid or neutral in pH. It is unlikely to tolerate frost or continuous cold weather. Winter rainfall climates are likely to cause fungal problems on the leaves. Use only light fertiliser on potted plants until requirements can be assessed.

Propagation Seed In the wild, it sets many seeds that should respond to soaking overnight in warm to hot water to enhance germination. *Cutting* Untested. Should strike well from firm, young growth. *Grafting* Untested. *G. robusta* should be a good rootstock.

Horticultural features *G. glabrescens* is an attractive plant with pale green leaves and masses of white flowers, borne prominently over the shrub in early winter. The species would have some appeal in mass plantings or roadside situations where colour was not a priority. The white flowers are sweetly scented, noticeable mostly around the middle of the day. The prickly leaves preclude close plantings beside walking paths or near access areas.

General comment The distribution commences over 100 km SW of the most southerly collection of *G. angulata*.

Grevillea glauca J.Banks & D.Solander ex J.Knight (1809)   Plate 145

On the Cultivation of the Plants belonging to the Natural Order of Proteeae: 121 (England)

Bushman's Clothes-Peg; Nut Wood;
Kawoj (NW New Guinea)

Aboriginal: 'nalgo' (Cardwell)

The specific epithet is derived from the Latin *glaucus* (greyish), in reference to the leaf colour.
GLOR-KA

Type: near Endeavour R., [Qld], June–July 1770, J.Banks & D.Solander (lectotype: BM) (McGillivray 1993).

Single-stemmed **tree** or large **shrub** 3–15 m high with dark-grey, furrowed bark. **Branchlets** ± terete, silky to tomentose. **Leaves** 6–20 cm long, 1.5–6.5 cm wide, ascending, petiolate, simple, ovate to elliptic, pliable; apex acuminate with non-pungent mucro; upper and lower surfaces similar, visibly penninerved, midvein and intramarginal veins prominent, silvery-grey pubescent, slightly more appressed on undersurface; juvenile leaves silky bronze especially towards apex; margins entire, flat, undulate. **Conflorescence** decurved to pendulous, shortly pedunculate, terminal, usually branched; unit conflorescence 8–18 cm long, 2–3 cm wide, dense, cylindrical, opening from apex to synchronously; peduncle bracteate, pubescent; rachis pubescent; bracts 0.5–1 mm long, ovate, pubescent outside, persistent to anthesis. **Flower colour**: creamy white or greenish white. **Flowers** transversely oriented, the ventral sutures of each pair back-to-back; pedicels 2.5–3.8 mm long, pubescent; torus c. 1 mm across, oblique; nectary conspicuous, V-shaped, sparingly toothed; **perianth** c. 4 mm long, 1 mm wide, narrowly oblong, silky outside, glabrous inside, cohering except along dorsal suture; limb revolute, ovoid to spheroidal; **pistil** 14–16.5 mm long, glabrous; stipe 0.6–1 mm long; ovary globose; style at first exserted from dorsal suture and looped upwards, straight to slightly inflexed at the middle after anthesis, filiform, style end slightly expanded; pollen presenter straight or slightly oblique, elongated conical with basal collar. **Fruit** 2.5–4 cm long, 2.5–4.5 cm wide, oblique, persistent, ± round to irregularly lens-shaped, rugose; pericarp 13–16 mm thick. **Seed** 7–15 mm long, 7–12 mm wide, broadly elliptic, ± flat, very broadly and membranously grey-winged all round, the raphe evident within the wing, the surface of both testa and wing bearing numerous short grooves resembling brush marks.

Distribution Qld, from Cape York Peninsula S to Bowen; also in Papua-New Guinea, along the southern section opposite Cape York; and on the Aru Islands, Indonesia. *Climate* Summer hot, humid, wet; winter mild, dry. Rainfall 500–1500 mm mainly in summer.

Ecology Widespread in many soil types and habitats, both coastal and inland, from open, grassy woodland to gravelly ridges, in open eucalypt forest or savannah woodland, sometimes in waterlogged soils at low altitudes. Flowers mainly winter–spring, occasionally in summer. After fire regenerates from seed and epicormic buds. Pollinator unknown.

Variation A fairly stable species with little variation.

Major distinguishing features Tall tree habit; leaves simple, entire, dorsiventral but similar on both sides, the margin flat; conflorescences branched, cylindrical; perianth zygomorphic, hairy outside, glabrous inside; pistil glabrous; ovary very shortly stipitate; pollen presenter erect, elongate-conical; fruit thick-walled; seed flat, broadly winged all round with brush-like markings.

Related or confusing species Group 4, especially *G. donaldiana* and *G. myosodes*, both of which may be distinguished by the perianth being glabrous outside.

Conservation status Not presently endangered.

Cultivation *G. glauca* grows well in warm to hot, tropical to subtropical climates at least as far south as Sydney, N.S.W., where it has proved hardy, and adaptable. Light frosts are tolerated provided the general climate is warm/hot. A seedling grafted onto *G. robusta* rootstock was successfully grown to c. 2.5 m high at the Royal Botanic Gardens, Sydney, before eventually blowing over in strong winds.

G. glauca

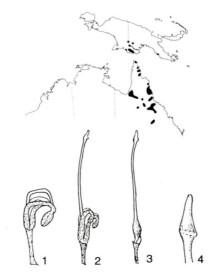

145A. *G. glauca* Fruits. (P.Olde)

Beautiful grafted plants are being grown in frost-free sites in a number of gardens in northern Vic. Given a suitable climate, it will grow into a low, silvery-grey tree, looking far more attractive in cultivation than in the wild. It succeeds best in a well-drained sunny position and responds favourably to summer watering. Soil should be neutral to acidic but most growers find that it is not fussy about soil type, growing equally well in sandy loam and heavy soil. Seed suppliers and nurseries, mainly in Qld, occasionally list this species.

Propagation Seed Mainly grown from seed sown fresh and preferably peeled or nicked to allow better water penetration. Soaking in extremely hot water for c. 24 hours is also reputed to be useful. *Cutting* Not generally tested. *Grafting* Has been successfully grafted onto a number of rootstocks including *G. robusta*, *G. banksii* and *G.* 'Poorinda Anticipation' using cotyledon or top wedge grafts. The unions appear to be compatible although several grafted plants have died without obvious cause. Unions should be made as close to the ground as possible so that the boldly-furrowed trunk may be admired without the distraction of a starkly-contrasting rootstock.

Horticultural features *G. glauca* grows into an attractive small tree with some potential for use in street landscapes or parklands. The foliage is bluish-grey or silvery-grey but rusty brown when young, the coppery tones of which are most decorative on a vigorously-grown plant. The adult foliage is quite dense, making it an excellent screen plant and, in its colouration, serving as a delightful contrast to green foliage in the tropical landscape. Scented, white racemes, which resemble bottlebrushes, dangle in profusion in winter but are not always conspicuous because, although terminal, they are usually borne on short side branches within the broad-leaved foliage and are therefore sometimes concealed. Nonetheless, they are attractive and lure nectar-loving birds and insects. The pendulous golf-ball size fruits may also be prominent when green or when dark-grey and spent, clustered around the base of the plant.

Uses The fruits were reputedly used by the early pioneers as rudimentary clothes' pegs because, when they drop their seed, they remain in a partially-open position, resembling a half-open mouth. Today, they are sometimes used in bush craft. The seeds were reportedly eaten raw by the Koka-amura tribe in New Guinea.

General comments The wood is described as dark-brown, prettily marked, close-grained and hard, but its greasy nature prevents it shining when polished (Bailey). Remnant specimens from the original bush have been retained in the median strip on the main Port Douglas Road. These beautiful old plants, with furrowed black trunks and contrasting silvery leaves, attract many honey-eating birds when in flower.

145B. *G. glauca* Flowers and foliage (M.Hodge)

145C. *G. glauca* Tree habit, cultivated at Brisbane, Qld (P.Olde)

Grevillea globosa C.A.Gardner (1964)
Plate 146

Journal of the Royal Society of Western Australia 47: 55 (Australia)

The specific epithet alludes to the globular flower heads and is derived from the Latin *globosus* (spherical, globular.) GLOW-BOE-SA

Type: 32 km N of Pindar, W.A., January 1963, F.Lullfitz 2241 (holotype: PERTH).

Spreading, domed **shrub** c. 2 m high, 3 m wide. **Branchlets** terete, (brown-) silky. **Leaves** 4–18 cm long, ascending, shortly petiolate, pinnatipartite; leaf lobes 3–9, 1.5–13 cm long, c. 1 mm wide, dipleural, subterete to narrowly linear, sometimes trigonous, firmly pliable, softly pungent; upper surface silky, midvein evident; margins revolute to midvein below; lower surface bisulcate, silky in grooves, midvein prominent. **Conflorescence** erect, long-pedunculate, terminal or axillary, simple or rarely few-branched; unit conflorescences 2 cm long, 2–3 cm wide, globose, dense; peduncle 2.5–8 cm long, brown-silky; rachis villous, broader than peduncle; bracts 2–4 mm long, ovate-acuminate, rusty villous outside, falling before anthesis. **Flower colour**: perianth green beneath a white to rusty indumentum outside, green inside, turning glossy black soon after anthesis; style green. **Flowers** adaxially oriented; pedicels 2.5–4 mm long, villous; torus c. 1.5 mm across, ± straight; nectary U-shaped, smooth; **perianth** 7–10 mm long, 2 mm wide, oblong, slightly dilated at base, villous-tomentose outside, sparsely pilose inside near base, otherwise glabrous, at first separated along dorsal suture, the tepals then separating to base, the dorsal tepals reflexing and exposing inner surface in upper half before anthesis; limb revolute, globular to spheroidal, densely white/brown villous, enclosing style end before anthesis, not immediately relaxed afterwards; **pistil** 13–22 mm long; stipe 1.4–2.5 mm long, villous; ovary densely villous; style exposed above ovary except style end before anthesis, afterwards strongly incurved, pilose, sparsely villous at base; style end flattened; pollen presenter oblique, obovate, flat or slightly convex. **Fruit** 9–12.5 mm long, 6.5–8 mm wide, erect, oblong-ellipsoidal, faintly ribbed, villous, especially base, soon glabrous, rugose; pericarp c. 1 mm thick. **Seed** 9–10 mm long, 3.8–4.8 mm wide, oblong, winged all round; outer face convex, rugulose; inner face slightly convex, channelled around the margin.

Distribution W.A., in the semi-arid region near Pindar, Paynes Find and Yalgoo. *Climate* Hot, dry summer; mild to warm, moist winter. Rainfall 200–250 mm.

Ecology Grows in Mulga shrubland in red loam or in mallee woodland in yellow, lateritic sand.

G. globosa

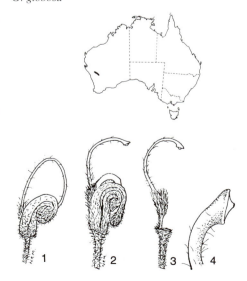

Flowering somewhat irregular, usually late spring–summer but at other times in response to rainfall. Regenerates from seed after fire. Pollinator unknown.

Variation This species is known from only 3 populations which differ in flower size (McGillivray 1993). In the Yalgoo area, plants generally grow to c. 1 m tall and have smaller flowers.

Major distinguishing features Leaves pinnatipartite, the lobes very narrow; conflorescences in dense, globose clusters; perianth zygomorphic, hairy on both surfaces; ovary villous, stipitate; fruit oblong-ellipsoidal; seed winged all round.

Related or confusing species Group 29. An unusual species of undetermined affinities.

Conservation status 3E. Three small populations known.

Cultivation *G. globosa* has rarely been cultivated but has been established in deep sand at Kings Park in Perth. In Sydney, where it was introduced in 1988, grafted plants have been held successfully in pots. In the wild, it tolerates light frosts and extended dry periods and is found scattered among low trees in partial shade or in full sun in very well-drained sand or gravelly loam. In cultivation, it would probably resent summer watering if grown on its own roots and, although it grows naturally into a compact shrub, occasional trimming of straggly branches may prove necessary. At present, it is unavailable from seed suppliers or from native plant nurseries but members of the Grevillea Study Group are attempting to increase numbers in cultivation. A specimen grafted onto *G. robusta* has succeeded and flowered for several years in the Blue Mountains of Sydney.

Propagation *Seed* Sets small quantities of seed which should germinate after pre-treatment. Average germination time 47 days (Kullman; sample size: 3 seeds). *Cutting* Strikes only with difficulty from cutting, although more testing is needed using cultivated plants. *Grafting* Has been grafted successfully onto *G. robusta*, G. 'Bronze Rambler' and G. 'Poorinda Royal Mantle', using top wedge and whip-and-tongue grafts.

Horticultural features *G. globosa* is a naturally rounded, dense shrub ideally suited to screening or massed planting along boundaries and fences. The young branchlets, foliage and floral stems have a quite rusty indumentum which conveys to the outer edge of the plant a brownish hue, in contrast to the otherwise dull green, finely divided foliage. The whitish-green globular flower heads are never displayed en masse but seem to be dotted around the shrub over a long period like balls of cotton, becoming possibly more profuse after rain. They terminate long, stiff stems, and while the flowers tend to blacken quickly there is some potential for the cut flower trade as the flowers can hold for a relatively long time without wilting. They also have a sweet, spicy smell and produce a reasonable drop of nectar that should interest honey-eating birds. It could be used widely in the general landscaping of dry, inland areas and, due to its rarity, should be widely planted for conservation purposes. While not a spectacular species, in many ways it is unique and should be greatly valued by lovers of the genus.

Grevillea glossadenia D.J.McGillivray (1975) Plate 147

Telopea 1: 21 (Australia)

The specific epithet is derived from the Greek *glossa* (a tongue) and *aden* (a gland), in reference to its tongue-like nectary. GLOSS-A-DEE-NE-AR

Type: near Bakerville, Qld, 17 Mar 1972, B.P.M.Hyland 5927 (holotype: NSW).

A compact, dense, greyish-brown **shrub** 1–2 m high, 1–2 m wide. **Branchlets** angular to terete, silky. **Leaves** 5–12 cm long, 2–3.5 cm wide, ascending, petiolate, simple, ± elliptic, obtuse, mucronate to acute; upper surface ± flat, sparsely silky with white or reddish-brown hairs; midvein and intramarginal veins conspicuous, lateral venation acutely angled; margins shortly recurved; lower surface silky; venation brochidodromus, prominent; juvenile leaves distinctly rusty-silky especially at apex and on underside. **Conflorescence** 2.5–3 cm long, 6–8 cm wide, erect, shortly pedunculate, terminal or axillary, simple, umbel-like to globose, few-flowered (6–10), not exceeding foliage; peduncle and rachises silky; floral bracts 0.5 mm long,

146A. *G. globosa* Plant in natural habitat, near Paynes Find, W.A. (J.Cullen)

146B. *G. globosa* Close-up of conflorescence (N.Marriott)

146C. *G. globosa* Flowering branches (P.Olde)

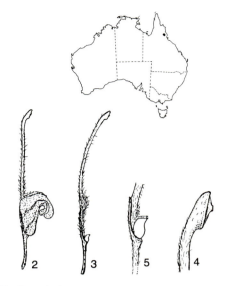

G. glossadenia

147. *G. glossadenia* Flowers and foliage (N.Marriott)

ovate, villous outside, falling before anthesis. **Flower colour**: perianth dull orange becoming bright at anthesis due to brilliantly coloured inner surface of tepal lobes above curve; style orange. **Flowers** basiscopic, abaxially oriented; pedicels 7–9 mm long, silky; torus 7–8 mm across, almost lateral; nectary conspicuous, erect, tongue-like, toothed; **perianth** 7–9 mm long, 6–10 mm wide, strongly zygomorphic, ovate in side-view, irregularly dilated, sharply constricted in middle, silky outside, pubescent inside, separated along dorsal suture from base to apex, the style end exposed before anthesis, tepals conspicuously narrowed at curve; limb revolute, spheroidal, the inner surface ± glabrous, brightly coloured; **pistil** 26–32 mm long, villous; stipe 7.5–9 mm long, broader than style, adnate to torus along most of its length; ovary globose, villous; style exserted from curve and looped upwards before anthesis, gently incurved afterwards, stout, slightly flattened, becoming glabrous at flattened, enlarged style end; pollen presenter lateral, obovate, convex. **Fruit** 12–14 mm long, 8 mm wide, erect, persistent, obovoidal, silky; pericarp c. 1.5 mm thick. **Seed** (per McGillivray) 7–8 mm long, 3.5–4 mm wide, oblong to narrowly ovate, convex and smooth on both surfaces, membranously bordered on the inner face.

Distribution Qld. Occurs in a small locality in the Atherton area near Herberton, Watsonville and Irvinebank, and W of Chillagoe. *Climate* Hot, wet summer; cool to mild, dry or wet winter. Rainfall 1000–1300 mm, mainly in summer.

Ecology Generally found in skeletal, stony soil of decomposed granite in open eucalypt woodland. Flowers mainly late autumn–winter but usually some flowers all year. Fire response unknown. Pollinated by birds.

Variation A uniform species with no apparent variation.

Major distinguishing features Leaves simple, entire with conspicuous venation; conflorescence umbel-like, 6–10-flowered; torus almost lateral; nectary conspicuous, erect, linguiform; perianth very zygomorphic, hairy on both surfaces; pistil villous; stipe adnate to torus, elongate; fruit obovoid.

Related or confusing species Group 19, especially *G. decora* which can be readily distinguished by its longer, many-flowered conflorescences.

Conservation status 2V. An endangered species with a restricted distribution.

Cultivation G. glossadenia has been in cultivation for more than 20 years and grows satisfactorily in a wide range of conditions, tolerating considerable extremes of climate. It has been grown successfully right down the east coast of Australia and inland and is known to tolerate frosts as low as –8°C. It will also tolerate dry spells but the soil should not be allowed to dry out completely. This reliable grevillea succeeds in most garden conditions, in either sandy or heavy soil. Pruning does not appear necessary as it is naturally a compact, rounded shrub. Although it will respond vigorously to summer watering and applications of most low-phosphorus fertiliser, these are rarely necessary. It is fairly easily obtainable in nurseries, especially in Qld.

Propagation Seed Grows readily from seed sown fresh on a capillary bed and will germinate freely if given the standard peeling treatment. *Cutting* Plants can be grown from cutting but it is generally very slow to produce roots. Hormone treatment and wounding improve the strike rate. Best results are achieved with half-ripe material taken in early summer. *Grafting* Has been grafted successfully onto *G. robusta* and G. 'Poorinda Anticipation', using both approach and top wedge methods.

Horticultural features G. glossadenia is a decorative, small to medium-sized shrub which, in some areas, has a greyish-white tomentum on its leaves and branchlets. The broad, green leaves are strongly veined, a feature particularly evident when backlit by the sun, and when young have a noticeable rustiness. The small clusters of orange flowers, while few in number, are boldly coloured and stand out against the foliage, attracting many honeyeaters throughout the cooler months. The fruits also persist on the plant over many months. This long-lived and hardy plant has proved useful as a structure plant and, having foliage dense to the ground, is ideal for low screening. In summer-rainfall climates, it is quite vigorous and pest-free while in cooler climates it tends to grow more slowly.

Hybrids No hybrids with other wild species have been collected but an attractive garden hybrid with *G. venusta* (G. 'Orange Marmalade') is known.

Grevillea goodii R.Brown (1810)

Plate 148

Transactions of the Linnean Society of London 10: 174 (England)

Good's Grevillea

Aboriginal: 'burrun burrun' (Gupapuynga). Nectar is shaken into the hand and sucked by the people of the Milingimbi area.

The specific name commemorates Peter Good (d. Sydney 1803), an assistant to Robert Brown during the voyage of the *Investigator* 1801–1803, and foreman gardener at Kew prior to the trip. GOOD-EE-EYE

Type: Carpentaria and Arnhem Land, [N.T.], 1803, R.Brown '3348' (possibly collected by Good himself) (lectotype: K) (McGillivray 1993).

A lignotuberous **shrub** with prostrate, trailing stems, sometimes sprouting annually, spreading 1–2 m. **Branchlets** angular, silky or sparsely so. **Leaves** 5.5–25 cm long, 1.5–5 cm wide, spreading, petiolate, simple, oblong-elliptic or narrowly ovate, obtuse or acute; upper surface glabrous or sparsely silky, penninerved, midvein conspicuous; margins entire, flat to undulate; lower surface glabrous to sparsely silky, midvein prominent, intramarginal veins and penninervation evident. **Conflorescence** 5–10 cm long, erect on upcurved peduncle, axillary or terminal, unbranched, secund, the apical flowers opening first; peduncle and rachis silky, sometimes glabrous; bracts 1 mm long, ovate, silky outside, falling before anthesis. **Flower colour**: perianth green; style pink or red. **Flowers** basiscopic; pedicels 4.5–7 mm long, silky; torus c. 5 mm long, very oblique to almost lateral; nectary inconspicuous, U-shaped; **perianth** 10–12 mm long, 4–6 mm wide, markedly zygomorphic, irregularly dilated with an oblique central constriction, abruptly narrowed at neck, ribbed, sparsely silky outside, pilose inside; dorsal tepals separated from base to limb before anthesis, the limb strongly reflexed just before and after anthesis; limb erect, conspicuous in young bud, soon revolute, spheroidal, the apex depressed, the margins slightly impressed; style end enclosed before anthesis; **pistil** 36–42 mm long; stipe 5–6 mm long, lateral, adnate to torus; ovary villous; style exserted first at curve and looped upwards, gently incurved after anthesis, villous at base, ± glabrous above; style end flattened, scarcely dilated; pollen presenter ± lateral, obovate, convex. **Fruit** 15 mm long, 10 mm wide, erect, ovoid to subglobose, tomentose with red-brown markings; pericarp 1–2 mm thick. **Seed** 8–9 mm long, 5 mm wide, obovoid with oblique, short wing at apex; outer face convex, smooth; inner face flat, smooth; margin recurved with prominent waxy border.

Distribution N.T., widespread from Darwin and the Top End inland to c. 100 km S of Adelaide River. *Climate* Summer hot, wet, humid; winter warm, dry. Rainfall 250–1200 mm.

Ecology Grows in brown gravelly loam or white quartzitic sand, in open eucalypt forest, low heath and savannah woodland, often scrambling among grass. Flowers summer–autumn, sometimes extending to winter, but dependent on length of wet season. After fire sprouts from lignotuber or regenerates from seed. Pollinated by birds.

G. goodii

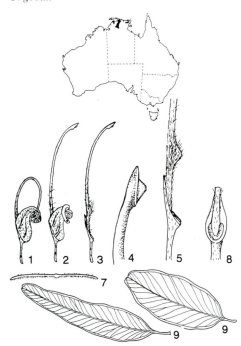

148A. *G. goodii* Close-up of conflorescence (K. & L. Fisher)

148B. *G. goodii* Flowers and foliage (J.Brock)

Variation A fairly uniform species.

Major distinguishing features Habit lignotuberous, with trailing branchlets; leaves glabrous or sparsely silky, simple, entire; conflorescence usually unbranched, secund to cylindrical, opening from apex first, the flowers reversely oriented; torus very oblique; perianth markedly zygomorphic; ovary villous; style glabrous in upper half; fruit with red-brown markings.

Related or confusing species Group 19, especially *G. decora* and *G. pluricaulis*. *G. pluricaulis* is an erect shrub with generally broader leaves, more sharply angular branchlets, a silky ovary, the upper part of the style loosely villous. *G. decora* has some or all conflorescences branched, densely silky leaves, an erect, single-stemmed bushy or tree habit and the upper half of the style is usually sparsely hairy.

Conservation status Not presently endangered.

Cultivation *G. goodii* has been grown for many years in the Darwin area but rarely further south, where it appears to be intolerant of the cooler winters even in areas as warm as Brisbane. Plants in the Grevillea Study Group collection have survived in pots as far south as Wollongong for a few years but overwintering in a glasshouse is essential. A well-drained, sunny to semi-shaded position in well-drained, sandy, acid soils should suit this species. It grows principally in the wet season and would clearly respond well to copious summer watering. Removal of leafless stems from the previous season will improve appearance. Specialist nurseries mainly in the Darwin area sometimes carry this species.

Propagation *Seed* Grows reasonably well if pretreated by nicking or peeling. *Cutting* Firm, young growth should be used but is sometimes difficult to strike. *Grafting* Has been grafted successfully onto *G. robusta* using the whip graft.

Horticultural features *G. goodii* is a dainty grevillea with a great deal of horticultural appeal. The toothbrush racemes of green and pink flowers on long, ground-hugging stems are bright and prominent and attract nectar-seeking birds. These stems arise from the lignotuber each season and the bluish-green leaves are reminiscent of eucalypts. These sometimes attract leaf-chewing caterpillars but damage is usually restricted to a few leaves. Leaf drop in winter may be seasonally unattractive as the plant retains only sparsely leaved branchlets into the new season. There is some potential as a rockery plant, open ground cover or low specimen plant for tropical landscapes.

Grevillea gordoniana C.A.Gardner (1964)
Plate 149

Journal of the Royal Society of Western Australia 47: 56 (Australia)

'This remarkable plant commemorates the name of Mr D.N.Gordon of Myall Park, Glenmorgan, Qld, who has cultivated numerous Western Australian plants in his garden, and who was instrumental in providing the first specimens of this species through one of his collectors (Mr. A.J.Gray)' (C.A.Gardner, 1964). GOR-DOH-NEE-ARE-NA

Type: near No. 1 Tank, 40 km N of Murchison R. on Carnarvon road [North West Coastal Hwy], W.A., 21 Dec. 1962, C.A.Gardner 14273 (PERTH).

Erect to spreading, irregular **shrub** to 5–7 m tall with long, almost leafless, flowering branches and bushy, vegetative branches. **Branchlets** erect, terete, silky. **Leaves** 15–36 cm long, 1–1.5 mm wide, ascending, sessile, simple, terete, leathery, silky, soon ± glabrous, the tips sometimes decurved; juvenile leaves sparingly divided, brown-tomentose; midvein not evident. **Conflorescence** erect, usually paniculate, axillary or terminal; unit conflorescence 3–4 cm wide, the axis 3–4 mm long, globose in bud becoming umbel-like, dense, shortly pedunculate; peduncle silky; rachis glabrous; bracts 5.5–9.5 mm long, imbricate, obovate to orbicular, brown-appressed-villous outside, glabrous inside, falling before anthesis. **Flower colour**: perianth yellow to orange; style similar, becoming red on the dorsal side. **Flowers** adaxially oriented; pedicels 7–12 mm long, glabrous; torus 2–2.5 mm across, slightly oblique, cup-like; nectary scarcely visible, adnate to inner face of torus; **perianth** 5 mm long, 2–3 mm wide, zygomorphic, dilated at base, strongly curved, oblong-acuminate, glabrous outside, densely bearded inside about level of ovary, at first separating along dorsal suture, all tepals separating to base yet cohering at limb, the dorsal tepals reflexing laterally to reveal inner surface before anthesis; limb conspicuous, revolute, angularly spheroidal; **pistil** 15–17 mm long, glabrous; stipe 2.3–2.6 mm long, stout; ovary globose; style slightly refracted above ovary, exposed over most of its length before anthesis, afterwards strongly incurved becoming hooked at the end; style end flattened, slipper-like; pollen presenter oblique, obovate, convex with a surrounding flange, umbonate. **Fruit** 23–28 mm long, 6.5–11 mm wide, erect, bean-pod-like, obovoid–S-shaped, flared at apex after de-hiscence, viscid; pericarp 1–2.5 mm thick. **Seed** 11.5–13 mm long, 1.8–2 mm wide, subterete to linear, winged all round, longitudinally lined; inner face channelled.

Distribution W.A., in inland areas from Yuna N almost to Exmouth Gulf. *Climate* Hot, dry summer; mild, moist winter. Rainfall 250–350 mm.

Ecology Grows on plains and sand hills, in red sometimes yellow, calcareous sand in tree and shrub steppe. Flowers spring to early summer. Fire response unknown. Pollinator unknown, probably birds.

Variation A stable, uniform species.

Major distinguishing features Leaves terete, long, leathery, simple; conflorescence ± globose to umbel-like, much-branched; floral bracts large, villous; torus straight, cup-like; perianth zygomorphic, glabrous outside, hairy within; pistil glabrous; style attached at base of pollen presenter; fruit obovoid to S-shaped, viscid, elongate.

Related or confusing species Group 17; this distinctive species is unlikely to be confused with any other.

Conservation status Not presently endangered.

Cultivation Not commonly cultivated, *G. gordoniana* has been successfully grown in warm-dry climates in western and north-eastern Vic., at Kings Park, Perth, and for a short while by Dave Gordon at Glenmorgan, Qld. It appears difficult to cultivate, demanding a deep, well-drained, acidic to alkaline, sandy soil in a warm to hot position, preferably in full sun. It can succeed in higher rainfall areas provided it is given perfect drainage and a sheltered, sunny site. It does not require watering once established and is hardy to medium frosts and extended dry conditions. Pruning the old, leafless flowering branchlets would keep the plant tidy but might reduce flowering. Unavailable at present in nurseries or from seed suppliers.

Propagation *Seed* Sets good quantities of seed that germinate well if pretreated. Average germination time 25 days (Kullman). Seedlings rapidly develop a very long and penetrating tap root. *Cutting* Difficult; the warmer months are suggested using firm, new growth. *Grafting* Has been grafted successfully onto *G. robusta*, G. 'Poorinda Anticipation' and G. 'Poorinda Royal Mantle', using both the top wedge and side wedge methods. Most grafted plants have proved vigorous and are growing rapidly but long-term compatibility is uncertain.

Horticultural features *G. gordoniana* is a large, robust, somewhat untidy shrub with leafless flowering branches waving amongst the upper branches and beyond the attractive pine-like foliage. The flowers, conspicuously bronze-brown in bud

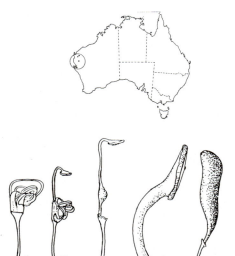

G. gordoniana

149A. *G. gordoniana* Shrub in natural habitat, near Denham, W.A. (P.Olde)

149B. *G. gordoniana* Flowers and buds (P.Olde)

and surrounded by large, papery bracts which fall off as the flowers open, are predominantly yellow except for the style which turns red on the dorsal side. The rounded heads of flower are produced quite densely along the stems and are most decorative in full bloom. The unusual bean-like seed pods which perch on the plant like miniature birds are an added feature both during the flowering season and after it has finished. This species would make an excellent screen or feature plant for spacious inland areas with deep sandy soil and should be long-lived in the right conditions.

Grevillea granulifera (D.J.McGillivray) P.M.Olde & N.R.Marriott (1994)   Plate 150

Telopea 5: 729 (Australia)

Based on *G. obtusiflora* subsp. *granulifera* D.J.McGillivray (1986), *New Names in* Grevillea *(Proteaceae)* 11 (1986, Australia).

The specific epithet refers to the finely granular upper leaf surface and is derived from the Latin *granulus* (a small grain) and *ferre* (to bear). GRAN-YOU-LIFF-ER-A

Type: 1–2 km W of Mt George, N.S.W., 26 Jan. 1980, L.A.S.Johnson 8518 (holotype: NSW).

Dome-shaped **shrub** to 1 m or an erect virgate shrub to 4.5 m with ascending branches. **Branchlets** angular to terete, tomentose. **Leaves** 1.5–6 cm long, 0.2–1.1 cm wide, ascending, petiolate, simple, elliptic, obtuse, mucronate; upper surface convex, finely or rarely coarsely granulose, faintly wrinkled, silky, soon glabrous, midvein and intramarginal veins evident to obscure; margins entire, revolute or recurved; lower surface silky, midvein prominent. **Conflorescence** strongly decurved or deflexed at base of peduncle, shortly pedunculate, usually terminal on short branchlet, sometimes axillary or cauline, simple or few-branched; development basipetal; unit conflorescence secund, rarely exceeding foliage, dense, 12–16-flowered; peduncle 3–10 mm long, bracteate, tomentose-silky; rachis 2–13 mm long, tomentose; bracts 1.5–2.5 mm long, narrowly triangular, tomentose outside, falling just before anthesis. **Flower colour**: perianth pale to dark pinkish red or pinkish purple at the base with a creamy white limb; style burgundy-red or brownish red with green pollen presenter. **Flowers** acroscopic; pedicels 3–7 mm long, loosely villous; torus 1.5–2 mm across, straight to slightly oblique, squarish; nectary subreniform; **perianth** 8–9 mm long, 3–4 mm wide, oblong–ovoid, dilated at base, detaching soon after anthesis, ribbed, sparsely tomentose outside, densely bearded inside just above dilation, sparsely villous to limb, cohering except along dorsal suture before anthesis; limb revolute, loosely villous, subcubic, the segments keeled; **pistil** 21–24 mm long; ovary sessile, oblong–ovoid, scarcely wider than style, white appressed-villous; style slightly inflexed just above ovary, sparsely tomentose, at first exserted towards base on dorsal side and looped upwards before anthesis, afterwards straight to gently incurved; style end broad, flattened; pollen presenter

150A. *G. granulifera* Close-up of flowers (M.Keech)

2.5 mm long, 2 mm wide, lateral, flat, obovate with a short basal attenuation; stigma prominent, distally off-centre. **Fruit** 14 mm long, 5.5 mm wide, erect, narrowly ellipsoidal, prominently ribbed, sparsely tomentose; pericarp c. 0.3 mm thick. **Seed** 8 mm long 2.5 mm wide, oblong-ellipsoidal with an apical wing 2 mm long; upper surface convex with subapical dilation, minutely pubescent; lower surface flat; margin revolute, narrowly winged along one side.

Distribution N.S.W., from near Wingham S to Polblue Creek, Barrington Tops State Forest, extending W to Wollomombi Falls. *Climate* Hot, dry summer; very cold, wet winter. Rainfall 1000–1300 mm.

Ecology Inhabits ridge tops and hillsides in open forest. Most collections have been from poor, stony serpentine soil, but at Polblue Creek the soil is decomposed granite sand. Flowers Sept.–Jan. Regenerates from seed after fire. Wattlebirds, Friarbirds and Noisy Miners have been seen probing flowers at Mt George and are presumed pollinators.

Variation There appear to be at least two distinct forms of *G. granulifera*.

Specimens from the type locality and elsewhere have strongly erect branches, a floral rachis 10–13 mm long with an average 12–14 flowers and a densely silky leaf undersurface, whereas specimens from Wollomombi have slender, weeping branchlets, a floral rachis (2–)3–5 mm long with an average 6–10 flowers and a moderately dense, slightly loose, subsilky leaf undersurface. Field study of all populations is necessary before infraspecific ranking of the forms can be considered.

Major distinguishing features Habit single-stemmed; leaves simple, entire, the upper surface finely granulose, the undersurface silky; conflorescence usually terminal, sometimes axillary or cauline, many-flowered, usually on decurved peduncle; floral rachis silky; perianth zygomorphic, hairy on both surfaces, the limb obtuse to emarginate, prominently keeled; pistil 21–24 mm long; ovary sessile, densely hairy, prominently ribbed; style loosely tomentose; pollen presenter flat with short basal attenuation and prominent stigma; fruit c. 14 mm long.

Related or confusing species Group 25; especially *G. guthrieana* which differs in its coarsely granular upper leaf surface, its villous lower surface, its glabrous to sparsely hairy peduncles and its longer (20 mm) fruits. *G. obtusiflora* and *G. kedumbensis* are also close but differ in their rounder ovary with coarser hairs, their pollen presenter lacking a basal attenuation, and their growth habit, the former being suckering and the latter lignotuberous.

Conservation status 3RC. Rarely collected and poorly known.

Cultivation *G. granulifera* has not been widely cultivated but is being grown by a number of collectors in Qld, northern N.S.W. and Vic. In these areas it has proved easy and reliable, producing abundant flowers and appearing to be free of insect and root pests. It should prove extremely adaptable in cultivation as it tolerates cold in the wild (to –4°C) and should perform well in similar climates such as in Tas. and New Zealand. A well-drained, acid to neutral soil in full or partial sun should be ideal. Pruning will improve shape but should rarely be necessary as it grows naturally into an attractive compact shrub. Low-phosphorus fertiliser applied in late spring will produce abundant new growth. Long, dry periods should also be well tolerated. Allow up to 1 m between plantings. Occasionally available from specialist nurseries in Brisbane.

Propagation *Seed* Untested. Sets good quantities of seed which should germinate readily. *Cutting* Grows readily from cuttings of half-ripe new growth at any time of the year. *Grafting* Has been grafted successfully onto *G. robusta* using whip and mummy grafts.

Horticultural features *G. granulifera* has considerable horticultural merit, being an attractive foliage or screen plant with interesting greyish leaves that have a bright silvery reverse. In addition, its pinkish red and cream flowers are displayed

G. granulifera

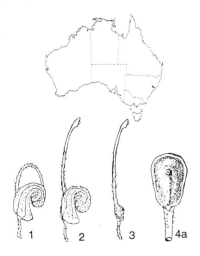

150B. *G. granulifera* Typical form, Mt George, N.S.W. (P.Olde)

150C. *G. granulifera* Habit of plants at the type locality, Mt George, N.S.W.

prominently over a long period and offer a bright contrast to the foliage. Plants from the Mt George area have a strongly erect, narrow-growing habit and would be useful for planting beside driveways and in narrow gardens. This species is useful in attracting nectar-feeding birds.

Grevillea granulosa D.J.McGillivray (1986) Plate 151

New Names in Grevillea *(Proteaceae)* 7 (Australia)
The specific epithet is derived from the Latin diminutive *granulum* (a small grain), in reference to the granulose upper surface of the leaves.
GRAN-YOU-LOW-SA

Type: half-way between Mullewa and Pindar, W.A., 2 Aug. 1965, A.M.Ashby 1574 (holotype: PERTH).

A compact, dome-shaped **shrub** c. 1 m high, 1 m wide with erect branches. **Branchlets** terete, silky. **Leaves** 5–16 cm long, 1.5–2.5 mm wide, ascending to erect, shortly petiolate, simple, linear, leathery; apex often slightly expanded, usually obtuse with a short, inflexed point; upper surface silky becoming glabrous and granulose, sometimes remaining sparsely silky, midvein obscure; margins entire, smoothly strongly revolute, lower surface usually uni-(bi-)sulcate, in broader leaves silky, the midvein prominent. **Conflorescence** sessile to shortly pedunculate, decurved, simple or few-branched, usually cauline, sometimes axillary; unit conflorescence c. 2 cm long, the axis 3–6 mm long, 3–8-flowered, secund, umbel-like; peduncle and rachis tomentose; bracts 0.7–1.3 mm long, narrowly triangular, appressed-villous outside, falling before anthesis. **Flower colour**: perianth and style red, pale red to orange, sometimes pure yellow. **Flowers** acroscopic; pedicels 5–8.5 mm long, silky-tomentose; torus 1.5–2 mm across, oblique; nectary U-shaped, thick and smooth, lobed; **perianth** 10–12 mm long, 2–3 mm wide, ovoid-attenuate, ribbed, sparsely silky outside, sparsely pilose inside in lower half, cohering except along dorsal suture before anthesis, ventral and dorsal tepals rolling back as opposed pairs after anthesis; limb densely silky, revolute, round with a pointed apex to subpyramidal; style end partially exposed before anthesis; **pistil** 20–23 mm long; stipe 1.6–2.2 mm long, villous on dorsal side, glabrous on ventral side; ovary villous; style exserted near base on dorsal side and looped upwards before anthesis, afterwards straight to slightly incurved, sparsely villous at base, mostly glabrous, minutely papillose or hairy on style end; pollen presenter lateral, obovate, convex, umbonate. **Fruit** 10.5–14 mm long, 4 mm wide, erect, narrowly ovoid, ridged, villous; pericarp c. 0.3 mm thick. **Seed** 11–12 mm long, 2 mm wide, narrowly oblong with a short apical and basal wing; outer face convex, minutely hairy; inner face flat, glabrous; margin revolute.

Distribution W.A., between Wubin and Mt Magnet, extending W to Yuna. *Climate* Hot, dry summer; mild, moist winter. Rainfall 200–350 mm.

Ecology Grows in open sandplain, Salmon Gum woodland, mallee broombush, and near granite outcrops in gravelly or loamy yellow sand, granite sand or red clay. Flowers winter–spring. Fire response unknown. Bird-pollinated.

Variation A fairly uniform species.

Major distinguishing features Leaves simple, entire, linear with revolute margins, granulose above; conflorescence few-flowered, axillary or cauline; torus oblique; perianth hairy on both surfaces, the ventral and dorsal tepals rolling back as two opposed pairs, the limb apiculate or pyramidal; ovary densely villous, stipitate; style end minutely hairy.

Related or confusing species Group 25, especially *G. pityophylla* which differs in its densely hairy outer perianth surface and hairy style. *G. extorris* differs in its perianth cohering at anthesis, its leaves ridged on the upper surface and its longer pistil (28–38 mm).

Conservation status 3R.

Cultivation Introduced to cultivation in the 1980s. Although grown by only a few people to date, *G. granulosa* has proved an adaptable, hardy species in dry, inland areas, resisting frost to at least –5°C

G. granulosa

and extended drought. It has been successfully in Perth, widely in Vic. and at Burrendong Arboretum, N.S.W. A very well-drained, acidic to neutral sand or gravelly loam in full sun supports it best; growers report that it is intolerant of poorly drained, cold and shady positions. Pruning is rarely required except to maintain shape. For established plants summer watering is inadvisable. Light applications of low-phosphorus fertiliser assist its vigour, especially if grown in light, sandy soil. It should make a nice pot plant if grown in a sunny situation in a well-drained, gravelly mix. Native plant nurseries occasionally sell it, usually as *G. extorris*, a species with which it has previously been confused.

Propagation *Seed* Sets numerous seed that germinate well when peeled or nicked. *Cutting* Firm to semi-hard young wood strikes readily especially in spring or early autumn. *Grafting* Untested.

Horticultural features *G. granulosa* is an ornamental, medium-sized shrub with bold, erect, grey-green leaves, making a good accent or contrast plant in the landscape. Flowers are usually a very attractive bright orange to red but tend to be partially hidden by the foliage when plants are young. As plants mature, they tend to lose leaves from the lower branches, exposing the masses of flowers on the old wood, making a very showy display. It appears to be relatively long-lived in the garden and would be ideal as a low screen or feature plant in dry regions. The flowers are extremely popular with honey-eating birds.

General comment A specimen with yellow flowers has been collected by the authors near Mullewa.

Grevillea guthrieana P.M.Olde & N.R.Marriott (1994)   Plate 152

Telopea 5: 731 (Australia)

The specific epithet honours Christine Guthrie, for many years the Secretary and Treasurer of the Grevillea Study Group and Editor of its Newsletter. GUTH-REE-ARE-NA

Type: 3 km E of Booral, N.S.W., 20 Sept. 1992, P.M.Olde 92/96 (holotype: NSW).

Shrub 1.5–2(–4.5) m high with spreading, pendulous branchlets. **Branchlets** terete, villous when young, soon ± glabrous. **Leaves** 2–6 cm long, 0.4–0.9 cm wide, ascending, oblong-elliptic, obtuse, mucronate, shortly petiolate, simple, soft; upper surface coarsely granular, faintly wrinkled, glabrous, the midvein evident; margins entire, loosely revolute; lower surface loosely villous, the lamina clearly visible beneath indumentum, the midvein prominent. **Conflorescence** decurved, pedunculate, terminal, simple or 1-branched; unit conflorescence 2–6(–10)-flowered, loose; development basipetal; peduncle slender, 7–22 mm long, 0.3–0.4 mm thick, glabrous or sparsely tomentose; rachis 2–6 mm long, glabrous or sparsely tomentose; bracts 1–1.5 mm long, narrowly triangular, glabrous to loosely tomentose outside, falling soon after bud formation. **Flower colour**: perianth green; style maroon; ovary and pollen presenter green. **Flowers** acroscopic; pedicels 6 mm long, with sparse spreading, crisped hairs; torus 2 mm across, slightly oblique, undulate; nectary reniform to cushion-like, prominent; **perianth** 9 mm long, 4–5 mm wide, oblong with a cons-picuous subannular dilation in basal 2 mm, sparsely tomentose outside, bearded inside c. 2 mm from base where reflexed hairs crowded for c. 2 mm, elsewhere with scattered, appressed hairs or glabrous; tepals ribbed, falling soon after anthesis; limb subcubic, sparsely tomentose, apically depressed, the segments prominently keeled; **pistil** 25–26 mm long; stipe absent to 0.2 mm long, glabrous; ovary 2–2.5 mm long, oblong-ovoid, angular with prominent ribs, pubescent to

151A. *G. granulosa* Plant in natural habitat, near Wubin, W.A. (P.Olde)

151B. *G. granulosa* Flowers and foliage (N.Marriott)

151C. *G. granulosa* Yellow-flowered variant, near Mullewa, W.A. (P.Olde)

152A. *G. guthrieana* Habit of plant, Booral, N.S.W. (P.Olde)

152B. *G. guthrieana* Flowers and foliage (P.Olde)

tomentose-villous; style straight, loosely tomentose in lower third, glabrous or almost so in distal third; pollen presenter 2.2–2.5 mm long, obovate with short, basal attenuation, flat to convex with prominent stigma. **Fruit** 20 mm long, 5 mm wide, narrowly ellipsoidal, attenuate, prominently ribbed, sparsely tomentose; pericarp c. 0.4 mm thick. **Seed** 10 mm long, 2–2.5 mm wide, oblong-ellipsoidal with subapical dilation and apical wing 3–4 mm long; upper surface strongly convex, faintly pubescent, otherwise smooth; margins revolute with narrow waxy border along one side extending into an excurrent apical elaiosome 3–4 mm long and a short basal wing c. 0.5 mm long; inner face obscured by margin.

Distribution N.S.W., known from only two locations, at Booral Creek and on the W edge of the Carrai Plateau. *Climate* Warm to hot, wet or dry summer; cool to cold, usually dry winter. Rainfall c. 1000 mm.

Ecology Grows in sandstone derived loam on creek lines in moist eucalypt forest. Flowers spring.

Variation A uniform species.

Major distinguishing features Leaves simple,

G. guthrieana

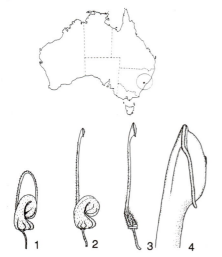

entire, villous below, coarsely granular above; conflorescence few-flowered; peduncle glabrous or sparsely hairy, slender, up to 20 mm long; perianth zygomorphic, with a subannular basal dilation, hairy on both surfaces; ovary pubescent-tomentose, sessile; pollen presenter lateral, flat with short basal attenuation; fruit c. 20 mm long.

Related or confusing species Group 25; especially *G. granulifera* which differs in its leaves which have a silky undersurface and finely granular upper surface, its thicker, much shorter, strongly decurved, densely hairy peduncle, in its shorter pistil and in its smaller fruits.

Conservation status Recommended 2E. There has recently (1993) been considerable inadvertent damage to the population at Booral through road works.

Cultivation *G. guthrieana* was introduced to cultivation in 1992 in Brisbane and north-eastern N.S.W. where grafted plants have grown readily in pots and large gardens. Judging from its natural habitat, it requires a sandy, well-drained yet moist soil in either open or semi-shaded positions. A neutral to slightly acid soil is preferred. Frost and drought tolerance are untested but the species will almost certainly tolerate quite cold conditions as regular frosts and even snow occur in its natural habitat.

Propagation *Seed* Sets plentiful seed in the wild that should germinate readily when fresh or with some pretreatment. *Cutting* Strikes well from cuttings of firm, new growth at most times of the year. *Grafting* Has been grafted successfully onto *G. robusta* using the mummy and whip techniques.

Horticultural features *G. guthrieana* is not an especially attractive plant from a horticultural viewpoint, as it tends to hide its flowers within the foliage and in any event its green colour tends to blend with the foliage, but it has an attractive, weeping habit and makes an interesting subject for the large shrubbery where it should attract nectar-feeding birds. One for the collector!

Grevillea hakeoides C.F.Meisner (1848)

in J.G.C.Lehmann (ed.), *Plantae Preissianae* 2: 252 (Germany)

Hakea Leaf Grevillea

The specific epithet is a reference to the resemblance of this species to a terete-leaved Hakea. HAY-KEE-OY-DEES

Type: Swan River, W.A., 1844, J.Drummond coll. 2 no. 325 (lectotype: NY) (McGillivray 1993).

Spreading, compact, mid-dense **shrub** 1.5–3 m high, 2–3 m wide. **Branchlets** sharply to slightly angular becoming terete, ribbed, silky, soon glabrous. **Leaves** 3–11 cm long, 0.5–2 mm wide, ascending, the juvenile divided, the adult simple, ± sessile, subterete to narrowly linear, firm, pungent, glabrous and glaucous; juvenile leaves silky; upper and lower surfaces without obvious midvein, either delineated by a longitudinal groove each side packed with hairs, or dorsiventral, the upper surface channelled or smooth and convex, the lower bisulcate, when fresh the lamina sometimes exposed. **Conflorescence** erect, shortly pedunculate, usually terminal, simple or few-branched; unit conflorescence 7–10 mm long, c. 10 mm wide, dome-shaped, dense; peduncle < 1 cm long, silky-tomentose; rachis 2–5 mm long, tomentose, noticeably broader than peduncle; bracts c. 0.8 mm long, broadly ovate, tomentose outside, falling before anthesis. **Flowers** adaxially oriented; pedicels 2–5 mm long, silky, rarely almost glabrous; torus 0.5–1 mm across, straight to slightly oblique; nectary conspicuous, cushion-like; **perianth** 2–4 mm long, 0.5–1.5 mm wide, narrowly oblong, undilated, ribbed, glabrous or sparsely to densely silky outside, papillose or shortly bearded inside just above level of ovary, scattered hairs sometimes extending to the curve; tepals first separating along dorsal suture to base, the tepals below limb then separated, the dorsal tepals reflexing laterally and exposing inner surface before anthesis, afterwards all free to base, soon falling; limb revolute, subcubic-spheroidal, the segments ribbed; style end partially exposed; **pistil** 5–12.5 mm long, rarely 2 per flower, usually glabrous; stipe and style with scattered hairs on dorsal side; stipe 0.6–2 mm long, ventrally channelled, sometimes broader than style; ovary globose; style exserted below curve before anthesis, strongly incurved afterwards; style end flattened; pollen presenter oblique, obovate, flat to slightly convex with prominent stigma. **Fruit** 7–12 mm long, 3–7.5 mm wide, slightly oblique, oblong-ellipsoidal, rugose, ribbed, glabrous; pericarp c. 0.5 mm thick. **Seed** 5–7.5 mm long, 2 mm wide, narrowly oblong; outer face convex, minutely pubescent; inner face channelled, obscured by strongly revolute margin; margin on one side with a waxy border drawn to a short apical wing.

Major distinguishing features Leaves simple, terete or linear, with a prominent groove each side or 2-grooved below; conflorescence regular, dome-shaped; perianth zygomorphic, undilated, glabrous or silky outside, bearded or sparsely hairy within; pistil usually glabrous; ovary stipitate; fruit rugose; seed pubescent with strongly revolute margins, a waxy border on one side and short apical wing.

Related or confusing species Group 16, especially *G. argyrophylla* and *G. commutata*, both of which differ in their broader leaves with more conspicuous venation and exposed undersurface. *G. commutata* usually has an ovoid conflorescence, a longer pistil and larger fruit. A number of populations with dome-shaped conflorescences and divided leaves are included within *G. commutata* but may be distinguished by their long pistil.

Variation Two subspecies are recognised.

Key to subspecies

Pistil < 8 mm long; most leaves < 5 cm long
                                        subsp. **hakeoides**
Pistil ≥ 9 mm long; most leaves 4–11 cm long
                                        subsp. **stenophylla**

Grevillea hakeoides C.F.Meisner subsp. hakeoides   Plate 153

Leaves dipleural, 3–5 cm long, 0.5–0.7 mm wide, subterete, straight to slightly curved. **Floral rachis** 2–3 mm long; pedicels 2–2.5 mm long. **Flower colour**: perianth creamy grey or greenish white, sometimes pale pink; style white or pale pink with green tip. **Perianth** 2–3 mm long, 0.5 mm wide, glabrous or silky outside, bearded or sparsely hairy inside; pistil 5–7 mm long; stipe 0.6–1 mm long. **Fruit** 7–8 mm long, 3–5 mm wide, faintly rugose. **Seed** 5–6 mm long.

Distribution W.A., in scattered populations from Moora and Goomalling SE through Tammin to Lake Grace. *Climate* Warm to hot, dry summer; cool to mild, wet winter. Rainfall 450–500 mm.

Ecology Grows in sandy to gravelly loam, usually over clay, in scrub, eucalypt woodland and open forest. Flowers winter to spring. Fire response unknown. Small, mosquito-like insects have been seen in large numbers seeking nectar from the flowers and may be the pollinator.

Conservation status Not presently endangered, although it never occurs in large numbers. Conserved in Charles Gardner Reserve, near Tammin.

Variation The type specimen has a glabrous perianth. Several collections, mainly from the northern part of the range, are sparsely silky on the outer surface of the perianth and bearded or sparsely hairy inside. These should be studied to determine whether taxonomic separation is warranted.

Cultivation Subsp. hakeoides has been grown successfully since about 1970 at Glenmorgan, Qld,

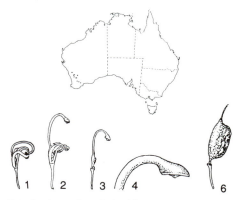

G. hakeoides subsp. *hakeoides*

153A. *G. hakeoides* subsp. *hakeoides* Plant in natural habitat, near Tammin, W.A. (P.Olde)

153B. *G. hakeoides* subsp. *hakeoides* Flowering branches (P.Olde)

153C. *G. hakeoides* subsp. *hakeoides* Pinkish flowers (P.Olde)

by Dave Gordon and more recently at Stawell and other locations in western Vic. Some grafted plants have appeared lately in cultivation in the Brisbane area, but in general it is rarely grown. It appears to be extremely hardy and adaptable, being resistant to frosts to at least −3°C and probably hardy in dry summer conditions. In the wild, G. hakeoides enjoys an acidic, sandy or gravelly loam or well-drained clay soil in a sunny position. As it forms naturally into a dense, spreading shrub, pruning should be rarely required unless it is grown in shady conditions. Although not really suited to pot culture, it can nonetheless be grown satisfactorily in a tub for a number of years, using a free-draining potting mix and slow-release fertiliser. It is not presently available at nurseries.

Propagation *Seed* Sets many seeds that should germinate well if peeled or nicked before sowing. *Cutting* Cuttings of firm, new growth, especially from cultivated plants, strike readily, particularly in spring and autumn. *Grafting* Has been grafted successfully onto *G. robusta* using the top wedge technique.

Horticultural features Subsp. *hakeoides* is a dull green shrub bearing crowded short, narrow leaves and is ideally suited to screening or as a foil to more spectacular and long-flowering species. In late winter or early spring it produces impressive numbers of delicate creamy-white or pink conflorescences that emit a sweet perfume and attract many insects. Although the individual flowers and conflorescences are quite small, they are crowded in almost every upper leaf axil or at the ends of the branches and, as the plant tends to bloom all at once, with few flowers out of season, it produces a spectacular effect. The light scent is especially noticeable in the middle of sunny days. It has no known pests or diseases and is relatively long-lived in the wild.

General comments We have only partially accepted the treatment of *G. hakeoides* by McGillivray (1993); more research is required into variation in this species.

Grevillea hakeoides subsp. stenophylla
(W.V.Fitzgerald) D.J.McGillivray (1986)

Plate 154

New Names in Grevillea *(Proteaceae)* 7 (Australia)

Based on *G. stenophylla* W.V.Fitzgerald, *Journal of the West Australian Natural History Society* 2: 30 (1905, Australia).

Subspecific epithet from the Greek *stenos* (narrow) and *phyllon* (a leaf), in reference to the leaves.
STEN-OWE-FILL-A

Type: Mingenew, W.A., Sept. 1903, W.V.Fitzgerald (holotype: NSW).

Leaves dipleural or dorsiventral, 4–11 cm long, 0.7–2 mm wide, subterete to linear, straight to very wavy; upper surface of dorsiventral leaves channelled; leaf margins sometimes slightly exposing undersurface (Wubin area). **Floral rachis** 3–5 mm long; pedicels 2.5–5 mm long. **Flower colour**: perianth silvery grey to whitish; style pink to white with green tip. **Perianth** 3.5–4 mm long, silky outside, bearded inside; **pistil** 9–12.5 mm long; stipe 1–2 mm long. **Fruit** 8–12 mm long, 5–7.5 mm wide, rugose. **Seed** c. 7.5 mm long.

Distribution W.A., widely distributed. The Slender-leaved form occurs on Dirk Hartog Is. and from the lower Murchison R. and Geraldton S to Cowcowing and Watheroo and inland to Paynes Find. The Stiff-leaved form occurs from Indarra and Latham S to Tammin. *Climate* Hot, dry summer; warm to cool, wet winter. Rainfall 330–500 mm.

Ecology Grows in various habitats from low heathland to tall, open shrubland and from streamsides to the tops of sandy rises. Soils range from yellow to red sand to sandy loam and are often poorly drained. Flowers winter–spring. Regeneration after fire is from seed. Native bees, flies and other insects have been seen attending the flowers but the pollinator is uncertain.

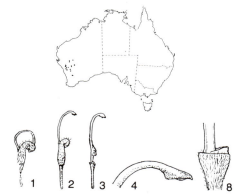

G. hakeoides subsp. *stenophylla*

Variation There are at least 2 forms in this subspecies.

Stiff-leaved (typical) form This form occurs in scattered populations between Wubin and Tammin. Its leaves are c. 4–11 cm long, 1–2 mm wide, rather rigid, relatively straight or smoothly curved, the upper surface channelled. Pistil length is 9–11 mm. Plants from near Wubin included in this form sometimes have the leaf undersurface partly exposed.

Slender-leaved form Leaves of this form closely approach those of subsp. *hakeoides* in structure (i.e. grooved along each side and subterete) but in their flower structure and size they are closer to the typical, Stiff-leaved form. The longest leaves in this form are up to c. 11 cm long and very wavy. It occurs over a wide range from the Murchison R. to Paynes Find and S to Cowcowing. This form has a longer pistil than subsp. *hakeoides* (10–12.5 mm) and may represent a separate infraspecific taxon.

Conservation status Not presently endangered.

Cultivation Subsp. *stenophylla* has rarely been cultivated although it has been grown successfully in Kings Park, Perth, and in western Vic. as well as at Dave Gordon's arboretum at Glenmorgan, Qld. Given its wide natural distribution and its occurrence often in poorly drained sites, it should prove quite adaptable and hardy. Natural habitat suggests that it likes an acidic to neutral sand or sandy loam in full sun. Selected forms may tolerate poor drainage. Pruning would rarely be necessary as it is naturally a compact shrub. Tip pruning when young would improve shape. Although not really suited to pot culture, it could be grown in a well-drained, sandy mix with slow-release fertiliser. Seed is sometimes available from specialist suppliers, usually under its old name *G. stenophylla*.

Propagation *Seed* Untried. Should respond to nicking or peeling treatment. *Cutting* Firm, new growth would undoubtedly strike well especially from cultivated plants. *Grafting* Has been grafted successfully onto G. 'Poorinda Royal Mantle' and *G. robusta* using the top wedge technique but long-term compatibility is uncertain.

Horticultural features Subsp. *stenophylla* has an attractive, dense, spreading habit making it suitable for screening or as a structure plant in large-scale gardens. Its attractive, fine foliage contrasts with its conspicuous white or pink conflorescences which at times are borne prolifically over the plant; it can be spectacular when in full flower. The flowers are lightly scented, especially during the middle of the day. It may be useful in landscaping for poorly drained areas and it appears to be fairly long-lived.

General comments A number of populations closely approach this subspecies but have some features of subsp. *hakeoides* and cannot be placed clearly (McGillivray 1993). Further studies may increase our understanding of this diverse and variable subspecies.

Grevillea halmaturina R.Tate (1883)

Plate 155

Transactions of the Royal Society of South Australia 6: 141 (Australia)

The specific epithet is derived from the Latin *halmaturus*, a name proposed originally for a genus of small wallabies, now included in *Macropus*, and *-inus*, belonging to, an allusion to Kangaroo Island

154. *G. hakeoides* subsp. *stenophylla* (P.Olde)

where the type was collected. HAL-MAT-YEW REE-NA

Type: Kangaroo Is., S.A., date unknown, F.G.Waterhouse (holotype: K).

Synonym: *G. parviflora* var. *acuaria* F.Mueller ex G.Bentham (1870).

Spreading to erect, single-stemmed, extremely prickly **shrub** 0.5–1.4 m high, 0.5–1.2 m wide. **Branchlets** angular, stout, silky with prominent glabrous ribs decurrent from leaf bases. **Leaves** 1–2.5(–3) cm long, 0.8–1.2 mm wide, sessile, simple, entire, spreading to patent, strongly recurved just above base, rigid, subterete to narrowly linear, subulate, extremely pungent with a rigid spiny tip to 1.5 mm long; upper surface glabrous, smooth to 7-ribbed, the ribs smooth, obscure to prominent, sometimes extending to apex, the channels usually darker than ribs; margins smoothly revolute to angularly refracted, enclosing undersurface; lower surface bisulcate, silky in grooves; midvein glabrous, prominent with prominent pulvinus at the base; point of attachment much broader than leaf base. **Conflorescence** 1 cm long, 1 cm wide, globose in bud, erect, sessile, simple, terminal, usually on short branchlet extending down main branch, umbel-like, open, enclosed in foliage; rachis 0.5–2.2 mm long, brown-pubescent; bracts 1.5 mm long, not exceeding buds, ovate with incurved apex, brown-silky outside, the basal bracts (?juvenile leaves) with a glabrous tip, falling before anthesis. **Flower colour**: perianth white to pale pink; style white to pink. **Flowers** acroscopic; pedicels 4–5.5 mm long, silky, sometimes becoming glabrous at base, ?markedly expanded at apex; torus 0.5–1 mm across, square, straight to slightly oblique (rarely very oblique); nectary obscure, arcuate to U-shaped with toothed ends; **perianth** 3.5–4 mm long, 0.5–0.8 mm wide, undilated, oblong-cylindrical, white-silky to sparsely so outside, conspicuously bearded inside in lower half adjacent to ovary, the hairs c. 0.4 mm long, patent, with scattered hairs above, elsewhere glabrous; tepals separating to base before anthesis, curling back to beard at anthesis; limb nodding to revolute, angularly spheroidal with depressed apex, the dorsal side noticeably depressed around the suture, silky; **pistil** 6.5–8.5 mm long; stipe 0.8–1.3 mm long, glabrous; ovary narrowly ovoid, style exserted from lower half on dorsal side, angularly looped out, geniculate 1–2 mm below style end; style end abruptly divergent, granular or with minute erect hairs; pollen presenter orbicular to squarish, oblique, flat with prominent stigma. **Fruit** 10–14 mm long, 3–5.5 mm wide, erect to slightly oblique on swollen, slightly incurved stipe, narrowly ovoid, smooth to faintly warty; pericarp 0.2–0.4 mm thick. Seed 9 mm long, 2 mm wide, oblong with subapical pulvinus; outer face convex, slightly wrinkled, minutely pubescent; margins revolute with a waxy ridge along one side extending into a short basal and longer apical elaiosome to 2.5 mm long; inner face obscured.

Distribution S.A., on Kangaroo Is. and southern parts of the Eyre Peninsula. *Climate* Summer hot and dry; winter cold and wet. Rainfall c. 600 mm.

Ecology Grows in silty sand or gravelly loam in swampy sites in heath or woodland, rarely on ridges. Flowers late winter–spring. Fire response is unknown. Probably pollinated by insects.

G. halmaturina

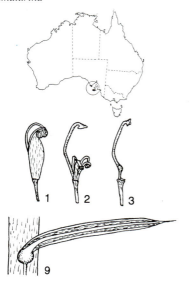

Variation Two distinct forms are recognised.

Typical form This form is distinguished by its longitudinally ribbed leaves and angularly refracted leaf margins. The ovarian stipe is c. 0.8 mm long, the pedicels glabrescent at the base. It is confined to creek lines or poorly drained areas on Kangaroo Is.

Eyre Peninsula form Recorded between Port Lincoln and Hundred of Lake Wangary. This form differs in its relatively smooth leaves with smoothly revolute margins and ovarian stipe c. 1.3 mm long. The pedicels remain silky to the base. It is confined to slopes and low-lying areas between Port Lincoln and a few kilometres inland.

Major distinguishing features Leaves simple, entire, patent, smooth, rigid, extremely pungent, the base broad, upper surface smooth or smoothly ribbed; margins smoothly to angularly refracted; undersurface not exposed; inflorescence terminal, usually on short axillary branchlet, simple, sessile, umbel-like; rachis 1–2 mm long, brown-pubescent; bracts ovate, not visible beyond line of subglobose buds; perianth zygomorphic, hairy outside, bearded within; pistil glabrous except a granular or minutely hairy style end; fruit faintly warted.

Related or confusing species Group 21, especially *G. parviflora* which differs in its slender branchlets, more strongly ascending, less pungent leaves, subsecund conflorescences, the leaves with granular veins and punctate upper surface.

Conservation status 3VC. Populations on Eyre Peninsula require assessment.

Cultivation *G. halmaturina* has not been widely cultivated, although some enthusiasts have grown it since at least the early 1980s. It grows quite readily in an open position in well-drained, acidic to neutral sand or sandy loam, doing best when not crowded by other plants. It tolerates quite heavy frosts (to at least -4°C) as well as extended dry periods, though subsoil moisture should never be too low. It has a naturally compact habit which is improved with occasional tip pruning. Once established, fertiliser and watering are unnecessary. Nurseries rarely stock this species, except occasionally in S.A.

Propagation Seed Sets reasonable quantities which should germinate if pretreated by nicking or peeling the testa. *Cutting* Strikes readily from cuttings of half-hardened new growth in early spring. *Grafting* Untested.

Horticultural features *G. halmaturina* is an extremely prickly low shrub that needs careful consideration about its placement in any landscape, especially near paths or high traffic areas. However, it has an attractive all-over flowering habit with delicate pink or white flowers appearing in profusion down the length of the branchlets in early spring. The prickly foliage affords ideal protection and nesting sites for small birds and has value when planted en masse as a people deterrent.

General comments McGillivray (1993) placed *G. halmaturina* in synonymy under a broadly circumscribed *G. linearifolia* which has similar flowers. While taxonomy and species boundaries in this group are at present being reconsidered, in our opinion the two forms to which it was referred (as Forms 'l' and 'm') should be recognised as a distinct species.

Grevillea haplantha F.Mueller ex G.Bentham (1870)

Flora Australiensis 5: 451 (England)

The specific epithet is derived from the Greek

155. *G. halmaturina* Typical form from Kangaroo Island, S.A., cultivated at Stawell, Vic. (N.Marriott)

haplos (single) and *anthos* (a flower), in reference to the solitary flowers found in the leaf axils of the type specimen and frequently on other collections. HAP-LAN-THA

Type: south-western W.A., J.Drummond (lectotype: MEL) (McGillivray 1993). The remaining syntype is a specimen of *G. disjuncta* F. Mueller.

Rounded, dense **shrub** with erect branches, 1.5–2 m tall, 1.5–3 m wide. **Branchlets** angular to terete, tomentose. **Leaves** 3–8 cm long, c. 1.5 mm wide, ascending to erect, sessile, simple, linear, pungent, leathery; upper surface convex, tomentose, soon glabrous, granulose, ribbed, midvein not evident; margins angularly revolute to midvein below; lower surface bisulcate, silky in grooves, midvein evident, sometimes recessed below margins. **Conflorescence** axillary or cauline, ± sessile, usually simple, often solitary or 2–4(–6)-flowered in umbel-like clusters; peduncle and rachis short (combined length c. 1 mm), villous; bracts 0.5–1 mm long, linear-ovate, villous outside, persistent to anthesis. **Flowers** abaxially oriented; pedicels 5–9 mm long, villous; torus 2–2.5 mm across, oblique; nectary V-shaped, entire; **perianth** 6–7 mm long, 3 mm wide, scarcely dilated at base, oblong-ovoid, strongly curved, conspicuously narrowed at throat, ribbed, villous outside, bearded inside, cohering except along dorsal suture; limb conspicuous, revolute, ovoid, villous, cohering after anthesis; **pistil** 18–25 mm long; stipe 0.5–1.3 mm long; ovary villous, sessile or almost so; style exserted at curve and looped upwards before anthesis, afterwards straight to gently incurved, villous in lower half, sometimes throughout; style end expanded, exposed before anthesis; pollen presenter lateral, elliptic, convex, umbonate. **Fruit** 10–13 mm long, 5 mm wide, erect, ellipsoidal, prominently ribbed, villous; pericarp c. 0.5 mm thick. **Seed** 8–10 mm long, 2.5 mm wide, oblong-elliptic with an apical wing 1.5 mm long, minutely pubescent; outer face convex, inner face flat; margin strongly revolute on one side, shortly recurved on the other.

Major distinguishing features Leaves simple, entire, ribbed and slightly granulose, the undersurface double-grooved; conflorescence few-flowered, axillary or cauline; torus oblique; perianth zygomorphic, villous outside, bearded within; ovary villous, shortly stipitate (stipe 0.5–1.3 mm); style hairy; fruit ribbed.

Related or confusing species Group 25, especially *G. disjuncta* and *G. dolichopoda* which differ in their leaf undersurface being single-grooved. The outer perianth surface of *G. haplantha* is more conspicuously villous.

Variation: Two subspecies are recognised.

Key to subspecies

Stylar indumentum extending onto back of pollen presenter, the hairs uniform in size; pedicels 7–9 mm long; pistil (22–) 24–25 mm long subsp. **haplantha**

Stylar indumentum either lacking in apical 2–3 mm or the hairs much reduced in size; pedicels 5–7 mm long; pistil 18–20 (–22) mm long subsp. **recedens**

Grevillea haplantha F.Mueller ex G.Bentham subsp. haplantha   Plate 156

Shrub 1.5–2 m tall, to 3 m wide. **Branchlets** erect, ± straight, with prominent axillary flowers. **Flower colour**: perianth dull pink to red, with a yellow limb; style red. **Perianth** dull pink with cream limb, with white indumentum; **pistil** (21–) 24–25 mm long; stipe c. 1 mm long; style villous throughout.

Distribution W.A.; in the Coolgardie area, extending from north of Koolyanobbing to W of Goongarrie and up to 50 km S of Coolgardie. *Climate* Hot, dry summer; cool to mild, wet winter. Rainfall 250 mm.

Ecology Grows in heathland and mallee shrubland, sometimes in tall shrubland, in gravelly loam, sometimes in dense laterite. Flowers winter–spring (autumn–early summer in cultivation). Appears to be killed by fire and regenerates from seed. Pollinated by birds.

Variation A uniform subspecies; some populations have markedly greyer foliage than others.

Conservation status Not endangered.

Cultivation *G. haplantha* subsp. *haplantha* has been in cultivation since the 1970s in Vic. where it has proved reasonably hardy. For best results, it

156A. *G. haplantha* subsp. *haplantha* Near Bullabulling, W.A. (N.Marriott)

156B. *G. haplantha* subsp. *haplantha* Plant in natural habitat, N of Koolyanobbing, W.A. (P.Olde)

G. haplantha subsp. *haplantha*

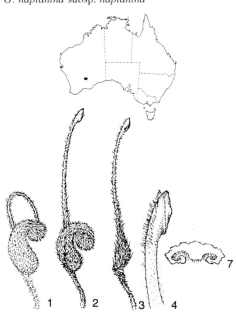

demands full sun in well-drained gravelly or gravelly loam, acidic to neutral soil. Summer-humid climates are not to its liking, but in dry climates with cool to cold winters, it is a most rewarding plant. Pruning is rarely required but may be necessary if grown in sandy soil. Fertiliser may produce short-term lushness at the expense of long-term health. It is hardy to frosts to at least –6°C and to extended dry periods. A few specialist nurseries sometimes stock this species.

Propagation Seed Not tested. Fresh seed should germinate readily if nicked or peeled before sowing. *Cutting* Strikes with some difficulty using firm, young growth in spring. *Grafting* Has been grafted successfully onto *G. robusta* and G. 'Poorinda Royal Mantle' using top wedge grafts. Long-term compatibility is uncertain.

Horticultural features Subsp. *haplantha* is an impressive, medium-sized shrub with bold, upright branches with bluish-grey leaves and dull pink flowers produced in profusion along the branchlets. New growth is bright purple and provides a harmonious counterpoint to its conspicuous late-winter flowers. The flowers are eagerly sought by honey-eating birds for many months of the year. This plant has high horticultural merit and has value as a winter-flowering feature plant or screen plant. It should be much more widely used in inland gardens.

General comments In a recent revision of the genus, McGillivray (1993) treated subsp. *haplantha* as the Coolgardie District race of *G. haplantha*.

Grevillea haplantha subsp. recedens
P.M.Olde & N.R.Marriott (1993)   **Plate 157**

Nuytsia 9: 290 (Australia)

The specific epithet is derived from the Latin *recedere* (to recede), in reference to the receding hairs on the style end. REE-SEED-ENZ

Type: near Manmanning, W.A., 6 July 1986, B.H.Smith 658 (holotype: NSW).

Shrub 0.6–1 m tall, 1 m wide. **Branchlets** divaricate. **Conflorescences** mostly concealed within the bush; pedicels 5–7 mm long. **Flower colour**: perianth including the limb brownish pink; style pinkish red with greenish yellow tip. **Pistil** 18–20(–22) mm long; stipe 0.5 mm long; style villous becoming glabrous or sparsely and minutely pubescent in apical 2–3 mm.

Distribution W.A.; from Mollerin to near Ballidu, extending S to Cunderdin and Merredin. *Climate* Hot, dry summer; cool to mild, wet winter. Rainfall 350 mm.

Ecology Grows in heavy clay loam often with a strong lateritic association in open shrubland or woodland. Flowers winter–spring (autumn–early summer in cultivation). Regenerates from seed after fire. Pollinated by birds.

Variation A uniform subspecies.

Conservation status 3EC-i recommended. Subsp. *recedens* has been collected only a few times in recent years, mainly from degraded road verges near Manmanning, and once from a nature reserve.

Cultivation Subsp. *recedens* has been in cultivation for a number of years, particularly in Vic. It is at times difficult to establish but then is hardy to frosts to at least –6°C and to extended dry periods. It requires a very well-drained situation in full sun either in heavy, acidic to neutral gravelly loam or in sandy loam. In warm, dry climates it will succeed in dappled shade. In cold, wet and poorly drained conditions, it quickly succumbs to fungal attack of the roots and foliage. Summer watering in any climate could be harmful. Being a naturally compact, dense shrub, pruning should rarely be required. It would make an excellent pot plant for a number of years using a well-drained mix and slow-release, low-phosphorus fertiliser, placed in full sun. A few specialist nurseries carry this subspecies but it is often incorrectly labelled as *G. disjuncta*.

Propagation Seed Not tested. Seed is rarely available. *Cutting* Strikes reasonably well from firm, young growth especially during spring or early autumn. It can be slow to root at times. *Grafting* Has been grafted successfully onto *G. robusta* and G. 'Poorinda Royal Mantle' using top wedge grafts. Long-term compatibility appears good.

Horticultural features Subsp. *recedens* is a dense, low shrub with dull green to greyish green leaves and pinkish brown flowers that are somewhat concealed within the foliage. Despite this, close inspection reveals a profusion of flowers that are eagerly sought by honey-eating birds over many months. This plant has value as a screen or contrast plant where winter flowers are desired.

General comments McGillivray (1993) treated this taxon as the Avon District race of *G. haplantha*.

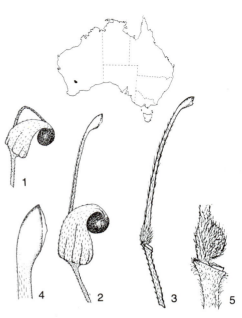

G. haplantha subsp. *recedens*

157. *G. haplantha* subsp. *recedens* (N.Marriott)

Grevillea heliosperma R.Brown (1810)
Plate 158

Transactions of the Linnean Society of London, Botany 10: 176 (England)

Red Grevillea, Rock Grevillea

Aboriginal: 'yilingbirradangwa yinungkwurra' (Anindilyakara, Groote Eylandt); 'djamudu' (Bardi); 'yalyana' (Yanyula); 'nyenyirri dirramu' (Gupapuynga); 'anbardbard', 'andjengerrer' (Mayali)—the Mayali acknowledged two forms of this species.

The specific name was derived from the Greek *helios* (sun) and *sperma* (a seed), in reference to the round seed with its encircling membranous wing looking like the sun. HELL-EE-OWE-SPERM-A

Type: Gulf of Carpentaria, [N.T.], 1802, R.Brown 3320 (lectotype: K) (McGillivray 1993).

Open, spreading to erect **tree** 4–8 m high. **Branchlets** terete, ridged, silky. **Leaves** 15–40 cm long, spreading, detaching readily, usually bipinnatipartite, rarely once or tripinnatipartite; lobes 3–13 cm long, 0.3–0.8(–1) cm wide, oblong-lanceolate to falcate, sometimes again lobed, concolorous or slightly discolorous, acute, the base obliquely and often unequally tapered to midvein, sometimes decurrent; upper and lower surfaces similar, usually glabrous and glaucous or with sprinkled appressed hairs, rarely silky; margins ± flat; venation fine, ± longitudinal, more prominent below, midvein inconspicuous. **Conflorescence** 10–25 cm long, 6–10 cm wide, decurved, pedunculate, axillary or sometimes terminal, usually branched; unit conflorescence conico-secund, open and lax, not exceeding foliage; peduncle silky or glabrous; rachis glabrous; bracts 1.5–3 mm long, ovate-acuminate, glabrous or silky, falling before anthesis. **Flower colour**: perianth pink or reddish, turning red at anthesis; style red with green tip. **Flowers** acroscopic, twisting on pedicels before anthesis; pedicels 5–12 mm long, glabrous, retrorse; torus 6–10 mm across, lateral or very oblique; nectary U-shaped, entire; **perianth** 8–11 mm long, 3–4 mm wide, oblong to ovoid, slightly dilated at base, strongly curved, glabrous and glaucous outside, bearded near base or sometimes glabrous inside, cohering except along dorsal suture; limb revolute, ovoid to spheroidal; **pistil** 34–46.5 mm long, glabrous; stipe 9.8–14 mm long, adnate to torus over part of its length; ovary conspicuous, globose; style exserted first just below curve, angularly looped out and upwards before anthesis, afterwards refracted from line of dorsal suture at anthesis and gently incurved, dilating a few mm before style end; pollen presenter oblique, obovate, convex. **Fruit** 18.5–35 mm long, 18–31 mm wide, oblique, round to ellipsoidal, glabrous, glaucous, granulose; pericarp 4–8 mm thick. **Seed** 7–12 mm long, 5–9 mm wide, obovate, central within an encircling membranous wing 5–6 mm wide with a conspicuous raphe; outer surface convex, smooth; inner face concave.

Distribution Qld, N.T. and W.A.; Qld, in the Gulf country, east to about 140°F longitude and south to about Mt Isa; N.T., in Arnhem Land and the Darwin area extending inland to Katherine; also on coastal islands such as Melville Is.; W.A., confined to the N Kimberley and possibly a few offshore islands.
Climate Summer hot, wet, rain sometimes falling in brief showers; winter mild, dry. Rainfall 350–1200 mm.

Ecology Grows in open woodland and eucalypt forest in gravelly loam, and on the escarpment plateau around sandstone outcrops, on ridges and slopes and on rocky cliffs, mainly in sandy soil. Generally flowers June–Sept. There are reports of a more extended flowering time on Melville Is. Usually killed by fire but may regenerate from epicormic growth after low-intensity fire. Pollinated by birds.

Variation A relatively uniform species.

Mt Isa form This form has smaller (c. 21 mm long) and more rounded fruits than the typical form and occurs in inland locations such as Mt Isa. The foliage is hairier and the inner perianth surface glabrous.

Major distinguishing features Small tree habit;

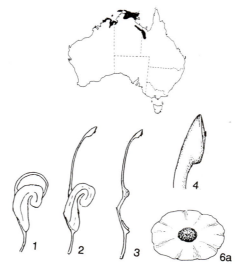

G. heliosperma

158A. *G. heliosperma* Flowers and foliage (M.Hodge)

158B. *G. heliosperma* Cultivated plant, Brisbane, Qld. (M.Hodge)

158C. *G. heliosperma* Fruits (P.Olde)

leaves usually bipinnate, the lobes lanceolate to elliptic, the base decurrent or not, the venation longitudinal and very fine; conflorescence secund, lax; flowers red, reversely oriented by twisting of the pedicels at anthesis; torus very oblique; perianth zygomorphic, glabrous outside, hairy or glabrous outside; limb round; pistil glabrous; ovary long-stipitate, the stipe adnate; fruit very large with thick pericarp; seed central within a broad, membranous wing.

Related or confusing species Group 5, especially *G. decurrens* which differs in its flower colour (cream to pale pink), its flowering period (November to April), and (usually) its leaves that are usually pinnate, the pinnae broader (up to 4 cm wide), more obtuse and decurrent, with coarser venation.

Conservation status Not presently endangered.

Cultivation *G. heliosperma* seems best suited to tropical, subtropical or monsoonal gardens and has been grown and flowered successfully as far south as Brisbane. It has grown well in Coffs Harbour and in pots in a frost-free area in Bulli south of Sydney. A robust grafted specimen is grown at the Royal Botanic Gardens, Sydney, and grafted plants are growing well even in frost-free sites in northern Vic. Further testing of this desirable species is a high priority although it would not probably be suitable for extremely cold areas. Although drought-resistant, it does best with some summer watering. A warm, sunny, open position in well-drained, preferably sandy or gravelly, slightly acidic soil produces best results in plants established on their own roots. It may be slow to establish and in its early years will thicken up its trunk before going into vigorous growth. Pruning is rarely necessary as it forms a naturally compact, slender tree. Not generally available in nurseries but seed is often listed in specialist catalogues.

Propagation *Seed* Usually grown from seed sown fresh and either peeled or soaked in hot water for 24 hours to improve germination. Seed is best sown in spring. *Cutting* Although it has been struck from cutting, it is quite difficult as a rule. The long trifoliate leaves are difficult to trim satisfactorily and the leaves generally fall off early. Cuttings taken and prepared absolutely fresh will stand a better chance, especially if the material contains no large compound leaves. Success is more likely if it is propagated in very hot, moist conditions. *Grafting* Has been grafted successfully onto *G. robusta* using the top wedge and whip-and-tongue methods and G. 'Poorinda Royal Mantle' using top wedge grafts. Early indications are that these combinations are compatible.

Horticultural features *G. heliosperma* is a slender, tree with an open, light crown, suitable for many applications in the landscape. The bright, blue-green to pale green leaves with their finger-like pinnae highlight the rather large, waxy red flowers that arise in profuse spikes at the ends of the branchlets during spring. Flowering continues spasmodically over a long period and is followed by large, pale green fruits, containing seeds with a very large papery wing. The fruits are quite ornamental in themselves and have a whitish bloom that rubs off when handled. The bright red flowers are very spectacular and attract nectar-feeding birds in great numbers. The landscaper in tropical climates could scarcely find a more novel and interesting specimen plant or background structure plant and thought should be given by civic bodies to using it in street plantings. Plants flower for a much longer time in the garden than in the wild.

Uses The Bardi people ate the seed when the fruits split open as well as the gum ('gugdju'). They also chewed the flowers for nectar. The branches were sometimes used to make windbreaks ('lungin') (Smith & Kalotas 1985). The Yanyula language people from Borroloola chop and boil the bark in water and use the liquid to treat infected sores and scabies (Smith 1991).

Hybrids Some hybridisation with *G. decurrens* has been observed on Melville Is (McGillivray 1993).

General comments The taxonomy of the *G. heliosperma*/*G. decurrens* complex is unresolved. More studies, using new characters, may provide a more acceptable basis for the separation of what are clearly distinct species.

Grevillea helmsiae F.M. Bailey (1899)
Plate 159

Queensland Agricultural Journal 4 (3): 195 (Australia)

Helms' Grevillea

The specific epithet refers to Sabine Helms (1866–1929), botanical illustrator and collector of the type. She was the wife of Richard Helms who collected during the Elder Exploring Expedition, 1891. HELMS-EE-EYE

Type: Childers, Qld, date uncertain, S.Helms (Mrs R.Helms) (isotypes: BRI, MEL).

A small, rough-barked **tree** 3–10 m tall. **Branchlets** angular to terete, silky or sparsely so. **Leaves** 5–20 cm long, 1–4 cm wide, spreading, shortly petiolate, simple, obovate to elliptic; juvenile leaves narrowly linear, pendulous; upper surface glabrous, glossy, silky when young, minutely pitted, with much reticulated brochidodromous venation; lower surface minutely silky to glabrous, the venation prominent; margins entire, flat or slightly recurved. **Conflorescence** 2–4.5 cm long, 3–4 cm wide, erect, shortly to prominently pedunculate, terminal or axillary, usually simple, rarely branched; unit conflorescence cylindrical, dense, not exceeding foliage; apical flowers opening first; peduncle and rachis silky to villous; bracts 0.5–0.8 mm long, obovate with incurved apex, villous both sides, falling before anthesis. **Flower colour**: perianth white; style green. **Flowers** basiscopic; pedicels 5–8.5 mm long, silky; torus 1.2–1.5 mm across, straight, square; nectary obconical, toothed; **perianth** 6–8 mm long, 1.5 mm wide, erect, oblong, silky outside, bearded inside in lower half, at first separating along dorsal suture, all tepals then free below limb, the dorsal tepals laterally reflexed and exposing inner surface before anthesis, afterwards free to base and independently rolled down below ovary; limb revolute, subcubic; **pistil** 12.5–20 mm long, glabrous; stipe 1.5–3.5 mm long; ovary ovoid; style exserted c. halfway and looped outwards before anthesis, afterwards strongly incurved, ridged along dorsal side; style end club-shaped; pollen presenter very oblique, obovate, convex. **Fruit** 20–25 mm long, 10–14 mm wide, horizontal to stipe, on thickened rachis, oblong-ellipsoidal, compressed, smooth; pericarp c. 1.5 mm thick. **Seed** 12.5 mm long, 6 mm wide, ovate, winged all round.

Distribution Qld, widespread near the coast from the Kennedy district N of Rockhampton S to Brisbane. *Climate* Hot, wet summer; mild, wet or dry winter. Rainfall 1200–1600 mm.

Ecology Occurs in dense softwood scrub, on rainforest margins, in semi-evergreen vine thickets along creek banks, and in dry rainforest on steep basalt slopes, in sandy loam soil derived from mixed acidic and basic rocks. Flowers mainly winter–spring but some flowers all year. Regenerates from seed but is not fire tolerant. Pollinator unknown.

Variation A uniform species.

Major distinguishing features Leaves simple, entire, broadly elliptic to obovate, glossy, penninerved; conflorescence shortly cylindrical, usually simple, not exceeding the foliage; torus straight; perianth zygomorphic, silky outside, bearded within; pistil glabrous; ovary stipitate; style dorsally ridged; fruit compressed, elliptic, transverse to the stipe, smooth; pericarp moderately thick-walled.

Related or confusing species Group 3. Closely related to *G. elbertii* from the Celebes which is distinguished by its branched conflorescence that exceeds the leaves and by its erect fruit with a thicker pericarp.

Conservation status Not presently endangered.

Cultivation *G. helmsiae* has been introduced to cultivation recently in Qld where it has flowered in a protected site. Like most rainforest species it should do well in a frost-free site; in nutritious, well-watered, acidic soil with a high humus level and good drainage it gives excellent results. Extended, dry conditions and drought are tolerated but a moist subsoil is best. It should be grown in a sunny or partially shaded position sheltered from hot winds. Grafted plants do well in a wider range of habitats and have been grown even in northern Vic. for some years. Denser flowering may be induced by tip pruning. Plants have been offered for sale in Qld nurseries but not commonly. Inquiries should be directed to rainforest nurseries.

Propagation *Seed* Plants have been raised from seed sown fresh. Germination is improved by soaking in hot water for 24 hours before sowing or by nicking or peeling the testa. *Cutting* Plants have been grown in limited quantities from cuttings of firm, young wood in spring. Young material strikes readily during the warmer months. *Grafting* Successfully grafted onto *G. robusta* and G.

G. helmsiae

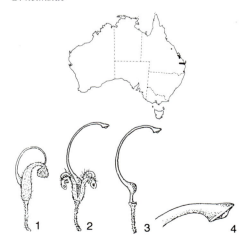

159A. *G. helmsiae* Conflorescence (M.Keech)

159B. *G. helmsiae* Fruits and foliage (P.Olde)

'Poorinda Royal Mantle' using top wedge, whip and mummy grafts.

Horticultural features *G. helmsiae* is a small, long-lived tree with potential in many situations such as parks, street plantings and rainforest gardens. The juvenile foliage and leaves from adventitious shoots from the trunk are markedly different from the adult foliage, being long, narrow and weeping, gradually passing to shorter and broader forms. Although the green and white flowers are not conspicuous, being somewhat hidden in the dense, dark green foliage, or out of sight in the crown of the tree, they have an interesting, spicy fragrance and would make an interesting vase specimen.

Grevillea hilliana F.Mueller (1858)

Plate 160

Transactions and Proceedings of the Philosophical Institute of Victoria 2: 72 (Australia)

White Silky Oak, Hill's Silky Oak, White Yiel Yiel

Aboriginal: 'yiel yiel' (northern N.S.W.)—a term applied in general to plants of the Proteaceae.

Named after the discoverer of the species, Walter Hill (1820–1904), who was director of the Brisbane Botanic Gardens and later the first Colonial Botanist for Qld. HILL-EE-ARE-NA

Type: Pine River, Moreton Bay, Qld, undated, W.Hill & F.Mueller (holotype: MEL).

A leafy, dense **tree** 10–30 m tall. **Branchlets** angular, silky to pubescent. **Adult leaves** 9–24 cm long, 2–6 cm wide, spreading, petiolate, simple, entire, ± oblong to obovate, leathery, undulate, obtuse or acute; juvenile leaves 28–40 cm long, 15–30 cm wide, deeply lobed to pinnatifid; lobes 1–10, 10–25 cm long, 1–5 cm wide, oblong to narrowly ovate; upper surface ± glabrous, sometimes sprinkled with appressed hairs, penninerved, the midvein, intramarginal veins and acutely angled lateral venation conspicuous; margins recurved; lower surface silky grey, penninervation prominent. **Conflorescence** erect to decurved, pedunculate, terminal or axillary, simple or rarely few-branched; unit conflorescence 9–22 cm long, 2–3 cm wide, cylindrical to S-shaped, dense; peduncle and rachis silky; bracts 0.5 mm long, oblong, villous, falling soon after raceme expansion. **Flower colour**: perianth green when young, maturing white; style white. **Flowers** oriented at right angles to rachis; pedicels 2.2–2.5 mm long, silky; torus 1–1.5 mm across, straight to slightly oblique, cup-shaped; nectary U-shaped; **perianth** 4 mm long, 1 mm wide, narrowly oblong, silky outside, glabrous or minutely pubescent in lower half inside, first separated along dorsal suture, the dorsal tepals then separated, reflexed laterally and exposing inner surface before anthesis, afterwards free to base and soon falling; limb revolute, spheroidal; **pistil** 13.5–16 mm long, glabrous; stipe 1.2–2.2 mm long, partially enclosed in torus; ovary ovoid; style exserted in lower half, strongly incurved to C-shaped after, dilating evenly at style end; pollen presenter very oblique, convex, ± elliptic. **Fruit** 17–26 mm long, 12–17 mm wide, erect, compressed-ovoid, apiculate, wrinkled; pericarp c. 2.5 mm thick. **Seed** 14 mm long, 7 mm wide, ovate, winged all round.

Distribution N.S.W., from about Burleigh Heads northwards, and Qld, from the N.S.W. border to about Cooktown, within 50 km of the coast. *Climate* Hot, wet summer; cool to mild, wet or dry winter. Rainfall 1000–1600 mm.

Ecology Occurs on the margins of tall mixed rainforest and coastal rainforest, sometimes around creeks in deep alluvial soils. Flowers spring–summer with a few flowers sometimes out of season. Regenerates from seed after fire. Pollinator unknown, probably insect.

G. hilliana

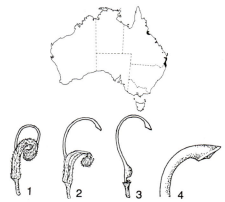

Variation A relatively uniform species. Plants from more southern populations have narrower, slightly more leathery, darker green leaves and more lobing of juvenile leaves.

Major distinguishing features Tree with juvenile, broadly lobed leaves with a silky-grey undersurface; conflorescence cylindrical, dense, usually simple; torus straight; perianth zygomorphic, silky outside, glabrous to pubescent inside; pistil glabrous, 13.5–16 mm long; pollen presenter very oblique; fruit compressed-ovate, apiculate; pericarp c. 2.5 mm thick.

Related or confusing species Group 3, especially *G. baileyana* which can be readily distinguished by the rusty-bronze indumentum of the undersurface of its juvenile leaves. In *G. baileyana* the conflorescences are many-branched and held erect above the foliage; the flowers are directed towards the apex, have shorter pistils (9–13.5 mm) and the follicles have thinner walls (c. 0.5 mm thick). *Buckinghamia celsissima* which is superficially very similar to *G. hilliana* has pedicels over 5 mm long.

Conservation status Not presently endangered.

Cultivation *G. hilliana* is relatively easy to cultivate on the east coast and has shown itself to be adaptable, being grown since 1858 as far south as Melbourne where large, old plants can be still be seen in the Royal Botanic Gardens. In unlikely climates it has succeeded in shadehouses and hothouses, such as at Burrendong Arboretum, N.S.W., as well as Canberra, and still grows at Edinburgh, Scotland. Light frosts are tolerated although it will be cut back severely by heavy frost and it does not endure very dry conditions for long. It prefers a well-drained position protected from strong winds in deep, slightly acidic, humus-rich loam or sandy loam. Light applications of nitrogenous fertiliser and regular watering especially in summer produce lush foliage. It does not require pruning, growing naturally into a well-shaped, small tree. The species is sensitive to *Phytophthora* fungus. When grown in full sun, it will not grow as tall but becomes more spreading and floriferous. This species is difficult to find in nurseries but is occasionally available from those specialising in rainforest plants.

Propagation Seed Propagated mainly from fresh seed which sets abundantly. Better germination will result from nicking or peeling. It will self sow in the garden. *Cutting* Difficult to strike from cutting, it has, nonetheless, been propagated using half-ripe wood taken in spring. *Grafting* Has been grafted successfully onto *G. robusta* using the top wedge graft.

Horticultural features G. hilliana is a densely crowned, small tree with lush, dark green leaves that are most interesting at their juvenile stage. They are large and deeply lobed, appealing for their luxuriant size and shape. The white flowers are individually quite small but cluster in pendulous, cylindrical racemes near the branch ends, highlighted by the dark green foliage. Flowering commences in late spring and continues through early summer, and although the season is relatively short, tends to be quite spectacular. This hardy and long-lived tree makes an interesting screen plant or feature plant if given a sheltered, warm position. For many years, it remains a large shrub and the flowers can be easily seen. It makes an ideal addition to the rainforest garden and is suited to street and park plantings in most climates. Young plants can be grown as attractive indoor plants for many years, being most hardy in a well-lit, cool to warm room.

History Right from its discovery, *G. hilliana* was considered worthy of cultivation and has been grown at Melbourne Botanic Gardens since 1858 (as *G. hillii*), and in Brisbane since 1875. Catalogues from botanic gardens overseas indicate widespread European cultivation in the late 19th Century and early this Century, in Italy and France and at Edinburgh, Scotland. William Macarthur sent two plants to Veitch & Son, nurserymen of London, in 1859, not long after the species was discovered.

General comments The hard, durable timber has attractive, dark brown, close-grained markings and was much sought for veneer and cabinet work in the early colonial days.

Grevillea hirtella (G.Bentham) P.M.Olde & N.R.Marriott (1994) **Plate 161**

The Grevillea Book 1: 182 (Australia)

Based on *G. pinaster* var. *hirtella* G.Bentham, *Flora Australiensis* 5: 427 (1870, England).

The specific epithet is derived from the Latin diminutive *hirtellus* (somewhat hairy), in reference to the leaf indumentum. HER-TELL-A

Type: near Champion Bay, W.A., date uncertain, P.Walcott 14 (holotype: MEL).

Synonym: *G. thelemanniana* subsp. *hirtella* (G.Bentham) D.J. McGillivray (1986).

A single-stemmed **shrub** c. 1 m high spreading up to 2.5 m wide. **Branchlets** terete, pubescent to villous, the hairs very fine and mostly > 1 mm long. **Leaves** 0.6–2 cm long, erect to spreading, usually crowded, shortly petiolate, obtuse, mucronate, simple, linear, or bi- or trifid, occasionally pinnatipartite with up to 5 lobes, both simple and divided leaves usually present; leaf lobes 1.5–8 mm long, 1–1.5 mm wide, strongly ascending; upper surface slightly convex, conspicuously pitted, loosely villous to glabrous, venation obscure; margins smoothly to angularly revolute, usually overtopping midvein below or partially so; lower surface obscured, villous in groove, midvein c. level with or recessed below marginal roll, glabrous. **Conflorescence** 1–1.2 cm long, decurved, pedunculate, terminal, simple, secund, relatively loose; peduncle and rachis glabrous or with scattered appressed hairs; bracts 0.5 mm long, ovate, pubescent outside, falling before anthesis; pedicels 4 mm long, glabrous or a few hairs at apex. **Flower colour**: perianth pale to deep pink-red; style red

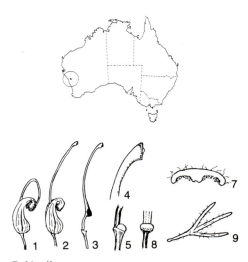

G. hirtella

160A. *G. hilliana* Flowers and foliage (M.Hodge)

160B. *G. hilliana* Young tree in cultivation at the Royal Botanical Gardens, Melbourne, Vic. (N.Marriott)

161A. *G. hirtella* Flowers and foliage (N.Marriott)

161B. *G. hirtella* Flowering habit of plant cultivated at Stawell, Vic. (N.Marriott)

with a green tip. **Flowers** acroscopic; torus c. 1 mm across, slightly oblique; nectary prominent, cushion-like to oblong; **perianth** 7.5–9 mm long, 3–4 mm wide, ovoid, dilated at base, glabrous and sometimes slightly glaucous outside, tomentose-pubescent inside, densely so about ovary, the hairs otherwise concentrated along tepal margins; tepals cohering except along dorsal suture; limb revolute, spheroidal-subcubic, emarginate in side-view, the segments carinate with keels c. 0.4 mm high, sparsely silky; **pistil** 21–23 mm long, glabrous; stipe 3.5–4 mm long, incurved, flattened; ovary triangular, stipitate; style before anthesis exserted from curve and looped strongly upwards, afterwards gently curved; style end abruptly expanded; pollen presenter 1.2–1.5 mm long, lateral to very oblique, oblong-elliptic, convex, flanged. **Fruit** 12–13 mm long, 5–6 mm wide, erect to oblique, oblong-acuminate to ellipsoidal, faintly ridged at base, ribbed, the ribs not extending beyond ventral plane, wrinkled; pericarp c. 0.5 mm thick. **Seed** not seen.

Distribution W.A., confined to several, small areas between Mingenew and Walkaway. *Climate* Summer hot, dry; winter mild, wet. Rainfall 450–500 mm.

Ecology Grows in open heathland in grey sand or gravelly to sandy loam. Flowers winter–spring. Pollinated by birds. Regeneration appears to be solely from seed.

Major distinguishing features Habit single-stemmed; branchlets pubescent to villous; leaves 0.6–2 cm long, sometimes all simple and linear, more usually mixed with some bifid or trifid; upper surface noticeably pitted, flat to convex, not channelled, venation obscure; lower surface obscured by strongly rolled margins, the midvein usually recessed; pedicels glabrous; perianth zygomorphic, glabrous outside except limb, hairy inside; limb glabrous or with a few scattered hairs, subcubic with keels on segments up to 0.4 mm high; pistil glabrous; stipe flattened, incurved; ovary triangular; pollen presenter oblique; fruit ridged at base.

Related or confusing species Group 14, especially *G. thelemanniana*, *G. delta* and *G. obtusifolia*. *G. thelemanniana* and *G. obtusifolia* have an appressed indumentum on the leaves and branchlets. *G. delta* differs in its mostly twice-divided leaves, pedicels sparsely silky all over and a more rounded perianth limb.

Variation Plants in the south of the species' range have mostly simple leaves, whereas those in the north have mostly divided leaves.

Conservation status 2E. An extremely limited distribution reduced to only a few populations in degraded road verges. Not believed to be conserved.

Cultivation *G. hirtella* has been cultivated for many years both in eastern Australia and W.A. It has proved to be a reasonably adaptable species that tolerates cool-wet and warm-dry climates as well as subcoastal summer-rainfall climates where humidity is not too high. It grows well in sandy soils of the Perth region. Extended dry conditions and frosts to at least –4°C are tolerated. Greatest success has been achieved when it is planted in full sun in well-drained acidic to alkaline (pH 6–7.5) sandy loam. It rarely requires any fertiliser and dislikes summer watering unless applied in the cool of evening. Due to its compact habit it rarely requires pruning. It makes a most attractive pot plant when planted in a free-draining, gravelly mix in a medium size tub. Occasionally available at specialist native plant nurseries as G. 'Walkaway Wanderer'.

Propagation *Seed* Untested. Seed should germinate well if the testa is nicked before sowing. *Cutting* Firm, young growth strikes well at most times of the year. *Grafting* Untested.

Horticultural features *G. hirtella* is a delightful, small shrub with a profusion of lovely, pink flowers and dainty small, hairy leaves. In cultivation, it is reasonably long-lived and makes a most decorative low, mounded shrub or groundcover. It is suitable for massed plantings and as a feature rockery plant in public, commercial and residential landscapes and is a popular plant with honey-eating birds.

General comments Given the diagnostic features that separate the populations formerly included in *G. thelemanniana*, *G. hirtella* stands out as very distinctive, notwithstanding a similar leaf size and arrangment to *G. thelemanniana*. It is especially distinguished by its punctate upper leaf surface, loosely villous branchlets, leaf indumentum and very prominently ribbed perianth limb.

Grevillea hockingsii W.Molyneux & P.M.Olde (1994) Plate 162

Telopea 5: 784 (Australia)

The specific epithet honours Mr F.David Hockings (1928–) who discovered this species. HOCK-INGZ-EE-EYE

Type: area of the Rockhole, off Tuckers Rd, c. 27 km W of Monto, Qld, 12 Oct. 1989, W.Molyneux & S.Forrester (holotype: MEL)

Dense, upright **shrub** 1.5–2 m high. **Branchlets** angular, silky. **Leaves** 4–14 cm long, 4–18 mm wide, shortly petiolate, simple, oblong to narrowly lanceolate, acute to obtuse with a short soft mucro, the indumentum purple-pink when young; upper surface glabrous, smooth to slightly pitted, shallowly concave with prominent venation; lower surface silky, midvein prominent; margins shortly recurved. **Conflorescence** usually decurved, occasionally erect, pedunculate, axillary, rarely terminal to short lateral branchlet; often cauline on older stems, simple, occasionally 1-branched; unit conflorescence 1.5–2.2 cm long, 2–3 cm wide, shortly cylindrical, open; apical flowers opening first; primary peduncle 4–15 mm long, terete or almost so, silky; secondary peduncle usually flattened or subterete, sometimes arising from old peduncle; rachis 2–8 mm long, silky to tomentose; bracts variable, 1–3.5 mm long, 0.4–1.6 mm wide, linear to lanceolate to narrowly triangular, margins revolute to strongly so, nearly concealing inner face, rusty-villous outside, falling early. **Flower colour**: perianth reddish-pink outside, mauve-pink inside; style reddish-pink. **Flowers** acroscopic; pedicels 3–7 mm long, sparsely silky; torus 1.5–1.9 mm across, square to slightly rectangular, oblique; nectary arcuate to semi-circular, flattened above; **perianth** 11–15 mm long, 2–2.5 mm wide, oblong, sparsely silky outside with scattered white and purple hairs, densely bearded inside 2–2.5 mm from base but clearly above ovary, sparsely hairy elsewhere, the hairs sometimes spathulate; limb subcubic, 2 mm long and wide, 1.8–2 mm deep, nodding to almost revolute, silky; tepals before anthesis separating along dorsal side, at anthesis free at apex

G. hockingsii

and all rolling back equally for 3–4 mm as 2 opposed pairs, the ventral tepals ultimately rolled down c. 2 mm further; **pistil** 13–17.5 mm long; stipe 1–2.5 mm long, glabrous, vertically grooved on ventral side; ovary 1.5–2 mm long, glabrous, ovoid, scarcely wider than style; style before anthesis bowed out through dorsal suture, sparsely pubescent but more densely so in apical 5–6 mm with short glandular hairs; style end c. 1 mm thick, sometimes emarginate; pollen presenter 2–2.5 mm long, 1.8–2 mm wide, lateral, obovate to cordate, flat; stigma slightly raised, distally off-centre. **Fruit** 20–36 mm long, 5–6.5 mm wide, 8–10 mm deep, strongly ribbed on dorsal side, ellipsoidal to ovoid-ellipsoidal, markedly swollen at each end, smooth to rugulose; pericarp 0.3–0.4 mm thick. **Seed** 11–13 mm long, 3–4.5 mm wide, 1.5–2 mm thick, narrowly ellipsoidal with an apical triangular elaiosome 2–3 mm long; margin recurved with a waxy wing 0.2 mm wide.

Distribution Qld, from the Coominglah State Forest between Monto and Biloela and on the Razor Back Range west of Mt Morgan. *Climate* Subtropical: summer wet, winter wet or dry. Rainfall 800 mm.

Ecology Grows in sandstone country, at least sometimes in red sand, either on the edges of breakaways or on sandy flats, occasionally on the edge of soaks containing *Callistemon* sp. Flowers (April–)June–December. Nectarivorous birds, possibly White-throated Honeyeaters, have been observed feeding at the flowers. Regeneration is from seed.

Variation A relatively uniform species.

Major distinguishing features Leaves simple, entire, silky on undersurface; conflorescence usually decurved, axillary, simple; perianth zygomorphic, bearded inside above level of ovary, the hairs of equal length; ovary glabrous, scarcely wider than style; style hairy; pollen presenter lateral; fruit ribbed with dilations at each end.

Related or confusing species Group 21, especially *G. victoriae* which differs in its mostly terminal, branched conflorescences and in its beard positioned lower on the inner perianth surface with hairs of unequal length.

Conservation status Suggested 3V. A thorough survey is required to establish the abundance of this species.

Cultivation *G. hockingsii* has been cultivated fairly widely in Qld gardens over the last few years. It has proved quite easy to grow in most situations that offer good drainage and full to partial sunlight and has flourished in both sandy and heavier neutral to slightly acidic soils. In Vic. it has so far proven slightly unreliable, often succumbing during cold, wet weather. Pruning is rarely required as it is a naturally compact species but some summer watering is appreciated, especially during extremely dry periods. A light annual dressing of all-purpose fertiliser low in phosphorus has proved beneficial in at least one Brisbane garden. Frost hardiness is untested but it is possible that it will tolerate short, cold spells. Some specialist nurseries in Qld occasionally stock this species.

162B. *G. hockingsii* Flowering habit of plant ex Biloela in cultivation near Brisbane, Qld (P.Olde)

162A. *G. hockingsii* Flowers and foliage (N.Marriott)

Propagation Seed Untested, but seedlings readily appear in gardens and it is assumed that fresh seed will germinate easily without pretreatment. *Cutting* Strikes readily from firm, new growth from spring to autumn. *Grafting* Grafts easily using mummy or side-wedge grafts onto *G. robusta* rootstock.

Horticultural features *G. hockingsii* is admired for both its relatively dense compact habit and its bright pinkish purple flowers which, although somewhat hidden within the foliage, appear over a long period. It makes an excellent screen or feature plant in the garden and the flowers attract many nectar-feeding birds that also use the foliage for shelter. The pinkish purple new growth is a notable feature. Has great potential in commercial and home landscapes especially in subtropical to mediterranean climates.

Grevillea hodgei P.M.Olde & N.R.Marriott (1994) Plate 163

The Grevillea Book 1: 185 (Australia)

Coochin Hills Grevillea

The specific epithet honours Mervyn Hodge (1933-), horticulturist, photographer, plant breeder, founding leader of the Grevillea Study Group of the Society for Growing Australian Plants. HODG-EYE

Type: East Peak, Coochin Hills, Qld, 17 July 1992, N.Marriott NM92/06 (holotype: NSW).

Erect **shrub** 1–4 m high, usually with a single main upright stem. **Branchlets** angular, ridged, tomentose. **Leaves** (6–)10–19 cm long, pinnatisect; leaf lobes 3–7 per side, 1.5–2.8 mm wide, (1.2–)5.5–12 cm long, narrowly linear, acute with a non-rigid, often curved, black mucro, decurrent at base, green to grey-green with an indumentum of curled tan to white hairs on upper surface, becoming sparse with age, although remaining dense along midvein; lower surface silky with white hairs and occasional erect, tan hairs; margins revolute, partially enclosing lower surface, the midvein prominent, silky; venation on upper surface consisting of a sunken midvein and inconspicuous intramarginal veins; petiole c. 1 cm long. **Conflorescence** 2–6(–8) cm long, 4 cm wide, terminal, erect, pedunculate, cylindrical, simple, dense, with development synchronous or on one side; peduncle 1–9 cm long, tomentose; rachis 1.5–3 mm thick at base, velvety; bracts 5–8 mm long, 0.5–1 mm wide, linear-lanceolate, villous outside, glabrous within, most falling well before anthesis. Flower colour: perianth cream with a dense indumentum of rusty hairs, concentrated especially on the limb and at the base of the perianth; style pale yellow, becoming pale green just below the yellow pollen presenter. **Flowers** basiscopic, crowded; pedicels 3–6 mm long, villous; torus c. 2 mm across, straight to very slightly oblique; nectary prominent, thin, erect, closely pressed to ovary, the margin dentate, erose; **perianth** 6–10 mm long, 2–3 mm wide, oblong-ovoid, villous-woolly outside with a mixture of short, white curled hairs and longer, rusty hairs, glabrous inside, at anthesis the dorsal and ventral tepals remaining united at limb while splitting widely into 4 separate segments from c. 5 mm from base to just above anthers in the limb, exposing inner surface and creating a broad platform, eventually separating into 4 segments; limb 3–4 mm long, 3–3.5 mm wide, rusty-villous, revolute, ovoid to subglobose, enclosing style end before anthesis; pistil 26–35 mm long; ovary sessile, villous; style glabrous, thickened in apical 2–3 mm, with a small hump on dorsal side below style end; pollen presenter slightly oblique, broadly conical, narrowly flanged at base. **Fruit** 13–14 mm long, 10–11 mm wide, oblique, ovate-elliptic, compressed, tomentose with mixed short and long hairs, grey with tan streaks on suture side particularly towards apex; pericarp c. 0.5 mm thick. **Seed** 9–12 mm long, 6–7 mm wide, oblong-elliptic, ridged around margin; outer face convex, smooth or slightly wrinkled; inner face flat; all encircled by a membranous wing.

Distribution Qld, confined to several small areas in the Beerwah area, notably Coochin Hills and Rupari Hill. *Climate* Summer hot, wet; winter mild to cool, wet or dry. Rainfall 700–900 mm.

Ecology Grows in low to tall scrub on exposed rocky hills, usually at base or on top of exposed rock platforms, in shallow sandy soil with *Eucalyptus curtisii*, *Leptospermum luehmanii*, *Keraudrenia collina*, *Rulingia* sp., *Acacia* spp. Flowers probably all year round, peaking March–Oct.

Variation A species with little variation.

Major distinguishing features Leaves pinnatisect, the undersurface enclosed by margins; conflorescence cylindrical, usually < 6 cm long; torus straight; nectary prominent, irregularly dentate; perianth zygomorphic, hairy outside with rusty hairs, glabrous within; pistil < 36 mm long; ovary sessile, densely villous, the style dilated in apical few mm and with a subapical dorsal hump; fruit with reddish stripes or blotches; seed winged all round.

Related or confusing species Group 35, especially *G. whiteana* and *G. banksii*. *G. whiteana* has a longer conflorescence (usually 10–12 cm long), pistil (> 40 mm long) and an indumentum of rusty hairs that mostly all fall before anthesis, (a few sometimes persistent on limb). *G. banksii* differs in having its leaf undersurface exposed, in its glandular perianth indumentum, and in its shorter floral bracts (< 2 mm long).

Conservation status Suggested 2V. Confined to a few hills in the Glasshouse Mountains; the major population on North Peak, Coochin Hills remains unprotected, being gazetted as a future quarry site.

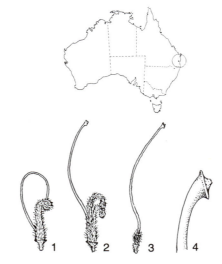

G. hodgei

163A. *G. hodgei* Comparison of conflorescences of *G. whiteana* (left) and *G. hodgei*

163B. *G. hodgei* Flowers and foliage (P.Olde)

Cultivation A widely grown species, *G. hodgei* has proved both long-lived and reliable, capable of enduring frosts and extended dry conditions with relative ease. It has been grown as far south as Ocean Grove, near Geelong, Vic., and inland at Rylstone, N.S.W., withstanding frost to -6°C. This subtropical plant does best, however, in a warmer climate and should be placed in full sun with some protection from strong winds as it has brittle branches and a shallow root sytem. It flourishes in acidic to neutral gravelly loam but will tolerate poor sandy soil. Once established pruning, fertilising and watering are rarely necessary, although light watering in long dry periods is beneficial. Available widely in general and native nurseries as Grevillea 'Coochin Hills', it has also been sold under the cultivar name Grevillea 'Honeycomb'.

Propagation Seed Sets prolific seed that germinates readily. Soaking in hot water improves germination. Cutting Strikes with some difficulty from cuttings of firm, new growth. Grafting Has been grafted successfully onto *G. robusta* using the top wedge, whip and mummy methods, but grafting is rarely necessary for this species.

Horticultural features *G. hodgei* is naturally bushy with dense, dark green foliage to the ground. The creamy brown flowers are prolific in early spring, contrasting with the foliage. Although most flowers are towards the top of the bush, enough form at eye-level to allow close inspection. The flowers attract birds strongly, and the foliage provides shelter and nesting sites. The species has been used successfully for both screening and feature planting over a wide climatic range and, except in areas with cold winters, can be strongly recommended for both beginners and experienced growers.

Grevillea hookeriana C.F.Meisner (1845)
Plate 164

in J.G.C.Lehmann (ed.), *Plantae Preissianae* 1: 546 (Germany)

Black Toothbrushes

The specific epithet honours William Jackson Hooker (1785 -1865), for many years director of the Royal Botanic Gardens, Kew, and to whom James Drummond sent many of his collections. HOOK-ER-EE-ARE-NA

Type: south-western W.A., J.Drummond 633 (lectotype: NY) (McGillivray 1993). A paratype—near the Swan R., L.Preiss 2626— is probably part of Drummond's collection.

Synonym: *G. pritzelii* L.Diels (1904).

G. hookeriana

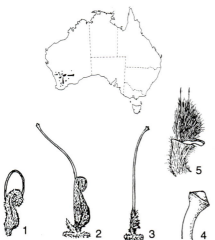

Spreading, dense **shrub** 1.5–2.5 m tall, 1.5–2.5 m wide with flowering racemes on branches ascending at c. 45° to the stem; branchlets angular, rarely terete, tomentose with curled hairs, silky or villous. **Leaves** 3–13.5 cm long, ascending to erect, sometimes clustered, ± sessile, leathery, simple, entire or deeply divided, either pinnatisect with 5–9 lobes or some or all leaves bi- or tripartite with the apical lobe subdivaricate, often with mixed simple and divided leaves; simple leaves sometimes curved; divided leaves sometimes subsecund or apical lobe subdivaricate; leaf lobes or simple leaves 0.3–13.5 cm long, 0.8–2.2 mm wide, linear to narrowly linear, sometimes wavy, basal lobes sometimes bi- or tripartite; juvenile leaves pinnatifid to pinnatisect; upper surface faintly granular to smooth, sometimes with faint longitudinal grooves, either with a persistent indumentum or glabrous, flat or convex, the venation obscure; lower surface bisulcate, the lamina not visible, hairy in grooves, the midvein prominent; margins smoothly, sometimes angularly revolute. **Conflorescence** 2.5–8 cm long, erect, shortly pedunculate, terminal, simple, conico-secund; peduncle villous; rachis angular, tomentose to villous, sometimes ± woolly, usually with some spreading hairs; bracts 1.5–3.5 mm long, ovate, sometimes the tip recurved, villous outside, a few usually persistent at anthesis, often spreading at early bud stage; pedicels 1–2 mm long, silky, curved; torus 1–1.2 mm across, ± straight; nectary patent, linguiform. **Flower colour**: perianth yellowish grey or yellowish green to pink; style black, maroon-black, red or yellow with green tip. **Flowers** acroscopic; **perianth** 6–8 mm long, 2.5 mm wide, ovoid, sometimes dorsally concave, dilated in lower half, silky to appressed-villous outside, glabrous inside; limb revolute, densely villous to silky, globular to ovoid; **pistil** 18–21.5(–23) mm long; stipe villous, c. 0.3 mm long; ovary subsessile, densely villous; style glabrous, sometimes with few erect hairs, gently curved but soon straight or slightly undulate, swept back with age; style end hoof-like; pollen presenter oblique, conical. **Fruit** 12–14 mm long, 8 mm wide, erect on curved pedicel, often persistent, oblong-ellipsoidal, tomentose with brown striping or blotches; pericarp 0.5–1 mm thick. **Seed** 9–11 mm long, 3.5–4.5 mm wide, oblong-ellipsoidal; outer face convex, smooth; inner face broadly gill-like around margin with a central smooth, flat section.

Distribution W.A., widespread in a roughly triangular area between Winchester, Dryandra and Newdegate. The distribution overlaps that of *G. apiciloba* but the two have not been seen growing together. *Climate* Hot, dry summer; cool to mild, wet winter. Rainfall 300-500 mm.

Ecology Grows in heath or mixed shrubland, usually in yellow sand but also in gravelly sand, rarely in laterite. Flowers early winter-early summer.

164A. *G. hookeriana* Long-lobed form, W of Wongan Hills, W.A. (P.Olde)

164B. *G. hookeriana* Simple-leaved form, Charles Gardner Reserve, W.A. (P.Olde)

Regenerates from seed after fire. Pollinated by nectar-feeding birds, especially mynahs.

Major distinguishing features Leaves subpinnatisect, rarely simple and entire; venation of upper surface obscure; margins smoothly rounded; undersurface not visible except two grooves beside midvein; conflorescence terminal, simple, conicosecund; pedicels short (1–2 mm); nectary tongue-like; perianth zygomorphic, hairy outside, glabrous inside; pistil 18–21.5 mm long; ovary sessile, densely hairy; style black or reddish black; fruit with reddish markings.

Related or confusing species Group 35, especially *G. apiciloba*, *G. armigera*, *G. calliantha* and *G. crowleyae*. *G. armigera* differs in its divaricately twice-divided, rigid, very pungent leaves. *G. calliantha* and *G. crowleyae* have a longer pistil (> 25 mm long). *G. apiciloba* differs in its obovate-cuneate leaves, the undersurface at least partially exposed.

Variation *G. hookeriana* is a variable species in need of further taxonomic research. There is variation in length, width, indumentum and degree of division of leaves as well as variation in indumentum of conflorescence and floral parts. See also flower colour.

Long-lobed (Typical) form This form occurs between Winchester and Wongan Hills. It generally has pinnatisect or tripartite leaves, the lobes 0.8–9 cm long, sometimes with secondary division, the upper surface usually persistently hairy, sometimes granular. Branchlet indumentum consists of curly hairs and the floral rachis is villous to subwoolly. Lobes on leaves are 5–10 mm apart. The perianth, especially the limb, is densely villous.

Simple-leaved form Merges with the typical form between Dowerin and Piawaning where the leaves are mostly tripartite with some pinnatisect. Between Kellerberrin and Cunderdin, the leaves are mostly all simple, although there are usually some tripartite leaves on all plants. Simple leaves and leaf lobes of this form are usually 1–2 mm wide, with lobes of divided leaves 1–3 cm long, greyish with a persistent indumentum. Branchlets also bear a spreading indumentum of wavy rather than curled hairs and the floral rachis is tomentose to villous. The perianth is silky-villous. An unusual, yellow-flowered plant of this form has been collected between Corrigin and Kondinin.

Narrow-lobed form This merges with the Simple-leaved form and extends south and south-west of Corrigin. In this area, the leaves and leaf lobes become gradually narrower (usually 1–1.2 mm wide). Between Corrigin and Kulin, the leaves are quite long and usually still either simple or tripartite. At its southern-most extremity, however, between Newdegate, Lake Grace and Nyabing, the leaves are mostly pinnatisect with the lobes slightly narrower again (0.8–1 mm wide) and closely aligned, more so than in the Long-lobed form. The midvein on the undersurface is prominently exserted beyond the margin. Simple leaves are also sometimes present.

Woodland form This form is closest to the Narrow-lobed form. Plants from two localities, at Boyagin and Dryandra, are from dry eucalypt woodland in association with dense laterite. In these collections, the conflorescences are sometimes quite long (to 8 cm) and the pistil is also slightly longer than in other forms (21–23 mm long). Yellow-styled plants also occur here. The pinnatisect leaves with lobes usually < 1 mm wide are hairier than similar leaves of the Narrow-lobed form and often have secondary division of the lower lobes. There is too much

164C. *G. hookeriana* Yellow-flowered variant, W of Kondinin, W.A. (P.Olde)

164D. *G. hookeriana* Flowering branch (N.Marriott)

overlap of these and other features to assign higher taxonomic status to this form until further research is completed. Plants of this form are closely related to *G. crowleyae*.

Conservation status Not endangered, although many populations are now restricted to degraded road verges.

Cultivation *G. hookeriana* has been widely cultivated for many years, mostly as *G. pritzelii*. It has proved hardy especially for dry-climate inland gardens, adapting well to a range of soils from sand to clay loam. For best results soil must be extremely well drained and the aspect warm and sunny. Once established, it will tolerate extremely dry conditions and regular frosts to at least -6°C. Wet-humid summer climates are to some extent untested with most cultivation having occurred to date in Vic., but grafted plants in Sydney and Brisbane have shown no sign of fungal disease on foliage or stems. Pruning improves the shape and density of the bush although it may temporarily reduce flowering. Fertiliser is not recommended. *G. hookeriana* is occasionally still sold in specialist nurseries as *G. pritzelii*.

Propagation *Seed* Untested. Sets moderate quantities of seed that should germinate well. *Cutting* Firm, young growth from plants in cultivation, taken during spring and early autumn, roots quite reliably. Cuttings from woody plants in the wild or slightly soft material can prove almost impossible to root. *Grafting* The most suitable rootstocks have been *G.* 'Poorinda Royal Mantle' on which plants grow quite vigorously and reliably. It has also been grafted successfully onto *G. robusta*, *G.* 'Poorinda Anticipation' and *G.* 'Bronze Rambler'.

Horticultural features *G. hookeriana* is an interesting, attractive plant with a rather spreading, dense habit and, usually, with greyish foliage. The toothbrush racemes poke from the tips of the branches, usually at an angle of c. 45° to the ground. Some might see the black-style flowers as rather dull, as they do tend to blend with the foliage. Most enthusiasts, however, delight in their subtle beauty, and as they are so numerous and extend over such a long period, most would agree that this is a very rewarding plant in the garden. The yellow-flowered selection from near Kondinin is extremely bright and floriferous. The dense foliage and nectar-filled flowers encourage many honeyeaters in search of both shelter and food. It is eminently suited to commercial landscapes for inland areas where an unusual utility plant is required. It is ideal as a screen or structure plant.

Hybrids Confusion exists in the horticultural trade where a presumed garden hybrid of uncertain origin has been sold for many years as *G. hookeriana*. One parent is almost certainly *G. tetragonoloba*. This is a robust shrub with wide-spreading to sprawling branches, rich green, pinnatisect leaves and bright red, terminal flowers that are usually sterile. This plant has been recently registered as G. 'Red Hooks'. Another plant often sold as *G. hookeriana* Fine Leaf Form is in fact *G. tetragonoloba*.

General comments Over the years, a number of species have been known incorrectly as *G. hookeriana*, most botanists relying on a description that could be (and was) applied to several, closely related species including *G. cagiana*, *G. concinna* and *G. tetragonoloba*.

Grevillea huegelii C.F.Meisner (1845)
Plate 165

in J.G.C.Lehmann (ed.), *Plantae Preissianae* 1: 543 (Germany)

Aboriginal: 'orrbi', 'urrbi' (Wirrung), S.A.

Named after Baron Karl A.A. von Hügel (1795–1870), an Austrian statesman, botanist and horticulturist with a large garden at Heitzing, near Vienna. Hügel visited Australia between 17 November 1833 and 6 October 1834, travelling to W.A., Tas. and N.S.W. HEW-GEL-EE-EYE

Type: near the Avon R., 6 miles (c. 10 km) from York, W.A., 10 Sept. 1839, L.Preiss 691 (lectotype: NY) (McGillivray 1993).

Synonym: *G. rigidissima* F.Mueller ex C.F.Meisner (1854).

A polymorphic **shrub** 0.3–3 m high, 1–3 m wide, sometimes prostrate or procumbent, sometimes irregular and straggly, sometimes bushy and erect, sometimes rounded, compact and

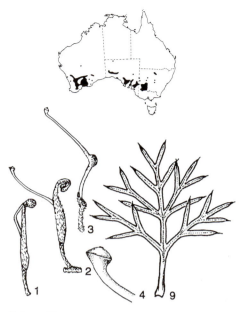

G. huegelii

165A. *G. huegelii* Close-up of conflorescence (P.Olde)

165B. *G. huegelii* Procumbent variant of the Pinnatisect-leaved form, N of Stirling Range, W.A. (N.Marriott)

165C. *G. huegelii* Prostrate habit of Glaucous-leaved form, near Marvel Loch, W.A. (P.Olde)

165D. *G. huegelii* Glaucous-leaved form (P.Olde)

dense. **Branchlets** terete, silky-tomentose when young, soon glabrous. **Leaves** 1–6 cm long, 2–6.5 cm wide, ascending to spreading, sessile to shortly petiolate, sometimes glaucous, pinnatipartite, sometimes pectinate; rachis straight to angularly or smoothly and strongly recurved; primary lobes (2–)5–11(–25), usually secund, sometimes flat, simple or with secondary or occasionally tertiary division; ultimate lobes 0.5–4 cm long, 0.7–2.2 mm wide, ascending to divaricate, linear, rigid, pungent; upper surface glabrous or sparsely silky, the venation obscure; margins smoothly revolute; lower surface mostly enclosed by margin and obscure, occasionally visible at sinuses where tomentose, midvein prominent and ± level with margin. **Conflorescence** erect, sessile to prominently pedunculate, terminal, axillary, or sometimes stem-borne on old wood, simple or branched at base; unit conflorescence 2.5–5 cm long, 2.5–5 cm wide, subglobose to umbel-like and many flowered, or few-flowered and irregular; peduncle 0.1–4 cm long, silky-tomentose; rachis silky-tomentose; bracts 0.4–1.2 mm long, 0.3–0.4 mm wide, ovate to narrowly triangular, silky-villous outside, glabrous inside, falling before anthesis. **Flower colour**: perianth bright to dull red, orange-red, apricot, burgundy, yellow, or rarely white—the glabrous inner surface of the perianth is more richly coloured; limb green to white; style red or yellow with green tip. **Flowers** acroscopic; pedicels 6–16.5 mm long, silky or sparsely so; torus 1.5–4.5(–6) mm across, oblique, cup-shaped; nectary mostly obscure and lining inside of torus, occasionally slightly raised above torus but following its outline, rarely the margin denticulate; **perianth** 17–22 mm long, 1.5–5 mm wide, cylindrical but sometimes protruding ventrally c. halfway along, densely to sparsely silky outside, sometimes with mixed, short, erect glandular trichomes, glabrous inside with raised longitudinal ridges evident in upper half; tepals at first splitting almost to base along dorsal side before anthesis, afterwards all separating to curve but otherwise cohering; limb nodding, globular to ovoid, densely silky, enclosing style end before anthesis; **pistil** 19.5–28.5 mm long, glabrous; stipe 7–14 mm long, inserted just within dorsal rim of torus and adnate to it at its base; ovary subglobose; style slightly exserted and bowed in upper half before anthesis, refracted almost at 90° about ovary, sometimes recurved from stipe, straight or incurved just after anthesis; style end abruptly dilated, slightly flanged; pollen presenter oblique, oblong-elliptic, convex to flat with obscure stigma. **Fruit** 10–12 mm long, 8–9 mm wide, erect, subglobose to oblong-ellipsoidal with attenuate apex, granulose; pericarp 0.6–1 mm across. **Seed** 8–12 mm long, 5–7 mm wide, 1–2 mm thick, elliptic to obovate; outer face slightly convex, wrinkled, the margin thickened into slightly raised ridge; inner face flat to slightly convex with raised margin, all encircled by an irregular, narrow to scarcely evident wing.

Distribution A disjunct transcontinental species found in lower rainfall areas across the southern half of Australia, but absent from the Nullarbor Plain. N.S.W.: far western plains from Wilcannia to Bourke and Byrock, S to c. Pooncarrie. Vic.: from Swan Hill W to the State border. S.A.: widespread in the south from the Vic. border to Maralinga and extending N to the Musgrave Ra. W.A.: widespread in drier parts of the southern half N to a line between Moora and Cundeelee. *Climate* Summer hot and dry; winter cool to mild and moist. Rainfall 200–700 mm depending on region.

Ecology Found in a wide range of habitats, but in the east usually in red, alkaline sand to sandy loam over limestone rubble. In the west, it is found in heavy clay-loam, gravelly loam and sandy loam. It is usually associated with mallee woodland or shrubland and sometimes open heathland. Flowers at any season but with a peak from winter to summer.

Variation *G. huegelii* is a highly variable species, some of which is discussed here. The informal names adopted combine those of McGillivray (1993) with our own continuing studies of this species. Further collections are needed, particularly of the Western Australian forms. Some of the forms here delimited may warrant formal names.

Maroon-flowered (typical) form This form seems to correspond in part to that group of plants north and east of 32°S and 118°E, noted by McGillivray as having bipinnate leaves. We have collected this form near Beverley, close to the type locality, and from near Manmanning, Moora, and Kellerberrin. It is distinguished by the following features: compact rounded, shrubby habit 1.5–2 m high, 1.5–2 m wide; leaves dull green, twice to three-times divided with distant (c. 5 mm apart) lobes, the rachis not usually recurved, the lobes ascending, rarely divaricate, up to 4 cm long; conflorescence enclosed by the foliage, usually well hidden within the plant, sessile, 8–10-flowered, usually cauliflorous, but sometimes axillary or terminal; perianth maroon to burgundy, densely silky outside with pinkish hairs. The distinctive flower colour is particular to this form.

Pinnatisect-leaved form The habit of this form varies from prostrate to decumbent, to irregular and erect or sometimes bushy and erect. It is distributed very widely over most of the species' range and is broadly distinguished in having simply pinnatisect leaves with a strongly recurved rachis, although some plants, mostly from south-west W.A., have secondary, divaricate bi- or tripartite division of the lower lobes. Plants in the south-western interzone have a prostrate to irregular, erect habit whereas those from eastern Australia have a more robust, shrubby habit, but prostrate forms occur in S.A. and western Vic. near Swan Hill. In some plants with shortened leaf lobes, some conflorescences at least

165E. *G. huegelii* Maroon-flowered form near Dalwallinu, W.A. (P.Olde)

165F. *G. huegelii* Close-up of conflorescence of Glabrous-flowered form, S of Norseman, W.A. (C.Woolcock)

165G. *G. huegelii* Pinnatisect-leaved form, near Swan Hill, Vic. (I.Evans)

partially exceed the foliage and are either sessile or borne on peduncles up to 2 cm long; they are densely flowered (up to 20 flowers), mostly terminal or axillary, occasionally cauline. The perianth is red, orange-red, yellow, apricot, pink or white, usually with dense to quite sparse appressed hairs. McGillivray (1993) stated that in plants from south-eastern Australia the perianth is usually 3–5 mm wide, whereas in the west it ranges from 1.5–3.5 mm wide.

Glaucous-leaved form This form typically has the decumbent to irregular habit of the Pinnatisect-leaved form from south-west W.A. but differs in its glaucous leaves. We have collected this form around Bullabulling and it is widespread from near Corrigin to Southern Cross. A completely prostrate, yellow-flowered form near Marvel Loch has been grown in cultivation.

Glandular-haired form McGillivray (1993) drew attention to plants from the Pioneer–Kalgoorlie–Southern Cross region which have glandular hairs on the perianth. Our early studies show that the appearance of this hair type in *G. huegelii* appears to be of little taxonomic significance. At one locality, some plants have glandular hairs, others do not.

Glabrous-flowered form Of all the variants in *G. huegelii*, this is the most distinct. It has the habit of the Pinnatisect-leaved form, growing as a sprawling, prostrate to decumbent shrub between Cape Arid, Esperance and N to Kumarl. The blue-grey leaves are short, 1–2 cm long, pectinate, the rachis smoothly and very strongly recurved. The lobes, usually 9–13, are closely aligned (not divaricate) and rarely exceed 1 cm in length. The bright red or yellow flowers are glabrous outside except a few hairs on the limb. In leaf morphology it approaches some plants E of Norseman which have a silky perianth. Horticulturally, this is an outstanding form.

Major distinguishing features Leaves pinnatisect, mostly with > 5 rigid, pungent lobes, the upper surface with obscure venation, the lower surface mostly enclosed by the margin, the rachis usually recurved; conflorescence subglobose to few-flowered, terminal, axillary or stem-borne; torus slightly to markedly oblique, cup-shaped; nectary within the torus; perianth zygomorphic, hairy outside at least on the limb, glabrous and ridged within; pistil glabrous; ovary long-stipitate; style reflexed about ovary at c. 45–70°; pollen presenter convex.

Related or confusing species Group 6. Closely related to *G. sarissa* which has very similar flowers but usually has simple, linear, straight leaves. Divided leaf forms of *G. sarissa* have mostly < 3 lobes.

Conservation status Not presently endangered.

Cultivation First introduced in 1858 at the Melbourne Botanic Gardens, *G. huegelii* has not proved adaptable in cultivation, growing reliably only in soils closely resembling those of its natural habitat. In practically all other soils it dies after one or two seasons. It requires a well-drained, alkaline sand or sandy loam in full sun or dappled shade for successful cultivation. Little experience has been gained with the western forms but there are indications that they may be a little more adaptable than eastern clones and should grow in neutral to acidic heavy loam or gravelly loam. Grafted plants grow slowly but steadily in an open sunny site but succumb in a cold, shaded one. It is not practical to continuously have to prune this species due to its prickly nature, although some of the more open forms would certainly require it. Once established, it is both frost- and drought-resistant and requires no summer watering. Cold, wet conditions and poor drainage are not tolerated. Most forms are unsuited to pot culture. It is occasionally available in native plant nurseries in S.A.

Propagation Seed Germinates well from seed that has been treated by peeling or nicking, but unless potted into very well-drained alkaline soil (eastern provenance) and placed in a warm, sunny position, seedlings rapidly succumb to fungal diseases. *Cutting* Difficult to strike from cuttings. Most success has been achieved using semi-hard wood during early autumn or late spring. Some forms strike more readily than others. *Grafting* Little success has so far been achieved although a few scions have been successfully grafted onto *G. robusta* and *G.* 'Poorinda Royal Mantle' using the top wedge technique.

Horticultural features *G. huegelii* is a most variable species bearing, in most forms, conspicuous, large, bright and attractive conflorescences that are strongly bird-attractive. Forms with terminal flowers should be selected for introduction and experimentation if flowers are the principal motivation. Both foliage and flowers are a feature of the Glabrous-flowered form, while the Glaucous-leaved form could be very attractive in some situations. The value of this species is limited by its very prickly foliage, although the Maroon-flowered form has possible value as a barrier or screen plant while other forms could be used most effectively in the shrubbery of inland gardens.

Grevillea humifusa P.M.Olde & N.R.Marriott (1994) **Plate 166**

The Grevillea Book 1: 182 (Australia)

Specific epithet from the Latin *humifusus* (spread along the ground), in reference to the habit. HEW-MEE-FEW-SA

Type: Cantabilling Rd, 26.5 km W of Brand Hwy, S of Eneabba, W.A., 15 Sept. 1991, P.M.Olde 91/96 (holotype: NSW).

Lignotuberous **shrub** with trailing stems to 3 m long. **Branchlets** angular, openly villous, the hairs to 2.5 mm long. **Leaves** 1.5–2 cm long, ascending to spreading, shortly petiolate, bipinnatisect; rachis straight to strongly recurved; lobes 0.5–1 cm long, 0.5 mm wide, narrowly linear, ascending to spreading; upper surface pilose, midvein evident to obscure; margins loosely revolute; lower surface partially exposed, pilose, midvein protuberant. **Conflorescence** 2 cm long, erect or decurved, pedunculate, terminal, simple, conico-secund, dense; peduncle and rachis pilose; bracts 1.5 mm long, ovate, acuminate, villous outside, falling before anthesis. **Flower colour**: perianth pink to pale red with cream limb; style pink, red or orange-red with yellow tip. **Flowers** acroscopic; pedicels 3–5 mm long, glabrous; torus c. 1 mm across, oblique; nectary cushion-like, prominent; **perianth** 5–7 mm long, 1.8–2 mm wide, ovoid, dilated at base, glabrous outside, pubescent inside near curve and along tepal margins, cohering except along dorsal suture; limb revolute, spheroidal-subglobose, silky, not ribbed; **pistil** 22–24 mm long, glabrous; stipe 3.5 mm long, flattened, incurved; ovary triangular; style before anthesis exserted at curve and looped upwards, afterwards gently incurved; style end slightly expanded, exposed before anthesis; pollen presenter 1–1.2 mm long, oblique, convex, ellipsoidal to orbicular. **Fruit** 12–15 mm long, 3–4 mm wide, erect, oblong, acuminate, with strong basal ridging, grooved; pericarp 0.5 mm thick. **Seed** not examined.

Distribution W.A., in a small area inland from Jurien. *Climate* Summer hot, dry; winter cold, wet. Rainfall c. 500 mm.

Ecology Grows in brown, gravelly loam in or near woodland. Flowers autumn–spring. Regenerates from seed or lignotuber. Presumably pollinated by birds.

Major distinguishing features Prostrate habit; branchlets angular, pilose with long white hairs; leaves bipinnatisect, pilose; conflorescence conico-secund; bracts > 1 mm long; perianth zygomorphic, glabrous outside except limb, hairy inside; pistil glabrous; ovary triangular on incurved, flattened stipe; pollen presenter oblique; fruit with strong basal ridging.

Related or confusing species Group 14, especially *G. delta* and *G. preissii*, neither of which has a prostrate trailing habit. *G. delta* also differs in its more hairy perianth and pedicels and in its less crowded flowers. *G. preissii* also differs in its glabrous to sparsely silky or densely tomentose-villous branchlets.

Variation A morphologically uniform species.

Conservation status 2E. Extremely rare, known from one population of c. 50 plants, beside a road in mostly cleared country.

Cultivation *G. humifusa* has been cultivated and appreciated widely since the 1960s (as *G. thelemanniana* Grey-leaf prostrate form). It appears to have been introduced by H.Demarz, collector for Kings Park, Perth, until recently the only collector of the species. It has proved easy to grow in drier, inland as well as coastal climates but is sometimes short-lived in summer-rainfall areas. It endures frost to at least -3°C and extended dry conditions without damage. It grows best in well-drained but moist acidic to slightly alkaline sand, sandy loam or gravelly loam in full sun. Partial shade is also tolerated. Rarely requires pruning except to restrict spread, and is an excellent pot plant using a standard, well-drained soil mix with light dressings of low-phosphorus, slow-release fertiliser. Native plant nurseries sometimes carry this species.

Propagation Seed Sets prolific seed in the wild. Germination is improved by nicking the testa before sowing. *Cutting* Grows readily from firm, young growth cuttings taken at most seasons. *Grafting* Untested.

Horticultural features *G. humifusa* is one of the most popular species in the *G. thelemanniana* complex and is valued for its dense, ground-covering habit, its hoary, grey-green foliage, and bright, pink-red, yellow-tipped flowers covering the plant in autumn and winter. Its trailing habit makes it an ideal spill-over plant for rockeries and walls and it is an excellent contrast or feature plant in the

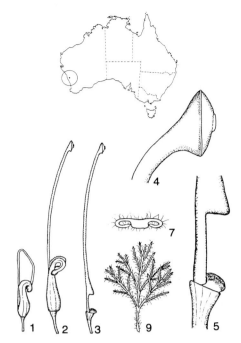

G. humifusa

166A. *G. humifusa* Flowers and foliage (N.Marriott)

166B. *G. humifusa* Trailing habit (M.Hodge)

landscape. It is both long-lived and attractive and could be used more frequently in landscaping than it currently is. It is popular in gardens of people interested in native plants.

General comments *G. humifusa* is recognised as distinct because of its unique habit and distinctive branchlet and leaf indumentum. It appears closely related to *G. preissii* but shares many important features with *G. delta*. Until its relationships can be properly assessed, it is here recognised as a distinct species. The name *G. humifusa* P.M.Olde & N.R.Marriott has no association with *G. humifusa* A.Cunningham, a nomen nudum which Bentham (1870: 436) placed under *G. laurifolia*.

Grevillea iaspicula D.J.McGillivray (1986)
Plate 167

New Names in Grevillea *(Proteaceae)* 7 (Australia)
Wee Jasper Grevillea

Specific epithet is a clever piece of linguistic humour using the Latin *iaspis* (jasper) and the diminutive *-ul* (small), to put the Wee into Jasper. EYE-ASS-PICK-YOU-LA

Type: Thermal Paddock, A.Howard's property, Wee Jasper, N.S.W., 16 May 1980, D.J.McGillivray 3962 (holotype: NSW).

Erect, dense, ± glabrous **shrub** 1–2 m high. **Branchlets** terete or slightly angular. **Leaves** 1–3.5 cm long, 4–10 mm wide, glabrous, crowded, ascending, sessile, simple, oblong-elliptic,

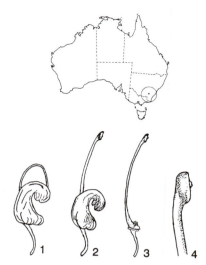

G. iaspicula

acuminate to obtuse, mucronate; margins entire, flat or shortly recurved; upper surface convex, smooth, slightly wrinkled above, midvein and lateral veins evident; lower surface paler, venation prominent. **Conflorescence** glabrous, decurved, sessile to shortly pedunculate, terminal or in upper axils, branched; unit conflorescences 2–3 cm long, 2–4 cm wide, partially secund, umbel-like to hemispherical; rachis glabrous; bracts 1–1.5 mm long, triangular, a few apical hairs sometimes evident, falling before anthesis, a few sometimes persistent at anthesis. **Flower colour**: perianth creamy white and pink; style pink. **Flowers** acroscopic; pedicels 4–6.5 mm long; torus c. 1.5 mm across, oblique; nectary conspicuous, arc-like, smooth; **perianth** 6–7 mm long, 3–4 mm wide, oblong to ovoid, markedly dilated at base, strongly curved, bearded inside above ovary, scattered hairs over remainder, cohering except along dorsal suture; limb revolute, spheroidal; style end partially exposed before anthesis; **pistil** 16–18 mm long; stipe c. 0.5 mm long, stout, with a tuft of hairs on ventral side, rarely glabrous; ovary glabrous; style exserted at curve and looped upwards before anthesis, afterwards straight to slightly incurved, inflexed above ovary; style end disc-shaped; pollen presenter lateral, round, ± flat to concave, umbonate. Fruit 12–17 mm long, 6–7 mm wide, oblique, ± cylindrical, longitudinally ribbed; pericarp c. 0.5 mm thick. Seed 8.5 mm long, 3.2 mm wide, narrowly elliptic with a short apical wing; outer face convex, minutely pubescent; inner face channelled at margin; margin revolute.

Distribution N.S.W., restricted to the Wee Jasper–Burrinjuck area. Climate Cool to cold, wet or dry winter; mild to hot, wet or dry summer. Rainfall 600–800 mm.

Ecology Grows in limestone-derived soil, in rock crevices around the entrance to caves, on steep rocky hillsides or at the base of cliffs. Flowers almost all year but mainly winter–spring. Pollinated by birds. Fire response is unknown but it is almost certainly seed-regenerative only.

Variation A morphologically uniform species.

Major distinguishing features A glabrous shrub; leaves simple, entire; conflorescence many-branched, glabrous; perianth zygomorphic, glabrous outside, hairy inside; stipe gibbous, usually hairy; ovary glabrous; pollen presenter lateral, flat.

Related or confusing species Group 25, especially *G. baueri* and *G. rosmarinifolia*. *G. baueri* has a hairy style and a villous ovary. *G. rosmarinifolia* has leaves < 4 mm wide, usually with hairs on the undersurface, and conflorescences mostly simple.

Conservation status 2E. A rare species with limited distribution. The recent discovery of a large population in the Burrinjuck area will assist its conservation.

Cultivation Introduced to cultivation c. 1984, *G. iaspicula* has proved fairly adaptable, growing well in most States of Australia. Well-grown plants can

167A. *G. iaspicula* Flowers and foliage (N.Marriott)

167B. *G. iaspicula* Limestone sinkhole habitat, Wee Jasper, N.S.W. (P.Hind)

be seen at the Mount Annan Botanic Garden, Sydney, the Australian National Botanic Gardens, Canberra, and in many private gardens on the east coast. It grows well in a well-drained acidic to alkaline soil in a sunny position and will tolerate heavy frosts and dry conditions, although it prefers some summer watering. It may occasionally become a little untidy in cultivation but regular pruning will assist to maintain shape. Although it tolerates summer-humid climates reasonably well, it does not like highly acidic sandy soil. Flowering may be affected by bud-drop psyllids and plants may occasionally be covered by white louse scale, although this is easily treated with white oil. Occasionally available in nurseries, especially in Vic.

Propagation Seed Untested. *Cutting* Easy to strike from firm, young growth at any time of the year. *Grafting* Grafts readily and reliably onto *G. robusta*.

Horticultural features *G. iaspicula* is a charming plant with pale green, smooth leaves and prominent, cream and red flowers displayed for most of the year. The flowers clustered in small umbel-like heads attract many nectar-seeking birds from late winter to summer. It forms an attractive, compact shrub and is useful as a feature or screen plant. The pale green foliage also has potential as a contrast feature in the landscape. Its erect habit can be accentuated by judicious pruning, making it suitable for planting beside paths and narrow spaces. Recent attempts by the Australian National Botanic Gardens to introduce it to wider cultivation are to be encouraged because of its many worthy horticultural features and its limited wild populations.

Grevillea ilicifolia (R.Brown) R.Brown (1830)

Supplementum Primum Prodromi Florae Novae Hollandiae exhibens Proteaceas Novas 21 (England)

Based on *Anadenia ilicifolia* R.Brown, *Transactions of the Linnean Society of London* 10: 167 (1810, England).

Holly Grevillea, Holly Bush

Aboriginal: 'gukwonbeyurgalk' Lake Hindmarsh, Vic. (Smyth).

Named for the resemblance of the foliage (Latin *folium*, a leaf) to that of *Ilex aquifolium* (holly). EYE-LISS-I-FOAL-EE-A

Type: near the sea-shore, Bay X [near Port Lincoln, S.A.], Feb./March 1802, R.Brown (holotype: BM).

Erect, dome-shaped **shrub** to 2 m high spreading to 3 m wide, or prostrate or decumbent. **Branchlets** angular, silky. **Leaves** 3–11 cm long, ascending, shortly petiolate, extremely variable, ovate to obovate in outline with markedly cuneate base, *either* toothed in upper half, sometimes with secondary toothing, the teeth few or many, broadly triangular and pungent, *or* (bi-)pinnatifid, kite- or wedge-shaped, regularly or irregularly lobed, the lobes narrowly triangular or linear, acute or obtuse, pungent, *or* irregularly once or twice divided into narrowly linear, elk-horn-like lobes, the segments distant, curved, pungent, *or* rarely, simple and elliptic; upper surface silky, venation evident; margins shortly recurved to revolute; lower surface sometimes obscured, usually visible, silky or tomentose with prominent venation. **Conflorescence** erect to decurved, terminal, shortly pedunculate, usually unbranched; unit conflorescence 2–5 cm long, secund; peduncle and rachis silky; bracts 0.3–0.9 mm long, ovate, silky outside, falling before anthesis. **Flower colour**: perianth pale green to grey; style red, pink, orange or pale yellow with green tip; S of a line between Arno Bay and Elliston, S.A., the style is always yellow (R.Mason pers. comm.). **Flowers** acroscopic; pedicels 1.5–4.5 mm long, silky-tomentose to almost glabrous; torus 1–1.8 mm across, ± straight to slightly oblique; nectary U-shaped; **perianth** 7–10 mm long, 2–3 mm wide, narrowly ovoid, sometimes dorsally concave, silky-tomentose outside, the hairs sometimes only sprinkled, glabrous inside, cohering except along dorsal suture; limb revolute, globular, enclosing style end before anthesis; **pistil** 19.5–25 mm long; stipe 2–4.4 mm long, silky; ovary silky; style glabrous or with basal hairs, exserted from near curve and looped upwards before anthesis, afterwards gently incurved, dilating smoothly and evenly into style end; pollen presenter oblique, elliptic, convex. **Fruit** 10.5–16.5 mm long, 5.5–7.5 mm wide, erect, narrowly ovoid, silky, reddish-striped, sometimes faintly ridged; pericarp 0.5–1 mm thick. **Seed** 10 mm long, 3.5 mm wide, ellipsoidal; outer face convex, smooth; inner face flat; margin slightly recurved with a membranous border.

Distribution N.S.W., Vic. and S.A. N.S.W.: in the south-west near Griffith but almost extinct there. Vic.: widespread in west and north-west with an isolated population near Dunolly, central Vic. S.A.: widespread on Eyre Peninsula, Yorke Peninsula, Kangaroo Is. and in the lower south-east from the Murray R. to Penola. *Climate* Hot, dry summer; cool to mild, wet winter. Rainfall 250–700 mm, usually 400–600 mm.

Ecology Grows in open eucalypt woodland, open heath, mallee heath, mallee woodland or scrub, in deep sand, sandy clay, limestone rubble, gravelly loam or sand over clay. Flowers spring–summer. Pollinated by nectarivorous birds. Regenerates from seed or sucker after fire.

Major distinguishing features Foliar and floral indumentum usually appressed; habit single-stemmed or suckering; leaves usually toothed or lobed, sometimes entire or deeply divided; conflorescences secund; floral bracts < 1 mm long, falling early; perianth zygomorphic, glabrous within; ovary stipitate, silky; dorsal side of style end convex; fruits erect.

Related or confusing species Group 35, especially *G. aquifolium* and *G. infecunda* (See *G. infecunda* for comments). *G aquifolium* differs in its broader style end, the spreading indumentum of the floral rachis and the decurved stipe of the fruit.

Variation An extremely variable species in habit, leaf shape and degree of leaf division. There are minor floral differences within forms which need further study to examine their correlation with foliar features. Two varieties are here accepted of which var. *ilicifolia* is the most morphologically diverse.

Key to varieties

Leaves entire, toothed or deeply divided with triangular to linear lobes .... var. **ilicifolia**

Leaves deeply divided with narrowly linear to subulate lobes .... var. **angustiloba**

Grevillea ilicifolia (R.Brown) R.Brown var. ilicifolia
Plate 168

Leaves entire, toothed or deeply divided with triangular to linear lobes.

Synonyms: *G. ilicifolia* var. *attenuata* R.Brown (1830); *G. ilicifolia* var. *dilatata* R.Brown (1830); *G. ilicifolia* var. *integrifolia* F.Mueller ex C.F.Meisner (1856); *G. ilicifolia* var. *lobata* (F.Mueller) G.Bentham (1870); *G. behrii* R.Schlechtendal (1847); *G. lobata* F.Mueller (1855); *G. dumetorum* C.F.Meisner (1956); *G. lobata* var. *sturtii* C.F.Meisner (1856); *G. approximata* M.Gandoger (1919).

Wedge-leaved form This form is relatively common through south-eastern S.A., as far west as Kimba on Eyre Peninsula, and western Vic. with a few collections from south-western N.S.W. Leaf shape is variable, usually obovate-cuneate but sometimes kite-shaped to diamond-shaped or ovate, and is distinguished by simple, triangular, shallow toothing around the upper margin and a markedly tapered leaf base. The leaf undersurface is tomentose with an indumentum of straight or curled hairs. Occasional simple, elliptic leaves are also encountered on plants of this form. The grey-green to blue-green leaves are often strongly convex with prominent reticulate venation of the upper surface. It ranges in size from prostrate to 1–2 m high.

Kangaroo Island form This form is not restricted to Kangaroo Is. where it is consistent in its leaf type but occurs also on southern Eyre Peninsula and Yorke Peninsula. Typically it has short, fan-shaped, prickly-toothed leaves that are olive green in colour and have a glabrous upper surface and a silky undersurface. Leaf toothing is not only simple but often bi- or tri-dentate, the toothing appearing crowded around the margin. Flowers are pale pink to bright red, usually on short, somewhat insignificant conflorescences. It is usually a dense shrub to 1.5 m high and up to 3 m wide, with occasional decumbent forms.

Lobed-leaf form This group of leaf forms typically has oak-leaf to herring-bone leaves and is fairly widespread and variable. Habit ranges from prostrate to upright, while leaves range from grey-green to a beautiful silvery blue-grey with white undersurface. The leaves may have only shallow dissection (in which case the lobes are broad and somewhat rounded at the end) or be quite deeply once or twice-divided (lobes linear or lanceolate). Plants currently grouped in this variety occur in S.A.

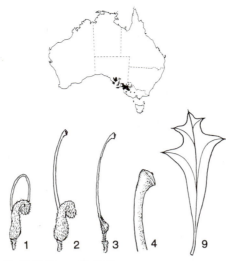

G. ilicifolia var. *ilicifolia*

from around the Murray R. S to near Penola and in Vic. from the Little Desert N to the Murray R., being particularly common in the Big Desert. There is also an extremely isolated population near Dunolly in central Vic. These forms are spectacular foliage plants. Plants from the Hattah area and Tutye, Vic., differ further in the spreading indumentum of the conflorescence and ovary.

Grevillea ilicifolia var. angustiloba
F.Mueller (1868)
Plate 169

Fragmenta Phytographiae Australiae 6: 122 (Australia)

Type: Mount Arapiles, Vic., 22 Sept. 1860, collector not cited (holotype: MEL).

Leaves deeply divided with narrowly linear to subulate lobes.

This variety has leaves very deeply once to thrice divided into one-sided, narrowly linear, curved lobes less than 1.5 mm wide. It is usually an attractive prostrate mat with greenish, grey-green or bronze-

168A. *G. ilicifolia* var. *ilicifolia* Yellow-flowered variant, Western Black Range, The Grampians, Vic. (N.Marriott)

168B. *G. ilicifolia* var. *ilicifolia* Flowering branch (N.Marriott)

168C. *G. ilicifolia* var. *ilicifolia* Form ex Little Desert, cultivated at Stawell, Vic. (N.Marriott)

169A. *G. ilicifolia* var. *angustiloba* Erect variant with unusual orange flowers, Little Desert, Vic. (N.Marriott)

169B. *G. ilicifolia* var. *angustiloba* Typical prostrate habit of this variety, Little Desert, Vic. (N.Marriott)

green leaves and branchlets bearing a brown indumentum. The style is usually red but sometimes is orange-red. Occurs in the Keith–Little Desert area and scattered localities in western Vic. and eastern S.A. Occasional erect plants to 1 m tall occur, especially in the Little Desert. This is the most distinct of the variants of the species, hence we retain the current name.

Conservation status Not presently endangered although in many areas populations have been all but wiped out by clearing for agriculture.

Cultivation *G. ilicifolia* has been in cultivation for many years during which time it has proven very hardy and reasonably adaptable, succeeding in cool-wet, warm to hot-dry, and even summer-rainfall climates. Many selections are in cultivation at Burrendong Arboretum, N.S.W. as well as the Australian National Botanic Gardens, Canberra, and it is today cultivated in hundreds of private gardens, mainly in Vic. and S.A. Two forms are cultivated in the U.S.A. It tolerates drought conditions and is hardy to frosts to at least -7°C. Best results have been obtained when planted in full sun in well-drained, neutral to alkaline, sandy loam. It does not require summer watering or fertilising but responds to regular tip pruning to maintain density and shape. Smaller forms make very attractive specimens when grown in tubs using a well-drained mix with light applications of slow-release fertiliser. Select forms are occasionally available in specialist native plant nurseries, especially in Vic. and S.A.

Propagation *Seed* Sets many seeds that germinate well if pretreated. *Cutting* Firm, young growth usually strikes well at most seasons, but material taken from wild plants sometimes strikes with considerable reluctance. *Grafting* Has been grafted successfully onto *G. robusta* using the approach and wedge methods.

Horticultural features *G. ilicifolia* is a most interesting, attractive plant with many diverse leaf variations. Plants of the Lobed-leaf form can be quite spectacular in both the shape and colour of the leaves. The flowers are showy without being brilliant and are popular with honey-eating birds. It is such a variable species that a whole garden could be created using this species alone with large shrubs, small to medium-sized shrubs and beautiful ground covers combining in a montage of unusual variety and contrast. All forms are long-lived in cultivation and make a valuable addition to the landscaper's armoury especially in warm and dry climates.

Hybrids A few hybrids of *G. ilicifolia* have arisen in garden situations, among them G. 'Old Gold' and G. 'Crimson Glory'. The beautiful hybrid, G. 'Pearl Light' was found in the wild among a patch of *G. ilicifolia*, and is a presumed F2 hybrid, having *G. lavandulacea*, *G. alpina* and *G. ilicifolia* influence.

History Brown described this species under the name *Anadenia ilicifolia* in 1810, but by 1830 regarded it as a *Grevillea*. It was in cultivation in Melbourne in 1858 but does not appear to have been grown in Europe. The name *G. ilicifolia* or *G. illicifolia* appears three times in Allan Cunningham's journal (Lee 1925: 297, 357, 359). These refer to two species, one of which was later named *G. angulata* by Brown. Neither applies to the true *G. ilicifolia*. Horticultural lists from 19th century gardens also include the name *G. illicifolia* (e.g. Salm-Reifferscheid-Dyck 1834, Jacques 1843) but the reference is almost certainly to another taxon, probably *G. acanthifolia* subsp. *acanthifolia*.

General comments In his revision of the genus, McGillivray (1993) did not accept any published varietal names. An extensive study of this species is needed prior to more formal treatment of the variation. Although we here continue to recognise two varieties on horticultural grounds, their taxonomic status needs verification.

Grevillea inconspicua L.Diels (1904)
Plate 170

Botanische Jahrbücher für Systematik 35: 153 (Germany)

Cue Grevillea

The specific epithet was derived from the Latin *inconspicuus* (not apparent, not remarkable), a reference to the small, rather hidden flowers. INN-KON-SPICK-YOU-A

Type: near Cue, W.A., June/July 1901, L.Diels 3277 (holotype: B).

Synonym: *G. brachyclada* W.V.Fitzgerald (1905).

A silvery, rounded, much-branched **shrub** to 1 m. **Branchlets** wiry, terete, silky, soon glabrous. **Leaves** 1–5 cm long, 0.8–1.8 mm wide, geniculate and somewhat twisted at base, patent to vertically descending, crowded, shortly petiolate, simple, narrowly linear, rigid, pungent to slightly uncinate; margins angularly refracted about an edge-vein; upper surface convex to subtriangular, silky to glabrous, midvein sometimes prominent; lower surface bisulcate, silky to villous in grooves, midvein evident but somewhat recessed. **Conflorescence** 1–1.5 cm long, 1 cm wide, erect, pedunculate, terminal on short branchlets, unbranched, usually with 6–8 flowers, umbel-like; peduncle silky; rachis short, villous; bracts not seen, falling very early. **Flower colour**: perianth silvery grey; style white or pinkish white. **Flowers** adaxially directed; pedicels 3.5–5 mm long, silky; torus 1 mm across, oblique; nectary U-shaped, entire; **perianth** 5–6 mm long, 1 mm wide, narrowly oblong, undilated, silky outside, bearded inside, at first separated along dorsal suture, then all tepals free below limb, the dorsal tepals reflexed laterally and exposing inner surface before anthesis, afterwards free to base and independently rolled down, quickly falling; limb revolute, ovoid; **pistil** 9.5–11.5 mm long, ± glabrous; stipe 1.1–1.5 mm long; ovary ellipsoidal, scarcely broader than style; style exserted from near curve and looped upwards before anthesis, angularly incurved, afterwards hooked near apex, short hairs at back of pollen presenter, grooved along ventral side; style end flat, plate-like; pollen presenter oblique, flat with conspicuous, pointed stigma. **Fruit** 7.5–14 mm long, 5–6.5 mm wide, almost perpendicular to stipe, ovoid-ellipsoidal, faintly ribbed, granulate; pericarp 0.3 mm thick. **Seed** not seen.

Distribution W.A., confined to a few isolated localities around Cue, Meekatharra, the Weld Ra. and NE of Sandstone. *Climate* Arid; very hot, dry summer; mild to warm, sporadically wet winter. Rainfall 210–220 mm.

Ecology In gravelly, red, clay-loam in drainage channels on hillsides, generally in treeless sparsely

G. inconspicua

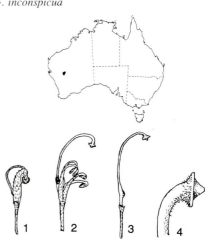

170A. *G. inconspicua* Plant in natural habitat, Cue Hill, W.A. (P.Olde)

170A. *G. inconspicua* Flowers and foliage (P.Olde)

vegetated areas. Flowers winter–spring. Pollinator unknown, probably insects. Fire response unknown.

Variation A morphologically uniform species.

Major distinguishing features Branchlets wiry; leaves simple, entire, mostly patent to vertical, twisted at base, the midvein sometimes evident on upper surface; conflorescence few-flowered; perianth zygomorphic, undilated, hairy outside, bearded within; pistil glabrous except style end; fruit faintly ribbed.

Related or confusing species Group 21, especially *G. costata* which has ascending leaves with the midvein on the upper surface not visible and conspicuously ribbed fruit.

Conservation status 3EC-i. Known from only a few small populations. Growing in at least one nature reserve.

Cultivation Introduced to cultivation in 1986, *G. inconspicua* is proving reasonably hardy in warm, dry inland climates and has been grown successfully in western Vic. Plants have survived frosts to at least -3°C but probably tolerate lower temperatures. Early results indicate a need for a well-drained, acid to slightly alkaline gravelly or clay loam in full sun for best growth. Its close relative *G. costata* is extremely long-lived in cultivation and the same may apply to this species. It is hardy to extended dry conditions and even drought but succumbs in wet, cold or summer-humid climates. Due to its naturally dense habit, it should not require pruning. It could be tried in a large earthenware tub using a well-drained mix with low-phosphorus, slow-release fertiliser, especially if situated in a sunny, sheltered courtyard. Not presently available commercially.

Propagation *Seed* Untested. *Cutting* Cuttings of firm, new growth have struck well in spring or early autumn. *Grafting* Has been grafted successfully onto *G. robusta* using the whip graft. Species in Group 21, especially *G. costata*, may prove better root-stocks.

Horticultural features *G. inconspicua* does not have showy, flowers and on no account could be recommended for this feature, but the shaggy appearance afforded by the close, vertically oriented, grey, linear leaves and the wiry, short branchlets give it unmistakeable character, suggesting use as a small feature or contrast plant, or as a low hedge. The inconspicuous white or pinkish flowers are sweetly scented, attracting butterflies and small moths. Its dense habit also provides shelter for small, insectivorous birds. Due to its endangered status, attempts should be made to introduce this species more widely among inland, dry climate growers.

Grevillea incrassata L.Diels (1904)
Plate 171

Botanische Jahrbücher für Systematik 35: 156 (Germany)

The specific epithet is derived from the Latin *incrassatus* (thickened), a reference to the leaves which are short and thick. INN-KRASS-ARE-TA

Type: near Parker Ra. S of Southern Cross, W.A., 1892, E.Merrall (holotype: B).

Synonym: *G. integrifolia* subsp. *incrassata* (L.Diels) D.J.McGillivray (1986).

An erect, bushy, silvery **shrub** 0.3–2 m tall with emergent floral branches. **Branchlets** silky, terete. **Leaves** 0.5–2.5 cm long, 0.5–1.5 mm wide, ascending to spreading, ± sessile, simple, terete, faintly ribbed, or narrowly linear with both surfaces the same but prominently channelled on undersurface, incurved, crowded, white-silky, shortly pointed. **Conflorescence** erect, pedunculate, terminal, usually branched; unit conflorescence 1–2 cm long, 1–1.5 cm wide, umbel-like or shortly cylindrical; peduncle and rachis silky; bracts 0.5 mm long, ovate-acuminate, strongly incurved, villous outside, falling before anthesis. **Flower colour**: bright yellow or yellow-gold, rarely pale yellow; flowers emit a spicy perfume. **Flowers** regular, glabrous; pedicels 2.5–3 mm long; torus c. 0.4 mm across, straight; nectary obscure; **perianth** 4 mm long, 0.6–0.8 mm wide, cylindrical, the tepals separating before anthesis; limb erect, ellipsoidal; **pistil** 7 mm long; stipe 0.5 mm long; ovary obovoid, conspicuously lobed on ventral side; style straight; pollen presenter cylindrical with its base straight, the apical, stigmatic opening expanded, slightly concave. **Fruit** 10–16 mm long, 2–3 mm wide, erect, usually persistent over one season, subcylindrical to obovoidal with two ventral horns at apex, smooth; pericarp c. 0.5 mm thick. **Seed** (immature) 6 mm long, 0.5 mm wide, narrowly obovate-cuneate with broad apical wing; outer face shiny with longitudinal wrinkles, convex; inner face smooth; margin recurved with a membranous border.

Distribution W.A., in inland regions from Narembeen E to Southern Cross and S to Lake King. *Climate* Hot, dry summer; cool to mild, wet winter. Rainfall 230–280 mm.

Ecology Often dominant in open sandplain and open shrub mallee, usually in sand, often over gravel or occasionally in clay-sand. Flowers spring to early summer. Pollinator unknown, probably insect; native bees have been seen at the flowers. Regenerates from seed after fire.

Variation A uniform species in homogeneous populations, with only minor variation in leaf shape and flower colour.

Major distinguishing features Leaves simple, usually terete, incurved, silvery; conflorescence shortly cylindrical to umbel-like, usually bright yellow-gold; floral rachis silky; nectary absent; flowers glabrous; perianth regular; ovary stipitate; fruit narrowly obovoid with two apical horns, persistent.

Related or confusing species Group 2, especially *G. incurva* which differs in its glabrous floral rachis and its much longer conflorescence (usually > 5 cm).

Conservation status A widespread, common species, secure in its inland, semi-arid habitat.

Cultivation *G. incrassata* is suited only to dry, arid climates and has been successfully cultivated at Stawell in western Vic. and at Glenmorgan, Qld, where it tolerates frost to at least -3°C. It requires a warm, dry climate with very low humidity, preferring full sun in very well-drained, open, acidic sandy or gravelly loam. Success is increased by planting on a mound built up substantially with gravel or sand. Very good results have been achieved in less than optimal climates by growing it in pots in a free-draining mix and an annual feeding of low-phosphorus slow-release fertiliser. It rarely requires pruning as the emergent, most visible branches are also the floral branches. Summer watering can be fatal to cultivated plants even if well-established. Rarely available from nurseries or seed suppliers.

G. incrassata

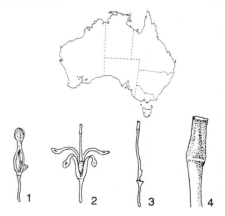

171C. *G. incrassata* Flowering branch, E of Southern Cross, W.A. (N.Marriott)

171B. *G. incrassata* Close-up of flowers and foliage (M.Keech)

Propagation *Seed* Sets many seeds that germinate freely even without pretreatment. Overnight soaking improves germination rate. Average germination time 40 days (Kullman). Seedlings have broader, flattened leaves. *Cutting* Difficult to strike. Best results are obtained in a low-humidity propagation frame using firm, young vegetative growth in spring or early autumn. *Grafting* Has been grafted successfully onto *G. robusta*, G. 'Poorinda Anticipation' and G. 'Poorinda Royal Mantle' using the top wedge technique. Plants on *G. robusta* have grown well but many have broken at the union in windy weather, possibly indicating long-term incompatibility.

Horticultural features This truly delightful small shrub with masses of bright, yellow, rounded heads of flowers over the upper parts of the bush and tiny, incurved, silver leaves give a first impression of a small *Acacia* or *Phebalium*, and at a distance it is hard to believe it is a grevillea. Even when not in flower, *G. incrassata* is an attractive shrub with its silvery stems and leaves and would make a superb display in raised beds or rockeries in the inland garden or landscape. It is pollinated by insects and as a result has a most attractive, delicate perfume described by some as smelling like an exotic chrysanthemum. This is undoubtedly one of the most exquisite of the small grevilleas, worthy of pride of place in any garden.

171A. *G. incrassata* Plant in natural habitat, E of Southern Cross, W.A. (N.Marriott)

Grevillea incurva (L.Diels) P.M.Olde & N.R.Marriott (1994)   Plate 172

The Grevillea Book 1: 177 (Australia)

Based on *G. integrifolia* var. *incurva* L.Diels, *Botanischer Jahrbücher für Systematik* 35: 157 (1904, Germany).

The specific epithet is derived from the Latin *incurvus* (incurved), a reference to the leaf curvature. IN-CUR-VA

Type: Avon District, W.A., Nov 1901, E.Pritzel 893 (holotype: B).

Erect **shrub** 1.5–2.5 m high. **Branchlets** silky, terete. **Leaves**: lower broadly elliptic (to 10 mm wide), successive leaves becoming narrowly elliptic or obovate, both surfaces similar, glabrous to sparsely silky with densely silky margin, the lower surface with more prominent midvein and slightly wrinkled lamina; adult leaves unifacial, 1–2(–4.5) cm long, 0.7–1 mm wide, ascending to erect, sessile, simple, subterete, strongly incurved, silky, the apex slightly recurved, non-pungent; margins entire; venation obscure, sometimes a longitudinal rib evident. **Conflorescence** erect, pedunculate, terminal or usually in upper axils, simple or branched; unit conflorescence (2.5–)4–5 cm long, 0.8–1 cm wide, cylindrical, the flowers opening ± synchronously, sometimes from apex first; peduncle silky; rachis silver-silky, usually brown (sometimes yellow) beneath indumentum, occasionally sparsely silky or glabrous; bracts 1.2–1.5 mm long, ovate, acuminate to caudate, silky-villous with white and red hairs, falling early. **Flower colour**: perianth and style creamy white. **Flowers** regular, glabrous, adaxially oriented; pedicels 1.5 mm long, 4-lobed; torus c. 1 mm across, straight; nectary absent; **perianth** (below limb) 2 mm long, 0.5–1 mm wide, oblong; tepals channelled inside, with obscure to prominently exserted V-shaped labia below anthers, before anthesis separating to base and bowed out below cohering limb, afterwards all free; limb 2 mm long, ellipsoidal, the segments flanged at margins, ribbed at base; distance from base of anthers to narrowed section of tepal 0.3–0.4 mm; **pistil** 5.5–6 mm long, glabrous; stipe 0.3–0.5 mm long; ovary 0.7 mm long, lateral, obovoid; style 4–4.5 mm long, 0.4 mm thick, smooth or granular, deeply grooved, before anthesis exserted from near base and folded outwards, after noticeably kinked about middle, becoming straight to undulate; style end 0.7 mm across, straight; pollen presenter 1.2 mm long, narrowly conico-cylindrical with cupped stigma. **Fruit** 7.5–10 mm long, 2.5–3 mm wide, often persistent, narrowly obovoid to obovoid, smooth; pericarp 0.2–0.5 mm thick. **Seed** not seen.

Distribution W.A., central wheatbelt from Meckering to Kellerberrin and S to Harrismith. *Climate* Summer hot, dry, winter wet, cool. Rainfall 400 mm.

Ecology Grows in yellow or white sand in open heath. Flowers late spring. Pollinated by insects. Regenerates from seed.

G. incurva

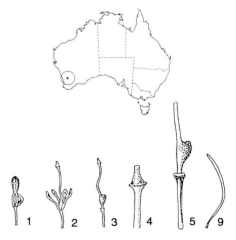

172. *G. incurva* Flowering branches (P.Olde)

Major distinguishing features Leaves short, incurved, subterete, silvery; conflorescence cylindrical; rachis silky; flowers regular, glabrous; pedicels c. 1.5 mm long; fruit obovoid, relatively short.

Related or confusing species Group 2, especially *G. biformis* (southern form) which differs in its longer, narrowly linear leaves, and *G. incrassata* which has a shorter floral rachis, golden-yellow flowers and longer, narrowly obovoid-cylindrical fruits.

Variation A relatively uniform species.

Conservation status Known from only a few populations. A code of 2EC is recommended.

Cultivation *G. incurva* is unknown in cultivation. A position in full sun is recommended in neutral to slightly acidic deep, sand or gravelly sand. Frosts as low as -5°C are tolerated in the wild. This naturally compact plant should require little maintenace once established. A cool-wet to hot-dry climate is recommended. Summer humid climates may reduce life expectancy and increase pest problems.

Propagation *Seed* Sets large quantities of seed that should germinate readily if pretreated by nicking. *Cutting* Untested. Should grow easily using firm, new growth in spring. *Grafting* Untested.

Horticultural features *G. incurva* is a beautiful silvery shrub with attractive, short, curved leaves and massed creamy-white flowers above. It has potential as a low screen or feature plant in suitable climates. The sweetly scented flowers attract insects. They are very bright and give a lift to the shrubbery. It has great potential as a foliage plant for the cut-flower trade.

General comments McGillivray (1993) treated this taxon as a form of *G. integrifolia* subsp. *biformis*. We consider that classification too broad as some formal recognition is warranted. It shares most features with *G. biformis* but there are important differences in leaf length and structure that suggest a close relationship with *G. incrassata*. It occurs in self-reproducing populations over a considerable area, and is here considered distinct, at least until the phylogenetic position of all taxa in the *G. integrifolia* group is clarified.

Grevillea infecunda D.J.McGillivray (1986)
Plate 173

New Names in Grevillea *(Proteaceae)* 7 (Australia)

Anglesea Grevillea

The failure of this species to set seed inspired the specific epithet which is derived from the Latin *infecundus* (unfruitful).   INN-FEK-UN-DA

Type: Gum Flat road between Harvester Co.'s concession and Wensleydale, Anglesea district, Vic., 25 Oct. 1969, J.H.Willis (holotype: MEL).

A suckering, open **shrub** 0.3–1.2 m high. **Branchlets** terete to angular, silky-tomentose. **Leaves** 3–7 cm long, 1.3–4 cm wide, ascending, shortly petiolate, ovate to oblong, simple, toothed

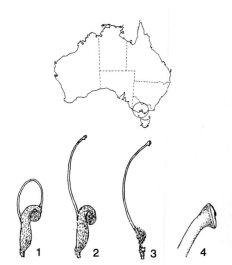

G. infecunda

with 5–11 broadly triangular, pungent lobes, the lower lobes sometimes bifid; upper surface sparsely pubescent becoming ± glabrous, midvein and lateral veins evident; margins shortly recurved; lower surface silky-tomentose, midvein and reticulate venation prominent. **Conflorescence** 2–4 cm long, erect or deflexed, shortly pedunculate, usually terminal, secund, unbranched; peduncle and rachis silky-tomentose; bracts 1.5–3 mm long, ovate, silky outside, usually persistent to anthesis. **Flower colour**: perianth greenish yellow; style dull pink to bright red; pollen brownish red. **Flowers** acroscopic; pedicels 1.5–2 mm long, tomentose; torus 1–2 mm across, straight; nectary U-shaped; **perianth** 7–9 mm long, 2–2.5 mm wide, narrowly ovoid, silky-tomentose outside, glabrous inside, cohering except along dorsal suture; limb revolute, globular, enclosing style end before anthesis; **pistil** 18–26 mm long, sometimes digynous; stipe 1.8–3.7 mm long, appressed-villous dorsally, almost glabrous ventrally; ovary silky to appressed-villous; style glabrous except near ovary, exserted at curve and looped upwards before anthesis, afterwards angularly incurved, noticeably so just below hoof-like style end; pollen presenter very oblique, oblong, broadly convex with prominent stigma. **Fruit** (immature ex cult. N.Marriott 1988) 10 mm long, 5 mm wide, on strongly incurved stipe, densely villous with white and reddish hairs; pericarp 0.5 mm thick. **Seed** not developed, oblong-ellipsoidal, outer face convex, smooth; inner face smooth; margin shortly recurved, bearing a short, membranous wing.

Distribution Vic., confined to the Anglesea area.

173A. *G. infecunda* Flowers and foliage of a shrubby variant, near Anglesea, Vic. (N.Marriott)

173B. *G. infecunda* Follicles developing on a plant cultivated at Stawell, Vic. (N.Marriott)

In 1852 Mueller collected it near Brighton (Melbourne) on 'scrubby, red sandhills' but it is presumed extinct there now.

Ecology Grows on undulating hills in dry sclerophyll eucalypt woodland in sandy, gravelly or clay loam. Flowers spring. The flowers are attended by birds. Regeneration after fire is exclusively by sucker. *Climate* Summer warm, dry; winter cold, wet. Rainfall 600–800 mm.

Variation A reasonably stable species with some variation in indumentum density on flowers and leaves. Habit ranges from decumbent to upright; leaves vary in size, shape and depth of lobing. Throughout its range, at least 3 populations can be separated on these features.

Major distinguishing features Habit consistently suckering; branchlet indumentum appressed; leaves toothed, the undersurface with an appressed indumentum; conflorescence simple, secund; floral bracts usually persistent, > 1.5 mm long; perianth zygomorphic, glabrous within; pollen sometimes misshapen; ovary stipitate, silky to appressed-villous; style end usually concave on dorsal side.

Related or confusing species Group 35, especially G. aquifolium, G. bedggoodiana and G. ilicifolia. G. aquifolium differs in the spreading indumentum on its foliage and floral parts. G. bedggoodiana has a pistil < 17 mm long and distinctive bi-coloured conflorescences. G. ilicifolia differs in its minute (< 1 mm long) caducous floral bracts, erect fruit and less conspicuously expaded style end.

Conservation status 2VC. Extremely rare, confined to a small area mostly within a coal-mining lease where it is threatened by gravel extraction, trail-bike damage and periodic fires. Urgent conservation measures are required.

Cultivation G. infecunda has been in cultivation for a number of years during which time it has proved reasonably hardy and adaptable, being grown successfully in Vic. and by members of the Grevillea Study Group, mainly in pots. It grows well in a variety of conditions, but greatest success is achieved when planted in the sun or dappled shade in acid, sandy or gravelly loam. It does not require summer watering or fertilising but responds to the occasional light pruning. Grows well in a medium to large tub in a well-drained, gravelly mix. In the wild it shoots vigorously from the base after fire and would probably respond similarly to hard pruning. Extended dry conditions are readily endured and it is recorded as frost-resistant at -5°C. Specialist nurseries in Vic. occasionally sell it.

Propagation *Seed* An infertile species that sets no viable seed. *Cutting* Firm, young growth generally strikes readily at most seasons. *Grafting* Has been grafted onto G. robusta using the whip graft. *Suckers* Once established, plants sucker lightly, especially when surface roots are scarred or damaged. Established suckers can be either carefully transplanted or treated as cuttings which give a very high strike rate.

Horticultural features G. infecunda is an interesting, low to medium-sized shrub with attractive 'holly' leaves and showy if not spectacular pink to red toothbrush flowers that turn dark red after anthesis. The low percentage of fertile pollen appears to have evolved in response to the development of non-sexual means of reproduction. Nonetheless, the flowers are popular with honey-eating birds. It is a long-lived plant, suited as part of a shrubbery or as a low feature plant. Due to its endangered position in the wild, it deserves to be more widely cultivated in public and private landscapes.

General comments As fruits have never been collected in the wild, specimens in cultivation should be closely monitored for follicle development as there is some possibility that the species could be cross-pollinated by birds carrying pollen from other grevilleas in the garden. Recently a plant in the author's garden at Stawell began to produce fruit but they aborted after several weeks for unknown reasons.

Grevillea infundibularis A.S.George (1974) Plate 174

Nuytsia 1: 371 (Australia)

Fan-leaf Grevillea

The specific epithet (Latin *infundibularis*, funnel-shaped) refers to the leaves which partially encircle the stem. INN-FUN-DIB-YOU-LARE-ISS

Type: W side of Middle Mt Barren, Fitzgerald R. NP, W.A., 16 July 1970, A.S.George s.n. (holotype: PERTH).

Spreading, sprawling decumbent **shrub** to c. 1 m. **Branchlets** angular when young, soon terete, sometimes ridged, glandular-pilose. **Leaves** 1–6 cm long, 1–4 cm wide, erect to spreading, ± glabrous, sessile, obovate to hemispherical, the base cuneate to amplexicaul; margins denticulate, entire or with broad, obtuse teeth, shortly recurved, the lobes scarcely pungent; upper surface funnel-like near base, midvein and acutely angled lateral veins visible; lower surface bluish, sometimes with a few scattered hairs, minutely pitted, venation prominent. **Conflorescence** erect, shortly pedunculate, terminal on short leafy branchlet or axillary, sometimes on sparsely leaved floral branch, usually simple, rarely few-branched; unit conflorescence 1.5–3 cm long, 4–8-flowered, umbel-like; peduncle glandular-pilose; rachis scarcely evident; bracts minute, ovate, sparsely villous outside, glabrous inside, falling before anthesis. **Flower colour**: Bright red. **Flowers** abaxially oriented; pedicels 5–8 mm long, sparsely glandular-pubescent; torus 1.6–2.2 mm across, reversely oblique, cup-shaped; nectary plate-like, the margin broadly toothed; **perianth** 7–12 mm long, 2–4 mm wide, oblong-ovoid, dilated at base, with mixed glandular/silky hairs outside, pilose inside, cohering except along dorsal suture; limb revolute, spheroidal to subcubic, the segments prominently keeled; **pistil** 18–20 mm long, glabrous; stipe 0.5–1.2 mm long; ovary ovoid with two ventral rounded protuberances; style exserted at curve and looped upwards before anthesis, afterwards gently incurved; style end flattened, exposed before anthesis; pollen presenter lateral, oval, flat. **Fruit** 13–15 mm long, c. 9 mm wide, erect, ovoid, ventrally ridged or with two protruding triangular eyes on ventral side, granulate; pericarp 1–1.5 mm thick. **Seed** 8 mm long, 3 mm wide, oblong; outer face convex, mottled, smooth; inner face with a central elliptic raised ridge surrounded by a broad channel, all covered (when immature) with a membranous skin drawn to an excurrent point 1.5 mm long at apex and sometimes at basal end.

Distribution W.A., confined to the central part of Fitzgerald R. NP. *Climate* Cold, wet winter; warm, dry summer. Rainfall 300–600 mm.

Ecology Occurs in heathland among quartzite boulders or in coastal sand. Flowers year-round peaking in winter–spring. Probably pollinated by birds. Fire response unknown.

Variation A species rarely seen in the wild because of its remote habitat. It has at least two known variants.

Typical form Known only from the base and lower slopes of several peaks in the Fitzgerald R. NP where it grows as a sprawling shrub to c. 60 cm. high. This form has been cultivated for a number of years and has leaves with toothed edges.

Entire-leaved form A form with entire or broadly lobed, obovate leaves has been collected from Fitzgerald R. NP just above the beach near Middle Mt Barren and at Twin Bays. This form is almost prostrate with many of the conflorescences terminating short leafy branches off a prostrate main stem. It grows in dune sand over orange clay.

Major distinguishing features Branchlets glandular-pilose; leaves sessile, ovate to round with marginal teeth or entire, funnel-like near base, minutely pitted on undersurface; conflorescence few-flowered, pedunculate; torus reversely oblique; nectary cup-shaped; perianth zygomorphic, hairy on both surfaces; pistil glabrous, ovary stipitate with two cushion-like swellings on ventral side, pollen presenter lateral; fruit ventrally horned, granulate.

Related or confusing species Group 12, especially G. nudiflora which differs in its linear leaves and non-glandular branchlet indumentum. For many years G. infundibularis was sold mistakenly as G. asteriscosa which differs in its smaller, pricklier leaves, oblique torus and hairy ovary.

Conservation status 2VC-t. Although locally frequent, it is restricted to a few peaks and nearby areas in Fitzgerald R. NP.

Cultivation Introduced in the 1970s, this hardy, adaptable plant grows readily in hot-dry and cool-wet climates, tolerating summer rain, extended dry conditions and frosts to at least -4°C. It has been grown successfully in Sydney and at Burrendong Arboretum, N.S.W., in many Vic. gardens and in

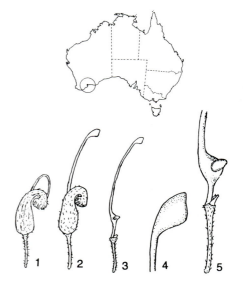

G. infundibularis

174. *G. infundibularis* Flowers and foliage (N.Marriott)

the U.S.A. *G. infundibularis* performs best in an open, full sun situation in well-drained, sandy or gravelly acidic loam, forming a low, sprawling, shrub. When planted among other shrubs it climbs through them, eventually covering a far larger area than when on its own. It dislikes excessive fertiliser but responds to summer watering, although this is rarely needed. Regular pruning creates a far denser plant. It makes an attractive cascading specimen for a large tub or pipe. Sometimes available at specialist native plant nurseries, especially in Vic.

Propagation Seed Sets few seeds but should germinate well if pretreated by peeling. *Cutting* Firm, young growth strikes well at most seasons, the highest strike rates in early autumn. *Grafting* Has been grafted successfully onto *G. robusta* and *G.* 'Poorinda Royal Mantle' using top wedge, whip and mummy grafts.

Horticultural features *G. infundibularis* is a most interesting species with its fan-shaped leaves, its glowing red flowers and its open, sprawling habit. The foliage is enhanced by turning red-green during dry or cold conditions. New growth is an attractive bronze. Although the flowers are not spectacular, they are very bright and dainty and can be found scattered over the bush practically all year round. They are strongly bird-attractive. It is a long-lived species that would make a useful addition to the armory of landscapers looking for something different. Its long, unbranched leafy stems make beautiful cut foliage for floral arrangements.

Grevillea insignis R.Kippist ex C.F.Meisner (1855)

Hooker's Journal of Botany and Kew Garden Miscellany 7: 76 (England)

Wax Grevillea

The specific epithet refers to the distinguished appearance of the plant and is derived from the Latin *insignis* (remarkable, extraordinary). INN-SIGG-NISS

Type: W.A., J.Drummond coll. 5 suppl. no. 12 (holotype: K). The Supplement to Drummond's Fifth Collection was gathered in July 1849, mostly from around Mt Stirling and Mt Caroline.

Erect, spreading, glabrous, glaucous **shrub** 2–4 m high, 3–5 m wide. **Branchlets** terete. **Leaves** 3–9 cm long, 3–4 cm wide, glabrous, glaucous, ovate, oblong to obovate, spreading, shortly petiolate, leathery to rigid; margin flat, sinuate, pungently toothed; rachis curved; upper and lower surface glaucous, slightly discolorous; midvein, edge-vein, lateral and penninervation prominent. **Conflorescence** erect to decurved, pedunculate, terminal on short branchlet, simple or few-branched; unit conflorescence 2.5–5 cm long, globose to shortly cylindrical, the apical flowers opening first; peduncle and rachis glabrous; bracts 1.8–2.5 mm long, triangular, glabrous except scattered apical and marginal hairs, falling before anthesis. **Flowers** acroscopic; pedicels 5–10 mm long, filamentous; torus 2–3 mm across, cup-shaped, funnel-like at base of stipe, oblique; nectary saucer-shaped, scarcely evident above toral rim; **perianth** 8–9 mm long, 3–5 mm wide, oblong-ovoid, narrowed towards throat, shiny, glabrous outside, white bearded inside about level of ovary, cohering except along dorsal suture and between dorsal and ventral tepals at the curve; limb revolute, spheroidal, the segments impressed at margin; style end partially exposed before anthesis; **pistil** 11–20 mm long; stipe 1.2–2.5 mm long, dorsally villous, ventrally glabrous, perpendicular to torus, adnate for c. 2 mm at base; ovary villous; style exserted at curve and looped upwards before anthesis, afterwards straight to slightly incurved, villous in lower half, otherwise glabrous, gradually dilated to style end; style end club-shaped, partly exposed before anthesis; pollen presenter lateral, oblong, flat. **Fruit** 10–13 mm long, 9 mm wide, erect, oblong-ellipsoidal, villous, soon glabrous, glaucous; pericarp 1.5–2 mm thick. **Seed** 8–9 mm long, 4 mm wide, oblong-ellipsoidal; outer face concave, smooth; inner face flat; margin shortly winged with an oblique apical attenuation.

Major distinguishing features A glabrous shrub; leaves oblong, sinuate, pungently toothed; conflorescence terminal, subglobose to cylindrical; torus oblique, funnel-like at base of stipe; perianth zygomorphic, glabrous outside, sparsely bearded within, the dorsal and ventral tepals separating at curve, limb with impressed margins; stipe inserted perpendicularly on toral rim; ovary villous; style end club-shaped.

Related or confusing species Group 26, especially *G. pilosa* which differs in its hairy outer perianth surface and hairy leaves.

Variation Two geographically disjunct subspecies are recognised.

Key to subspecies

Branchlets glaucous; most leaves with truncate to spreading base and shallowly arcuate sinuses   subsp. **insignis**

Branchlets not glaucous; most leaves with cuneate base and arcuate sinuses   subsp. **elliotii**

Grevillea insignis R.Kippist ex C.F.Meisner subsp. insignis   Plate 175

Spreading **shrub**. **Branchlets** white or pinkish, glaucous. **Leaves** 3–9 cm long, dentate with shallowly arcuate sinuses, bluish grey, slightly discolorous, rigidly leathery, the base usually cuneate to spreading, sometimes cuneate. **Flower colour**: perianth cream turning pink; style red. **Pistil** 15–20 mm long.

Distribution W.A., originally widespread in the central and southern wheatbelt regions between Tammin, Tarin Rock and Nyabing. *Climate* Hot, dry summer; cool to mild, wet winter. Rainfall 330–500 mm.

Ecology Grows in very well-drained, dry sites in sand over laterite, usually associated with low heath to thick scrub. Flowers winter–summer. Pollinated by birds. Regenerates from seed after fire.

Variation A uniform subspecies.

Conservation status 2RC. Uncommon and severely at risk over much of its range.

Cultivation Subsp. *insignis* does not adapt readily to cultivation, demanding perfect drainage in a warm, dry site. It resents summer humidity and cold-

175. *G. insignis* subsp. *insignis* Close-up of conflorescence, Tarin Rock, W.A. (P.Olde)

wet climates and may die after summer thunderstorms, even in inland gardens. It can withstand extended dry, harsh conditions and frosts to at least -6°C without signs of stress and has been successfully grown in most inland areas including Burrendong Arboretum and Rylstone, N.S.W., throughout northern Vic., at Glenmorgan, Qld, in S.A. and in Kings Park, Perth. It thrives in a hot, full-sun site in very well-drained sandy to gravelly, acidic to neutral loam that benefits from adding a good quantity of gravel. It will not tolerate excessively wet soil, and dislikes fertiliser and summer watering, both of which can kill the plant. Regular tip pruning is beneficial in increasing the density of the plant, but caution should be taken as it does not respond to pruning into the hard wood. It is generally too large for standard pot culture.

Propagation Seed Sets numerous seed that germinate although quite slowly. Average germination time 74 days (Kullman). Some pretreatment might improve germination time and rate. *Cutting* Young growth should be carefully selected to ensure that it is not too hard. This material generally gives good results, especially in early autumn. *Grafting* Essential for cultivation in east coast conditions. It has been grafted successfully and is compatible with *G.* 'Poorinda Royal Mantle' and *G. robusta*.

Horticultural features Subsp. *insignis* is a magnificent plant, with waxy, cream-pink flowers nestled impressively among its glaucous, blue-grey holly-like leaves. New growth is an attractive red that fades with maturity, leaving only the venation red, and that colour eventually disappearing. Although blooms continue for a long period in cultivation, this robust species continues to impress all year even when flowering has finished. In the right conditions it is long-lived, although grafted plants would be more reliable for most growers. It is a large shrub suited to feature plantings in large gardens or landscapes. Because of its prickly leaves, it should not be planted too close to pedestrian traffic although it could be an effective barrier plant. It is strongly attractive to both honeyeaters and nesting birds. Unfortunately, even specialist nurseries rarely carry it.

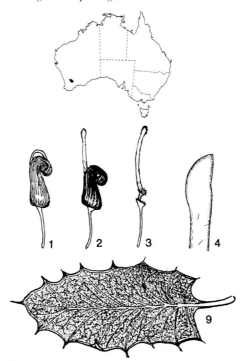

G. insignis subsp. *insignis*

Grevillea insignis subsp. elliotii
P.M.Olde & N.R.Marriott (1993)   **Plate 176**

Nuytsia 9: 285 (Australia)

The subspecific epithet honours Victorian nurseryman W.Rodger Elliot (1941–), horticulturist and author on the Australian flora. ELL-EE-OTT-EE-EYE

Type: Digger Rocks, E of Varley, W.A., 4 Oct. 1986, P.M.Olde 86/757 (holotype: NSW).

Bushy **shrub** 2–3 m wide. **Branchlets** non-glaucous. **Leaves** 3–4(–6) cm long, ovate to oblong, discolorous, green, firmly papery to leathery, sometimes faintly glaucous, ovate to oblong, with deep arcuate sinuses, the base cuneate or narrowly so, sometimes spreading. **Flower colour**: perianth cream turning rose-pink; style rose-pink. **Pistil** 11–17 mm long.

Distribution W.A., confined to a small area E of Varley. *Climate* Hot, dry summer; cool to mild, wet winter. Rainfall 330–500 mm.

Ecology Grows in very well-drained, dry sites in densely laterised loam in eucalypt woodland. Flowers winter–summer. Regenerates from seed after fire. Pollinated by birds.

Conservation status 2E suggested. The subspecies is fairly restricted in an area subject to intensive mineral exploration.

Cultivation Subsp. *elliotii* has been cultivated successfully for many years, especially in Vic. It prefers a warm, dry site in full sun in neutral to slightly acidic well-drained loam or sandy loam. While it is more tolerant of summer humidity than subsp. *insignis* it is not widely tested in these conditions. It withstands extended dry, harsh conditions and frosts to at least -6°C without signs of stress and has been successfully grown in most inland areas including Burrendong Arboretum, N.S.W. and widely in Vic. Regular pruning is seldom needed as the plant is naturally far more compact than subsp. *insignis*. Native nurseries sometimes stock this species, especially in Vic.

Propagation Seed As for subsp. *insignis*.

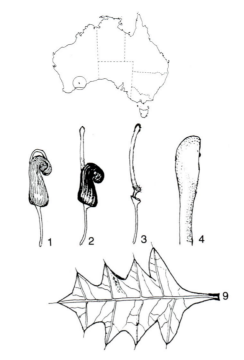

G. insignis subsp. *elliotii*

Horticultural features Subsp. *elliotii* is a desirable species with bright rose-red and cream flowers and bright green leaves with attractive bronze-red new growth. Flowering continues over many months, usually from spring to autumn but often all year given good conditions, attracting many nectar-feeding birds. Large landcapes or specialist gardens could utilise it to great advantage as it would make an ideal screen or specimen plant. Grafted plants should preferably be used in summer-wet climates, although it is far hardier on its own roots than is subsp. *insignis*.

176. *G. insignis* subsp. *elliotii* Digger Rocks, W.A. (N.Marriott)

Grevillea integrifolia (S.L.Endlicher)
C.F.Meisner (1856)   **Plate 177**

in A.P. de Candolle, *Prodromus* 14: 385 (France)

Based on *Anadenia integrifolia* S.L.Endlicher, *Stirpium Australasicarum Herbarii Hügeliani* 21 (1838, Austria).

Entire-leaved Grevillea

The specific epithet refers to the simple, entire leaves on this species and is derived from the Latin *integer* (entire) and *folium* (a leaf). INN-TEG-RI-FOAL-EE-A

Type: original collection gathered east of York, W.A., in 1836 by J.S. Roe has been lost. Neotype: Tuttaning Reserve, Possum Rd, E of Pingelly, W.A., 1 Oct. 1967, G.Heinsohn 14 (PERTH) (McGillivray 1993).

Erect, mid-dense **shrub** 1–2.5(–3) m high. **Branchlets** silky, terete. **Leaves** bifacial, 1–5 cm long, 2–10 mm wide, ascending, sessile, simple, obovate, acute to obtuse, sometimes recurved, usually with a short, blunt point; upper and lower surfaces similar, silky; margins entire, involute, often also undulate; venation obscure on upper surface, the midvein and usually 2 intramarginal veins prominent below, sometimes also acutely angled lateral veins evident. **Conflorescence** erect, pedunculate, terminal or usually in the upper axils, simple or branched; unit conflorescence 3–4.5(–7.5) long, 0.8 cm wide, cylindrical, the flowers opening ± synchronously; peduncle and rachis silky, occasionally sparsely so; bracts 1–1.5 mm long, ovate, acuminate, silky-villous with white and red hairs outside, falling early. **Flower colour**: perianth and style white to creamy white, perianth sometimes with pinkish tinges. **Flowers** regular, glabrous, strongly ascending; ovary adaxially oriented; pedicels 1.5–2.2 mm long, prominently 4-lobed; torus c. 1 mm across, straight. square; nectary absent; **perianth** below limb 2 mm long 0.5–1 mm wide, oblong; tepals channelled inside, obscurely V-shaped below anthers, before anthesis separating to base and bowed out below cohering limb, afterwards all free; limb 1.6 mm long, erect, ovoid to ellipsoidal, the segments slightly flanged at margins and each with a prominent to obscure midrib; distance from base of anthers to narrowed section of tepal 0.5 mm; **pistil** 5.5–6.5 mm long, glabrous; stipe 0.3–0.6 mm long; ovary 1 mm long, lateral, obovoid; style 4 mm long, 0.5 mm thick, smooth, before anthesis exserted near base and folded outwards, afterwards noticeably kinked about middle, becoming straight to undulate; style end 0.7–1 mm across, abruptly expanded, straight; pollen presenter 1.2 mm long, conico-cylindrical, erect with cup-shaped stigma. **Fruit** 9–11 mm long, 2.5–3.5 mm wide, oblique, narrowly obovoid, smooth, often persistent; pericarp 0.2–0.5 mm thick. **Seed** 4–8.5 mm long, 1.2–2 mm wide, obovoid or narrowly so, rugose to smooth; apex and base with a membranous band 0.5–1 mm wide; outer face convex; inner face flat.

Distribution W.A., Burracoppin, Quairaiding, Kukerin and Corrigin. *Climate* Hot, dry summer; cool to mild, wet winter. Rainfall 300–360 mm.

Ecology Usually in open heath in sand over laterite. Flowers spring–summer. Pollinated by insects. Regenerates from seed.

G. integrifolia

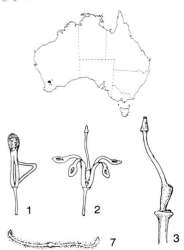

177A. *G. integrifolia* Flowers and foliage (N.Marriott)

177B. *G. integrifolia* Close-up of conflorescences (M.Pieroni)

177C. *G. integrifolia* Plant in natural habitat, Charles Gardner Reserve, near Tammin, W.A. (N.Marriott)

Major distinguishing features Leaves simple, entire but wavy, similifacial, obovate, silky on both surfaces; conflorescence cylindrical, < 1 cm wide; flowers glabrous, regular; nectary absent; ovary shortly stipitate; pollen presenter an erect cone or cylinder; fruit narrowly obovoid.

Related or confusing species Group 2, especially the hairy-leaf form of *G. shuttleworthiana* subsp. *obovata* which differs in its pale yellow flowers, wider fruits and flat leaf margins. *G. shuttleworthiana* differs in its glabrous to sparsely hairy, strongly wrinkled leaves with flat margins and its wider fruits. *G. didymobotrya* (Group 3) is also closely related but differs in its zygomorphic perianth with hairy outer surface.

Conservation status Known from a number of flora reserves; not presently endangered.

Cultivation *G. integrifolia* is not widely known in cultivation and is rarely available at commercial outlets. It has been successfully propagated by enthusiasts in western Vic. where it has proved both frost tolerant and hardy to extended dry periods. It requires a warm, sunny situation in a well-drained, sandy or gravelly loam, pH 5–7, and does not like fertiliser or summer watering once established. It is suited to cultivation in large pots or tubs in a sunny location. Pruning makes it more compact.

Propagation *Seed* Sets many fruits that often persist on the plant for long periods. Seed germinates readily if peeled or nicked before sowing. *Cutting* Firm, young growth strikes readily in the warmer months. Difficulties have been experienced with cutting frames using mists or high humidity. *Grafting* Has been grafted successfully onto *G. robusta* and *G.* 'Poorinda Royal Mantle' but may not be long-term compatible.

Horticultural features *G. integrifolia* is a most attractive medium-dense, upright shrub with pleasing, silvery, undulate leaves and masses of showy, creamy white flowers that prominently exceed the leaves. It has potential as a screen or feature plant in inland regions but requires considerable testing before widespread use is recommended. One for the enthusiast!

General comments McGillivray (1986, 1993) included *G. biformis*, *G. ceratocarpa*, *G. incrassata* and *G. shuttleworthiana* as subspecies of *G. integrifolia*. Other populations with distinctive morphological features were included informally in these subspecies. Several of these subspecies grow sympatrically. We have observed (in many places) sympatric occurrence of *G. shuttleworthiana* and *G. biformis*, *G. ceratocarpa* and *G. incrassata* as well as with our *G. incurva* and *G. eremophila*. *G. incrassata* also grows sympatrically with *G. ceratocarpa* E of Southern Cross. McGillivray's classification, based primarily on the morphological similarity of the flowers, is thus seen as too broad and ignores the biological implications of sympatric occurrence. In at least one situation, subtle factors, such as slightly different flowering times, appear to operate effectively. Thus these taxa which differ discontinuously in other important morphological characters such as foliage and fruit, maintain and reproduce themselves without interbreeding. Until relationships in this group are clarified, we propose to recognise population-based taxa exhibiting morphological discontinuity as species.

Grevillea intricata C.F.Meisner (1855)
Plate 178

Hooker's Journal of Botany and Kew Garden Miscellany 7: 74 (England)

The specific epithet is derived from the Latin *intricatus* (entangled), a reference to the leaf lobes which tend to become entwined with each other.
INN-TRICK-ARE-TA

Type: north of the Swan R., W.A., 1850–51, J.Drummond coll. 6 no. 189 (lectotype: NY) (McGillivray 1993).

Dense, intricately branched **shrub** 1–3 m high, 1–3 m wide. **Branchlets** angular, reddish, glabrous, lax. **Leaves** 8–16 cm long, ascending, sessile, pinnatisect, divaricately 2 or 3 times divided, the segments tripartite, distant, usually at right angles to each other, lax, glabrous or sparsely appressed-hairy; leaf rachis dipleural, the grooves hairy; lobes 3–8 cm long, linear-subulate to subterete, non-pungent; new growth silky, softly curled. **Conflorescence** erect, usually pedunculate, rarely sessile, usually terminal, branched; unit conflorescence 5–8 cm long, 3 cm wide, ovoid to conico-cylindrical, dense to open; peduncle almost glabrous or silky; rachis glabrous, sometimes with sprinkled, appressed hairs; bracts 1–1.7 mm long, triangular, glabrous with ciliate margins, falling before anthesis. **Flower colour**: buds yellow; perianth creamy white or greenish white; style white. **Flowers** acroscopic to almost transverse, glabrous; pedicels 3.5–8 mm long, filamentous; torus 0.5–0.7 mm across, square, ± straight; nectary obscure; **perianth** 1.5–2 mm long, 0.5 mm wide, strongly curled in bud, straight-sided, occasionally minutely papillose near base or with a few solitary trichomes inside; tepals ribbed, separated first along dorsal suture, free near limb, the dorsal tepals reflexed laterally before anthesis, afterward free to base and loose; limb revolute, spheroidal, the segments prominently ribbed, enclosing style end before anthesis; **pistil** 4.5–5.5 mm long, glabrous; stipe 1–1.2 mm long; ovary ovoid-ellipsoidal, style first exserted near curve on dorsal side and looped upwards, straight to slightly incurved after anthesis, dilated near apex; pollen presenter erect, truncate-conical with oblique, flanged base. **Fruit** 11–15 mm long, 5–7 mm wide, erect to very oblique on curved pedicel, oblong-ellipsoidal, rugose, glabrous; pericarp c. 1 mm thick. **Seed** 9–10 mm long, 3 mm wide, narrowly ellipsoidal, biconvex, slightly compressed at margin into an encircling flange; outer face smooth; inner face with a central oblong-elliptic section surrounded by a broad, smooth, membranous border.

Distribution W.A., confined to a limited area from just north of Geraldton to Ajana and inland to the Chapman East River. *Climate* Hot, dry summer; cool but mild, wet winter. Rainfall 460–475 mm.

Ecology Found in various soils from poor, white

G. intricata

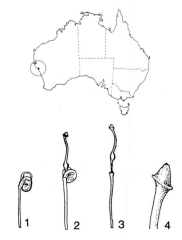

178A. *G. intricata* Close-up of flowers. Note the red stems. (N.Marriott)

178B. *G. intricata* Plant in natural habitat, N of Geraldton, W.A. (C.Woolcock)

sand to lateritic gravel, sometimes in loam or granitic soil, in heath, mallee woodland or thick shrubland. Flowers winter–spring. Pollinated by insects. Regenerates from seed and possibly lignotuber after fire.

Variation A uniform species.

Major distinguishing features Leaves divaricately pinnatisect, the lobes subterete, grooved along each side of rachis, patent, somewhat distant; young growth lax, curled; conflorescence conical, usually branched; torus straight; nectary not evident; perianth zygomorphic, glabrous, strongly curled in bud; pistil glabrous; pollen presenter conical with oblique base; fruit oblong-ellipsoidal, rugose.

Related or confusing species Group 8, especially *G. leucoclada* which differs in its stiffer leaf lobes, glaucous branchlets, shorter stipe (< 1 mm long) and consistently hairy inner perianth.

Conservation status 2V. Despite its limited distribution, and the fact that it comes from a region in which the natural vegetation has in most cases been long-since cleared, *G. intricata* is still quite common because of its ability to regenerate even in cleared, roadside habitats.

Cultivation *G. intricata* is a hardy, adaptable species that will grow in cold-wet or hot-dry climates as well as summer-rainfall regions. It withstands frosts to at least -6°C and extended dry conditions, although branchlets in the middle of the shrub tend to die if conditions are too harsh or too shady. Lovers of the genus now cultivate it in most areas of Australia as well as in the U.S.A. Best success is achieved when planted in a well-drained, sunny site in sandy or gravelly loam. Some summer moisture is beneficial in the first season. Under ideal conditions it can become a very large plant, spreading to 2–5 m, but more usually it is smaller and relatively compact. It grows successfully in slighty alkaline to acidic soil and tolerates higher amounts of fertiliser than most species, although this is rarely required. Pruning is generally unnecessary but may be required to reduce the size of the plant. Specialist native plant nurseries usually stock this species.

Propagation *Seed* Sets numerous seed that normally germinate readily when peeled or nicked. Average germination time 29 days (Kullman). *Cutting* Firm, young growth gives a very good strike at most seasons. Care must be taken when lifting cuttings as the intricate foliage tangles readily with neighbouring plants and can cause them to be lifted also if not first separated. *Grafting* Not usually necessary but has been grafted successfully onto *G. robusta* using the top wedge method.

Horticultural features Unique, densely tangled foliage reminiscent of tumble-weed, showy red branchlets and a massed display of creamy white flowers in short, erect spikes all over the upper part of the bush combine to make *G. intricata* a most desirable plant for the home garden, especially broad-acre type gardens. It is long-lived in cultivation, is valued as a dense screen or feature plant and is ideal in commercial, public or home landscaping, choking out all but the toughest weeds with its density. The flowers have a pleasant, sweet perfume and are attractive even in bud, being massed like small yellow rockets above the foliage.

History *G. intricata* was first introduced to cultivation in England about 1869 from seed collected by Oldfield and Burges. Seed was successfully germinated in the Royal Botanic Gardens, Kew, and raised to flowering in 1871 (*Curtis's Botanical Magazine* 1871, t. 5919). It was also reported in cultivation in Paris in 1900. The Adelaide Botanic Gardens first cultivated it in Australia in 1872.

Grevillea involucrata A.S.George (1974)
Plate 179

Nuytsia 1: 372 (Australia)

Lake Varley Grevillea

Specific epithet derived from the Latin *involucratus* (having a ring of bracts), in reference to the floral bracts that are conspicuous and persist after the perianth has fallen. INN-VOLL-YOU-KRAR-TA

Type: between Hyden and Lake Varley, W.A., 30 June 1970, A.S.George 9890 (holotype: PERTH).

Open, decumbent, spreading **shrub** prostrate to 0.5 m high and 2 m wide. **Branchlets** secund, slender, terete, silky. **Leaves** 2.5–4 cm long, 0.6–1 cm wide, ascending to patent, shortly petiolate, secund, almost pinnatisect; leaf lobes usually 9–15, 2.5–8 mm long, c.1 mm wide, linear, closely aligned, obtuse, mucronate with a short recurved point; upper surface glabrous, minutely granular, the midvein within a sunken channel; margins smoothly revolute; lower surface bisulcate, obscured, silky in grooves, glabrous or sparsely hairy; midvein prominent, ± level with margins. **Conflorescence** erect, sessile to shortly pedunculate, terminal or axillary, simple, 1–3-flowered with a basal whorl of imbricate bracts; peduncle silky or sparsely so; rachis scarcely evident, glabrous; bracts 5.5–9 mm long, elliptic, glabrous, persistent past fruiting. **Flower colour**: perianth pale to dark pink; style brownish red; floral bracts pink to dark red. **Flowers** abaxially oriented, glabrous outside; pedicels 8–11 mm long, glabrous; torus 2.8–4 mm across, cup-shaped, oblique; nectary conspicuous, U-shaped with a thick, entire margin; **perianth** 10–12 mm long, 5–6 mm wide, oblong, waxy and shiny, hirsute inside in lower half, elsewhere minutely papillose; tepals ridged, cohering except along dorsal suture and between dorsal and ventral tepals at the curve; limb revolute, spheroidal, prominently indented at tepal margins; **pistil** 23–25 mm long; stipe 1.8–2.4 mm long, refracted at toral rim, ± at right angles to torus; ovary white-villous; style exserted at curve and looped upwards before anthesis, afterwards straight to incurved, thick, brown-villous, dilating smoothly into style end; style end club-shaped, glabrous, partially exposed before anthesis; pollen presenter lateral, obovate, flat. **Fruit** 13–15 mm long, 6–8 mm wide, oblique, oblong or ellipsoidal, attenuate, loosely villous;

178C. *G. intricata* Flowering branches (M.Hodge)

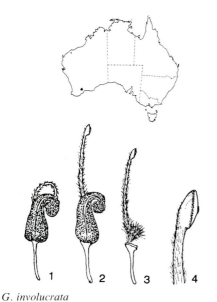

G. involucrata

pericarp c. 1 mm thick. **Seed** 8 mm long, 3–3.2 mm wide, ellipsoidal; outer surface strongly convex, slightly mottled; inner face flat with a raised central ridge and a raised submarginal, waxy border drawn at both ends into a short wing, the apical wing oblique, triangular, c. 1 mm long.

Distribution W.A., restricted to a few populations near Newdegate. *Climate* Summer hot and dry; winter cool and wet. Rainfall 300–400 mm.

Ecology Occurs on sandheath or low scrub in sandy loam over laterite. Flowers winter–spring. Pollinated by birds. Regenerates from seed after fire. This species may be short-lived in the wild.

Variation A uniform species with little variation apart from minor variation in flower and bract colour.

Major distinguishing features Leaves almost pinnatisect, the lobes short, closely aligned; conflorescence 1–3-flowered with persistent, conspicuous floral bracts; torus oblique, cup-shaped; perianth zygomorphic, oblong, glabrous and waxy outside, hairy within; ovary densely hairy, stipitate, the stipe inserted ± at right angles to torus; style end club-shaped.

Related or confusing species Group 26, especially *G. fulgens* which differs in its hairy pedicels and perianth limb.

Conservation status 2EC-i. Known from 10 populations, all on roadsides; one known to the authors has 17 plants.

Cultivation G. involucrata has adapted well to cultivation and has been grown widely since 1983 in the eastern States, especially N.S.W. and Vic. It grows extremely well at Mount Annan Botanic Garden and Burrendong Arboretum near Wellington, as well as in many private gardens, including the author's at Sydney. More recently, it has been grown in the U.S.A. It requires a warm, sunny, open position in a well-drained, acidic to neutral sand or gravelly loam. In pure sand it tends to become open and straggly though this can be remedied by adding humus and slow-release fertiliser. Plants respond vigorously to regular pruning, becoming denser and far more floriferous. Once established, summer watering is not required. Extended, dry conditions, summer-humid conditions and frosts to at least -6°C are tolerated well. In recent years, it has become more readily available from native plant nurseries.

Propagation *Seed* Sets small quantities of seed that germinate well when given the standard peeling treatment. *Cutting* Strikes readily from cuttings of firm, young growth at most seasons, although occasionally results are inexplicably poor. *Grafting* Grafts successfully onto a number of rootstocks including *G. robusta*, *G. macleayana* and G. 'Poorinda Royal Mantle', mostly using the top wedge technique.

Horticultural features G. involucrata is a low, domed shrub with distinctively arched branches and appealing bronze-red stems. The narrow, divided leaves give the plant an open appearance, allowing the flowers to stand out. Although the pink flowers and conspicuous red bracts are quite showy, flowering is not prolific and is often restricted to a few branches, some of which may have flowers in every leaf axil. These attract many honey-eating birds, resulting in many prominent fruits developing above the persistent, dark pink bracts. It is proving to be long-lived in cultivation and has value in the public landscape as a feature plant, low screen or spill-over plant for the large rockery. It has been grown successfully in open-ended pipes and tubs, its arching branches cascading to the ground. Worth planting for conservation alone, *G. involucrata* is a most rewarding and unusual grevillea which deserves wider acceptance.

Grevillea jephcottii J.H.Willis (1967)
Plate 180

Muelleria 1: 117 (Australia)

Green Grevillea, Pine Mountain Grevillea

The specific name honours members of the Jephcott family of Ournie, on the upper Murray R., with special acknowledgement to Sydney Wheeler Jephcott who first collected it in 1878. JEFF-COT-EE-EYE

179A. *G. involucrata* Flowering branch. (P.Olde)

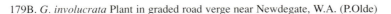

179B. *G. involucrata* Plant in graded road verge near Newdegate, W.A. (P.Olde)

G. jephcottii

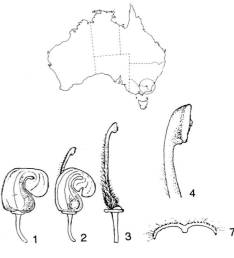

180. *G. jephcottii* Flowers and foliage (N.Marriott)

Type: SW slopes of Pine Mountain, c. 11 km SE of Walwa, upper Murray region, Vic., 17 Nov 1964, J.H.Willis (holotype: MEL).

Dense **shrub** 1 to 3 m tall with erect, subverticillate branches. **Branchlets** terete, villous. **Leaves** 1.5–3.5 cm long, 2.5–6(–8) mm wide, ascending, clustered, sessile, simple, ovate to oblong-lanceolate, often with a slight twist, obtuse with a fine point; upper surface villous when young, soon ± glabrous, scabrous, midvein scarcely prominent, intramarginal veins evident; margins entire, shortly recurved; lower surface glabrous, smooth, ± glaucous, midvein prominent. **Conflorescence** 2–3 cm long, 3 cm wide, erect, sessile, terminal, unbranched, usually 4–9-flowered, dense, umbel-like; rachis sparsely villous; bracts 0.8–2.4 mm long, triangular, villous outside, some usually persistent at anthesis. **Flower colour**: perianth pale green becoming translucent white, turning black; style dull maroon with a conspicuous green style end. **Flowers** acroscopic, glabrous outside; pedicels 4.5–6.5 mm long; torus c. 1.5 mm across, straight; nectary semi-circular, prominent, entire; **perianth** 4–5 mm long, 3–4 mm wide, strongly curved, triangular-ovate, saccate, ribbed, densely bearded inside above level of ovary, cohering except along dorsal suture, persistent to fruiting; limb revolute, spheroidal; **pistil** 9–10.5 mm long; ovary sessile, villous; style scarcely exserted at curve before anthesis, afterwards straight to gently incurved and scarcely longer than perianth, stout, villous on dorsal side, ± glabrous on ventral side; style end conspicuous, club-shaped, exposed before anthesis; pollen presenter lateral, round, flat. **Fruit** 10–15 mm long, 4–7 mm wide, erect, ovoid, ribbed, sparsely pilose; pericarp c. 0.5 mm thick. **Seed** 8–11 mm long, 2.3 mm wide, narrowly ellipsoidal with a cushion-like swelling below apical wing, minutely hairy; outer face convex; inner face flat, channelled around margin.

Distribution Vic., confined to several very small areas in the Burrowa–Pine Mountain area of north-eastern Vic., N.S.W. near border adjacent to Victorian distribution. *Climate* Hot, dry summer; cool, wet winter. Rainfall 700–800 mm.

Ecology Grows in open to closed eucalypt forest in a mixed, often fairly thick, scrub understory or as the dominant shrub in open woodland, often on steep hillsides and among granite boulders in gravelly or sandy clay-loam. Flowers winter–summer. Bird-pollinated. Regenerates after fire from seed only.

Variation A species with little variation. Plants in open, exposed situations are low and bushy whereas those in shaded forest are tall.

Major distinguishing features Branchlets villous; leaves simple, entire, glabrous on underside of adult leaves, scabrous above; perianth zygomorphic, strongly curved, saccate, glabrous outside, hairy within; ovary villous, sessile; style scarcely exceeding perianth, style end clavate; pollen presenter lateral.

Related or confusing species Group 25, especially *G. lanigera* which differs in its hairy leaf undersurface and longer pistil.

Conservation status 2RC. Habitat very restricted but well conserved within the Burrowa–Pine Mountain NP.

Cultivation *G. jephcottii* has been in cultivation for many years, during which it has proved extremely hardy and adaptable, succeeding in cold-wet to warm-dry, and even summer-rainfall climates such as Sydney. It has been grown extensively in Vic. and N.S.W. as well as inland Qld and around Brisbane. It has also been cultivated in the U.K. and U.S.A. There is a superb grove planting at Burrendong Arboretum, N.S.W. It tolerates a wide range of conditions including extended dry conditions and frosts to at least -6°C. It shows a preference for dappled shade to full sun in well-drained, acidic sand or gravelly loam. Once established, summer watering is not required. Regular tip-pruning is necessary to maintain density when grown in shade. If left unpruned, it grows tall and leggy with foliage crowding at the ends of the branchlets. It is not really suited to tub culture because of its subdued flowering and its lanky habit. Specialist and general nurseries occasionally stock this species, especially in Vic.

Propagation *Seed* Sets many seeds that germinate well if pretreated. Hybridises with other species in the garden, and wild-source seed should be used if possible. *Cutting* Firm, young growth strikes readily at most seasons. *Grafting* Although rarely necessary, it has been grafted successfully onto *G. robusta*.

Horticultural features *G. jephcottii* is an elegant, upright shrub with crowded, soft leaves and massed clusters of greenish cream flowers at the branchlet tips. The branchlets with their long white hairs are also most interesting and attractive. On close examination, the flowers display subdued delicate colours. It is highly regarded for its bird-attracting

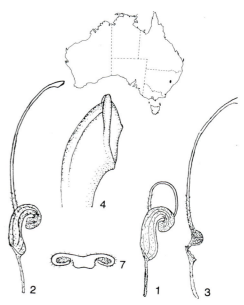

G. johnsonii

qualities. A large clump of this species at Burrendong Arboretum was almost continuously vibrating with the number of birds it attracted during spring, the flowers supplying copious nectar for many months. It is a very long-lived species, suitable as a screen or background shrub in the garden. It is ideal for narrow sites beside paths and driveways where spreading plants would be a nuisance.

Hybrids *G. jephcottii* hybridises with *G. lanigera* in the wild. Some of these intermediates are brightly flowered and attractive.

Grevillea johnsonii D.J.McGillivray (1975)   Plate 181

Telopea 1: 22 (Australia)

Johnson's Grevillea

The epithet honours Lawrence A.S. Johnson, Director of the Royal Botanic Gardens, Sydney, 1972–1985. JON-SON-EE-EYE

181A. *G. johnsonii* Flowers and foliage (P.Olde)

Type: Kerrabee Mountain, Kerrabee, N.S.W., 16 Oct. 1955, L.A.S.Johnson NSW 33695 (holotype: NSW).

Pine-like, spreading **shrub** 2–4.5 m high with a single, prominent trunk. **Branchlets** angular, ridged, rusty-tomentose. **Leaves** 9–25 cm long, ascending, ± sessile, subpinnatisect with 5–10 lax, acutely angled, elongate-linear lobes, 8–15 cm long, 0.7–2 mm wide; margins revolute; upper surface sprinkled with silky brown hairs when juvenile, soon ± glabrous, grooved beside raised midvein; lower surface mostly obscured, bisulcate, silky in grooves, the midvein prominent. **Conflorescence** 1–3(–5) cm long, 7–8 cm wide, erect, pedunculate, terminal or axillary, simple or rarely branched; unit conflorescence shortly cylindrical or umbel-like, lax, rarely exceeding foliage; development basipetal; peduncle and rachis brown-silky; bracts 1.5–2 mm long, narrowly cymbiform, ovate-acuminate, rusty villous outside, falling soon after expansion of buds. **Flower colour**: perianth pink with a creamy white to yellow limb, occasionally uniformly orange; style red. **Flowers** basiscopic; pedicels 8.5–15 mm long, sparsely silky; torus 3.5–4.5 mm across, very oblique to lateral; nectary long U-shaped, toothed at ends; **perianth** 9–12 mm long, 4–5 mm wide, oblong, strongly curved from c. half-way, sparsely silky outside, the tepals noticeably keeled and somewhat waxy, bearded at base around and below ovary inside, cohering except along dorsal suture and between dorsal and ventral tepals at the curve; limb revolute, spheroidal, the segments prominently keeled, impressed at margins; **pistil** 26–37 mm long; stipe 2.5–4.5 mm long, inserted perpendicular to torus but then continuing laterally and lining toral surface, silky; ovary subglobose, silky; style silky in lower half, glabrous above, exserted from c. halfway along dorsal suture and looped upwards before anthesis, afterwards gently incurved, refracted about ovary from line of stipe, gently curved, dilating evenly into enlarged, glabrous style end; style end not exposed before anthesis; pollen presenter very oblique, oblong-elliptic, convex. **Fruit** 12–16 mm long, 10–13 mm wide, very oblique to stipe, ± lateral to pedicel, subglobose to oblong-ellipsoidal, silky to pubescent; pericarp 1.5–1.8 mm thick. **Seed** 8–9 mm long, 4.5–5 mm wide, obovoid, granular; outer face convex; inner face slightly convex, channelled around margin; margin flat.

Distribution N.S.W., in the Goulburn R. catchment at Coxs Gap, Gungal, Widden and Kerrabee areas, and a probable southern disjunct occurrence at Brown Mountain. *Climate* Cool to cold, wet winter; hot, dry summer. Rainfall 800–1000 mm.

Ecology Grows at the base of sandstone cliffs or on the sides of steep gullies in woodland, usually in protected positions. Flowers spring. Pollinated by birds. Regenerates mostly from seed after fire, although we have observed new shoots emergent from roots close to the main stem below the soil surface.

Variation A species with little variation except in flower colour.

Major distinguishing features Shrub with a single, prominent trunk; leaves pinnatisect, the lobes elongate-linear with the undersurface double-grooved, lamina ± obscure beside midvein; conflorescence usually simple, outermost flowers opening first, the rachis 0.5–1.2 cm long, usually c. 10-flowered; torus very oblique; perianth zygomorphic; tepals prominently keeled, the outer surface sparsely silky, densely so at base and limb, the inner surface hairy at base; ovary stipitate, silky, the stipe inserted perpendicular to torus but lining it over much of its length; style hairy above ovary; fruit subglobose to oblong-ellipsoidal.

Related or confusing species Group 26, especially *G. longistyla* which differs in its multi-stemmed habit, its broader, linear leaves or leaf lobes with recurved margins (not concealing the undersurface), its longer, cylindrical conflorescences borne clear of the foliage; its style glabrous or nearly so.

Conservation status 2RC. Restricted, but conserved in Goulburn R. NP.

Cultivation *G. johnsonii* has been cultivated successfully over a wide climatic range from Qld to Vic., although it has a reputation for unreliability in coastal, summer-humid climates. It is both drought- and frost-resistant, tolerating light snow

181B. *G. johnsonii* Flowers and fruit (P.Olde)

181C. *G. johnsonii* Plant in cultivation at Burrendong Arboretum, N.S.W. (P.Olde)

and frosts to at least -11°C. Superb plants up to 3 m across can be seen growing at the Australian National Botanic Gardens, Canberra, and at Burrendong Arboretum, N.S.W., where they thrive in well-drained acidic sandy or loam in both full sun and partial shade. It does not require summer watering but occasional tip-pruning when young improves shape. Protection from strong winds is advisable as the species tends to be shallow-rooted and the limbs are somewhat brittle. It prefers a cool root run, either from deep soil or a shaded position, although a good layer of mulch is quite adequate. Available in many nurseries on the east coast.

Propagation *Seed* Readily germinates from fresh seed which has been subjected to the standard peeling treatment. High germination rates have also been achieved by soaking seed in hot water for 24 hours prior to sowing. *Cutting* Shy to strike from cutting. It is best propagated by this technique using fresh, semi-ripe wood in summer to early autumn. *Grafting* Grafts readily using the top wedge, approach or cotyledon methods onto *G. robusta* rootstock. Other compatible rootstocks are G. 'Poorinda Royal Mantle', G. 'Copper Rocket' and *G. shiressii*.

Horticultural features *G. johnsonii* is one of the most decoratively foliaged grevilleas, with long, finely divided leaves sometimes weeping in the manner of a willow. Foliage is dull to dark green and contrasts delightfully with the bright red branchlets which boast a terminal profusion of pink-cream or orange-red flowers over a long period from early spring. It makes a fine feature or screen plant in any open or semi-shaded situation but may need to be grafted when used as a structure plant in summer-humid, coastal gardens. It is strongly bird attractive. In recent years, both the general public and commercial landscapers have come to prize this species as one with great horticultural potential.

Hybrids Has hybridised with *G. longistyla* in cultivation, the resultant plants sold as G. 'Long John', and G. 'Elegance', the latter name inappropriate because of confusion with G. 'Poorinda Elegance'.

General comments Reports of a disjunct population from Brown Mountain and Brogo River near Bega in southern N.S.W. should not be dismissed lightly. Although no herbarium collection is known, the original collector (G.Althofer) was a reliable and knowledgeable enthusiast, who propagated and distributed material to other growers, and who clearly remembers the collection and location (near the river, below a lookout). Seed from that source sent to D.Gordon in Qld is still grown there today as the Brown Mountain form, and there is little substantial difference from plants from elsewhere. Searches in recent years have failed to locate the species on Brown Mountain.

Grevillea juncifolia W.J.Hooker (1848)

in T.L.Mitchell, *Journal of an Expedition into the Interior of Tropical Australia* 341 (England)

Honey-suckle Grevillea

Aboriginal names: 'jultukun' (Pintubi); 'ultukunpa' (Pitjantjatjara/Yankunytjatjara); 'umbagumba' (Blyth Range); 'erolunga' (Aranda); 'yuldigunyba', 'badubiri', 'yilyilba', 'badubadjalba' (Gugadja); 'walunari' (Walbiri); 'tarrakinea' (Alyawara).

The specific epithet refers to the long linear leaf lobes which reminded Hooker of a rush (*juncus*) with the Latin *folium* (a leaf). JUNK-IFF-OLE-EE-A

Type: probably near Mt Pluto, inland Qld, 184-, T.L.Mitchell 477 (holotype: K).

Erect to spreading, greyish **shrub** to c. 7 m. **Branchlets** terete, ridged, pubescent. **Leaves** 10–30 cm long, ascending, petiolate, leathery, simple and linear or pinnatipartite; lobes 0.5–2 mm wide, linear, pungent; upper surface pubescent or tomentose, sometimes glabrous, midvein scarcely visible; margins angularly revolute to midvein below; lower surface bisulcate, silky in grooves, midvein prominent. **Conflorescence** erect, pedunculate, terminal, axillary or occasionally stem-borne, usually branched; unit conflorescence 4–17 cm long, 6–10 cm wide, conico-cylindrical, loose; peduncle and rachis pubescent, sometimes with glandular hairs; bracts 2.1–6.5 mm long, obovate, villous outside, falling early. **Flowers** basiscopic, sometimes irregularly oriented; pedicels 8–20 mm long; torus c. 2 mm across, slightly oblique; nectary patelliform to broadly V-shaped, smooth; **perianth** 7–8 mm long, 4–7 mm wide, ovoid-saccate, much contracted at neck, the tepals ribbed, outside sparsely to densely silky or pubescent, sometimes the hairs glandular, inside glabrous; limb strongly revolute, ovoid, each tepal limb with a subterminal, villous appendage; **pistil** 18–27 mm long; stipe 0.5–2 mm long, appearing sessile by enclosure within torus; ovary sessile or appearing so, villous; style robust, glabrous or papillose, villous near ovary, gently curved, grooved on ventral side; style end scarcely dilated; pollen presenter lateral or oblique, oblong-elliptic, convex. **Fruit** 20–29 mm long, 11–15 mm wide, persistent, horizontal or oblique to pedicel, compressed-ovoid to ellipsoidal, pubescent; pericarp 0.5–1.5 mm thick. **Seed** 18–23 mm long, 8–11.5 mm wide, ± flat, ovate to oblong elliptic, broadly winged all round, smooth with slight marginal ridge.

Major distinguishing features Leaves or lobes elongate, narrowly linear, the under surface double-grooved, lamina not exposed; conflorescence conico-cylindrical to cylindrical, loose; torus oblique; nectary patelliform; pedicels elongate; perianth zygomorphic, orange, ovoid-saccate, hairy outside, glabrous within, the tepal limb with a subterminal appendage; ovary sessile or scarcely stipitate, densely villous; pollen presenter very oblique to lateral; fruit compressed.

Related or confusing species Group 35, especially *G. excelsior* which differs in its secund conflorescence, acroscopic flowers, shorter pedicels (< 6 mm long), and scarcely dilated perianth.

Variation Two subspecies are recognised.

Key to subspecies

Outer perianth surface, pedicels and rachis glandular-pubescent; leaves divided, occasionally simple; conflorescence conical
    subsp. **juncifolia**

Outer perianth surface, pedicels and rachis silky; leaves simple; conflorescence usually cylindrical   subsp. **temulenta**

Grevillea juncifolia W.J.Hooker subsp. juncifolia   Plate 182

Leaves divided, occasionally simple. **Conflorescence** conical. **Flower colour**: buds green; perianth yellow to orange; style yellowish-orange. Outer **perianth** surface, pedicels and rachis glandular-pubescent;

Synonyms: *G. sturtii* R.Brown (1849); *G. sturtii* var. *pinnatisecta* F. Mueller; (1863); *G. mitchellii* C.M.Lemann ex C.F.Meisner (1855).

Distribution A transcontinental species found in inland regions of N.S.W., Qld, S.A., W.A. and N.T. *Climate* Hot, dry summer; mild to warm winter with fleeting rain extending to winter-wet climates. Rainfall 150–500 mm.

Ecology Grows on sandplain, sand hills, stony rises and open plains often associated with light timber and scrub or spinifex. Soil ranges from deep red or yellow sand, sandy clay, gravelly loam, or granite sand/loam. Flowers usually winter–spring but occur be at any season following rain. Probably mammal or bird-pollinated but numerous insects, including ants, are also attracted to the nectar. Regenerates from seed.

Variation There is variation in leaf division and size of flowers and fruits.

Typical form This is widespread, extending over most of the distribution except south-western W.A. and the Eyre Peninsula, S.A. It is distinguished by its divided leaves. Most collections are relatively uniform but plants with larger flowers and fruits occur in the Simpson Desert. There is some variation in foliar indumentum and its degree of persistence.

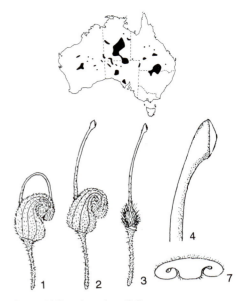

G. juncifolia subsp. *juncifolia*

182B. *G. juncifolia* subsp. *juncifolia* Flowering branches (M.Hodge)

182A. *G. juncifolia* subsp. *juncifolia* Conflorescence (M.Hodge)

Eyre Peninsula form Resembles the typical form in every way except that the leaves are predominantly simple.

Conservation status Not presently endangered.

Cultivation Subsp. *juncifolia* does not adapt well to cultivation, preferring a dry, inland climate only. In these conditions it has proved hardy to regular winter frosts and extended dry conditions and has been grown successfully at Dave Gordon's Glenmorgan Arboretum, Qld, as well as at Burrendong Arboretum, N.S.W. A perfectly drained, deep sand or sandy loam in full sun is demanded; in less than perfectly drained sites, it will succeed for a short time but only until a wet season arrives during which it is likely to die. It dislikes fertiliser and, once established, watering is unnecessary. Regular tip-pruning can be beneficial. Although too large for standard pot culture, it can be grown for a number of years in an open-ended pipe or drum filled with free-draining sandy soil. In these conditions, it has even survived and flowered for a number of years in summer-humid Sydney. It is sometimes available from specialist nurseries or from seed suppliers.

Propagation *Seed* Sets many seeds that normally germinate readily when pretreated by soaking in warm water for 36 hours. Kullman reports average germination time of 42 days (subspecies unknown). *Cutting* Difficult to strike unless vigorous half-hardened regrowth is used, preferably during the warmer months. *Grafting* Has been grafted successfully onto *G. robusta* using cotyledon, mummy and top wedge grafts.

Horticultural features Subsp. *juncifolia* is a spectacular, free-flowering, large shrub with beautiful silvery or grey-green foliage contrasted with bright orange flowers that cluster on the branch tips. Usually, the flowers are full of nectar which tends to flow and drip to the ground around the middle of the day, attracting a parade of ants and nectar-feeding birds. At this time the flowers are also redolent with a strong, honey scent, although, in dry conditions, both scent and nectar flow are

G. juncifolia subsp. *temulenta*

much reduced. It is a long-lived plant that would make a superb addition to gardens and landscapes in inland, arid regions.

History It was first introduced to cultivation in Qld in 1875.

Hybrids In the wild we have seen hybrids with *G. spinosa*.

General comments *G. juncifolia* was used by tribal Aborigines as a source of nectar, 'wama', for some their only source of sugar apart from the honey ant. They drew the conflorescence sideways through their mouths or simply placed it in the mouth for a few minutes to extract the nectar. A sweet drink was also made by soaking the flowers in water. This rapidly fermented and had an intoxicating effect. The shrub has also been used as a fodder tree by graziers in time of drought although it is reported to be low in protein and phosphorus and high in fibre.

Grevillea juncifolia subsp. temulenta
P.M.Olde & N.R.Marriott (1994)   **Plate 183**

The Grevillea Book 1: 185 (Australia)

The subspecific epithet is derived from Latin *temulentus* (intoxicated), in reference to the effect produced by consumption of the floral nectar after it has fermented.   TEM-YOU-LEN-TA

Type: 104 km N of Kalgoorlie on road to Menzies, W.A., 16 Sept. 1989, B.J.Conn 3439 & J.Scott (holotype: NSW).

Leaves simple. **Conflorescence** usually cylindrical. **Flower colour**: perianth and style orange. Outer **perianth** surface, pedicels and rachis silky.

Distribution W.A., widespread in the south-west; from Pindar and Perenjori to near Lake King and NE to Laverton and Queen Victoria Spring. *Climate* Hot, dry summer; cool, damp winter. Rainfall 200-400 mm.

Ecology Grows in yellow sand, sometimes with laterite, in open shrubland. Flowers spring–summer. Probably pollinated by nectarivorous bird. Regenerates from seed.

Variation A relatively uniform subspecies.

Conservation status Not presently endangered.

Cultivation Subsp. *temulenta* has not been cultivated widely to date, though a few grafted plants have appeared recently in Qld. A dry climate in deep, well-drained, sandy, neutral to slightly alkaline soil in full sun is strongly recommended. It will tolerate drought as well as extreme cold (to -6°C) as long as daylight temperatures rise and sunlight predominates. Summer-humidity is not tolerated well.

Propagation *Seed* Sets large quantities of seed that should germinate well if pretreated by soaking in hot water for 24 hours. *Cutting* Untested, but unlikely to be easy to strike. Grafting Has been grafted successfully onto *G. robusta* using mummy grafts but long-term compatability is untested.

Horticultural features Subsp. *temulenta* is an eye-catching robust shrub with bright orange flowers in great abundance in the flowering season. Tourists frequently stop at roadside stands to take advantage

183A. *G. juncifolia* subsp. *temulenta* Flower colour change, near Mt Jackson, W.A. (P.Olde)

183B. *G. juncifolia* subsp. *temulenta* Plants in natural habitat, N of Kalgoorlie, W.A. (M.Hodge)

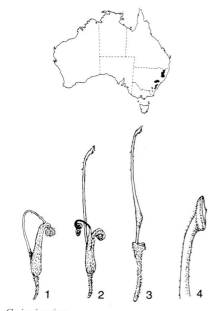

G. juniperina

of the excellent photo opportunities this species offers in the wild. It has potential as a feature or massed landscape plant, especially when combined with blue-flowering shrubs, and warrants close investigation for its horticultural potential.

Hybrids Hybrids with *G. excelsior* have been seen in the wild.

General comments Serious consideration was given to recognising subsp. *temulenta* at specific rank, but the occurrence of simple-leaved plants in several populations of subsp. *juncifolia* indicated that on present knowledge there is insufficient discontinuity. Field work and comparative studies of conflorescence morphology and new character states may show that such ranking is warranted.

Grevillea juniperina R.Brown (1810)
Plate 184

Transactions of the Linnean Society of London 10: 171 (England)

Juniper-leaf Grevillea

The foliage of this species was thought to resemble that of the genus *Juniper*, hence the specific epithet. JOO-NIP-ER-EE-NA

Type: about 11 km NW of Prospect, near Port Jackson, N.S.W., Oct. 1803, G.Caley & A.Gordon (lectotype: BM) (McGillivray 1993).

Synonyms: *G. sulphurea* A.Cunningham (1825); *G. acifolia* F.Sieber ex K.Sprengel (1827); *G. acicularis* J.H.Schultes & J.H.Schultes (1827); *G. trinervis* R.Brown (1830); *G. juniperina* var. *trinervata* J.H.Maiden & E.Betche (1899); *G. juniperina* var. *sulphurea* (A.Cunningham) G.Bentham (1870); *G. juniperina* forma *sulphurea* (A.Cunningham) I.K.Ferguson (1978).

A prostrate or decumbent, spreading **shrub** to 30 cm or an erect, dense **shrub** 0.5–2.5 m high. **Branchlets** terete, villous. **Leaves** 0.5–3.5 cm long, 0.5–6 mm wide, spreading to ascending, clustered, simple, sessile, narrowly linear to ± triangular, acute to subulate, pungent; margins entire, angularly recurved or revolute; upper surface glabrous, smooth, sometimes shiny, occasionally with scattered hairs, longitudinally 3–5-nerved, the venation obscure to prominent; lower surface obscured by the margin on narrowly linear leaves, silky or sparsely so, sometimes glabrous, midvein obscure. **Conflorescence** 1.5–3.5 cm long, erect or deflexed, sessile or shortly pedunculate, usually terminal on short, side branchlet or axillary, simple, rarely branched; unit conflorescence dense, umbel-like or irregular, usually secund; peduncle silky to villous, sometimes almost glabrous, rachis usually tomentose, sometimes villous; bracts 0.5–1 mm long, narrowly triangular, villous outside, falling before anthesis. **Flower colour**: perianth and style greenish yellow, lemon, yellow, pinkish yellow, apricot, orange or red with a brownish or greenish limb and a green style end. **Flowers** acroscopic; pedicels 2.5–5 mm long, tomentose, square in cross-section; torus c. 2 mm across, square, oblique; nectary conspicuous, broadly lunate to U-shaped, entire, crenate or toothed; **perianth** 11–15 mm long, 2–4 mm wide, ovoid to oblong-ovoid, slightly dilated at base, tomentose or sparsely so outside, bearded inside above and at level of ovary, cohering except along dorsal suture before anthesis, tepals separating and rolling back to beard after anthesis, the ventral tepals curled back furthest; limb erect in late bud, nodding before anthesis, ovoid to spheroidal, villous; **pistil** 18.5–27 mm long; stipe 1.8–4 mm long, glabrous; ovary glabrous, narrowly ellipsoidal, scarcely broader than style; style ± glabrous or sparsely hairy, exposed along dorsal suture, bowed out and up before anthesis, afterwards straight or gently incurved, channelled on ventral side; style end small, flattened, minutely hairy, not exposed before anthesis; pollen presenter

184A. *G. juniperina* Red-flowered variant, cultivated at Stawell, Vic. (N.Marriott)

184B. *G. juniperina* Corang River form in natural habitat, N.S.W. (P.Olde)

184C. *G. juniperina* Broad-leaved form in cultivation at Brisbane, Qld (N.Marriott)

very oblique, oblong-elliptic to obovate, convex. **Fruit** 10–18 mm long, 4–8 mm wide, erect or slightly oblique, narrowly ellipsoidal to fusiform, sometimes ribbed, usually smooth or faintly rugose; pericarp c. 0.5 mm thick. **Seed** 7.5–12 mm long, 2.2–3.3 mm wide, narrowly elliptic with an apical cushion-like swelling and short, oblique wing; outer face convex, minutely hairy; inner face channelled, minutely hairy; margin recurved, more strongly on one side.

Distribution Qld, N.S.W., A.C.T. Widespread in inland areas on tablelands and slopes from Stanthorpe, Qld, to Canberra. *Climate* Cool to cold, wet winter; hot, wet or dry summer. Rainfall 600–800 mm.

Ecology Occurs in many soils including sandy loam, granitic loam, shale and gravelly alluvium, in various habitats including creek banks, wet heath and open eucalypt woodland. Flowers spring to early summer. Pollinated by birds. Regenerates from seed after fire.

Variation There is considerable variation in this species although each population appears uniform.

Typical form (slender-leaved race of McGillivray 1993) Erect to spreading shrub with narrow, pungent leaves 0.5–1 mm wide with strongly revolute margins that mostly obscure lower surface including midvein; leaves usually crowded. *G. sulphurea*, which has pure yellow flowers and a robust erect habit to 2.5 m, appears to be this form. Flowers are usually yellow or pinkish, sometimes orange or red. Occurs between Windsor and St Marys (between Sydney and the Blue Mountains), in open eucalypt woodland.

Broad-leaved form Occurs in the northern part of the species' range, mainly on the Northern Tablelands and is characterised by a prostrate to decumbent habit with leaves 2–6 mm wide, broadly triangular, linear-elliptic or linear, with 3–5 prominent, longitudinal veins visible on the upper surface, the undersurface exposed and silky. Flower colour is yellow or red and various shades in between.

Canberra form Usually a robust shrub with rigid, crowded, pungent, linear leaves 1–2 mm wide with the undersurface completely enclosed by the revolute margins. Flowers are usually red, yellow or pink but not as villous as the Corang River form. It is found along waterways and creeks such as Paddys R., N.S.W., and Pine Island, A.C.T. Populations of prostrate plants occur near Braidwood and at Cullen Bullen, the northernmost population of this form which appears to link it with the Broad-leaved form.

Corang River form (villous-flowered race of McGillivray 1993) The bushy shrubs to about 1.5 m which grow along the banks of the Corang R. and on Mt Currockbilly may be distinguished by a larger pollen presenter and a villous perianth. Both red- and yellow-flowered forms occur. Foliage is similar to the Canberra form.

Major distinguishing features Branchlets terete, villous. Leaves simple, crowded, rigid, pungent, narrowly linear to triangular, mostly with 3–5 longitudinal veins; conflorescence irregular, usually secund; torus oblique, nectary prominent; perianth zygomorphic, hairy on both surfaces; limb nodding; ovary glabrous, stipitate; style end minutely hairy; pollen presenter very oblique.

Related or confusing species Group 21, especially *G. molyneuxii* and *G. speciosa*, both of which have angular branchlets. *G. speciosa* differs further in its broader leaves, usually with lateral venation, its less oblique pollen presenter and its longer pistil. *G. molyneuxii* differs further in the conspicuous midvein on the lower surface of the leaf.

Conservation status Not currently endangered.

Cultivation *G. juniperina* is a hardy, adaptable plant notable for its tolerance of climatic extremes. It performs well in an open, sunny position in a well-drained, neutral to acidic heavy or light soil but will also tolerate a degree of dappled shade. Summer watering is preferred but the species is drought-hardy. Plants seem to require a cold winter to initiate dense flowering and will tolerate extreme frost and snow without damage. Pruning may be needed to shape but this may be an unpleasant duty, considering the prickliness of the foliage of most forms. Most specialist nurseries stock at least one form of this species.

Propagation *Seed* Germinates readily from fresh seed usually without any treatment, but better after applying standard pre-treatments. *Cutting* Strikes readily from cuttings of healthy, half-hard wood at any season. *Grafting* Rarely necessary. Prostrate forms have been grafted successfully onto *G. robusta* for use as 'standards' but the resultant plants have limited appeal.

Horticultural features *G. juniperina* shows great variety of habit, leaf type and flower colour. Foliage is usually dark green and, through the variety of its leaf form, of great horticultural interest. By judicious selection of forms, prostrate broad-leaf groundcovers and robust, narrow-leaved screen plants can be grown together in such a way as to appear to be of different species. Using a variety of different flower colours can also add interest, as flowering is continuous over most of the year. Regardless of which form is selected, all are long-lived, hardy and attract nectar-seeking birds. Small birds in need of shelter value the shrubby forms. A bright, orange-flowered selection introduced to cultivation at Burrendong Arboretum, N.S.W., is outstanding and worthy of wider planting.

Hybrids *G. juniperina* is a parent of a number of hybrids of garden origin including: G. 'Australflora Canterbury Gold', G. 'Canberra Gem', G. 'Pink Lady', G. 'Pink Pearl', G. 'Priors Hybrid' and G. 'Molonglo', as well as the following Poorinda hybrids: 'Adorning', 'Annette', 'Beauty', 'Splendour', 'Wonder', 'Belinda', 'Leanne', 'Elegance', 'Constance', 'Hula', 'Queen', 'Rachel', 'Signet' and 'Pink Coral'.

History First cultivated in 1821 in England from seeds sent by Allan Cunningham in 1820. It is still grown there out of doors in favoured positions in the south-west and as far east as Surrey and Sussex. Baron von Hügel was growing two forms at Vienna in 1831 and it has since been widely grown in Europe. A number of forms are grown in N.Z. Eight colour forms are grown in the U.S.A. where it is well regarded in the nursery trade. It was first grown in Australia at the Sydney Botanic Gardens in 1857 and a fine collection is growing at both Mount Annan Botanic Garden, N.S.W. and at the Australian National Botanic Gardens, Canberra.

General comments The taxonomy of *G. juniperina* remains unresolved and requires further study.

Grevillea kedumbensis (D.J.McGillivray) P.M.Olde & N.R.Marriott (1994) Plate 185

Telopea 5: 727 (Australia)

Based on *G. obtusiflora* subsp. *kedumbensis* D.J.McGillivray (1986, Australia).

The specific epithet is derived from the locality at which the holotype was collected, Kedumba, with the Latin adjectival suffix -*ensis* indicating place. KED-UM-BEN-SISS

G. kedumbensis

Type: Kedumba Valley, 8 km beyond homestead, N.S.W., 7 Oct. 1977, A.M.Blombery NSW 117349 (holotype: NSW).

A lignotuberous, twiggy **shrub** 0.2–1 m high. **Branchlets** slender, terete to slightly angular, glabrous to silky or sparsely so. **Leaves** 1–3 cm long, 0.1–0.5 cm wide, ascending, shortly petiolate, simple, narrowly elliptic to obovate, obtuse, mucronate, the mucro c. 1 mm long; upper surface glabrous to sparsely silky, coarsely granulate, densely to sparsely distributed, midvein evident; margins entire, usually strongly revolute in dried specimens; lower surface silky, midvein prominent. **Conflorescence** erect, terminal or axillary in upper axil, subsessile to shortly pedunculate, simple or 1–3-branched at base; unit conflorescence 12–20-flowered, relatively loose, subglobose to subcylindrical; development basipetal; peduncle 0–3 mm long, silky; rachis silky to glabrous; bracts 1.5–3.5 mm long, narrowly triangular, sparsely silky to glabrous outside, glabrous inside, mostly persistent to anthesis; nectary conspicuous, reniform. **Flower colour**: perianth green to cream; style brownish red; pollen presenter green. **Flowers** adaxially oriented; pedicels (3–)5–6 mm long, loosely tomentose to glabrous, usually retrorse; torus c. 1 mm across, oblique at c. 40°; **perianth** 6–7.5 mm long, 2.5–3 mm wide, falling soon after anthesis, cohering except along dorsal suture, ribbed, oblong-ovoid, ± glabrous to loosely tomentose outside, bearded inside adjacent to and above ovary for c. 2 mm, glabrous or with scattered hairs elsewhere; limb strongly revolute, obtuse-subcubic, loosely tomentose or with scattered hairs only, the segments prominently keeled; **pistil** 12.5–17.5 mm long; stipe absent to 0.2 mm long; ovary villous; style loosely tomentose to loosely villous interspersed with several short, erect hairs;

185A. *G. kedumbensis* Close-up, Mt Cookem variant, N.S.W. (P.Olde)

185B. *G. kedumbensis* Flowering habit, Kedumba Valley, N.S.W. (P.Olde)

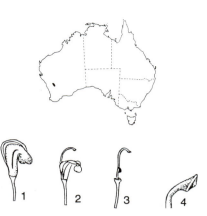

G. kenneallyi

style end discoid, not exposed before anthesis; pollen presenter lateral, flat to slightly convex, round to oblong to obovate-elliptic; stigma prominent, distally off-centre, slightly oblique. **Fruit** 14 mm long, 5–6 mm wide, erect, ovoid, sparsely tomentose, very faintly ribbed; pericarp 0.3–0.5 mm thick at suture. **Seed** 8.5 mm long, 2.5 mm wide, ellipsoidal with a subapical dilation and apical excurrent wing c. 1 mm long; outer face convex, smooth, minutely pubescent; inner face flat; margin revolute with a narrow, membranous wing along one side.

Distribution N.S.W., restricted to the Kedumba Valley and Yerranderie areas. *Climate* Cold, wet winter; cool to hot, wet summer. Rainfall 800–1200 mm.

Ecology Grows in sandy, gravelly loam, usually on decomposed sandstone, in dry sclerophyll forest. Flowers all year with a winter–spring peak. Regenerates from seed and lignotuber after fire. Pollinator uncertain, probably nectarivorous birds.

Variation Plants from the Kedumba Valley area grow to 1 m tall whereas those from the Mt Cookem area are rarely taller than 20 cm.

Major distinguishing features Habit lignotuberous; leaves simple, entire, upper surface coarsely granulate; lower surface silky; margins strongly revolute; conflorescence subsessile, erect; torus oblique; perianth zygomorphic, falling early, almost glabrous outside except limb, bearded inside; limb revolute, emarginate to obtuse, the segments strongly keeled; pistil 12.5–17.5 mm long; ovary subsessile to very shortly stipitate, densely villous; pollen presenter broadly convex; fruit ovoid, ribbed.

Related or confusing species Group 25, especially *G. floribunda* subsp. *tenella* and *G. mucronulata*. *G. rosmarinifolia* is also superficially similar but differs in its glabrous style and ovary. *G. floribunda* subsp. *tenella* differs in its globose perianth. *G. mucronulata*, especially the Picton form, differs in its longer leaf mucro (1–2 mm vs < 1 mm long), its fruit with an inflexed style, its finely granulose leaves and in its more conspicuously keeled perianth limb. *G. obtusiflora* differs in its generally longer pistil, its more densely hairy, persistent perianth and in its suckering habit.

Conservation status 2VC. This species is very restricted in its distribution. The Kedumba Valley population is threatened by flooding if the proposed enlargement of the Burragorang water storage area proceeds.

Cultivation *G. kedumbensis* has been introduced to cultivation only recently but has adapted to most situations encountered in the garden on the east coast. It tolerates frosts to at least -4°C and snow, as well as dry periods but may not be strongly drought-tolerant. It responds favourably to occasional waterings during long, dry summers. A sunny or partially shaded position in acidic to neutral well-drained soil is required, possibly in gravelly or rocky loam for best results. Plants at Mount Annan Botanic Garden are growing well in sandstone-derived, imported soil. Allow up to 1 m between plantings for best results. Not available from nurseries at present.

Propagation *Seed* Sets small quantities of seed that should germinate readily given standard pre-treatment. *Cutting* Strikes readily using firm, young growth. *Grafting* Untested.

Horticultural features *G. kedumbensis* is a low to medium shrub with granular leaves, small clusters of creamy white flowers and dull red styles. Flowering does not stand out on wild plants, being somewhat obscured by the foliage, but the nectar-filled flowers nonetheless strongly attract honey-eating birds. It is suitable as a low screen or fill-in shrub or could be successful in a rockery. Cultivated plants in the right conditions are more floriferous. This species should be long-lived both on the coast and inland. One for the collector!

Grevillea kenneallyi D.J.McGillivray (1986)
Plate 186

New Names in Grevillea *(Proteaceae)* 8 (Australia)

Named in honour of Kevin Kenneally, botanist with the Western Australian Herbarium, who has collected a number of new species of *Grevillea* and is an expert on the flora of the Kimberley. KEN-EE-LEE-EYE

Type: 18.5 km NW of Wongan Hills towards Piawaning, W.A., 27 Aug. 1976, R.Coveny 7839 & B.R.Maslin (holotype: NSW).

Spreading, dense **shrub** to c. 3 m high, 3 m wide. **Branchlets** terete, silky. **Leaves** 4–8 cm long, ascending, shortly petiolate, pinnatisect, divaricately once to three-times divided; leaf lobes 1–3.5 cm long, 1 mm wide, subterete, dipleural, the grooves hair-filled, rigid, pungent; upper surface glabrous or almost so, grooved down the middle; lower surface consisting entirely of the raised midvein; margins rounded. **Conflorescence** erect, spreading, sometimes decurved, pedunculate, axillary or terminal, usually branched; unit conflorescence 2–3 cm long, conico-secund, crowded, dense; peduncle and rachis silky; bracts 0.5–1 mm long, ovate, glabrous with ciliate margins, falling before anthesis. **Flower colour**: perianth and style white with a pink tinge on young buds. **Flowers** acroscopic; pedicels 4–5 mm long, glabrous; torus c. 0.5 mm across, square, straight; nectary U-shaped; **perianth** 2–3 mm long, 0.5 mm wide, strongly curled in bud, oblong, ribbed, glabrous outside, bearded inside in lower half, separated first along dorsal suture and dorsal tepals reflexing slightly, loose to free below limb before anthesis, afterwards free to base; limb revolute, spheroidal, the segments prominently ribbed; **pistil** 5–6 mm long, glabrous; stipe 0.7–0.9 mm long; ovary compressed-globose; style at first exserted from near curve, strongly incurved after anthesis; style end flanged, enclosed before anthesis; pollen presenter oblique, elliptic, convex. **Fruit** 9–11 mm long, 5–6 mm wide, erect on curved pedicels, oblong-ellipsoidal, glabrous, rugulose; pericarp c. 1 mm thick. **Seed** not seen.

Distribution W.A., confined to the Wongan Hills area. *Climate* Hot, dry summer; cool to mild, wet winter. Rainfall 300–400 mm.

Ecology Confined to strongly lateritic, sandy loam in valleys, slopes and plains among grevillea and dryandra shrubland. Flowers late winter to early spring. Probably pollinated by insects. Regenerates from seed after fire.

Variation A uniform species.

Major distinguishing features Branchlets silky; leaves divaricately pinnatisect with rigid, pungent, subterete, dipleural lobes; conflorescence secund; perianth zygomorphic, strongly curled, glabrous outside, bearded within in lower half, limb subrectangular; pistil glabrous; pollen presenter oblique, convex; fruit oblong-ellipsoidal.

Related or confusing species Group 8, especially *G. subtiliflora* which differs in its longer (> 6 cm long), cylindrical conflorescence.

Conservation status 2EC. Restricted to a very small area near Wongan Hills.

Cultivation *G. kenneallyi* was introduced to cultivation in 1986 but since it grows naturally in lateritic soil it can be difficult to establish and maintain. It tolerates frost to at least -4°C and is drought-resistant. Most plants in cultivation at present have been grafted onto hardy rootstocks, and therefore performance in the ground on its own roots is speculative. Success should be expected if planted in an open, sunny position in a deep, acidic gravelly loam. The incorporation of gravel into the soil is strongly recommended. It naturally forms a compact, dense shrub without pruning. Summer watering is not recommended once established. Not currently available from nurseries.

Propagation *Seed* Untested but likely to be successful if pretreated by nicking or peeling. *Cutting* Firm, young growth from wild plants in

186A. *G. kenneallyi* Close-up of flowers and foliage (N.Marriott)

186B. *G. kenneallyi* Flowering branches (M.Hodge)

186C. *G. kenneallyi* Plant in natural habitat near Wongan Hills, W.A. (P.Olde)

spring has been struck. Material from cultivated plants should prove easier. When trimming, do not cut across the leaf lobes as this can cause the remainder of the lobe to die. Difficulties have been experienced in maintaining struck cuttings. *Grafting* Has been grafted successfully onto *G. robusta* using whip and mummy grafts.

Horticultural features During early spring, *G. kenneallyi* is a most floriferous shrub festooned with one-sided white-flowered racemes perched densely all over the upper branches. Although flowering is not continuous, the green, prickly foliage is concealed in a mantle of white, indicating that this species could prove suitable as a screen or feature plant in inland gardens. Foliage would afford protection to small birds, but these do not visit the flowers which are lightly scented and attract insects including butterflies.

Grevillea kennedyana F.Mueller (1888)
Plate 187

Transactions & Proceedings of the Royal Society of Victoria 24: 172 (Australia)

Flame Spider Flower

The specific epithet honours Mrs M.B.Kennedy who collected specimens for Mueller in north-western N.S.W. KEN-ED-EE-ARE-NA

Type: Grey Range, N.S.W., c. 1887, W.Baeuerlen (holotype: MEL).

Diffuse, much-branched, lignotuberous, suckering **shrub** to 0.7–1.5(–2) m high. **Branchlets** angular to terete, silky to tomentose. **Leaves** 0.7–3.3 cm long, 1.2–2 mm wide, often fasciculate, ascending to patent, ± sessile, simple, linear, rigid, pungent with a long aristate point; upper surface silky becoming sometimes almost glabrous, midvein evident only as a faint groove; margins angularly revolute, entire; lower surface bisulcate, midvein prominent. **Conflorescence** 2–4 cm long, 3–5 cm wide, erect, ± sessile, terminal, simple, umbel-like; rachis silky; bracts 0.7–1 mm long, narrowly triangular, villous, falling before anthesis. **Flower colour**: perianth pink, red or orange red with an enamel green limb; style red. **Flowers** abaxially oriented; pedicels 6.5–11 mm long, usually glabrous, filamentous; torus 4.4–5.5 mm long, very oblique, cup-shaped; nectary long U-shaped, entire, inconspicuous; **perianth** 14–21 mm long, 3–6 mm wide, elongate-ovoid, glabrous or with sprinkled silky hairs outside, villous inside, tepals ribbed, cohering except along dorsal suture before anthesis, afterwards all free to curve and rolling back, the ventral tepals curling back further than dorsal tepals; limb nodding, ovoid-apiculate; **pistil** 27–31 mm long, glabrous; stipe 4–5.5 mm long, partially united with torus; ovary ovoid, scarcely broader than style; style exserted just below curve and looped upwards before anthesis, afterwards gently incurved, slightly inflexed above ovary; style end flattened, not exposed before anthesis; pollen presenter very oblique to lateral, oblong, convex with a surrounding flange. **Fruit** 15–19 mm long, 5.5–7 mm wide, erect, obovoid-ellipsoidal, granular; pericarp c. 0.5 mm thick. **Seed** 7.5–10.5 mm long, 2.7–3.1 mm wide, linear to narrowly ellipsoidal with a short, apical wing; outer face convex; inner face channelled.

Distribution N.S.W., in the north-western corner on off-shoot ranges of the Grey Ra., McDonald Peak, Olive Downs escarpment, Mount Wood and Mount Wood Hills and on isolated hills on Onepah station. Qld, at Naryilco on the Bygrave Ra. *Climate* Hot, dry summer; cool to mild, wet or dry winter. Rainfall 150–200 mm.

Ecology Grows on gentle mesa slopes with silcrete scree and on steep jump-ups as well as in creek beds, in small groups. Flowers winter–spring. Flowers attended by birds. Plants occasionally arise from seed but mostly regenerate from lignotuber and sucker, either naturally or in response to fire.

Variation There is considerable variation in perianth indumentum and flower size but whether these differences are random or population-related requires further study.

Major distinguishing features Leaves linear, pungent, silky-grey, the midvein on upper surface obscure; conflorescence usually terminal, sessile, many-flowered, umbel-like; pedicels filamentous; torus long, very oblique, cupped; perianth elongate-ovoid, the limb ovoid, nodding, the outer surface glabrous to silky, the inner villous; ovary long-stipitate, glabrous, scarcely wider than style; pollen presenter lateral; fruit obovoid, minutely granular.

Related or confusing species Group 15, especially *G. acuaria* which differs in its prominent ovary, its smaller torus and perianth and in its fruit. There is some similarity to some members of Group 21,

G. kennedyana

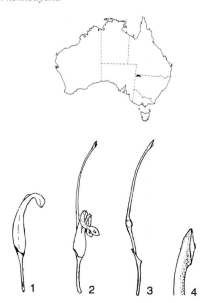

187A. *G. kennedyana* Flowers and foliage (N.Marriott)

187B. *G. kennedyana* Flowers before anthesis showing green limb (N.Marriott)

187C. *G. kennedyana* Plants in natural habitat, Sturt NP, N.S.W. (N.Marriott)

notably *G. juniperina* and *G. victoriae*, both of which differ in their hairy style and less oblique torus.

Conservation status 3VC. Recent surveys have revealed plant numbers upward of 7000 individuals.

Cultivation *G. kennedyana* has been established successfully at Mount Annan Botanic Garden, near Sydney, where it has thrived and flowered for a number of years. Plants at the Australian National Botanic Gardens, Canberra, have not responded as favourably except potted specimens in glasshouses. More recently, plants grafted onto *G. robusta* have been tried but results have been mixed. Members of the Grevillea Study Group, from Qld to Vic., have been testing a number of rootstocks, some of which have produced superb results. A well-drained gravelly loam in full sun, protected from strong winds, is recommended for successful cultivation. Soil should be neutral to slightly acidic. Regular tip pruning will overcome a natural tendency to stragginess. Occasional summer watering appears to be beneficial but caution is recommended if watering after very hot weather. A dry climate is preferred but to date it appears unaffected by summer humidity. Potted plants have thrived in standard commercial potting mixes, principally of pine bark, peat and coarse sand with nitrogenous fertiliser and trace elements. Periodic treatment for fungal attack may be necessary. It should establish well in cultivation as it will tolerate extended dry conditions and frosts to at least -4°C.

Propagation *Seed* Untested. Sets limited quantities of seed in the wild. *Cutting* Strikes readily from cuttings of firm, young growth from cultivated plants taken in spring. *Grafting* Successfully grafted onto *G. robusta*, G. 'Poorinda Royal Mantle', G. 'Poorinda Anticipation' and G. 'Ruby Clusters' using the top wedge, whip and mummy methods. Grafts appear to be compatible but resultant plants are often slow-growing.

Horticultural features *G. kennedyana* is a handsome species with its silvery-grey foliage in striking contrast to its brilliant orange or orange-red flowers. The flowers have long, prominent styles and a 'metallic', enamel-green limb that further enhances their beauty. They produce copious clear, sweet nectar and are strongly bird-attractive. The conflorescences tend to be scattered through the foliage, sometimes quite near the base of the plant. Even when not in flower, the foliage of this species is pleasing and provides a picturesque contrast to green-leaved plants. This species, so recently introduced and so beautiful, suggests itself for use as a feature plant or rockery plant, of equal value to collectors and horticulturists.

Grevillea lanigera A.Cunningham ex R.Brown (1830)   Plate 188

Supplementum Primum Prodromi Florae Novae Hollandiae exhibens Proteaceas Novas 20 (England)

Woolly Grevillea

The specific epithet alludes to the hairy branchlets and leaves and is derived from the Latin *lana* (wool) and *-ger*, (bearing).   LAN-IDG-ER-A

Type: bed of Murrumbidgee or Lachlan R., SW from Lake George, N.S.W., April 1824, A.Cunningham 42 (lectotype: BM) (McGillivray 1993).

Synonyms: *G. ericifolia* R.Brown (1830) (published with *G. lanigera*); *G. lanigera* var. *planifolia* C.F.Meisner (1856); *G. lanigera* var. *revoluta* (1856); *G. ericifolia* var. *muelleri* C.F. Meisner (1856); *G. scabrella* C.F.Meisner (1856); *G. baueri* var. *pubescens* G.Bentham (1870); *G. ericifolia* var. *scabrella* (C.F.Meisner) G.Bentham (1870); *G. lanigera* var. *ericifolia* (R.Brown) A.Ewart, Maiden & Betche (1916).

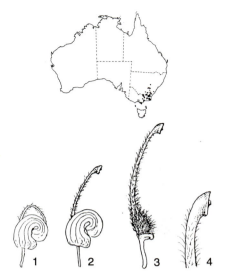

G. lanigera

A variable species, usually suckering; sometimes an erect, dense **shrub** to 1.5 m with erect branches; sometimes a low, weak, few-branched **shrub** to 30 cm; sometimes a small, dense, grey **shrub** to 30 cm; sometimes prostrate. **Branchlets** ± terete, pubescent to villous, thin or stout. **Leaves** 1–4 cm long, 1–5 mm wide, ascending to spreading, crowded, sessile, simple, narrowly oblong, obtuse-mucronate, sometimes acute, non-pungent; upper surface villous, sometimes the indumentum persistent, otherwise scabrous, midvein obscure; margins entire, smoothly revolute; lower surface villous, midvein prominent. **Conflorescence** decurved, usually shortly pedunculate, terminal and axillary in upper axils, simple or few-branched; unit conflorescence 1–3 cm long, 2–3 cm wide, secund, umbel-like, dense, apical flowers opening first; peduncle villous; rachis glabrous or with a few scattered hairs; bracts 1–3.5 mm long, narrowly triangular, villous outside, usually falling before anthesis. **Flower colour**: perianth bright pink or red-pink and cream, sometimes creamy green or yellow; style dull red, sometimes yellow-green. **Flowers** acroscopic; pedicels 3–5.5 mm long, glabrous; torus 1–2 mm across, slightly oblique; nectary conspicuous, saucer-shaped, lipped over rim of torus; **perianth** 5–6 mm long, 2–5 mm wide, ovoid, dilated at base, strongly curved, glabrous and often glaucous outside, bearded inside about level of ovary, silky above, cohering except along dorsal suture; limb revolute, spheroidal; **pistil** 13.5–19.5 mm long; ovary ± sessile (stipe absent to 0.8 mm long), villous; style villous, exserted below curve and looped outwards before anthesis, afterwards straight to slightly incurved, kinked forward above ovary, villous, ± glabrous on ventral side, gently curved to straight, stout; style end conspicuous, glabrous, flattened, partially exposed before anthesis; pollen presenter lateral, oblong-elliptic, flat, umbonate. **Fruit** 10–15 mm long, 4.5–7 mm wide, oblique, cylindrical to narrowly ovoid, ribbed, sparsely villous; pericarp 0.3–0.6 mm thick. **Seed** 7.5–8.5 mm long, 2.2–2.4 mm wide, narrowly elliptic with a short apical wing, minutely pubescent; outer face convex; inner face flat, channelled around margin; margin revolute with a membranous border on one side.

Distribution N.S.W., from Green Cape near Eden to the high country in the Snowy Mountains, usually at lower elevations and extending north through the A.C.T. to the Bathurst region. Vic., from Wilsons Promontory through Gippsland to the NE corner and into the high country. *Climate* Mild, warm to hot, dry summer; cool to cold, wet winter; subalpine populations have occasional snowfalls. Rainfall 600–1600 mm.

Ecology Grows in a wide range of habitats, primarily eucalypt woodland and medium scrub, sometimes in coastal heath, in skeletal rock scree, grey, shallow sand, granitic loam, shallow clay and deep sand, often beside the ocean. Flowers all year, principally in winter and spring. Pollinated by birds. Regenerates from seed or sucker.

Variation An extremely variable species with a wide natural distribution. Plants vary in habit, leaf size and shape, flower size and indumentum. A full study of variation in wild populations is still needed. The following forms are quite distinctive.

Grenfell form A low form to c. 30 cm with narrowly linear, grey, pubescent leaves and brilliant pink and cream flowers. This most delightful shrub was first grown at Burrendong Arboretum.

Robust form A shrubby form to 1.5 m sometimes spreading to 1.5 m with erect, stout branches bearing broad, oblong leaves longer than other forms. It has showy though less profuse pink-red and cream or sometimes pure creamy white flowers. It grows on the Suggan Buggan R., Vic., as well as on Black Mountain, A.C.T., and widely at lower elevations in the Snowy Mountains.

Upper Murray form In the high country around Cooma, extending to the Cotter R. and the hills of the upper Murray R., populations occur which sucker vigorously but tend to be low, weak-looking plants to c. 50 cm with narrowly oblong dull green leaves and thin branchlets. Flowers are usually a brilliant rich pink and cream but some are entirely yellow.

Coastal form A low, few-branched shrub to c. 30 cm with small oblong leaves c. 5 mm long and 2 mm wide. This strongly suckering form occurs in sandy heath at Green Cape and Nadgee Fauna Reserve, N.S.W., extending into coastal, eastern Vic. It tends to be somewhat inconspicuous in flower although some plants can be quite showy. We have seen attractively flowering plants growing among rocks overlooking the sea, almost in the salt spray, at Green Cape.

Prostrate form An outstanding form introduced to horticulture by the late Bill Cane from a location unknown to the authors. It forms a dense, decumbent to prostrate mat and has small oblong leaves and prominent, attractive pink and cream flowers. This very hardy form, which grows equally well in Brisbane and Tas., was thought to have been collected at Mt Tamboretha, Vic., and while *G. lanigera* does occur in this area it is not the beautiful prostrate form so well known in cultivation. Recent observations indicate that this form may have been selected from plants in the Yanakie–Wilsons Promontory area of Vic.

Wilsons Promontory form Similar in foliage and flower to the Prostrate form but with a dense, much-branched habit. It grows to c. 1 m and can be found almost to the high tide mark in granitic sand. It does not appear to sucker strongly.

Major distinguishing features Leaves simple, entire, pubescent like the branches, usually oblong and obtuse; conflorescence a short, one-sided cluster; torus slightly oblique; floral rachis glabrous; perianth zygomorphic, glabrous outside, hairy

188A. *G. lanigera* Upper Murray form, Mt Cudgewa, Vic. (N.Marriott)

188B. *G. lanigera* Flowers and foliage of the Prostrate form from Yanakie, cultivated at Stawell, Vic. (N.Marriott)

188C. *G. lanigera* Grenfell form, cultivated at Burrendong Arboretum, N.S.W. (P.Olde)

188D. *G. lanigera* Yellow-flowered plant of the Robust form, cultivated at Stawell, Vic. (N.Marriott)

within; ovary ± sessile, densely hairy; style villous to pubescent; pollen presenter lateral.

Related or confusing species Group 25, especially *G. rosmarinifolia* which has a glabrous ovary and style and a dense tuft of hairs on the stipe just below the ovary.

Conservation status Not presently endangered.

Cultivation A very hardy, adaptable species that has been in cultivation for many years in cold-wet, warm-dry and even summer-rainfall climates provided humidity is not excessive. *G. lanigera* grows best in a cool climate when planted in full sun or partial shade in well-drained, acidic sandy or gravelly loam or friable clay. It does not need summer watering although occasional deep watering during extended dry conditions will improve its general appearance. Fertiliser is not usually necessary. Occasional light pruning enhances shape and density. The smaller prostrate forms make delightful specimens in medium to large tubs. Flowering is improved if grown in winter-cold conditions. In humid areas leaves may drop from the lower branches, leaving plants very spindly and leggy. A number of forms are available from nurseries.

Propagation *Seed* Sets good quantities of seed that germinates well when pretreated. Great care needs to be taken to ensure that seedlings are pure, especially seed of garden origin as the species hybridises readily. *Cutting* Firm, young growth strikes readily at most seasons. *Grafting* Has been grafted successfully onto *G. robusta* using the whip graft.

Horticultural features *G. lanigera* is an attractive species distinguished by its soft, pubescent, grey or grey-green leaves that are often crowded on the branchlets, and its prominent clusters of pink and cream or pure creamy-yellow flowers. Some forms are particularly showy and provide a profuse display of delicate flowers nearly all year round. It is very popular with honey-eating birds and is extremely useful in landscaping in all climates, especially prostrate forms. Most forms are lightly suckering and will clump up if given space to grow.

Hybrids Natural hybrids have been recorded between *G. lanigera* and *G. rosmarinifolia* which often grow in close association with each other and share a similar diversity of habit and leaf shape. Populations along the Yarra, Goulburn and upper Murray Rivers, Vic., are sometimes of hybrid origin and are often more robust. In this form, the leaves are grey to grey-green, revolute to the midvein or nearly so, very narrowly linear with acute, pungent tips and ± glabrous. The pale pink and cream flowers appear to be smaller and are borne in open racemes. There are also hybrid populations in several areas of N.S.W. In north-eastern Vic., two other very interesting hybrid populations occur. At Granya Gap is a swarm of hybrids with *G. polybractea*; the lovely G. 'Granya Glory' is one of these. In the Pine Mountain-Burrowa NP, hybrids with *G. jephcottii* are sometimes found. The strongly suckering plant known as G. 'Little Thicket' is probably a hybrid between *G. lanigera* and *G. arenaria* and, although not appropriately named, is a useful plant for landscaping large open spaces.

In horticulture, the hybrid G. 'Poorinda Signet' is derived from *G. lanigera*. G. 'Clearview John' is believed to have been selected by the late Bill Cane from a wild population of *G. lanigera*, probably in the Gippsland area.

History Introduced into cultivation in England in 1822 as *G. baueri* var. *pubescens*, it is still grown there today by enthusiasts. Three 'forms' are in cultivation in the U.S.A. and it is cultivated in New Zealand.

General comments *G. lanigera* appears to be taxonomically unresolved at the infraspecific level. Further studies may clarify the various elements.

Grevillea latifolia C.A.Gardner (1923)
Plate 189

Forests Department Bulletin Western Australia 32: 43 (Australia)

The specific epithet is derived from the Latin *latus* (broad) and *folium* (a leaf), in reference to the very broad leaves. LAT-I-FOAL-EE-A

Type: plateau between Lawley and King Edward Rivers, W.A., 30 July 1921, C.A.Gardner 1498 (holotype: PERTH).

A multi-stemmed, lignotuberous **shrub** 1–2 m high. **Branchlets** compressed-angular, silky to glabrous. **Leaves** 6–15 cm long, 5–15 cm wide, ascending, petiolate, very broadly ovate to obovate, emarginate to obtuse with short mucro; upper and lower surfaces similar, sparsely silky, soon glabrous; venation prominent; lateral veins closely aligned, spreading at c. 45° to midvein, terminating in a conspicuous intramarginal vein 1–2 mm from margin, tertiary venation evident; margins entire, flat, undulate. **Conflorescence** erect, pedunculate, terminal, simple or, more usually few-branched; unit conflorescence 1–2.5 cm long, secund, 6–16-flowered, the flowers sometimes arising singly along rachis; peduncle and rachis silky, the rachis slightly decurved; bracts 2.7–3.2 mm long, narrowly ovate, silky outside, falling soon after expansion of conflorescence. **Flower colour**: perianth pale pink turning red from the base, rarely pure white with mauve tinge; style creamy white with tip turning red. **Flowers** acroscopic; pedicels 3.5–8 mm long, glabrous; torus 2–3 mm across, oblique; nectary high-walled, U-shaped, the margin sometimes recurved, wavy; **perianth** 9–12 mm long, 3–3.5 mm wide, oblong-ovoid, slightly recurved, dilated at base, glabrous outside, hairy inside with a narrow transverse line of hairs between ovary and curve, tepal margins slightly flanged, cohering except along dorsal suture, flared open at curve where dorsal tepals separate and reflex before anthesis, afterwards free to base and quickly detaching; limb erect and very conspicuous in bud, nodding to revolute before anthesis, ovoid, noticeably thickened over anthers; tepal margins prominently flanged; **pistil** 12–19.5 mm long, glabrous; stipe 2–5 mm long, inserted ± perpendicular to torus; style not exserted before anthesis, afterwards incurved to straight, not far exceeding perianth; style end gradually swollen, exposed before anthesis; pollen presenter ± lateral, oblong-elliptic, flat to slightly convex; stigma prominent. **Fruit** 16–19 mm long, 8–9 mm wide, erect,

G. latifolia

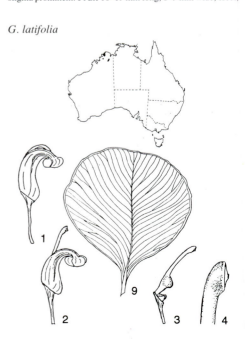

189A. *G. latifolia* Close-up of flowers (M.Hodge)

189B. *G. latifolia* Habit, near King Edward River crossing, W.A. (P.Olde)

189C. *G. latifolia* Flowers and foliage (P.Olde)

compressed-elliptic but attenuate, ridged on dorsal side, granular, glabrous; pericarp c. 0.5 mm thick. **Seed** 13 mm long, 6 mm wide, obovoid with triangular apical wing c. 2 mm long; outer face convex, pitted, with a thickened marginal border; inner face slightly convex, channelled inside thickened border; margin inconspicuously and membranously winged all round.

Distribution W.A., in the Kimberley, in scattered populations on the Mitchell and Gardner Plateaus, the King Edward and Lawley Rivers, extending E almost to Wyndham. *Climate* Summer monsoonal, hot and wet; winter warm, dry. Rainfall 800–1000 mm.

Ecology Grows on flat plateaus, often along creek lines, in yellow-brown sandy soil, white alluvial sand or lateritic gravelly soil, usually in eucalypt woodland and grassland. Flowers autumn–winter. Regenerates from seed and lignotuber after fire. Pollinator unknown, probably birds.

Variation A relatively uniform species.

Major distinguishing features Leaves simple, broadly ovate, entire, ± the same both sides, prominently and closely penninerved, the margins flat; conflorescence on decurved rachis, the flowers usually solitary (not in twinned pairs); torus very oblique; nectary prominent; perianth zygomorphic, glabrous outside, transversely narrow-bearded within, the limb 4-winged; pistil glabrous; ovarian stipe at 90° to torus, style end swollen; fruit compressed with apical attenuation.

Related or confusing species Group 5. *G. versicolor* seems most closely related but differs, among other things, in its toothed leaves.

Conservation status 3R. Occurs in fairly large numbers.

Cultivation To our knowledge, *G. latifolia* has not yet been cultivated on its own roots, but there are several grafted specimens (on *G. robusta*) in the Brisbane area and northern N.S.W. These are flourishing in well-drained, deep soil with good summer rain. In this area, the species seems to be free of pests, especially those that eat leaves as often occurs in the wild. Cold or frosty weather seems to cause leaf drop, hence this species is not recommended for cold-winter climates but should do reasonably well in mild, temperate climates (especially if grafted). Grafted plants have been grown to flowering in tubs in northern Vic. but must be overwintered in a glasshouse. Full sun is strongly recommended, although partial shade should be acceptable in tropical climates, where it should also do well on its own roots. May be pruned heavily. Grafted plants may soon be available commercially in the Brisbane area.

Propagation *Seed* Sets reasonable quantities of seed. Germination untested. *Cutting* Should strike from cuttings of firm, young growth during the warmer months. Has proved difficult to date, especially from the wild. *Grafting* Has been grafted successfully onto *G. robusta* using the whip and mummy grafts.

Horticultural features *G. latifolia* is something of a novelty for enthusiasts of the genus, mainly for its bright green, extremely broad, shapely leaves. The creamy pink flowers, while individually attractive, are never prominent, being usually concealed by the leaves, but they have curiosity value for the shape of the limb. Birds are attracted to the flowers. The beautifully veined, bright green circular leaves make this a delightful feature plant, best placed where they can be highlighted by backlighting. Pruning or regular burning improves its density and hence its horticultural appeal.

Grevillea laurifolia F.Sieber ex K.Sprengel (1827) Plate 190

Systema Vegetabilium 16th edn, *Curae Posteriores*: 46 (Germany)

Laurel-leaf Grevillea

The specific epithet alludes to the elliptic leaves which were thought to resemble those of the Laurel, *Laurus nobilis*. LOR-I-FOAL-EE-A

Type: probably in the Blue Mountains, N.S.W., c. 1823, F.Sieber 26 (holotype: S).

Synonyms: *G. laurifolia* F.Sieber ex J.H.Schultes & J.H.Schultes f. (1827); *G. amplifolia* M.Gandoger (1919); *G. cordigera* M.Gandoger (1919).

A prostrate, trailing **shrub** forming mats to 4.5 m across. **Branchlets** secund, terete, faintly ribbed, silky to glabrescent. **Leaves** 3–10(–16) cm long, 2.5–4.5(–6) cm wide, spreading, petiolate, simple, ovate or elliptic, usually obtuse, mucronate, sometimes acute; upper surface glabrous, penninerved, midvein and acutely angled lateral veins prominent, midvein sprinkled with silky hairs; margins entire, undulate to flat, shortly recurved; lower surface silky, sometimes sparsely so, venation prominent. **Conflorescence** 4–8 cm long, erect, pedunculate, terminal, unbranched, secund, rarely almost cylindrical; peduncle and rachis silky; bracts 1.5–2.3 mm long, ovate, silky outside, falling before anthesis. **Flower colour**: perianth grey or reddish grey; style red or dark red with green tips. **Flowers** acroscopic; pedicels 2–4 mm long, silky; torus c. 1.5 mm across, ± straight; nectary semi-circular, wavy; **perianth** 7–10 mm long, 1.8–2.5 mm wide, oblong-ovoid, dorsally incurved, silky outside, glabrous inside, cohering except along dorsal suture; limb revolute, globular; **pistil** 13–25 mm long; stipe 1.3–2.6 mm long, villous; ovary villous, globular; style glabrous, at first exserted below curve on dorsal side and looped upwards and outwards before anthesis, afterwards straight to incurved, ultimately swept back; style end hoof-like, not exposed before anthesis; pollen presenter oblique, oblong, convex. **Fruit** 9–13 mm long, 5 mm wide, oblique on curved stipe, ellipsoidal, tomentose with red-brown stripes or blotches; pericarp c. 0.3 mm thick. **Seed** 10 mm long, 4.2 mm wide, compressed-ellipsoidal, narrowly winged all round or with a waxy border, the wing drawn out obliquely at each end; outer face convex, rugose; inner face flat, rugose; margin shortly recurved.

Distribution N.S.W., widespread in the Blue Mountains and beyond to the Wombeyan Caves. *Climate* Cold, wet or dry winter; mild to hot, wet summer. Rainfall 800–1200 mm.

Ecology Grows in open heath or eucalypt woodland in dry sclerophyll forest, usually in moist to wet, sandy soils. Flowers spring–summer. Pollinated by nectarivorous birds. Regenerates only from seed after fire.

G. laurifolia

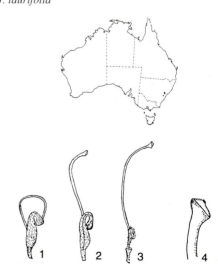

190. *G. laurifolia* Flowers and foliage (R.Page)

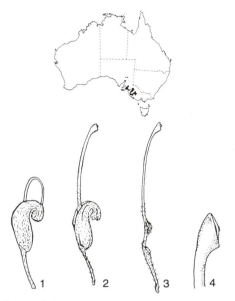

G. lavandulacea

Variation A species with considerable variation in leaf size and degree of leaf undulation as well as conflorescence shape and flower size. A wavy leaf and large leaf selection are grown in cultivation.

Smaller-flowered form In the lower elevations of the Blue Mountains, between Wentworth Falls and Valley Heights, plants with a compact, sub-cylindrical conflorescence and a pistil 13–15 mm long are found.

Major distinguishing features Prostrate, trailing habit; leaves simple, entire, ovate to broadly elliptic, the margins shortly recurved, undulate, undersurface silky; conflorescence secund to almost cylindrical; pedicels silky, 2–4 mm long; torus straight to slightly oblique; perianth zygomorphic, S-shaped, silky outside, glabrous within; ovary villous, shortly stipitate; fruits with reddish blotches or stripes.

Related or confusing species Group 35, especially *G. macleayana* which has similar leaves and flower structure but differs in its erect habit and leaf undersurface with an indumentum of curled hairs. Its flowers have pink styles.

Conservation status Not presently endangered.

Cultivation *G. laurifolia* grows best in areas with winter frosts and summer rainfall. It has been cultivated widely in private and public gardens in Tas. and Vic. and grows vigorously in the Australian National Botanic Gardens, Canberra, and elsewhere. Three leaf forms are grown in the U.S.A. It is not long-lived in winter-warm, summer-humid climates such as Sydney, even though its natural habitat is not far away. It tolerates hot-dry climates well as long as there is subsoil moisture. Cold winters are necessary for flower buds to initiate in large numbers. An open, sunny, exposed position or partial shade best suits this species and a well-grown plant will cover up to 3 m across. It likes a moist, well-drained, acidic sandy soil and is long-lived and hardy. Pruning is usually not necessary. Occasionally available from specialist native plant nurseries mainly in N.S.W., Vic. and Tas.

Propagation Seed Untested. Sets limited quantities of seed in the wild. *Cutting* Grows readily from cuttings at any season, with best results in spring–summer. *Grafting* Grafts readily onto *G. robusta* and has been sold as a weeping standard for many years. It can also be readily grafted to *G.* 'Poorinda Royal Mantle'.

Horticultural features *G. laurifolia* is an outstanding groundcover with large, oval, dark green leaves and masses of contrasting dark red flowers. These attract many honey-eating birds that eagerly hop around the ground seeking out the nectar-filled blooms. Good plants in suitable climates develop into very large, dense, long-lived plants that spread up to 4 m across. For large gardens and public areas, these are extremely valuable in softening bare areas, keeping soil cool and suppressing weeds. It also looks impressive when allowed to cascade down walls and banks, softening harsh lines and creating a veil of rich, green foliage.

Hybrids Forms natural hybrids with *G. acanthifolia*, such as the well-known *G.* × *gaudichaudii* which has very bright flowers and attractive divided leaves. Usually these hybrids are prostrate but sometimes erect forms are found. May also be a parent of *G.* 'Poorinda Royal Mantle'.

Grevillea lavandulacea
D.F.L.Schlechtendal (1847) **Plate 191**

Linnaea 20: 586 (Germany)

Lavender Grevillea

The foliage reminded Schlechtendal of lavender, hence the specific epithet. LAV-AND-YOU-LACE-EE-A

Type: near Port Adelaide, St Vincent Gulf, S.A., Dec. 1844–45, H.H.Behr (lectotype: HAL40153) (McGillivray 1993).

Synonyms: *G. rosea* J.Lindley (1851); *G. ramulosa* F.Mueller ex C.F.Meisner (1854); *G. lavandulacea* var. *latifolia* C.F.Meisner (1856); *G. lavandulacea* var. *lanceolata* C.F.Meisner (1856); *G. lavandulacea* var. *angustifolia* C.F.Meisner (1856); *G. lavandulacea* var. *sericea* G.Bentham (1870).

A compact, rounded, sometimes suckering **shrub** 0.2–1 m high. **Branchlets** terete, pubescent or silky. **Leaves** 0.5–4 cm long, 0.5–10 mm wide, ascending to spreading, sessile, simple, narrowly linear to narrowly obovate, acute, pungent, sometimes obtuse, mucronate; upper surface convex, glabrous and granulose or silky, midvein obscure; margins entire, usually smoothly revolute to midvein, sometimes shortly recurved; lower surface usually obscured, unisulcate, when exposed pubescent to tomentose, the midvein prominent. **Conflorescence** 1–5 mm long, 1–5 cm wide, erect, sessile to shortly pedunculate, mostly terminal, umbel-like to secund, 6–10-flowered, unbranched; peduncle tomentose-villous; rachis tomentose-villous; bracts 0.5–6.5 mm long, narrowly triangular, villous outside, usually falling before anthesis. **Flower colour**: perianth pink, white, red, mauve-pink, red and white, pink and white; style pink, red or rarely white. **Flowers** acroscopic; pedicels 4–8 mm long, tomentose; torus 1.5–2 mm across, very oblique; nectary U-shaped, the margin thick, smooth or toothed at ends; **perianth** 8–10 mm long, 3.5–5 mm wide, narrowly ovoid, strongly recurved, slightly dilated at base, ribbed, silky to almost glabrous outside, bearded about level of ovary inside, cohering except along dorsal suture; limb revolute, globular to spheroidal, silky to villous; **pistil** 21.5–28.5 mm long; stipe 1–2.5 mm long, partially adnate to torus; ovary silky; style hairy at base, glabrous above, exserted near curve and looped upwards before anthesis, afterwards gently incurved, dilating evenly upwards; style end discoid, partially exposed before anthesis; pollen presenter oblique, orbicular to oblong, convex. **Fruit** 11–15 mm long, 4.5–7 mm wide, erect, narrowly ovoid, pubescent; pericarp c. 0.5 mm thick. **Seed** 8–12 mm long, 2–3 mm wide, linear-elliptic, pubescent, with subapical cushion-like swelling and short wing; outer face convex; inner face channelled around border; margin strongly recurved with narrow wing along one side.

191A. *G. lavandulacea* Black Range form, cultivated at Burrendong Arboretum, N.S.W. (P.Olde)

Distribution S.A., widespread in drier and/or sandy areas in the south-east, extending north into the Flinders Ranges. Vic., in similar situations in the west and south-west. ***Climate*** Hot, dry summer, cool to mild, wet winter, except in the Flinders Ranges where winter can be dry with fleeting showers. Rainfall 200–700 mm.

Ecology Grows in various habitats including open heath, mallee heath, open eucalypt woodland, or low, dense scrub in shallow to deep, acidic to alkaline sand, sandy loam, sandy clay, limestone rubble or sandstone rubble. Flowers winter–spring.

Variation An extremely variable species with major differences between homogeneous populations in characters such as leaf shape, size, texture and colour, habit and flower colour. Almost all populations are sufficiently distinct to enable recognition by horticulturists. The following major forms will give a good guide for the present.

Victor Harbor form When not in flower, this form could be confused with *G. rosmarinifolia* as it has glabrous green leaves unlike any other form of the species. It makes an attractive, rounded shrub up to 1 m tall with a good display of showy red and white flowers. It grows naturally around the Victor Harbor area and in several other areas on the Fleurieu Peninsula. Many populations have been wiped out by clearing for agriculture.

Adelaide Hills form A most distinct and lovely form with broad, soft, grey-green leaves 5–10 mm wide, with a white pubescent undersurface and conspicuous smoky pink, mauve-pink or bright pink flowers. It is a small shrub to c. 50 cm and grows on the dry, stony hills E of Adelaide, especially around the Athelstone–Black Hill area.

Flinders Ranges form A population with clustered, small, grey, rigid leaves and pink, red or creamy white flowers. It is a very attractive open shrub to 60 cm high and grows in extremely dry, stony soil in several areas in the Flinders Ranges including Wilpena Pound. A white-flowered form is known from Mt Arkaroola.

Aldinga form An open to dense, low, mounded shrub to 0.8 m high which originated in the Aldinga area. This form, which has silver-grey elliptic leaves and masses of large pinkish red flowers, is one of the loveliest forms in cultivation.

Tanunda form A beautiful form with felted, silver-grey leaves and large clusters of distinct mauve-pink flowers. It makes a dense mounded shrub to 1 m and is confined to a number of areas in the Barossa Valley in particular near Tanunda. The form of *G. lavandulacea* in cultivation as the Tanunda form is of uncertain origin, but is suspected to have originated at Aldinga in S.A.

Black Range form A very open shrub with narrow, grey leaves and spectacular clusters of fiery red flowers. It is a lanky plant to 1 m in the wild but is more compact when regularly pruned in cultivation. It occurs naturally in the Black Ra. to the W of The Grampians where it grows in acidic, sandy loam to sandy clay.

191A. *G. lavandulacea* Adelaide Hills form, S.A. (N.Marriott)

191B. *G. lavandulacea* Cream-flowered plant ex Mt Arkaroola, Flinders Ranges, S.A., in cultivation at Melbourne, Vic. (R.Elliot)

191C. *G. lavandulacea* Victor Harbor form, cultivated at Burrendong Arboretum, N.S.W. (P.Olde)

191D. *G. lavandulacea* Desert form, N of Bordertown, S.A. (N.Marriott)

Billywing form This is a delightful, dwarf shrub which is lightly suckering, rarely more than 20 cm high, with leaves similar to the Black Range form. Its large, showy clusters of pinkish red to red flowers are also similar to the Black Range form. It grows in flat, treeless, seasonally swampy sites in the Billywing area of The Grampians in acidic sandy clay. A very similar, low to medium form with slightly broader leaves is widespread in low, sandy areas near Penola in S.A. and across the border in western Vic.

Desert form A very attractive shrub with tightly revolute, small silver leaves and showy pink to bright red flowers. It grows from 30 cm to 1 m and spreads by suckers. Soil ranges from acidic to alkaline light, sandy loam. It occurs in mallee heath in the Big Desert, Vic., continuing across the border into the desert areas of south-eastern S.A. Similar to the Monarto form. Bentham referred to this form as var. *sericea*.

Major distinguishing features Leaves simple, entire with granular upper surface; conflorescence terminal in short, one-sided clusters; torus very oblique; nectary prominent; perianth zygomorphic, sparsely hairy outside, bearded within; ovary silky, stipitate; style hairy just above ovary, otherwise glabrous; style end scarcely expanded; pollen presenter oblique.

Related or confusing species Group 25, especially *G. rogersii* which differs in its narrower perianth (2–3 mm wide), noticeably scabrous leaves and 1–4-flowered conflorescence.

Conservation status Not presently endangered although many populations have been cleared for agriculture.

Cultivation *G. lavandulacea* is a hardy, adaptable plant that has been in cultivation for many years in both private and public gardens especially in S.A. It succeeds in cool-wet, warm-dry and hot-dry climates but dislikes wet, summer-humid conditions, although the Victor Harbor form appears to be more tolerant in this respect. It is therefore wise to grow it in an open position where air circulation is plentiful, allowing for growth up to 1.5 m across. Inland forms tolerate extended dry or drought conditions and are hardy to frosts to at least -7°C. Best results are achieved when planted in full sun in well-drained, acidic to alkaline sand or sandy to gravelly loam. It does not require summer watering or fertiliser but most forms respond to regular tip pruning or annual pruning. In coastal, summer-rainfall climates it may be short-lived and is often affected by white scale although this occasionally affects the species even in the wild. This can cause leaf drop and results in unattractive, leggy plants. Smaller forms make ideal pot plants, creating spectacular displays when planted either singly or contrasting with other species in a medium to large tub.

Propagation Seed Sets many seeds that germinate well if pretreated. Garden hybrids with this species are common. Cutting Most forms strike well from firm, young growth at almost any season. Some inland forms are more difficult to strike, especially from material collected in the wild. Grafting Has been grafted successfully onto *G. robusta*, G. 'Poorinda Royal Mantle' and G. 'Canberra Gem' using top wedge grafts or approach grafts.

Horticultural features *G. lavandulacea* is a brilliant, free-flowering species providing splashes of intense colour in the garden. Its profuse display of flowers massed towards the ends of the branches often conceals the foliage. There is a large range in flower colour, the brilliant deep pink colours being the most eye-catching. Even when not in flower, it is an attractive foliage plant with some grey-leaved forms providing pleasing contrast in the landscape. The impact when in flower can be stunning. Extremely attractive and popular with honey-eating birds. In suitable, dry climates it is a long-lived species suitable for use in landscaping especially where a spectacular feature plant is required or as a low border of massed specimens. Nurseries frequently stock at least one form of this species.

Hybrids *G. lavandulacea* is known to hybridise with *G. alpina* in the wild, especially in the Black Ra. to the W of The Grampians where many beautiful hybrids have been collected and introduced to cultivation. The form sold in nurseries as the Penola form is of garden origin and closely resembles G. 'Poorinda Illumina'. The true Penola form is a scrambling, low shrub and not as bold as the usurper of its name. Known parent of a number of excellent hybrids including G. 'Poorinda Rondeau', 'Poorinda Illumina', 'Poorinda Ruby', 'Poorinda Ensign', 'Poorinda Tranquillity', 'Crosbie Morrison', 'Clearview David', 'Clearview Robyn' and 'Evelyn's Coronet'.

History First introduced to cultivation in Australia at the Melbourne Botanic Garden where it was recorded growing in 1858. Paxton indicated that two forms were cultivation in England in 1850 (as *G. lavendulaca* [sic!] and *G. rosea*).

General comments The taxonomy of *G. lavandulacea* appears to be unresolved and will require careful field and herbarium study to determine any formal division.

Grevillea leiophylla F.Mueller ex G.Bentham (1870) **Plate 192**

Flora Australiensis 5: 471 (England)

The specific epithet is derived from the Greek *leio* (smooth) and *phyllum* (a leaf), an allusion to the smooth leaves. LIE-OH-FILL-A

Type: Glasshouse Mountains, Moreton Bay, Qld, F.Mueller; probably from the same area, L.Leichhardt. (syntypes K, MEL).

Suckering **shrub** with mostly single stems 0.3–1 m high. **Branchlets** angular, slender, glabrous to silky, with prominent glabrous ribs decurrent from leaf bases. **Leaves** 2–5(–11.5) cm long, 1–3 mm wide, sessile, simple, entire, ascending to erect, pliable, straight, linear to narrowly elliptic or obovate, acute, non-pungent; upper surface glabrous, smooth, the midvein and intramarginal veins prominent, smooth; margins shortly but angularly refracted about intramarginal vein; lower surface clearly exposed, glabrous or a few scattered hairs; midvein glabrous; point of attachment scarcely broader than leaf base. **Conflorescence** usually decurved, shortly pedunculate, usually 1-branched, terminal, not extending down main branch; unit conflorescence 1–2 cm long, 1 cm wide, secund, moderately condensed, usually enclosed in foliage; peduncle 0–12 mm long, usually erect, sparsely silky to glabrous; rachis 5–12 mm long, usually recurved, sparsely silky, the hairs mostly white intermixed with scattered brown; bracts 1.5–2 mm long, usually exceeding ovoid buds, narrowly ovate-acuminate with straight apex, tomentose outside, usually falling before anthesis, sometimes persistent. **Flower colour**: perianth dark to pale-pink; style pink. **Flowers** acroscopic; pedicels 5.5–7 mm long, silky to tomentose, markedly expanded at apex; torus 1 mm across, square, straight to slightly oblique; nectary obscure, U-shaped; **perianth** 4–5 mm long, 0.7–0.8 mm wide, undilated, oblong, silky or sparsely so, sometimes with many short erect trichomes intermixed, conspicuously bearded inside in lower half adjacent to upper half of ovary, the hairs c. 0.5 mm long, spreading to slightly reflexed with scattered hairs above, elsewhere glabrous; tepals separating to base before anthesis, curling back to beard at anthesis; limb nodding to revolute, spheroidal to ovoid with depressed apex, silky to villous; **pistil** 7–13 mm long; stipe 1.2–1.5 mm long, glabrous; style angularly refracted about ovary, exserted from lower half on dorsal side, angularly looped out, geniculate 1–2 mm below style end; style end abruptly divergent, sparsely granular or with minute erect trichomes, rarely glabrous; pollen presenter 1–1.3 mm long, 1 mm wide, square to oblong-obovate, oblique, flat with prominent stigma. **Fruit** 9 mm long, 4 mm wide, erect to slightly oblique on swollen, slightly incurved stipe, narrowly ovoid, faintly verrucose; pericarp 0.2–0.4 mm thick. **Seed** not seen.

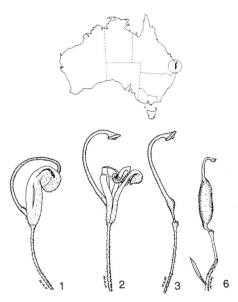

G. leiophylla

Distribution Qld, from the Caloundra–Beerwah area, Glasshouse Mountains, to North Stradbroke Is. *Climate* Summer hot and wet; winter mild, wet or dry. Rainfall 1200 mm.

Ecology Grows in silty sand or gravelly loam in swampy sites, in heath or woodland, rarely on drier sites. Flowers all year with a spring peak. Regeneration after fire is from seed or sucker. Pollination is almost certainly by insects.

Variation Typically, this grows to 30 cm but plants up to 1 m have been recorded. Plants from the Caloundra area have the longest pistils with flowers reminiscent of *G. sericea* but this is likely due to convergence.

Major distinguishing features Leaves simple, entire, smooth, non-pungent, the undersurface exposed and glabrous or a few scattered hairs, the upper surface smooth, the veins smooth, margins shortly recurved; conflorescence terminal, shortly pedunculate, usually 1–3-branched, secund; rachis with mostly appressed indumentum, 5–12 mm long; bracts narrowly ovate-acuminate, visible beyond line of buds; perianth zygomorphic, silky outside, conspicuously bearded within; pistil glabrous except a granular or minutely hairy style end; style sharply and angularly refracted at ovary; fruit faintly verrucose.

Related or confusing species Group 21, especially *G. parviflora* which differs in its narrower leaves, shorter ovarian stipe and pistil, and *G. linearifolia* which differs in its leaves mostly punctate on the upper surface, silky on the lower surface, the intramarginal veins granular, sometimes inconspicuously so.

Conservation status Not presently at risk, though its habitat is continually under threat from development pressures.

Cultivation *G. leiophylla* has been widely cultivated in Qld gardens at least since the 1960s, although some enthusiasts may have grown it earlier. It adapts readily to cultivation where it is both hardy and long-lived, even in frosty conditions (to at least -4°C). More recently it has been grown widely along the east coast of Australia. It flourishes in a moist but well-drained sandy acidic soil in partial shade to full sun and also tolerates heavier loam. It benefits

192A. *G. leiophylla* Cultivated at Rylstone, N.S.W. (N.Marriott)

192B. *G. leiophylla* Cultivated near Brisbane, Qld (P.Olde)

from occasional tip pruning and summer watering, especially when conditions become very dry. Fertiliser is unnecessary. Qld nurseries frequently stock this species as it makes a delightful potted plant in full flower.

Propagation *Seed* Sets reasonable quantities that should germinate if pretreated by nicking or peeling the testa. *Cutting* Strikes readily from cuttings of half-hardened new growth in early spring. *Grafting* Untested.

Horticultural features *G. leiophylla* is a delightful border or rockery plant that suckers sparsely through the garden, without becoming a nuisance. Larger-flowered plants in cultivation are extremely attractive as they bear prominent racemes of pale pink flowers. Its hardiness and longevity suggest possibilities for wider application than in the gardens of enthusiasts as it will also tolerate poorly drained conditions. Some landscapers have used it in commercial landscapes with considerable effect.

General comments McGillivray (1993) referred *G. leiophylla* to synonymy under *G. linearifolia* as form 'b'. While taxonomy and species boundaries in this group are at present being reconsidered, this taxon should continue to be recognised as a distinct species.

Grevillea sp. aff. leiophylla   Plate 193

Synonym: *G. linearifolia* form 'a' of McGillivray (1993).

Suckering **shrub** to 0.3 m high. **Branchlets** angular, slender, glabrous to silky with prominent glabrous ribs decurrent from leaf bases. **Leaves** (4–)6–12(–17) cm long, 1–3 mm wide, sessile, simple, entire, ascending to erect, pliable, straight, linear to narrowly elliptic, acute, non-pungent; upper surface glabrous, smooth, the midvein and intramarginal veins obscure, smooth; margins smoothly to angularly revolute; lower surface usually exposed, silky, sometimes enclosed by margins; midvein glabrous, prominent; point of attachment scarcely broader than leaf base. **Conflorescence** usually decurved, conspicuously pedunculate, usually 1–3-branched, terminal, not extending down main branch; unit conflorescence 1–2 cm long, 1 cm wide, secund, moderately condensed, usually exceeding foliage; peduncle 5–60 mm long, tomentose; rachis 5–12 mm long, usually recurved, tomentose; bracts 1.2 mm long, usually exceeding ovoid buds, narrowly ovate-acuminate with straight apex, tomentose outside, usually falling before anthesis, sometimes persistent. **Flower colour**: perianth dark to pale-pink; style pink. **Flowers** acroscopic; pedicels 5.5–7 mm long, tomentose; torus 0.7 mm across, square, straight to slightly oblique; nectary obscure, U-shaped; **perianth** 4–5 mm long, 0.7–0.8 mm wide, undilated, oblong, villous, conspicuously bearded inside about the middle, adjacent to upper half of ovary, the hairs c. 0.5 mm long, spreading to slightly reflexed with scattered hairs above, elsewhere glabrous; tepals separating to base before anthesis, curling back to beard at anthesis; limb nodding to revolute, spheroidal to ovoid with depressed apex, villous; **pistil** 7–9 mm long; stipe 1 mm long, glabrous; style angularly refracted about ovary and exserted from lower half on dorsal side, angularly looped out, geniculate 1–2 mm below style end; style end abruptly divergent, sparsely granular or with minute erect trichomes, rarely glabrous; pollen presenter 1 mm long, 1 mm wide, square to oblong-obovate, lateral, flat with prominent stigma. **Fruit** 9 mm long, 4 mm wide, erect to slightly oblique on swollen, slightly incurved stipe, narrowly ovoid, faintly warted; pericarp 0.2–0.4 mm thick. **Seed** not seen.

Distribution Qld, from Burrum Point to Tewantin. ***Climate*** Summer hot, wet; winter mild, wet or dry. Rainfall 1600 mm.

Ecology Grows at the margins of wallum in gravelly loam or sandy clay flat in woodland. Flowers all year with a spring peak. After fire regenerates from seed or sucker. Almost certainly pollinated by insects.

Variation A relatively uniform species.

Major distinguishing features Leaves linear, smooth, the venation obscure, the margins smoothly revolute, silky on undersurface; conflorescence conspicuously pedunculate, usually branched; perianth zygomorphic, villous outside, bearded about middle inside; pistil glabrous except minute trichomes on style end; style angularly refracted about ovary.

Related or confusing species Group 21, especially *G. leiophylla* which differs in its glabrous leaf undersurface, silky perianth and usually shorter peduncle and conflorescences less often branched. *G. linearifolia* differs in its shorter leaves with more prominent, granular veins, punctate upper surface, less conspicuously pedunculate conflorescences and silky perianth.

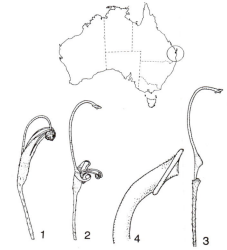
G. sp. aff. *leiophylla*

193. *G.* sp. aff. *leiophylla* Flowers and foliage, S of Maryborough, Qld (M.Hodge)

Conservation status The habitat of this species is continuously under pressure from coastal development.

Cultivation *G.* sp. aff. *leiophylla* has been cultivated by local enthusiasts around Rainbow Beach at least since 1980, although some may have grown it earlier. It adapts readily to cultivation where it is both hardy and long-lived, though its tolerance of frost and extended dry conditions is untested. It should flourish in a well-drained, moist, acidic sandy loam and is ideally suited to many coastal gardens. Heavier loams should also be tolerated. It grows in partial shade naturally. Watering should be unnecessary once established. Regular tip pruning might improve the appearance of this otherwise straggly species.

Propagation *Seed* Sets reasonable quantities that should germinate if pretreated by nicking or peeling the testa. *Cutting* Strikes readily from cuttings of half-hardened new growth in early spring. *Grafting* Untested.

Horticultural features *G.* sp. aff. *leiophylla* is a delightful border or rockery plant that will sucker lightly through the garden. It has some potential because its flowers are very attractive, but its straggly habit does not command attention. Potted plants should do well. One for the collector at present.

General comments McGillivray (1993) referred *G.* sp. aff. *leiophylla* to synonymy under *G. linearifolia* form 'a'. While taxonomy and species boundaries in this group are at present being reconsidered, this taxon is very distinctive and should be recognised as a distinct species.

Grevillea leptobotrys C.F.Meisner (1848)   Plate 194

in J.G.C.Lehmann (ed.), *Plantae Preissianae* 2: 256 (Germany)

Tangled Grevillea

The specific epithet is derived from the Greek *leptos* (fine, slender, weak) and *botrys* (in botany, an inflorescence), in reference to the small, delicate conflorescence.   LEP-TOE-BOT-RISS

Type: Swan River colony, W.A., 1844, J.Drummond coll. 3 no. 268 (lectotype: NY) (McGillivray 1993).

Synonym: *G. leptobotrys* var. *simplicior* F.Mueller (1868).

A low, clumping, dense **shrub**, or a prostrate, mat-forming shrub sometimes lignotuberous, sometimes suckering, 0.1–0.4 m high, 1–3 m wide. **Branchlets** secund, terete, pubescent,

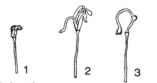

G. leptobotrys

silky or glabrous. **Leaves** 3–29 cm long, ascending to patent, sessile or shortly petiolate, the rachis sometimes flexuose, extremely variable in shape and degree of leaf division, obovate-cuneate in gross outline, rarely simple and entire, sometimes dentate, usually 2- or 3-times pinnatifid to almost pinnatisect; leaf lobes 1–10 mm wide, linear to triangular, acuminate or obtuse, mucronate or pungent; upper surface pubescent, silky,

194A. *G. leptobotrys* Brookton form, Boyagin Rock, W.A. (P.Olde)

194C. *G. leptobotrys* Tutanning form, in cultivation at Stawell, Vic. (N.Marriott)

194B. *G. leptobotrys* Commonly cultivated form (N.Marriott)

194D. *G. leptobotrys* Dryandra form, Dryandra Forest, W.A. (C.Woolcock)

tomentose, sometimes glabrous, midvein an impressed groove; margins usually flat or shortly recurved to revolute; lower surface pubescent, silky to glabrous, midvein prominent. **Conflorescence** erect, pedunculate, terminal, simple or branched; unit conflorescence 3–5 cm long, secund, open, apical flowers opening first; peduncle and rachis silky or glabrous; bracts 0.5–1 mm long, narrowly ovate, silky to glabrous-ciliate, usually persistent to anthesis. **Flower colour**: pink. **Flowers** acroscopic to transverse, often irregularly directed; pedicels 3–7.5 mm long, pubescent, silky or glabrous; torus c. 0.5 m across, straight; nectary obscure; **perianth** c. 2 mm long, 0.5–1 mm wide, cylindrical, strongly curved in young bud, pubescent, silky or glabrous outside, at first separating along dorsal suture, the dorsal tepals then reflexing laterally to reveal inner surface before anthesis, the dorsal tepals aligning with but slightly separate from ventral tepals to form a broad elliptic platform below style; limb revolute, globular or ovoid, the segments lightly ribbed, the lamina white-spotted; **pistil** 4–6 mm long, glabrous; stipe 0.9–1.8 mm long; ovary ellipsoidal; style exposed before anthesis, afterwards strongly retrorse and sigmoid, sharply inflexed c. 0.5 mm from slightly expanded style end; style end not exposed before anthesis; pollen presenter straight to slightly oblique, conical with basal collar. **Fruit** 9–14 mm long, 6 mm wide, erect, obovoid, red, smooth to faintly wrinkled, persistent; pericarp c. 0.5 mm thick. **Seed** 6–7 mm long, 3.5–4 mm wide, 3 mm deep, obovoid; outer face strongly convex, smooth, slightly flanged at rim; lower face with a prominent, convex central section, deeply and intramarginally channelled, the border waxy.

Distribution W.A., widely distributed but not common on the eastern side of the Darling Ra. from the Brookton Hwy to the Gordon R. *Climate* Hot, dry summer; cool, wet winter. Rainfall 600–800 mm.

Ecology Occurs in various soils from sandy loam to clay, always gravelly, usually on lateritic ridges, occasionally in flat areas. It is usually associated with eucalypt woodland or forest. Flowers late spring to summer in the wild, longer in cultivation. Pollinated by insects, probably native bees. Regenerates from lignotuber after fire.

Variation A polymorphic species that varies from population to population especially in leaf and indumentum features. Floral features are relatively uniform. The forms listed below cover some of the variation but further herbarium and field work is needed to resolve the taxonomy of this species.

Tutanning form A most beautiful form that grows in spreading mats. It is distinguished by its stout, pubescent branchlets and broad, grey-pubescent leaves 10–16 cm wide with flexuose rachis, the lamina divided into broad, linear lobes c. 5 mm wide with triangular apices. Flowers and rachis have a silky to pubescent indumentum. It is confined to a few lateritic ridges in and near Tutanning Nature Reserve and is one of the most distinctive forms.

Brookton form Fairly widespread in this area, this form grows in open to dense cushion-like clumps 1–3 m across and has glabrous, sharply angular branchlets, glabrous green leaves divided at the apex into linear lobes with a long basal stalk, the margin revolute almost to the midvein. Flowers and rachis are silky or glabrous. This robust shrub has attractive, red new growth and occurs in the Boyagin Rock to Brookton area. One collection has some simple leaves with a few divided into linear lobes.

Dryandra form A form similar in habit to the Brookton form but differing in the shape of its leaf lobes, which are generally shorter and triangular with a subterete basal stalk. The upper leaf-lobe surface is usually concave and the leaves are an attractive bluish grey.

Williams form A lignotuberous shrub with prostrate, trailing branches and relatively small (bi-)pinnatifid leaves with oblong lobes. This form displays youthful vigour, responding to disturbance or fire but dying back to a few short branchlets in old forest. Some collections from near Cranbrook have a similar habit but differ in their very finely divided leaves that are usually grey-green.

Major distinguishing features Leaves usually much divided; conflorescence open, secund; torus straight; nectary absent; perianth pink, zygomorphic, silky or glabrous outside, papillose inside, the tepals cohering at the limb but becoming everted before anthesis; pistil short, glabrous; ovary stipitate; style retrorse to S-shaped; pollen presenter erect, conical; fruit red, persistent, obovoid.

Related or confusing species Group 8. An unrelated species, *G. cirsiifolia*, has a similar perianth development at anthesis but differs in its yellow flowers, its hairy ovary and flat pollen presenter.

Conservation status 3EC. Confined to small populations scattered over a wide area.

Cultivation An extremely unreliable, touchy species that grows successfully in very few places away from its native habitat, where it is also sometimes short-lived. Some forms (Boyagin, Tutanning) have been cultivated successfully at Burrendong Arboretum in raised rockeries. It is mainly cultivated in Vic. among enthusiasts but at least 2 forms are grown in the U.S.A. Hardy to both extended dry conditions and to frosts to at least -6°C but requires perfect drainage and warm, dry, sheltered conditions, preferably in dappled shade. This can be critical during very hot, dry weather. If planting into a rockery, the soil may need total modification. Best results are achieved by planting on top of a good depth (at least 0.5 m) of almost pure coarse gravel or stones. The soil should be kept slightly moist, never waterlogged or completely dried out. Once established, plants respond to a light annual dressing of slow-release fertiliser. It can be successfully grown in pots and in a large pipe or drum or in a length of hollow log in a coarse, free-draining, neutral to acidic soil. Well-grown plants in tubs make prized specimens and can be moved readily to display flowers. Tubs should be kept fairly dry during the summer (not completely dry), as sudden collapse can even occur in tub-grown plants if allowed to get too wet. Occasionally available from specialist native plant nurseries but may be misidentified as *G. flexuosa*.

Propagation *Seed* Sets small quantities of seed that normally germinate readily when given the standard peeling treatment. Seedlings are extremely prone to damping off. Average germination time 69 days (Kullman). *Cutting* Firm, young growth strikes readily at most seasons. *Grafting* Successfully grafted onto *G. robusta*, *G. shiressii*, G. 'Poorinda Anticipation' and G. 'Poorinda Royal Mantle', mostly using the top wedge or approach technique. Unfortunately there have been many cases of rootstocks (especially *G. robusta*) showing incompatibility after 1 or 2 years, indicated by the leaves becoming blotched or discoloured; branches die and the whole plant succumbs soon after. More recently G. 'Bronze Rambler' has shown promise as a rootstock. An interstock, such as *G. flexuosa* might be worth trying.

Horticultural features *G. leptobotrys* is an amazingly diverse foliage plant with so many different forms that a small garden of them could be planted with none the same. In some forms new growth is a bright red and flowering is prolific, generally from late spring to the end of summer, in dense, delicate racemes of lightly perfumed, pink flowers. It makes a beautiful rockery or specimen plant. Growers should not count on more than three years' life span although a few instances of successful, longer survival have been noted in recent years, possibly due to ideal, local microclimates. A most desirable species unsuitable at present except for the collector, although a reliable rootstock could change that.

General comments Previously misidentified as *G. flexuosa* because, in some forms, it shares a similar, flexuose leaf rachis, but *G. flexuosa* is more closely related to *G. synapheae*.

Grevillea leptopoda D.J.McGillivray (1986) Plate 195

New Names in Grevillea *(Proteaceae)* 8 (Australia)

This species has fine, slender pedicels and the specific epithet is derived from the Greek *leptos* (fine, slender) and *podion* (a foot). LEP-TOE-POE-DA

Type: c. 11 km NW of Carnamah on Geraldton Hwy, W.A., 18 July 1953, R.Melville 4126 & J.Calaby (holotype: NSW).

A spreading, dense, prickly **shrub** c. 1 m tall. **Branchlets** angular, sparsely silky. **Leaves** 4–8 cm long, ascending, petiolate, pinnatisect, divaricately biternate; lobes 1–3.5 cm long, c. 1 mm wide, narrowly linear to subulate, dipleural with hair-lined grooves, rigid, pungent; upper surface channelled beside prominent midvein; margins not evident. **Conflorescence** erect, sessile or pedunculate, crowded, terminal and in upper axils, usually branched; unit conflorescence 4–6 cm long, dense, conico- to oblong-secund; peduncle and rachis silky to glabrous; bracts 1.5–2.2 mm long, obovate, glabrous or sparsely silky, the margins ciliate, falling before anthesis. **Flower colour**: perianth creamy white with pinkish tones on young buds; style white. **Flowers** acroscopic; pedicels 6–8.5 mm long, slender, glabrous; torus c. 0.5 mm across, square, slightly oblique; nectary arc-like, smooth; **perianth** c. 3 mm long, 0.5 mm wide, strongly curled at young bud stage, oblong and undilated, glabrous outside, hairy at base inside; tepals longitudinally ribbed, separating first along dorsal suture, the dorsal tepals then separating below limb, reflexing and exposing inner surface before anthesis, afterwards all free to base and detaching; limb revolute, depressed, spheroidal, the segments ribbed; **pistil** 7.5–9 mm long, glabrous; stipe 1.5–2 mm long; ovary subglobose, slightly compressed; style exserted first through dorsal suture, strongly incurved when free; style end abruptly expanded, not exposed before anthesis; pollen presenter oblique, oblong, convex. **Fruit** 9–10.5 mm long, 4 mm wide, erect on curved pedicel, oblong-ellipsoidal, rugulose; pericarp c. 0.5 mm thick. **Seed** 6.5 mm long, 3 mm wide, obovoid-ellipsoidal, biconvex; outer face dull, minutely tesselated and with a few asperities; inner face with an obovate central section encircled by a membranous waxy film or shallow channel, drawn at both ends into an excurrent wing c. 0.3 mm long.

Distribution W.A., scattered from Kalbarri NP, south to Moora. *Climate* Hot, dry summer; cool, wet winter. Rainfall 250–600 mm.

Ecology Grows in heath or shrubland in lateritic loam or sand. Flowers winter–spring. Pollinated by insects. Regenerates from seed after fire.

Variation A uniform species.

Major distinguishing features Branchlets angular, sparsely silky; leaves divaricately pinnatisect, the

G. leptopoda

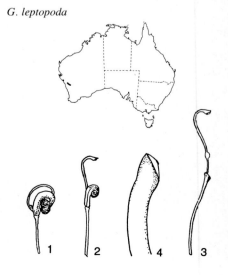

195A. *G. leptopoda* Plant in natural habitat, S of Mullewa, W.A. (N.Marriott)   194B. *G. leptopoda* Flowering branch (P.Olde)

lobes dipleural, pungent; conflorescence secund, 4–6 cm long; nectary obscure; perianth zygomorphic, strongly curled in young bud, glabrous outside, hairy at base within, the tepal segments medially ribbed; pistil glabrous; pollen presenter oblique, convex; fruit oblong, faintly rugose.

Related or confusing species Group 8, especially *G. teretifolia* which differs in its shorter conflorescence (c. 2 cm long), longer pistil (10–17 mm long) and the position of the beard on the inner perianth surface (most prominent in upper half).

Conservation status 3VC. Conserved in Kalbarri NP. Much of its former habitat has been cleared for agriculture.

Cultivation *G. leptopoda* was introduced to cultivation in 1986 and plants are presently growing rapidly in well-drained, neutral, sandy loam and acidic clay-loam in full sun. Plants are hardy to extended dry conditions and to frosts to at least -4°C, but dislike cold-wet or excessively humid conditions which may cause sudden death. Summer watering is deleterious. As it is a naturally compact species, pruning is unnecessary. Generally too prickly for pot culture, but plants grow well in these conditions using an open, well-drained mix and slow-release, low-phosphorus fertiliser. Not presently available from commercial sources but members of the Grevillea Study Group are growing this species.

Propagation *Seed* Untested, although seedlings are common in the wild. Seedlings have broad, pinnatifid leaves. *Cutting* Strikes fairly readily from cuttings using firm, new growth. *Grafting* Has been grafted successfully onto *G. robusta* using whip, top wedge and mummy grafts.

Horticultural features *G. leptopoda* is a spectacular plant with dense, white racemes of flowers smothering the plant and almost completely concealing the foliage. In habit, it is somewhat flat-topped with flowers presented at the top of the plant in a layer. The prickly, deeply divided leaves preclude its use near pathways or access areas but it makes an ideal barrier plant or border around buildings. Its spectacular flowering habit also suggests uses as a feature or accent plant in small gardens or rockeries, especially for inland, dry-climate gardens.

Grevillea leucadendron A.Cunningham ex R.Brown (1830)   **Plate 196**

Supplementum Primum Prodromi Florae Novae Hollandiae exhibens Proteaceas Novas: 25 (England)

Aboriginal: 'bambra' (Kalumburu); 'yanandi', 'nyindilba', 'malba' (Gugadja).

The specific epithet is derived from the Greek *leucon* (white) and *dendron* (a tree), due to the silvery-white appearance. LOO-KA-DEN-DRON

Type: Cambridge Gulf, W.A., Sept. 1819, A.Cunningham (lectotype: BM) (McGillivray 1993).

Synonyms: *G. obliqua* R.Brown (1830) (published with *G. leucadendron*); *G. viscidula* C.A.Gardner (1923).

Slender, open **shrub** or small **tree** 2–3 m high with lignotuber; also suckers. **Branchlets** stout, terete, silky. **Leaves** silvery-white, 15–30 cm long, unifacial, (seedlings bifacial with obvious midvein), ascending, sessile, leathery, ribbed, densely silky or moderately so, venation not evident, most leaves pinnatipartite, sometimes with secondary or tertiary division of the lower lobes, sometimes a few leaves simple; rachis terete; lobes 1–5(12) mm broad, narrowly linear to obovate, the lobes sometimes curved, acute to obtuse; upper edge channelled or fused into a pale stripe. **Conflorescence** erect, paniculate, terminal or in upper axils; unit conflorescence 6–13 cm long, 1–1.5 cm wide, conico-cylindrical, dense; primary peduncle with mixed silky

195C. *G. leptopoda* Close-up of conflorescences (K.Alcock)

G. leucadendron

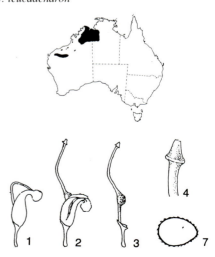

196A. *G. leucadendron* Flowers and foliage (P.Olde)

196B. *G. leucadendron* Flowering habit (A.Smith)

and glandular hairs at base; secondary and tertiary peduncle glabrous, sometimes triangular at base; rachis glabrous; bracts 1.5–5 mm long, narrowly ovate, imbricate in bud, glabrous, the margins glandular-ciliate, some usually persistent. **Flower colour**: white. **Flowers** acroscopic, glabrous; pedicels 1.2–2 mm long, stout; torus 1–1.5 mm wide, oblique, slightly concave; nectary U-shaped, inconspicuous, smooth; **perianth** 3–5 mm long, 1–1.5 mm wide, oblong to cylindrical, straight, papillose inside, first separated along dorsal suture, the dorsal tepals then reflexed and exposing inner surface at or before anthesis, afterwards free to the base and soon detaching; limb strongly recurved, globular; **pistil** 5–7 mm long, glabrous; stipe 1.5–2.5 mm long; ovary globose; slightly bowed out before anthesis, afterwards straight or slightly inflexed; style end slightly expanded; pollen presenter slightly oblique to straight, round, conical. **Fruit** 18–24 mm long, 15–18 mm wide, oblique to almost perpendicular to stipe, compressed-obovoidal to round, glabrous, viscid; pericarp 1–1.5 mm thick. **Seed** 10–13 mm long, 10 mm wide, ± flat, round to elliptic with an apical attenuation, both sides smooth, winged all round.

Distribution W.A. and N.T., widespread from Katherine W to Cambridge Gulf, S to Halls Creek and W almost to the coast. *Climate* Summer hot, wet (rain falling intermittently); warm to hot, dry winter. Rainfall 250–750 mm.

Ecology Usually found in gravelly loam in open eucalypt woodland or grassland. Flowers winter. Reshoots from epicormic buds at base or below ground following low-intensity fire. Pollinated by insects, especially flies, native bees and wasps.

Variation Two forms are recognised. Leaf lobes vary in width; narrow-leaved forms usually have more lobes.

Typical form *G. leucadendron* is extremely widespread in its Typical form, ranging from just W of Katherine almost to the W coast of the Kimberley. It is characterised by its silvery, silky divided leaves with lobes 1–12 mm broad.

Kalumburu form This form has sparsely silky to glabrous pinnatisect leaves with almost filiform lobes (< 1 mm wide). It scarcely differs from other narrow-lobed forms of *G. leucadendron* except in its indumentum and lobe width. The type of *G. viscidula* belongs with this form.

Major distinguishing features Branchlets glandular-silky or glabrous; leaves unifacial, silvery, densely to moderately silky, the rachis terete; most leaves pinnatisect with up to three orders of division, leaf lobes elongate, narrowly linear to narrowly linear-obovate, mostly 1–12 mm broad; conflorescence paniculate; primary peduncle usually with glandular hairs at base; secondary peduncle with a dark, triangular base; perianth zygomorphic, glabrous outside, papillose within; pistil glabrous; pollen presenter conical; fruit viscid.

Related or confusing species Group 3, especially *G. erythroclada*, *G. pyramidalis* and *G. mimosoides*, all of which have ± glabrous leaves.

Conservation status Not presently endangered.

Cultivation *G. leucadendron* has not been cultivated widely to date but has succeeded as far south as Brisbane where it has grown reliably in a few gardens for several years. Tropical or sub-tropical climates with a defined dry season would be preferred for this species. It should tolerate drought but not frost or extended cold weather. An open, sunny position in well-drained, sandy soil or gravelly loam should suit it best and it should respond to limited summer watering. Tip pruning may make the shrub more shapely. In the wild, regular burning causes it to shoot from below the ground, forming several branches but its natural tendency is to form a single-stemmed shrub with foliage at the branchlet ends. It is not usually sold in nurseries but seed may be obtained from commercial suppliers.

Propagation *Seed* Fresh seed should germinate if placed in hot water and soaked overnight. In cold areas, seedlings are subject to damping off and should be germinated only in the warmer months. Seedlings are large and quite robust. *Cutting* The thick, woody stems have proved impossible to strike, but young sucker growth is usually thinner and may give positive results. *Grafting* The same problems as for cuttings confront the grafter, hence there has been little success with the species. One of the authors used seedlings to produce a strong cotyledon graft onto *G. robusta*. These are growing well in a frost-free site in western Vic.

Horticultural features *G. leucadendron* is a beautiful, open silvery plant with a massed display of prominent, branched conflorescences in winter. Its habit of producing a dense mop of leaves atop long, leafless stems gives the impression of a small silver-leaved palm. Many insects are attracted to the flowers and the species would be useful in landscapes as either a contrast or feature plant. Potential growers should beware the caustic fruits which can cause serious, painful blistering. Aborigines used the caustic fruits to make tribal marks. Gardner (1923: 44) referred to the caustic exudate being used to prevent healing of tribal marks.

General comments Until further studies proceed in this group of related species (*G. mimosoides*, *G. pyramidalis*, *G. leucadendron*), *G. leucadendron* is here held to be separate from *G. pyramidalis* principally on the basis of its leaf morphology, persistent indumentum, rachis shape and to some extent lobe width which is consistent in this taxon over more than 500 km. Over most of its distribution, leaf lobes are relatively narrow (1–5 mm wide) and densely silky. *G. pyramidalis* is here restricted to plants with broadly lobed, ± glabrous, green leaves (leaf lobes 5–20 mm wide). Floral morphology does not appear to be a basis for separation.

Grevillea leucoclada D.J.McGillivray (1986) Plate 197

New Names in Grevillea *(Proteaceae)* 8 (Australia)

The specific epithet alludes to the whitish branchlets and is derived from the Greek *leucon* (white) and *clados* (a branch). LOO-KO-CLADE-A

Type: c. 23.5 km from North West Coastal Hwy on road to Kalbarri, W.A., D.J.McGillivray 3348 & A.S.George, 13 June 1976 (holotype: NSW).

Erect to spreading, mid-dense **shrub** 1–2 m tall. **Branchlets** terete, glabrous or sprinkled with appressed hairs, glaucous. **Leaves** 6–14 cm long, ascending, petiolate, pinnatisect, divaricately tripartite to biternate; lobes 2–6 cm long, 1–2 mm wide, narrowly linear, straight, rigid, pungent; upper surface glabrous or sometimes a few hairs evident, ribbed, channelled beside prominent midvein; margins angularly revolute; lower surface bisulcate, lamina not visible, lying in grooves, midvein prominent. **Conflorescence** erect, pedunculate, terminal or in upper axils, branched, rarely simple; unit conflorescence 4–6 cm long, 2 cm wide, conico-cylindrical, dense to open; peduncle and rachises sprinkled with few silky hairs or glabrous; bracts 0.6–0.9 mm long, ovate, glabrous with ciliate margins, falling before anthesis. **Flower colour**: white. **Flowers** mostly acroscopic, sometimes oblique to rachis; pedicels 4–6 mm long, glabrous; torus c. 0.5 mm across, straight to slightly oblique, square; nectary V-shaped, scarcely evident; **perianth** 1.5–2 mm long, 0.5 mm wide, strongly curled in bud, glabrous outside, prominently bearded inside near base, soon falling; tepals ribbed, opening first along dorsal suture, the dorsal tepals reflexing laterally and separating from ventral tepals before anthesis, afterwards all free to base and loose; limb revolute, spheroidal, the segments faintly to prominently ribbed, enclosing style end before anthesis; **pistil** 5.5–6 mm long, glabrous; stipe 0.6–0.9 mm long; ovary ellipsoidal; style first exserted near curve on dorsal side and looped upwards, straight to slightly incurved after anthesis, dilated abruptly into a flanged style end; pollen presenter erect, conical to truncate-conical with an oblique, flanged base. **Fruit** 10.5–13.5 mm long, 6–7.5 mm wide, erect, oblong/ellipsoidal with slightly upturned

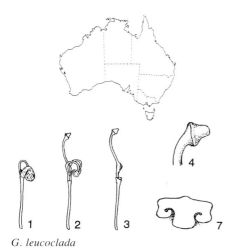

G. leucoclada

apex, rugulose; pericarp c. 1 mm thick. **Seed** 9 mm long, 3.5 mm wide, biconvex, ellipsoidal with a slightly lipped margin; outer face smooth; inner face with a central oblong section encircled by a smooth membranous border.

Distribution W.A., in the Kalbarri area. *Climate* Hot, dry summer; mild to cool, wet winter. Rainfall 300–450 mm.

Ecology Grows on clifftops, open plains, in ravines and on rocky hillsides, sometimes beside the river, usually in sandy or gravelly soil. Flowers late winter–spring. Probably pollinated by insects. Regenerates from seed.

Variation A species with little variation.

Major distinguishing features Branchlets glaucous; leaves rigid, divaricately divided into pungent, narrowly linear lobes with channelled upper surface; conflorescence conico-cylindrical; torus slightly oblique; nectary evident; perianth strongly curled, glabrous outside, bearded within, the tepal-limbs with scarcely prominent medial ribs; pistil glabrous; ovary shortly stipitate; pollen presenter oblique, conical; fruit oblong-ellipsoidal, faintly rugose.

Related or confusing species Group 8, especially *G. intricata* which differs in its non-glaucous branchlets, its non-rigid, pliable leaf lobes and its perianth glabrous within.

Conservation status 2RC. Although the distribution is small, most known populations are reserved in Kalbarri NP.

Cultivation *G. leucoclada* grows readily and rapidly in cultivation, preferring an open, sunny site in well-drained, acid to neutral sand or gravelly loam. It has been grown for many years at Kings Park, Perth, (as *G. intricata*) and more recently in northern Vic., N.S.W. and Qld. It is a relatively hardy species, withstanding hot, dry summers and frosts to at least -4°C. It has a rather untidy habit that would be improved by pruning. Summer watering is unnecessary and may be fatal. Plants are intolerant of poor drainage and succumb in cold, wet or very humid conditions unless grafted. It makes an excellent pot plant for a number of years using a well-drained mix with annual applications of slow-release fertiliser.

Propagation *Seed* Germinates readily especially if pre-treated. Seedlings have large, bipinnatifid leaves. *Cutting* Grows readily from cuttings of firm, new growth during spring or early autumn. *Grafting* Has been grafted successfully onto *G. robusta* using top wedge, whip, whip-and-tongue and mummy grafts.

Horticultural features During the flowering season, *G. leucoclada* is an extremely beautiful, free-flowering plant boasting a profusion of large white conflorescences, so crowded that the foliage is almost obscured. The prickly, bluish-green foliage with its long, stiff and wide-spreading lobes is interesting and contrasts attractively with the white, glaucous branchlets. The flowers have a sweet, honey scent, especially noticeable during the middle of the day, and attract many flying insects including butterflies. In the home garden, its flowering habit suggests a prominent, feature position or it could be considered as an accent or contrast plant. In large landscapes, massed plantings would be spectacular along a boundary or beside north- or west-facing buildings. The prickly foliage prevents its use near pathways or access areas. In cultivation, plants tend to grow larger (to 2 m), particularly when grown in sandy soil.

Grevillea leucopteris C.F.Meisner (1855)
Plate 198

Hooker's Journal of Botany and Kew Garden Miscellany 7: 76 (England)

Old Socks, White Plume Grevillea

The specific epithet alludes to the whitish wing surrounding the seed and is derived from the Greek *leucon* (white) and *pteron* (a wing). LOO-KOP-TER-ISS

Type: N of Swan R., W.A., 1850–51, J.Drummond coll. 6 no. 188 (holotype: NY).

Synonym: *G. segmentosa* F.Mueller (1863).

Dense, rounded **shrub** to 3 m high, 2.5–3 m wide, with emergent, arching, floral branches to 5 m. **Branchlets** terete, hoary-pubescent. **Leaves** 12–35 cm long, ascending, subsessile, broadly winged at base, leathery, coarsely pinnatipartite, or simple on floral branches; leaf lobes 10–20, 10–20 cm long, 1–10 mm wide, linear, acuminate, flexible, the seedling lobes tending to be broader; upper surface pubescent becoming ± glabrous, venation obscure; margins revolute, sometimes to midvein; lower surface bisulcate or lamina slightly exposed, pubescent, midvein prominent. **Conflorescence** erect, long-pedunculate, terminal, many-branched; unit conflorescence 10–15 cm long, 5–8 cm wide, cylindrical, dense; peduncle

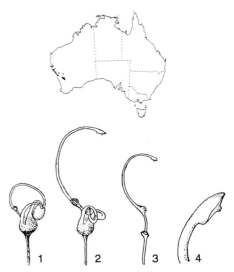

G. leucopteris

197A. *G. leucoclada* Flowering habit (P. Olde)

197B. *G. leucoclada* Close-up of conflorescences (P.Olde)

198A. *G. leucopteris* Flowering branch (M.Hodge)

198B. *G. leucopteris* Plant in natural habitat, Kalbarri NP, W.A. (P.Olde)

glandular-pubescent; rachis glabrous; conflorescence bracts large, leafy, conspicuous in bud, ovate; floral bracts 6.5–11 mm long, trullate, pink, glandular-pubescent, falling before anthesis. **Flower colour**: creamy white. **Flowers** acroscopic; pedicels 4–10 mm long, glabrous; torus c. 2 mm across, straight to slightly oblique; nectary annular, entire; **perianth** 5–7 mm long, 2–3 mm wide, cylindrical to narrowly ovoid, contracted at curve, slightly dilated at base, glabrous outside, bearded inside below level of ovary, at first separating along dorsal suture, the dorsal tepals reflexing laterally and exposing inner surface before anthesis, afterwards all free and soon detached; limb conspicuous, revolute, spheroidal; **pistil** 25–33 mm long, glabrous; stipe 4.5–7 mm long, slender, terete, centrally inserted; ovary obovoid; style refracted above ovary and first exserted below curve on dorsal side, looped upwards before anthesis, afterwards strongly incurved to C-shaped, flattened, stout, becoming slightly dilated below enlarged style end; pollen presenter oblique, obovate, broadly conical, flanged. **Fruit** 20–24 mm long, 14–15 mm wide, persistent, erect to very oblique to stipe, ellipsoidal, smooth, shiny; pericarp 1–1.5 mm thick.

198C. *G. leucopteris* Pink-bracted buds (N.Marriott)

Seed 9–11 mm long, 4.5 mm wide, grey, compressed-ellipsoidal, shiny, smooth or with occasional irregular wrinkles, broadly winged all round; outer face slightly convex, faintly lipped at margin, smooth; inner face similar but flatter with a faint submarginal channel.

Distribution W.A., widespread between the lower Murchison R. and Marchagee. *Climate* Hot, dry summer; cool to mild, wet winter. Rainfall 430–600 mm.

Ecology A pioneer species that grows in sand or sandy loam, sometimes over limestone, in open heath and tall shrubland. It quickly colonises disturbed sites in its natural habitat. Flowers spring–summer. Pollinator uncertain, possibly small mammals such as the honey possum (*Tarsipes rostratus*) or large insects attracted at night to the strong floral odour and associated nectar flow. Killed by fire but regenerates readily from seed with potential as a weed away from its natural habitat. Capable of selfing (see also Lamont 1982).

Variation A morphologically uniform species.

Major distinguishing features Shrub with emergent floral branches; leaves with winged base, pinnatipartite except floral leaves; conflorescence cylindrical, much-branched; peduncle glandular-hairy; rachis glabrous; torus slightly oblique; nectary ring-like; perianth zygomorphic, glabrous outside, hairy at base within; pistil glabrous; ovary long stipitate; style refracted above ovary, strongly incurled; pollen presenter broadly conical; fruit compressed: seed winged all round.

Related or confusing species Group 3. *G. annulifera* appears superficially similar but is glabrous and differs also in its fruit, seed and divaricately divided, pungent leaves.

Conservation status Not presently endangered; a coloniser of disturbed road verges and burnt areas.

Cultivation *G. leucopteris* has been widely and successfully cultivated in many areas of Australia but does best in subtropical to mediterranean climates in both high-rainfall coastal and drier, inland situations, where it is hardy to extended dry conditions and frost to at least -6°C. For best results, it should be planted in a warm, full-sun position in deep and well-drained sand, sandy loam or gravelly loam and given plenty of room to grow without crowding lest it become spindly and unattractive. There is a tendency to wind-throw in unprotected situations and staking may be necessary. It does not require fertiliser unless in hungry, sandy soil, and then only as an annual top-dressing. It will succeed in alkaline soils when drainage is good and the topsoil is slightly acidic. Once established, supplementary summer watering is unnecessary. The floral branches can be pruned off to improve appearance but it should be remembered that flowers are produced annually from the same canes. Seed is available from specialist suppliers and plants can be purchased at some native plant nurseries.

Propagation *Seed* Sets copious seed over a long period; sometimes germinates erratically even when given standard peeling or soaking pretreatments. Average germination time 30 days (Kullman). *Cutting* Seldom propagated by this technique due to the large leaves. Firm, young growth generally strikes well in the warmer months. *Grafting* Successfully grafted onto *G. banksii*, *G. shiressii* and *G. robusta* but there are signs of incompatibility as many plants snap at the union. More testing is required, however, as many have also grown strongly and the problem may be due to technique rather than any incompatibility.

Horticultural features *G. leucopteris* is a large, quick-growing species that makes a spectacular show in full flower with abundant, leafless spikes of cream flowers waving above the grey-green foliage. The beautiful, soft-pink developing conflorescences are borne for many months and give the impression of an extended flowering period. The seed pods are also attractive and retained for a long period. The whole plant, reminiscent in its habit of John Wyndham's triffids, has an attractive, dense form and can be used as a low screen plant or as a spectacular feature plant. The flowers have an strong perfume, foetid to some though pleasant to others, which can permeate the surroundings for long distances on still nights and for this reason the plantings should not be too close to the house or outdoor recreation areas. During the day this odour is less noticeable and in fact becomes an attractive, spicy smell. Apart from this drawback, *G. leucopteris* could be considered ideal for landscaping in parklands and large public areas.

Grevillea levis P.M.Olde & N.R.Marriott (1994)

Plate 199

The Grevillea Book 1: 175 (Australia)

The specific epithet is derived from the Latin *levis*, smooth, and refers to fruit surface. LEE-VISS

Type: Near Mt Churchman on track from Bimbijy Road, W.A., 23 Sept. 1991, P.M.Olde 91/188 (holotype: NSW).

Synonym: *G. paniculata* Forms 'g' & 'k' of McGillivray (1993).

Dense, intricate, suckering, mostly glabrous **shrub** 1–2 m high. **Branchlets** terete, glabrous, sometimes slightly glaucous. **Leaves** 2–3(–6) cm long, ascending to spreading, crowded, petiolate (the distance from point of attachment to first (lateral) lobes 0.7–1.5 cm), mostly divaricately biternate to triternate, sometimes tripartite or with the lateral lobes bi- or tripartite; primary leaf lobes 3 (rarely 5); ultimate lobes 0.5–2(–4) cm long, 0.3–0.5(–0.8) mm wide, linear-subulate with a lateral groove along each side, (trigonous in cross-section), rigid to slightly flexible, pungent, occasionally secondary lateral lobes bi- or tripartite, the lobes c. same length as petiole; upper surface glabrous, midvein prominent and channelled on either side, edge veins also prominent; margin angularly revolute; lower surface bisulcate, the grooves glabrous, the lamina enclosed by margins, midvein very prominent. **Conflorescence** ascending, pedunculate, usually axillary, sometimes terminal, simple or 2- or 3-branched, usually c. equal with or slightly enclosed within the leaves at anthesis; unit conflorescence 1–2 cm long, 1.5–2 cm wide, open, subglobose to ovoid; development acropetal; peduncle glabrous glaucous, sometimes loosely silky; rachis glabrous; bracts 0.5 mm long, 0.5 mm wide, ovate, glabrous with ciliate margins, falling very soon after bud elongation. **Flower colour**: perianth and style white or creamy white, frequently with pink tinges on the tepals. **Flowers** regular, ascending, glabrous; ovary adaxially oriented; basal pedicels 5–7 mm long, filamentous, much longer than perianth; torus c. 0.5 mm across, straight, square with prominently lobed corners; nectary prominent, erect, fan-shaped to U-shaped or arc-like, the margin uneven; **perianth** 3.5–4 mm long, 0.8 mm wide at base, oblong to obovoid below limb, ellipsoidal just before anthesis; tepals strap-like with medial rib, narrowed above base and just below limb, separating below limb and bowed out before anthesis, free to base and rolled down after anthesis, soon falling; limb 1.2 mm long, 1.2 mm wide, erect, subglobose, the segments slightly flanged at margin with faint midrib; **pistil** 3.5 mm long; stipe 1.5 mm long, straight to flexuose; ovary globose, smooth; style erect, constricted above ovary, the zone of constriction incurved, 0.05–0.2 mm long, dilated above, the dilation ovoid, c. 0.5 mm wide, c. same width as to slightly narrower than ovary, narrowed below expanded style end; style end 0.6–0.8 mm wide, straight; pollen presenter 0.8 mm long, conical with narrow basal rim. **Fruit** 6–9 mm long, 3–5 mm wide, ± horizontal to stipe; oblong-ellipsoidal to obovoidal, the base retrorse, smooth or obscurely colliculate; pericarp 0.5–0.7 mm thick. **Seed** 7 mm long, 2.8 mm wide, ellipsoidal, compressed at margin, biconvex; outer face mottled, longitudinally wrinkled; inner face partially overlain by a translucent membrane, ridged all round near outer edge.

Major distinguishing features Branchlets glabrous; leaves divaricately biternate or triternate, prominently stalked below first lobes; leaf lobes linear-subulate, < 1 mm wide, the undersurface not exposed; conflorescence enclosed within foliage; rachis glabrous; floral bracts < 1 mm long, glabrous, falling early; flowers regular, glabrous; style constricted above ovary, again dilated; pollen presenter conical; fruit oblong-ellipsoidal, smooth.

Related or confusing species Group 1 but especially *G. paniculata* which differs in its deeply wrinkled fruit and mostly less divided leaves with longer leaf lobes and mostly longer pedicels. *G. xiphoidea* differs in its more robust single-stemmed habit, broader leaf lobes with the undersurface usually partly exposed and shorter pedicels.

Distribution W.A., widespread in the areas Ajana, Murchison R., Lake Monger, Dalwallinu, Wubin, Morawa, Three Springs, Coorow, Mullewa, Mt Churchman, Bullfinch, Coolgardie. *Climate* Mild to hot, dry summer; cool, wet winter. Rainfall 200–400 mm.

Ecology Grows in sandy gravelly loam, often in depressions, in open heath or shrubland, occasionally in sandy loam around granite outcrops. Flowers late autumn–early spring. Regenerates from seed or sucker. Pollination is probably by insect.

Variation *G. levis* is a relatively uniform species over most of its range. Populations from the south-eastern part of its range, from Bullfinch to near Southern Cross, tend to have slightly coarser leaf lobes. A number of collections in the Wongan Hills area with tripartite leaves, very long lobes (to 4 cm long) and small, ellipsoidal fruit may warrant separation as a distinct species.

Conservation status Not presently endangered.

Cultivation Most plants grown as *G. paniculata* are in fact *G. levis* and have been grown successfully over a wide area in eastern Australia, mainly by enthusiasts. It is a very hardy, adaptable plant that grows readily in cold-wet or hot-dry climates and even succeeds in summer-rainfall climates. It is hardy to extended frosts to at least -6°C and grows well in most soil conditions that have reasonable drainage. It grows best in sunny to semi-shaded sites in free-draining, acidic to neutral sand, loam or gravelly clay. Fertiliser is not needed even when grown in hungry, open sand. Pruning may be required to maintain shape and density, especially when planted in semi-shade. Summer watering is not required once established. Native nurseries sometimes stock this species, usually as *G. paniculata*.

Propagation Seed Sets copious quantities of seed that germinate readily if pretreated by nicking the testa. *Cutting* Strikes well using firm, young growth of cultivated plants. Material from the wild can be difficult especially if taken while the species is in flower. Best propagated during autumn using fairly high hormone levels (8,000–16,000 ppm IBA powder). *Grafting* Untested.

Horticultural features Despite the prickly foliage, *G. levis* is a beautiful garden plant that smothers itself in white from mid-winter to mid-spring. A spectacular, pink-flowered form of unknown origin

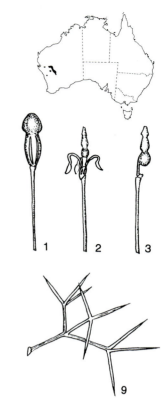

G. levis

199A. *G. levis* Pink-flowered variant (N.Marriott)

199B. *G. levis* Habit of plant, near Mt Churchman, W.A. (P.Olde)

199C. *G. levis* Flowers, foliage and fruit (P.Olde)

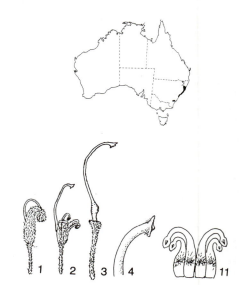

G. linearifolia

has been widely sold in Vic. It has some potential in horticulture as a feature or low-barrier plant but is grown mainly by enthusiasts. It should be planted away from pathways or in areas that have little traffic. The flowers usually have a sweet, spicy or vanilla perfume that is best appreciated about the middle of the day.

Grevillea linearifolia (A.Cavanilles) G.Druce (1917)   Plate 200

The Botanical Exchange Club and Society of the British Isles Report for 1916 4 (5), Suppl. 2: 625 (England)

Based on *Embothrium linearifolium* A.Cavanilles, *Icones et Descriptiones Plantarum* 4: 59, t. 386 f. 1 (1798, Spain).

Linear-leaf Grevillea

The specific epithet is derived from the Latin *linearis* (linear) and *folium* (a leaf) in reference to the leaves of the type collection. LINN-EE-ARR-IF-OL-EE-A

Type: near Port Jackson, N.S.W., April, 1793, L.Née (MA).

Synonyms: *Embothrium sericeum* var. *angustifolium* J.Smith (1794); *Embothrium lineare* J.Kennedy (1803); *Lysanthe lineariifolia* J.Knight (1809); *Grevillea linearis* R.Brown (1810); *G. linearis* var. *alba* W.Page (1818) nom. nud.; *G. linearis* var. *rubra* W.Page (1818) nom. nud.; *G. linearis* var. *alba* C.Loddiges (1824); *G. linearis* var. *incarnata* J.Sims (1826).

A robust, single-stemmed **shrub** to 3.5 m with open foliage, or a suckering **shrub** prostrate to 1 m high with crowded foliage. **Branchlets** angular, stout, silky with prominent glabrous ribs decurrent from leaf bases. **Leaves** 1–9 (–11.5) cm long, 1.5–5 mm wide, sessile, ascending to erect, pliable, straight to wavy, simple, entire, linear to narrowly elliptic or obovate, acute, non-pungent; upper surface glabrous or sparsely silky, usually punctate, sometimes smooth, the midvein and intramarginal veins prominent, prominently to faintly granular; margins angularly refracted about intramarginal veins, sometimes loosely; lower surface usually clearly exposed, silky, the midvein glabrous, the point of attachment scarcely broader than leaf base. **Conflorescence** usually decurved, pedunculate, usually 1-branched, terminal on long branchlet, not extending down main branch; unit conflorescence 1–2 cm long, 1 cm wide, secund, moderately condensed, usually extending well beyond foliage; peduncle 1–20 mm long, usually erect, sometimes refracted at base, silky-tomentose to sparsely silky or glabrous; rachis 5–12 mm long, usually recurved, subvillous to sparsely silky, the hairs mostly white intermixed with scattered brown; bracts 1.5–2 mm long, usually exceeding ovoid buds, narrowly ovate-acuminate with straight apex, tomentose outside, usually falling before anthesis, sometimes persistent. **Flower colour**: perianth white to pale pink; style white or pink. **Flowers** acroscopic; pedicels 5.5–7 mm long, silky to tomentose, markedly expanded at apex; torus 0.7–1 mm across, square, straight to slightly oblique; nectary prominent, U-shaped; **perianth** 4–5 mm long, 0.7–0.8 mm wide, undilated, oblong, white-villous to sparsely silky outside, sometimes intermixed with many short erect hairs, conspicuously bearded inside about the middle, adjacent to ovary, the hairs c. 0.5–1 mm long, spreading to slightly reflexed with scattered long hairs above, elsewhere glabrous; tepals separating to base before anthesis, curling back to beard at anthesis; limb nodding to revolute, spheroidal to ovoid with depressed apex, silky to villous, the hairs mostly brown; pistil 7.5–13 mm long; stipe 1.2–2.5 mm long, glabrous; ovary narrowly ovoid; style angularly refracted about ovary and exserted from lower half on dorsal side, angularly looped out, geniculate 1–2 mm below style end; style end abruptly divergent, sparsely granular or with minute erect trichomes, rarely glabrous; pollen presenter 1–1.3 mm long, 1 mm wide, square to oblong-obovate, oblique, flat with prominent stigma. **Fruit** 9 mm long, 4 mm wide, erect to slightly oblique on swollen, slightly incurved stipe, narrowly ovoid, faintly warted; pericarp 0.2–0.4 mm thick. **Seed** not seen.

Distribution N.S.W., from Sydney to Lismore. Queensland in the Darling Downs district. *Climate* Summer hot, wet; winter mild, wet or dry. Rainfall 1200 mm.

Ecology Grows in sandy loam with sandstone in eucalypt woodland, sometimes in low, wet heath. Flowers most of the year except winter. Regenerates after fire from seed or sucker. Almost certainly pollinated by insects. Introduced honey bees (*Apis mellifera*) regularly attend the flowers. Native pollinators unknown.

Variation Although the species has been more narrowly defined in this work, it remains variable and further separations are likely. A number of collections are unassigned and need further field work and investigation.

Type form (Form 'e' of McGillivray) Gosford area south to the Parramatta R. with an early collection from Watson Bay. This is a robust single-stemmed shrub to c. 3 m with open foliage. Leaves are undulate, linear to narrowly elliptic, 3.5–9 cm long, 2–4 mm wide, with most of the undersurface exposed; margins refracted; peduncle 1–20 mm long; flowers pink or white; pistil 7–12.5 mm long; fruit 13–15 mm long. Plants from the Mosman–Balgowlah area have the most robust flowers and may be hybridising with *G. sericea*. Plants from the Epping area have the smallest flowers. Plants from the Nowra area, included in form 'e' by McGillivray, are excluded here. They have not been assigned pending field work. They have much

200A. *G. linearifolia* Torrington form in natural habitat (P.Olde)   200B. *G. linearifolia* Typical form (P.Brady)

200C. *G. linearifolia* Angourie form (N.Marriott)

narrower, leathery leaves with the undersurface mostly enclosed.

Form 'c' of McGillivray Plants of this form sucker and rarely grow more than 1 m high. Fruits are generally more noticeably warted and rarely exceed 10 mm long. There appear to be at least four variants in this broadly circumscribed form:

1. *Coastal form* (N.S.W.) Occurs from near Gosford to Lismore in near-coastal heaths. Generally a low shrub with shorter, narrower leaves than the type form (1.5–5.5 cm long, 1.2–4 mm wide). Flowers pink to white.

2. *Torrington form* (N.S.W.) This form has yellowish flowers and greyish, strongly ascending leaves 4–6 cm long, and fruits prominently warted. Pistil 8–9 mm long; stipe 0.5–0.7 mm long. It occurs in granitic sand in heath.

3. *Darling Downs form* (Qld) Collected near Stanthorpe in Girraween NP. Leaves are somewhat shorter than the Torrington form. It also occurs in granitic sand. Pistil 7–7.5 mm long, stipe 0.5 mm long.

4. *Angourie area form* (N.S.W.) This is very variable. Prostrate plants with short, elliptic leaves have been collected from a few headlands, notably Shelley Beach. Generally it is a low but erect plant, difficult to distinguish from the Torrington form, but the leaves a little shorter and wider. A single collection with leaves resembling those of *G. speciosa* (c. 1 cm wide, 2.5 cm long) was made by the authors at Angourie. Flowers white.

Major distinguishing features Leaves simple, entire, ascending to erect, linear-elliptic to narrowly obovate, scarcely to not pungent, the upper surface usually punctate, the longitudinal veins granular; the undersurface exposed, silky; conflorescence terminal, pedunculate, usually branched, secund; rachis 5–12 mm long, the indumentum mostly appressed; bracts narrowly ovate, acuminate, visible beyond line of buds; perianth zygomorphic, hairy outside, bearded about middle within; pistil glabrous except a granular or minutely hairy style end; fruit warted.

Related or confusing species Group 21, especially *G. australis*, *G. leiophylla*, *G. parviflora* and *G. sericea*. *G. leiophylla* differs in its glabrous, smooth leaves. *G. parviflora* differs in its narrowly, linear leaves and its shorter pistil. *G. australis* has a glabrous inner perianth surface, smoothly rounded leaf margins, and the midvein on the undersurface obscure or densely hairy. *G. sericea* has a longer pistil and usually a more extensive indumentum on the upper style.

Cultivation *G. linearifolia* has been cultivated since the colonisation of Australia, both overseas and locally. It is extremely adaptable and hardy, tolerant of heavy frosts, sometimes even sea spray, extended dry periods but possibly not drought. It is suited both to well-drained and poorly drained moist, sandy or heavy loam in full sun or partial shade. Pruning will make plants more compact. Nurseries in the Sydney region sometimes carry this species.

Propagation Seed Sets reasonable quantities that should germinate if pretreated by nicking or peeling the testa. *Cutting* Strikes readily from cuttings of half-hardened new growth in early spring. *Grafting* Untested.

Horticultural features In its typical form, *G. linearifolia* is a graceful shrub with long weeping branchlets, the white flowers abundant and delicately veiling the green foliage. Its robustness has led to its frequent usage in both public and home garden landscapes. The prostrate habit of the Shelley Beach form is genetically fixed and it makes a delightful rockery plant, the bright white flowers blending with the leaves which have a prominent white undersurface. All forms are fairly long-lived and most are well suited to far wider use in horticulture, particularly in temperate regions. They are seldom troubled by pests and when in flower are popular with bees and other insects.

History Seed was sent to Lee & Kennedy, London, by William Paterson in 1790 and it was propagated successfully. It was cultivated widely throughout Europe for many years into the early part of the 19th century. It was cultivated at Sydney and Melbourne Botanical Gardens in 1857.

General comments *G. linearifolia*, as treated here, remains a variable species and awaits a comprehensive field and herbarium survey. Some elements included here may be separated. Other forms defined by McGillivray (1993) have been included in our treatment of *G. alpivaga*, *G. halmaturina*, *G. leiophylla*, *G. micrantha*, *G. neurophylla*, *G. parviflora* and *G. patulifolia* or as related unnamed taxa.

Grevillea linsmithii D.J.McGillivray
(1986) **Plate 201**

New Names in Grevillea *(Proteaceae)* 8 (Australia)

Named in honour of Lindsay Stuart Smith (1918–1970), highly regarded botanist who worked at the Qld Herbarium, and who was himself fond of using compound nominal epithets (McGillivray 1993). LINN-SMITH-EE-EYE

Type: Mt Greville, 20 km SW of Boonah, Qld, 27 Sept. 1973, I.R.H.Telford 3226 (holotype: BRI).

Open to mid-dense **shrub** 0.5–3 m high with spreading, drooping branches. **Branchlets** terete, pale-brown villous. **Leaves** 4–9 cm long, 4–10 mm wide, ascending, shortly petiolate, pliable, simple, oblong to obovate, obtuse, mucronate; upper surface villous to almost glabrous, midvein, lateral and intramarginal veins evident; margins entire, undulate, flat or shortly recurved to loosely revolute; lower surface pale-brown villous, midvein prominent. **Conflorescence** decurved on slender, pubescent peduncle, simple or few-branched, usually terminal, sometimes axillary; unit conflorescence loose, 2–4(6) flowered; peduncle and rachises villous; bracts c. 0.5 mm long, linear, densely villous, glabrous inside, falling before anthesis. **Flower colour**: perianth green with an orange-pink to red limb; style greyish pink. **Flowers** acroscopic; pedicels 2.5–4 mm long, tomentose; torus 1.5 mm across, oblique, square; nectary broadly U-shaped, spreading; **perianth** 8–11 mm long, 2 mm wide, oblong-ovoid, dilated at base, villous outside, bearded inside about and above ovary to the curve, cohering except along dorsal suture before anthesis, afterwards free to curve, the dorsal and ventral tepals rolling back as independent pairs, the ventral tepals strongly and further curled past curve, all detaching soon after anthesis; limb nodding to revolute, globular, tomentose-villous; **pistil** 10–16 mm long; stipe 0.7–1.2 mm long, glabrous; ovary ellipsoidal, scarcely wider than style, glabrous; style sparsely pubescent, slightly bowed out from lower half on dorsal side before anthesis, afterwards gently incurved to straight, somewhat flattened, ridged and channelled along dorsal side; style end conspicuous, flat, glabrous; pollen presenter lateral, obovate, slightly convex with surrounding flange. **Fruit** 10–18 mm long, 4–6 mm wide, oblique, subcylindrical, glabrous, faintly ridged; pericarp c. 0.5 mm thick. **Seed** 11 mm long, 3.5 mm wide, obovate-ellipsoidal with short apical wing; outer face convex, minutely pubescent; inner face channelled around margin; margin recurved.

Distribution Qld, on Mt Greville and Mt Maroon. N.S.W., near Wauchope, with large populations in Werrikimbe NP. *Climate* Hot, wet summer; cool to mild, moist to dry winter. Rainfall 800–1000 mm.

Ecology Grows on rocky ridges and ledges on south-facing lower slopes, often near rivers, in scrub and dry sclerophyll forest. Flowers late winter–spring. Probably pollinated by birds. Fire response unknown but probably killed with regeneration from seed.

Variation A relatively uniform species despite a large disjunction between populations in Qld and

G. linsmithii

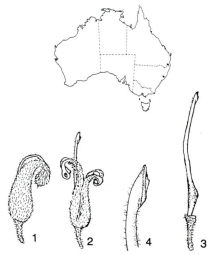

201. *G. linsmithii* Flowers and foliage (N.Marriott)

N.S.W., where populations sometimes have smaller flowers (pistil 11–12 mm).

Major distinguishing features Leaves simple, entire, oblong-obovate, thin, pliable, villous on undersurface; conflorescence few-flowered, decurved on slender pubescent peduncle; torus oblique; perianth zygomorphic, hairy on both surfaces; ovary stipitate, glabrous; style sparsely pubescent; pollen presenter lateral, 2.2–2.6 mm wide.

Related or confusing species Group 21, especially *G. mollis* which differs in its 6–20-flowered conflorescence, larger red flowers with broader pollen presenter (2.6–3 mm wide), and *G. victoriae* which differs primarily in its many-flowered conflorescence and leathery leaves with silky undersurface.

Conservation status 3RC-a. Scattered small populations, some in conservation reserves.

Cultivation *G. linsmithii* has not been cultivated widely and more trials are needed. It has been grown for many years at Pullenvale, near Brisbane, where it has self-sown and proved fairly hardy and adaptable. It has also been cultivated successfully in Sydney and western Vic. for a few years. The species tolerates light frosts and extended dry periods but prefers summer watering. A well-drained, acidic to neutral sandy or clay-loam soil produces best results in full sun or semi-shade. It may become a little straggly with age but this can be remedied by regular tip-pruning when young. We recommend planting 2 m apart. Not usually available from nurseries.

Propagation *Seed* Germinates readily from fresh seed, that generally requires no pretreatment. *Cutting* Easy to strike from cuttings of half-ripe wood at most seasons. *Grafting* Has been grafted successfully onto *G. robusta* using whip and mummy grafts.

Horticultural features *G. linsmithii* is grown mainly as a foliage plant because of its weeping branchlets and rather dense, spreading habit. Flowering is rather insignificant, the conflorescences containing usually 2–4 flowers and being well hidden inside the foliage. On close inspection, the flowers are attractive without being eye-catching. The green perianth with orange-pink limb is unusual and the flowers attract many honey-eating birds to the garden. It would be of interest to collectors or people seeking a shrub to add variety to the native garden. It has some potential as a soft-foliage screen plant and could be useful for planting under trees.

Grevillea lissopleura D.J.McGillivray (1986)
Plate 202

New Names in Grevillea *(Proteaceae)* 9 (Australia)

The specific epithet alludes to the leaf veins which are smooth and rounded. It is derived from the Greek *lissos* (smooth) and *pleuron* (a rib). LISS-OH-PLOO-RA

Type: about 25 km NNW of Mt Holland, W.A., 23 Aug. 1979, K.Newbey 5808 (holotype: PERTH).

A low **shrub** 0.5–1.2 m high, 1 m wide. **Branchlets** terete, silky-tomentose. **Leaves** 1–7 cm long, c. 1 mm wide, ascending, sessile, simple, narrowly linear, usually curved, with a short, recurved point, stiff and leathery, crowded; upper surface convex, glabrous, sometimes with scattered hairs, the base pubescent, longitudinally smooth-ribbed; margins entire, smoothly revolute to midvein; lower surface bisulcate, glabrous, midvein prominent, slightly depressed below level of margin. **Conflorescence** c. 1 cm long, 1 cm wide, erect, ± sessile, terminal or in upper axils, unbranched, umbel-like, not exceeding foliage; rachis 1–2 mm long, tomentose; bracts 1.5 mm long, oblong-obovoid, silky-tomentose, variably persistent, sometimes to anthesis. **Flower colour**: perianth and style white to creamy white. **Flowers** adaxially oriented; pedicels 2.8–3.6 mm long, silky; torus 0.8–0.9 mm across, oblique, cup-shaped; nectary inconspicuous, semicircular, smooth; **perianth** 2–4 mm long, 1 mm wide, slender, oblong-cylindrical, silky outside, bearded inside near level of ovary, first separated along dorsal suture, the dorsal tepals then separated from ventral tepals, reflexed laterally and exposing inner surface before anthesis, soon afterwards free to base and detaching; limb revolute, globular, the segments faintly ribbed; **pistil** 7–8 mm long; stipe 0.8–1 mm long, channelled ventrally, glabrous; ovary ovoid-ellipsoidal, silky-tomentose; style exposed from the base, scarcely exserted before anthesis, afterwards angularly incurved, geniculate in apical 2 mm, glabrous except hairs at base and apex; style end partially exposed before anthesis; pollen presenter oblique, round, with a central subconical stigmatic area encircled by a flat basal flange. **Fruits** 6–9 mm long, 3–3.5 mm wide, erect, ovoid, minutely silky, glabrescent; pericarp 0.2–0.4 mm thick. **Seed** not seen.

Distribution W.A., between Southern Cross and Mt Holland. *Climate* Hot, dry summer; cool, wet winter. Rainfall 300 mm.

Ecology Grows in open scrub in well-drained, stony loam on a slight ridge of banded ironstone. Seedlings observed emergent from an ant nest. Flowers winter–spring. Pollinators unknown, not attending the flowers in mid-morning. Ants appear to be responsible for the spread of coccid scale prevalent on most plants in the wild but with no apparent detrimental effect. Regenerates entirely from seed.

Variation A morphologically uniform species.

Major distinguishing features Leaves simple, entire, stiff but often slightly curved, narrowly linear, bisulcate below, smoothly ribbed; conflorescence sessile, umbel-like, unbranched; torus oblique; nectary obscure; pedicels silky; perianth zygomorphic, undilated, hairy on both surfaces; pistil 7–8 mm long; ovary silky above a glabrous stipe; style sparsely hairy at apex; pollen presenter oblique, flanged at base.

Related or confusing species Group 27, especially *G. scabrida* which differs in its scabrous leaves.

Conservation status 2E. This species was re-discovered by the authors in 1991 at two locations within 6 km of each other. Survey of the population size and distribution is now being undertaken by

G. lissopleura

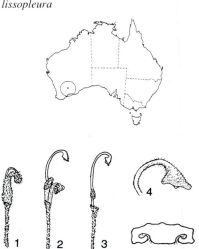

202A. *G. lissopleura* Flowers and foliage (P.Olde)

202B. *G. lissopleura* Habit and habitat, NNW of Mt Holland, W.A. (P.Olde)

the Department of Conservation and Land Management.

Cultivation *G. lissopleura* was introduced to cultivation in 1991 and is currently being grown at Mount Annan Botanic Garden, N.S.W., and by a few collectors in western Vic. To date, it has proved quite hardy in cultivation, tolerating light frosts in winter and some summer humidity without apparent ill-effect. It is also hardy to extended, dry conditions, even when quite young. A well-drained, acidic to neutral gravelly or loamy soil in full sun or partial shade has proved successful. Not commercially available at present.

Propagation *Seed* Sets prolific quantities of seed that should germinate well with some pretreatment. *Cutting* Strikes fairly well from cuttings of firm, new growth in spring. *Grafting* Has been grafted successfully onto *G. robusta* using whip and mummy grafts, but long-term compatibility is yet to be assessed.

Horticultural features *G. lissopleura* is a delightful rounded shrub with small heads of creamy white flowers dotted over the whole plant in a massed early spring display. Flowering appears to occur over a relatively short period, about 4–6 weeks. The leaves are bright green when young, dull green when mature, sometimes crowding the upper branches but not intruding on the floral display even though it is mostly enclosed in the foliage. There is potential as a rockery or small, low screen plant in the landscape if this species becomes more widely available commercially.

Grevillea longicuspis D.J.McGillivray (1986) Plate 203

New Names in Grevillea *(Proteaceae)* 9 (Australia)

The specific epithet is derived from the Latin *longus* (long) and *cuspis* (a point), in reference to the long pointed teeth around the margins of the leaves.
LON-GEE-KUSS-PISS

Type: near Port Darwin, N.T., Nov. 1943, T.Black (holotype: MEL).

Multi-stemmed, lignotuberous, **shrub** to c. 0.5 m high, 0.5 m wide. **Branchlets** slender, compressed slightly to terete, ridged, sparsely glandular-pubescent. **Leaves** 2.5–8 cm long, 2–6 cm wide, ascending, shortly petiolate, concolorous, ± ovate to oblong; upper and lower surfaces similar, concolorous, usually sparsely silky and with glandular hairs near margin, sometimes almost glabrous; midvein, reticulum and lateral veins prominent, the lateral veins leading to excurrent brittle spines to 4 mm long; margin a prominent, rounded edge-vein, flat or shortly recurved, undulate, prominently toothed or lobed (2–6 per side) with arcuate sinuses, sparingly ciliate with glandular hairs extending sometimes onto lobe spines. **Conflorescence** 1.5–4 cm long, erect, usually pedunculate, terminal or axillary, simple or few-branched; unit conflorescence c. 1 cm long, subglobose, mid-dense; peduncle and rachis silky or glandular-pubescent or often mixed; bracts 1.5–2.2 mm long, ovate, villous outside with ciliate margins, falling before anthesis. **Flower colour**: perianth bright red; style red or pinkish cream. **Flowers** adaxially acroscopic; pedicels 2–3.5 mm long, glabrous or sprinkled with silky hairs; torus 1.5–2 mm wide, slightly oblique, shallow-cup-shaped, markedly elongate on dorsal side; nectary prominent, U-shaped; **perianth** 3–4 mm long, 2–2.5 mm wide, ovate, irregularly dilated at base, strongly curved, glabrous outside, villous inside, cohering except along dorsal suture; limb revolute in young bud, angularly spheroidal, the segments ribbed; style end partially exposed before anthesis; **pistil** 6.5–9.5 mm long; stipe 1.5–2.2 mm long, fused and lateral with torus; ovary ovoid; style refracted from line of stipe, slightly exserted at curve before anthesis, afterwards angularly incurved, geniculate below style end, glabrous or sparsely papillose, sometimes a few scattered erect glandular hairs on dorsal side, slightly swollen beneath a flattened style end; pollen presenter lateral, convex, obovate to elliptic. **Fruit** 9.5–12.5 mm long, 5.5–7.5 mm wide, oblong-ellipsoidal with a short recurved apical point, coarsely granular, glabrous; pericarp 1–1.5 mm thick. **Seed** 6–8 mm long, 3–4 mm wide, ellipsoidal; outer face strongly convex, smooth, mottled; inner face convex in middle, the margin recurved, with a narrow, wavy and waxy wing all round, the wing drawn to a short, slightly oblique, excurrent point at both ends.

Distribution N.T., at Woolaning, Manton R. and Darwin. *Climate* Monsoonal. Hot, wet and humid summer; warm, dry winter. Rainfall 1200 mm.

Ecology Grows in open eucalypt woodland among grasses, in sand or lateritic gravelly soil or on slopes in quartzite scree. Flowers autumn–spring, sometimes in summer. Pollinator unknown. Regenerates from lignotuber and seed after fire which occurs frequently.

Variation Minor variations in leaf shape, otherwise relatively uniform.

Major distinguishing features Lignotuberous habit; leaves bifacial, concolorous, sparsely hairy, the margin flat or shortly recurved, pungently toothed; conflorescence erect, usually pedunculate,

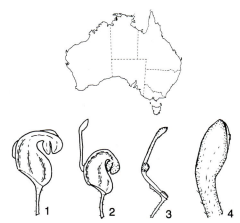
G. longicuspis

globose; torus oblique, elongate on dorsal side; nectary U-shaped; perianth small, zygomorphic, glabrous outside, hairy within; ovary stipitate, the stipe adnate to torus and ± at right angles to pedicel, glabrous; style glabrous or a few hairs on dorsal side, scarcely longer than perianth; pollen presenter lateral.

Related or confusing species Group 10, especially *G. cunninghamii* which differs in its stem-clasping leaves, and *G. adenotricha* which has longer, narrower leaves with more leaf lobes and a conspicuous spreading leaf and branchlet indumentum.

Conservation status 3E recommended. Known from only a few small populations.

Cultivation Little is known about the cultivation of this tropical species although it has been grown successfully since 1986 in the ground in Brisbane and as a potted plant as far south as Bulli, south of Sydney. Indications are that it will tolerate a cooler, mediterranean-type climate provided there is sufficient warmth in the non-summer period. Frost is likely to be deleterious. It would prefer a position offering full or dappled sun in well-drained, acidic sandy or loamy soil. Once established, supplementary watering will keep this plant happy if conditions

203A. *G. longicuspis* Habit and habitat, Pethericks Nature Reserve, N.T. (P.Olde)

203B. *G. longicuspis* Close-up of flowers (K. & L. Fisher)

become too dry. Annual vegetative stems sprout from the base and usually die after fruiting; an occasional fire over the plant may invigorate it if it begins to look sickly. Its general appearance would undoubtedly be improved by pruning off dead branches but no other maintenance is required. Not currently available in nurseries.

Propagation *Seed* Not tested. *Cutting* To date most plants in cultivation have been propagated by cuttings of half-ripe wood in summer. *Grafting* Has been grafted successfully onto *G. robusta* using top wedge, whip and mummy grafts.

Horticultural features *G. longicuspis* is an interesting though unspectacular plant with potential as a rockery plant in tropical and subtropical gardens. It has attractive bright green leaves and bright red clusters of flowers that nestle delicately on long, slender peduncles among the outer leaves. The flowers produce a reasonable drop of nectar but we have observed no potential pollinators probing for it. The floral structure is unusual and although individually fairly small the flowers are nonetheless quite attractive. More trials are needed to establish the climatic limits that this species will tolerate. One for the collector.

Grevillea longifolia R.Brown (1830)

Plate 204

Supplementum Primum Prodromi Florae Novae Hollandiae exhibens Proteaceas Novas: 22 (England)

Long-leaf Grevillea

The specific epithet alludes to the long leaves and is derived from the Latin *longus* (long) and *folium* (a leaf). LON-GEE-FOAL-EE-A

Type: 'above the Cataract', Port Jackson, N.S.W., July 1807, G.Caley (holotype: BM).

Synonym: *G. aspleniifolia* var. *longifolia* (R.Brown) K.Domin (1921).

Erect to spreading, dense **shrub**, 1–5 m tall, 1–5 m wide. **Branchlets** reddish brown, angular, silky, glabrescent. **Leaves** 7–22 cm long, 5–25 mm wide, simple, ascending, petiolate, broadly linear to narrowly triangular, acute, mucronate; margins shortly recurved, usually regularly serrate, often deeply so, rarely ± entire; upper surface silky, glabrescent, midvein a conspicuous groove, lateral veins evident; lower surface silky, midvein prominent. **Conflorescence** 4.5–7.5 cm long, erect, pedunculate, terminal, unbranched, secund; peduncle and rachis silky; bracts 1 mm long, ovate-acuminate, villous outside, falling before anthesis. **Flower colour**: perianth pinkish white; style bright pink, rarely creamy orange, with green tip. **Flowers** acroscopic; pedicels 1.5–3 mm long, silky; torus 1–1.5 mm wide, slightly oblique; nectary U-shaped, toothed; **perianth** 7.5–9 mm long, 1–2 mm wide, S-shaped-cylindrical, mainly silky-white outside, glabrous inside, cohering except along dorsal suture; limb revolute, ellipsoidal, brown-silky; **pistil** 21–24 mm long; stipe 0.7–1 mm long, silky; ovary subglobose, silky; style silky at base, otherwise glabrous, exserted near base on dorsal side and looped upwards before anthesis, afterwards ± straight but swept back strongly; style end hoof-like, not exposed before anthesis; pollen presenter slightly oblique, oblong, conical. **Fruit** 13–16 mm long, 6 mm wide, erect on curved pedicel, compressed-ellipsoidal with an apical attenuation, silky with brownish striping; pericarp c. 0.5 mm thick. **Seed** (n.v.) 11 mm long, 4.5 mm wide, oblong, rugose (McGillivray 1993).

Distribution N.S.W., from the Springwood area to Woronora and south to the Cataract R. In 1898 it was collected from the Parramatta R. near Burwood. *Climate* Cool to mild, wet or dry winter; warm to hot, wet or dry summer. Rainfall 1000–1200 mm.

Ecology Grows on river banks and nearby slopes in eucalypt woodland, usually in grey sand or gravelly sandstone-derived soil. Flowers late winter to summer. Pollinated by birds. Regenerates from seed after fire.

Variation A fairly uniform species with little variation other than leaf width and degree of serration. Wild plants occasionally have pale flowers.

Major distinguishing features Branchlets angular; leaves elongate, broadly linear, toothed, rarely entire, the undersurface silky; conflorescence

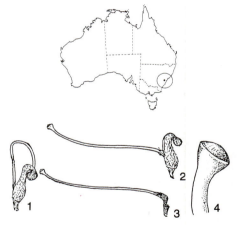

G. longifolia

terminal, simple, secund; torus slightly oblique; perianth zygomorphic, S-shaped, slender, silky outside, glabrous within; ovary silky, stipitate; style glabrous, pink; pollen presenter slightly oblique, conical; fruit with reddish stripes or blotches.

Related or confusing species Group 35, especially *G. aspleniifolia* and *G. wilkinsonii*. The latter differs in its shorter pistil (14–15 mm long). *G. aspleniifolia* differs in its terete branchlets, the matted indumentum of its leaf undersurface, the shape of its pollen presenter and its narrower, few-toothed leaves.

Conservation status 2RC. Restricted to creek lines in Royal, Heathcote and Blue Mountains NPs.

Cultivation Cultivated successfully near the coast in all eastern States from Brisbane south to Tas. as well as in S.A. Overseas it is also successfully cultivated in Europe, France, the south of England, as well as in California, U.S.A., and New Zealand. It tolerates both subtrobical and cold climates and is relatively hardy to extended, dry periods, even in shallow, sandy soils. *G. longifolia* is a compact, hardy plant in cultivation, growing equally well in full sun and semi-shade. It prefers a well-drained, acidic sand or sandy loam and responds to summer watering. Tip pruning makes it even more compact, especially as it gets older. Fertiliser is usually not required. Nurseries occasionally stock this species.

Propagation *Seed* Sets few seed in the wild but often large quantities in cultivation. Fresh seed sown after soaking in warm water germinates well. *Cutting* Sometimes slow to root. Usually successful from semi-ripe new wood between spring and autumn. *Grafting* Has been grafted successfully onto *G. robusta*, but this technique is rarely necessary for this species.

Horticultural features *G. longifolia* is an outstanding foliage plant with elongate, narrow leaves that in some forms are neatly and regularly serrate, making them ideal as a fill-in for floral arrangements. Foliage is cultivated for use in floral art in Europe. With the bonus of attractive and prolific pink toothbrush flowers over a long period in spring, attracting many honey-eating birds, the species can be widely recommended for horticulture and landscape work. It makes an excellent, dense shelter plant or structure plant in the landscape, growing reliably in a wide variety of conditions. It should be more widely planted for screening, especially as it is also long-lived in cultivation. It is not attacked by pests other than leaf caterpillars and occasionally wood grubs. Care should be exercised in handling the foliage as the leaf hairs are reported to cause irritations to sensitive skins.

Hybrids Often regenerates naturally in the garden from seed and hybridises with similar, closely related species. Known parent of G. 'Poorinda Anticipation', G. 'Poorinda Peter', G. 'Poorinda Blondie', G. 'Poorinda Enchantment' as well as G. 'Ivanhoe', G. 'Australflora Fanfare' and probably G. 'Boongala Spinebill.'

History *G. longifolia* was first recorded in cultivation in 1844 when William MacArthur of Camden sent 6 plants to Lee's nursery at Hammersmith, England. Both the Sydney and Hobart Botanic Gardens were growing it in 1850. Shepherd's Nursery, Sydney, listed the species in its 1851 catalogue. Charles Moore, then director of the Sydney Botanic Gardens, sent plants all over the world between 1851 and 1857, including Singapore, India, France and Holland. In 1864, Robert Henderson, Camellia Grove Nursery, Newtown, won an award of the Horticultural Society, Sydney, with a display that included this species, one of the first in the colony to include a native species.

204A. *G. longifolia* Close-up of conflorescences (P.Olde)

204B. *G. longifolia* Flowering branch (J.Noble)

204C. *G. longifolia* Well-grown specimen in cultivation, Seaford, Vic. (N.Marriott)

Grevillea longistyla W.J.Hooker (1848)
Plate 205

in T.L.Mitchell, *Journal of an Expedition into the Interior of Tropical Australia* 343 (England)

Long-style Grevillea

The specific epithet is derived from the Latin *longus* (extended, long) and *stylus* (a style), in reference to the prominent, long style. LON-GEE-STY-LA

Type: probably from the sandstone ranges near Mount Pluto and The Pyramids, Qld, 24 Sept. 1846, T.L.Mitchell (lectotype: K) (McGillivray 1993).

Synonym: *G. neglecta* R.Brown (1849).

Open, lignotuberous **shrub** with many erect stems and no conspicuous main trunk, 1.5–5 m tall. **Branchlets** terete or angular, silky. **Leaves** 15–25 cm long, 1.5–4 mm wide, ascending to spreading, shortly petiolate to subsessile, leathery, usually simple, linear, sometimes pinnatipartite; leaf lobes 3–6, 4–17 cm long, 1.5–4 mm wide, narrowly linear, longitudinally concave; upper surface silky, soon ± glabrous, midvein evident; margins shortly recurved; lower surface white tomentose-villous, midvein prominent. **Conflorescence** erect, pedunculate, usually terminal or in upper axil, branched, usually exceeding foliage; unit conflorescence 8–15 cm long, 6 cm wide, loose, cylindrical, apical flowers opening first; peduncle silky; rachis glandular-pubescent, viscid; bracts 2.2–3.7 mm long, ovate-acuminate, sparsely silky outside with ciliate margins, falling before anthesis. **Flower colour**: perianth bright red, pinkish red, creamy pink or orange; style red. **Flowers** basiscopic; pedicels 3–10 mm long, silky, sometimes glandular-pubescent, expanded near apex; torus 2–3 mm across, oblique; nectary U-shaped, entire; **perianth** 14–15 mm long, 3–4 mm wide, oblong, strongly curved from c. halfway, sparsely silky outside, usually with mixed glandular hairs, the tepals prominently keeled, somewhat waxy, bearded inside mostly above ovary; cohering except along dorsal suture and between ventral and dorsal tepals at the curve; limb revolute, subcubic, depressed, the segments conspicuously keeled, impressed at margins; **pistil** (28–)40–52 mm long; stipe (2–)3–4.5 mm long, glabrous or almost so, inserted obliquely to almost perpendicular on upper edge of torus and lining torus; ovary white-silky, subglobose; style glabrous except basal hairs, exserted from below the curve on dorsal side and looped upwards before anthesis, afterwards gently incurved; style end evenly dilated; pollen presenter very oblique, ± elliptic, convex. **Fruit** 10–13 mm long, 8–10 mm wide, oblique on erect stipe, round to oblong-ellipsoidal with short refracted apiculum, rugulose, shiny, sparsely sprinkled with hairs; pericarp c. 1 mm thick. **Seed** 7–8 mm long, 5 mm wide, compressed ellipsoidal-obovoid, narrowly winged all round; outer face convex, dull, minutely tesselated, irregularly wrinkled; lower surface flat faintly grooved around margin.

Distribution Qld, confined to sandstone ridges on the Blackdown Tableland, Isla Gorge and Gurulmundi areas. *Climate* Warm to cool, dry or wet winter; hot, dry or sometimes wet summer. Rainfall 600–700 mm.

Ecology Grows on sandstone ridges and elevated plains in sandy or gravelly soil, in open places or in tall, open eucalypt forest or woodland. Flowers late winter–early summer. Pollinated by birds. Regenerates by seed or from lignotuber after fire.

Variation There is some variation in flower colour and size. McGillivray (1993) reported a plant with mostly entire leaves, short conflorescences and small flowers, collected NE of Jandowae towards Kingaroy.

Blackdown form McGillivray (1993) observed that plants from this area lack the glandular hairs usually present on plants from the rest of its range. Plants with pale or orange flowers are also found in this area.

Major distinguishing features Lignotuberous habit; leaves linear, usually simple, sometimes divided, the undersurface exposed, white-hairy; conflorescence cylindrical, 8–15 cm long, usually branched, the apical flowers opening first; torus very oblique; perianth zygomorphic, oblong, usually glandular pubescent outside; tepals prominently keeled, hairy within above ovary; ovary silky, the stipe inserted obliquely on upper toral rim; style glabrous except some basal hairs.

Related or confusing species Group 26, especially *G. johnsonii* which differs in its shorter conflorescence, its perianth with non-glandular hairs, and the leaf undersurface enclosed by the margins.

Conservation status Not presently endangered.

Cultivation *G. longistyla* has been cultivated successfully mainly in tropical and subtropical climates and at least as far south as Vic., but there it often succumbs in the extended cold winters. It tolerates extreme drought and light frost and will probably survive most garden conditions except extreme, continuing cold or wet weather. It prefers an open, sunny position in well-drained, acidic to neutral, sandy gravel soil. Pruning the old flowering branchlets back to the nearest vegetative leaf axil will improve density and compactness and lead to

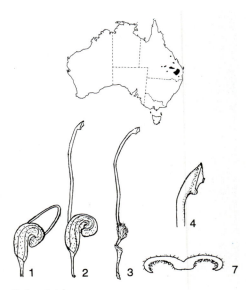

G. longistyla

205A. *G. longistyla* Flowering branches (M.Hodge)

205B. *G. longistyla* Flowering plant, in cultivation, Brookvale Park, Qld (M.Hodge)

205C. *G. longistyla* Pale-flowered variant, ex Isla Gorge, Qld (P.Olde)

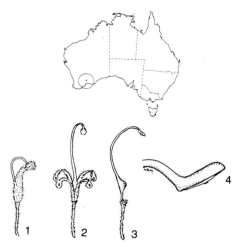

G. lullfitzii

more flowering branches the next season. *G. longistyla* is not widely available in nurseries but seed can often be obtained from specialist suppliers. The species was introduced to cultivation in 1875 in Qld and is widely grown there today by members of the Society for Growing Australian Plants.

Propagation Seed Easy to germinate from seed which sets prolifically. Seed should be sown fresh and preferably nicked or soaked for better results. In cultivation the species hybridises, hence seed may not come true. *Cutting* Strikes with reluctance usually from half-ripe wood in late spring or early summer. *Grafting* Successfully grafted onto *G. robusta* using mummy, top wedge, whip and whip-and-tongue techniques. The combination appears to be long-term compatible.

Horticultural features G. longistyla is a somewhat straggly, open plant in cultivation and is not recommended for landscapes where habit and foliage are of prime importance. When it bursts into flower, however, the conflorescences are both spectacular and prolific, borne continuously and prominently in the upper foliage over a long period. The individual flowers are mostly red and contain a copious flow of nectar that is strongly bird-attractive. It makes an excellent feature plant when combined with other plants in dense plantings and is both hardy and reliable. It is recommended for mixed plantings in public or private landscapes.

Hybrids Plants available commercially in Melbourne under this name are almost certainly a hybrid, between this species and *G. johnsonii*, which arose in Dave Gordon's arboretum at Myall Park, Qld. These plants share features of both species in their conflorescence and foliage. The hybrid is also grown in the U.S.A. under the name *G. longistyla*. An appropriate name for it is G. 'Long John'.

Grevillea lullfitzii D.J.McGillivray (1986)
Plate 206

New Names in Grevillea *(Proteaceae)* 9 (Australia)

The specific epithet honours Fred Lullfitz (1914–1983), West Australian nurseryman, botanical and seed collector for Kings Park and enthusiast for the Australian flora. LULL-FITS-EE-EYE

Type: Digger Rocks, W.A., 10 Dec. 1964, F.Lullfitz 4001 (holotype: KINGS PARK).

A spreading, prickly, ?suckering **shrub** to 2 m in height, 2 m wide. **Branchlets** sharply angular, ribbed, sparsely silky. **Leaves** 3–6 cm long, ascending, sessile, divaricately pinnatisect usually glaucous; leaf lobes 3–6, 1.5–3 cm long, 1.2–2 mm wide, rigid, pungent, linear; upper surface glabrous or with a few sprinkled hairs, smooth with obscure venation, the midvein sometimes a faint impressed groove; margins angularly revolute; lower surface obscured, bisulcate, the grooves silky, midvein prominent. **Conflorescence** erect, shortly pedunculate, terminal, simple or few-branched; unit conflorescence c. 2 cm long, 2 cm wide, dense, dome-shaped or umbel-like; peduncle and rachis silky, the rachis stouter than peduncle; bracts 0.5–0.8 mm long, ovate, silky outside, falling before anthesis. **Flower colour**: perianth greyish cream outside, creamy white inside; style pinkish brown with green tip. **Flowers** adaxially oriented; pedicels 2.5–3.5 mm long, sparsely silky; torus c. 0.8 mm across, slightly oblique, square; nectary arc-like, scarcely evident above toral rim; **perianth** 4–5 mm long, 0.8–1 mm wide, oblong-cylindrical, narrowing slightly at curve, silky outside, bearded inside about ovary, at first separated to base along dorsal suture, the dorsal tepals then separating and reflexed laterally, exposing inner surface before anthesis, soon afterwards free to base and detaching; limb erect in late bud, nodding through recurvature of upper perianth before anthesis, subcubic, silky; **pistil** 8–10.5 mm long, glabrous; stipe 1–1.7 mm long; ovary ovoid; style refracted above ovary, exposed to base and scarcely exserted before anthesis, afterwards angularly incurved, geniculate below apex; style end shoe-like; pollen presenter very oblique to almost lateral, elliptic, slightly convex. **Fruit** 14–17 mm long, 5–8 mm wide, erect, ellipsoidal/obovoidal, with a small obtuse apical attenuation or knob, granular; pericarp c. 0.5 mm thick. **Seed** not seen.

Distribution W.A., known from only a few collections between Digger Rocks and Hatters Hill. *Climate* Hot, dry summer; cool, wet winter. Rainfall 320 mm.

Ecology Grows in densely lateritic loam in eucalypt woodland, usually on rises. Flowers spring–summer. Regenerates from seed and possibly also from suckers after fire. Pollinator unknown.

Variation A relatively uniform species. The foliage is sometimes green rather than the usual blue-green.

Major distinguishing features Leaves divaricately pinnatisect, the lobes rigid, linear, pungent; conflorescence dome-shaped, terminal; torus oblique; nectary obscure; perianth zygomorphic, undilated at base, hairy on both surfaces, the limb nodding; pistil glabrous; style attached at base of pollen presenter; fruit with an obtuse, apical knob.

Related or confusing species Group 16. Unlikely to be confused with other species, it seems related most closely to *G. brachystachya* which differs in its simple, linear leaves.

Conservation status 2E recommended. Known only from a few collections. Mining activity is a serious threat.

Cultivation G. lullfitzii was introduced to cultivation in 1986 but plants did not prosper. Growers report that it succeeds in a well-drained, sunny or partially shaded site in deep, gravelly loam. It is likely to be both drought- and frost-tolerant but prefers a dry climate. Cold, wet conditions cause sudden losses. Grafted plants have succeeded in and near Brisbane. Young plants are growing strongly in western Vic. in well-drained sandy loam in full sun. Soil should be neutral to slightly acidic. Pruning will improve the shape as it tends to be open and rather untidy with age. Summer watering is not recommended. Unavailable from nurseries.

Propagation *Seed* Sets very few seeds and germinability is untested. *Cutting* Strikes reluctantly from cuttings of wild material, but firm, young growth from cultivated plants in spring or autumn will certainly improve success rates. More trials are required. *Grafting* Has been grafted successfully onto *G. robusta* and G. 'Poorinda Royal Mantle' using top wedge, whip, whip-and-tongue or mummy grafts.

Horticultural features *G. lullfitzii* is a curiously attractive plant with striking blue-green foliage. Many whitish-yellow racemes, produced in spring, are reasonably interesting and quite prominent, but this species is probably best grown by the collector or broad-scale landscaper. Prickly foliage limits its use to background situations or as a people stopper. Notwithstanding its limited horticultural appeal, it should be grown as widely as possible, especially by public institutions and in botanic gardens to ensure its survival. Large-scale, dry-area gardeners should be able to grow this species well. It would provide good shelter for small birds.

206A. *G. lullfitzii* Plant growing in gravel pit, Digger Rocks, W.A. (P.Olde)

206B. *G. lullfitzii* Close-up of flowers (P.Olde)

206C. *G. lullfitzii* Flowers, fruit and foliage (N.Marriott)